Springer Proceedings in Materials 61

Series Editors

Arindam Ghosh, *Department of Physics, Indian Institute of Science, Bengaluru, India*

Daniel Chua, *Department of Materials Science and Engineering, National University of Singapore, Singapore, Singapore*

Flavio Leandro de Souza, *Universidade Federal do ABC, Sao Paulo, Brazil*

Oral Cenk Aktas, *Institute of Material Science, Christian-Albrechts-Universität zu Kiel, Kiel, Germany*

Yafang Han, *Beijing Institute of Aeronautical Materials, Beijing, China*

Jianghong Gong, *School of Materials Science and Engineering, Tsinghua University, Beijing, China*

Mohammad Jawaid, *Laboratory of Biocomposite Technology, INTROP, Universiti Putra Malaysia, Serdang, Malaysia*

Springer Proceedings in Materials publishes the latest research in Materials Science and Engineering presented at high standard academic conferences and scientific meetings. It provides a platform for researchers, professionals and students to present their scientific findings and stay up-to-date with the development in Materials Science and Engineering. The scope is multidisciplinary and ranges from fundamental to applied research, including, but not limited to:

- Structural Materials
- Metallic Materials
- Magnetic, Optical and Electronic Materials
- Ceramics, Glass, Composites, Natural Materials
- Biomaterials
- Nanotechnology
- Characterization and Evaluation of Materials
- Energy Materials
- Materials Processing

To submit a proposal or request further information, please contact one of the following Springer Publishing Editors according to your affiliation location:

European countries: **Mayra Castro** (mayra.castro@springer.com)

India, South Asia and Middle East: **Swati Meherishi** (swati.meherishi@springer.com)

South Korea: **Smith Chae** (smith.chae@springer.com)

Southeast Asia, Australia and New Zealand: **Ramesh Nath Premnath** (ramesh.premnath@springer.com)

The Americas: **Michael Luby** (michael.luby@springer.com)

China and all the other countries or regions, as well as topics in materials chemistry: **Maggie Guo** (maggie.guo@cn.springernature.com)

This book series is indexed in **SCOPUS** and **EI Compendex** database.

Lech Czarnecki · Andrzej Garbacz · Ru Wang ·
Mariaenrica Frigione · Jose B. Aguiar
Editors

Concrete-Polymer Composites in Circular Economy

Proceedings of the 17th International Congress on Polymers in Concrete (ICPIC 2023)

Springer

Editors
Lech Czarnecki 🆔
Scientific Secretary
Building Research Institute (ITB)
Warsaw, Poland

Andrzej Garbacz 🆔
Faculty of Civil Engineering
Warsaw University of Technology
Warsaw, Poland

Ru Wang 🆔
School of Materials Science and Engineering
Tongji University
Shanghai, China

Mariaenrica Frigione 🆔
Department of Engineering for Innovation
University of Salento
Lecce, Italy

Jose B. Aguiar 🆔
Department of Civil Engineering
University of Minho
Guimarães, Portugal

ISSN 2662-3161   ISSN 2662-317X (electronic)
Springer Proceedings in Materials
ISBN 978-3-031-72954-6   ISBN 978-3-031-72955-3 (eBook)
https://doi.org/10.1007/978-3-031-72955-3

This work was supported by Warsaw University of Technology and Building Research Institute.

This Springer imprint is published by the registered company Springer Nature Switzerland AG
The registered company address is: Gewerbestrasse 11, 6330 Cham, Switzerland

If disposing of this product, please recycle the paper.

Preface

The 17th International Congress on Polymers in Concrete has taken place in Warsaw, Poland, on September 17–20th, 2023. It was organized by the International Congress on Polymers in Concrete, the Warsaw University of Technology and the Building Research Institute (ITB) under patronage of ACI, RILEM, the Rector of Warsaw University of Technology, Polish Association of Civil Engineers, the Section for Building Materials Engineering of the Committee on Civil Engineering of the Polish Academy of Sciences and Polish Association of Civil Engineers and Technicians.

The ICPIC congresses are cyclic events organized since 1975 by the International Congress on Polymers in Concrete. This non-profit organization gathers specialists in research and the practical use of polymers in concrete.

The aim of the 17th ICPIC 2023 was to create a forum for the exchange of multidisciplinary knowledge referring to the application of polymers "in concrete" and "on concrete": from the modification of concrete composition with modern admixtures and additives, through alternative binders (like geopolymers, sulphur concrete, and others), polymer composites for reinforcing concrete (including fibres, FRP bars and FRP strengthening systems), improvement of concrete surface properties (by impregnation, hydrophilization and coatings) to unique properties like self-healing, self-cleaning or energy consumption control with PCM (Phase Changing Materials).

The main issues of the 17th ICPIC held at Warsaw were challenges in the field of Concrete-Polymer Composites (C-PC), concerning the implementation of a circular economy. Therefore, an essential part of the 17th ICPIC 2023 was the Special Symposium entitled "C-PC in Circular Economy: Searching for a New Paradigm", organized by the Building Research Institute (ITB) under the chairmanship of Professor Lech Czarnecki.

The Part I of the proceedings contains the papers deal with the main topics of the Special Symposium. Part II of the proceedings was divided into seven chapters containing the papers deal with issues related to application and challenges for c-pc in circular economy, alternative binders, admixtures and additives, repair and protection of concrete structures, reinforcement and strengthening, improvement of the C-PC properties, special properties of concrete. At the 17th ICPIC Congress the papers have been presented by authors from 19 countries. We are very grateful to all authors for their participation, especially for exciting and fruitful discussions during the 17th ICPIC Congress on the new paradigm for Concrete-Polymer Composites, which will meet the challenges of circular economy implementation. The content of this proceedings clearly indicates a necessity for such discussion. I hope that the papers presented in the proceedings will be a source of inspiration and a guide for new ideas that have been outlined during the 17th ICPIC in Warsaw in the field of sustainable concrete-polymer composites.

On behalf of the Organizing Committee

Andrzej Garbacz
President of the International Congress on
Polymers in Concrete

Contents

Alternative Binders

Admixtures and Additives

Repair and Protection of Concrete Structures

Reinforcement and Strengthening

Improvement of the C-PC Properties

Special Properties of Concrete

C-PC in Circular Economy: Searching for a New Paradigm

Searching for a New C-PC Development Paradigm

Lech Czarnecki[1]([✉]) [iD], Dionys Van Gemert[2] [iD], Ru Wang[3] [iD],
and Mahmoud Reda Taha[4] [iD]

[1] Building Research Institute, ITB, Filtrowa 1, 00-611 Warsaw, Poland
l.czarnecki@itb.pl

[2] KU Leuven, Leuven, Belgium

[3] Key Laboratory of Advanced Civil Engineering Materials of Ministry of Education, School of Materials Science and Engineering, Tongji University, 4800 Cao'an Road, Shanghai 201804, China

[4] Department of Civil, Construction & Environmental Engineering, University of New Mexico, MSC 01 1070, Albuquerque, NM, USA

Abstract. In less than a century, concrete has become the most commonly used construction material worldwide. Today, it is difficult to imagine concrete entirely devoid of polymers. The implantation of polymers into concrete has taken effect in the form of Concrete Polymer Composite [C-PC = PMC + PCC + PIC + PC]. Several milestones are recognised in the development of C-PC. They are discussed here with particular emphasis on the innovative milestones that shaped the use of polymers in concrete. As the difference between polymer cement concrete and ordinary concrete diminishes, the question: "What should the paradigm of C-PC development be?" arises.

Keywords: Concrete Polymer Composites · Polymer Modified Concrete · Polymer Cement Concrete · Polymer Aggregate in Concrete · Polymer Fibres in Concrete · Nanomodified Polymers · Circular Economy · Development Paradigm · Sustainable Development

1 Introduction: Need for a New Paradigm

It is symbolic that our considerations take place one hundred years after the first patent (L. Cresson, 1923) on polymers in concrete was received [1]. Simultaneously, more than fifty years ago, the concept of "polymer in concrete" began to be thoroughly developed [2, 3]. Gradually, the general material concept (Fig. 1) has been transformed into various kinds of concrete – Polymer Composites C-PC = MPC + PCC + PIC + PC, see Table 1. There is fascination surrounding the fact that such a small amount of polymer can have such a great effect on concrete properties, see Table 2.

Polymers in concrete serve in a multitude of ways as significant modifiers in the processing ("binders") of concrete. Functional polymers are used due to their particular native properties and synergy, created by their own: bonding, adhesion, chemical

© The Author(s) 2025
L. Czarnecki et al. (Eds.): ICPIC 2023, 61, pp. 3–21, 2025.
https://doi.org/10.1007/978-3-031-72955-3_1

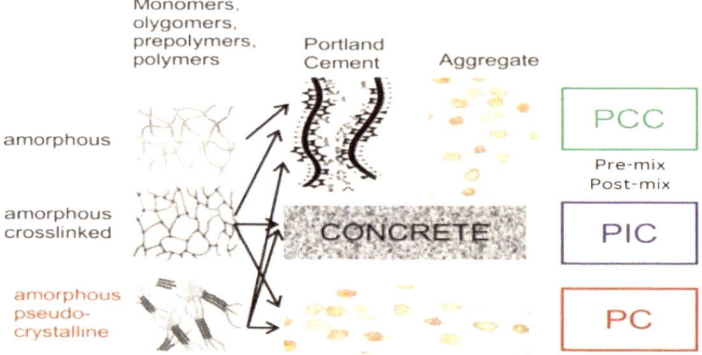

Fig. 1. General concept of Polymers in Concrete [4]

Table 1. Scope of Concrete-Polymer Composites C-PC [5]

Polymers		
in	on	for
Concrete		
Polymer Concrete, PC	Polymer overlays	Polymer repair mortars
Polymer modified Concrete, PMC	Polymer coatings and	Polymer crack repair
Polymer Cement Concrete, PCC	waterproofing materials	
Polymer-impregnated concrete,	Polymer used for bonding	
PIC	materials to concrete	
Polymer Fibers in Concrete, PfiC	Fiber-reinforced polymers for	
Polymer Aggregate in Concrete	strengthening Concrete, FRP-C	

Table 2. Content Polymer in Concrete, P/Con. %

Type of Polymers	P/Con. %
Nano PCC	<0.10
PMC	<0.15
PCC pre-mix	1–3
PCC post-mix	1–3
PIC	3–8
PC	8–12

reactivity, viscosity, regulation, lubrication, hydrophyllic/hydrophobic, water repellent, thermo-setting, thermo-plasting, chemo-setting and so on.

The results of the main research gradually revealed a mechanism of polymer in concrete modification (Fig. 2). The basis has been the utilisation of the synergy of polymers in concrete because their influence on the properties of the product is much higher than that declared from its mass share.

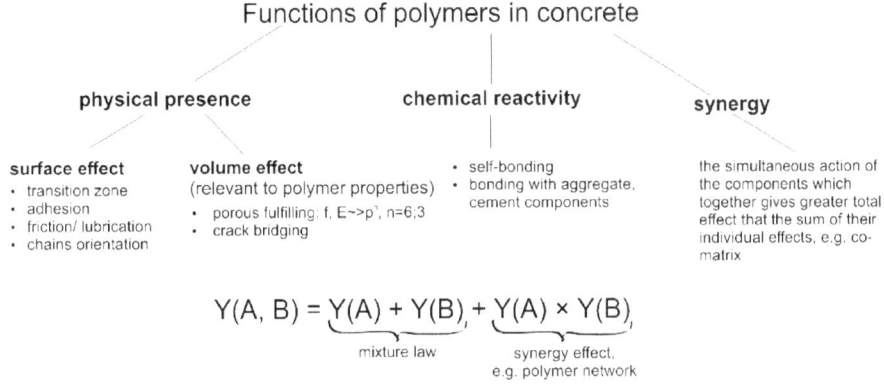

Fig. 2. Mechanism of polymer in concrete modification

The sixteen International Congresses since 1975 have marked the milestones in the C-PC development. The rapid development of the discipline and collection of new knowledge followed the incubation period. At a certain point, publications started to bring in less new knowledge (involving mainly confirmations or revisions). The literature also expressed doubts and negations, secondary discoveries or even repetitions. The maturity period was followed by "fatigue with the topic". The end of the cycle marked the beginning of new hysteresis (Fig. 3a), typically with a new paradigm, which is initiated on a much higher level of knowledge [6].

"Delusion of hysteresis" intrinsically includes a negative arrow of time and is not possible in physics categories. The "time trap" can be bypassed by introducing the discontinuity between the displaced logistic curves (Fig. 3b). The sense of hysteresis remains and corresponds to the researcher's state of mind in the reference period. The authors feel that this point is being addressed in terms of Concrete-Polymer Composites. In the field of C-PC, we have experienced the entire scientific cycle. This statement is the consequence of our question, which we asked on the 16th ICPIC: "Is polymer still the factor that contributes to progress in concrete technology?" [5].

The same Congress, however, retains and confirms the Congresses' motto "polymers for resilient and sustainable concrete infrastructure" [7]. The Concrete-Polymer Composites research area needs a new paradigm of development. Certainly, this is an arbitrary statement. Usually, that is a conclusion ex-post – it just happened. As authors, we would like to emphasise strongly that the basis of our statement is not pessimism; on the contrary, we would like to declare our deep confidence in the C-PC development. What is more, we are not alone in this belief. In 2022, a systematic review on C-PC was published [8] by three scholars from Melbourne University. Furthermore, the topic, similar to the ICPIC 2018 topic, was oriented towards "durable and resilient infrastructure" (!). In this paper, it is duly noted that the annual number of papers on C-PC is increasing (Fig. 4). According to the conclusions of the authors [8], the following particularly promising directions can be mentioned: corrosion inhibitors, electrical insulators, energy absorbers (vibration-damping), waterproofing, durable and resilient (high durability and ductile) means for infrastructure. Those "promises" are already pre-established applications. The future is always more unpredictable than it seems, which does not distract from the importance and need of forecasting.

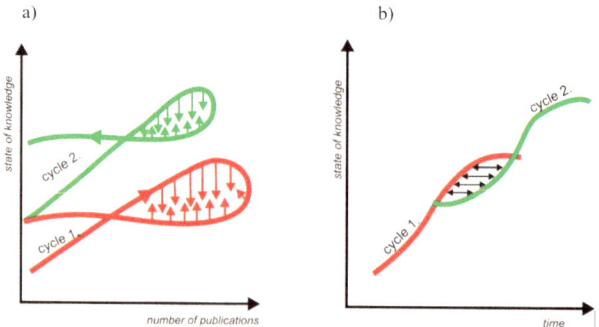

Fig. 3. Development of the state of knowledge according to the: a) apparent hysteresis; b) displaced logistic curves [6]

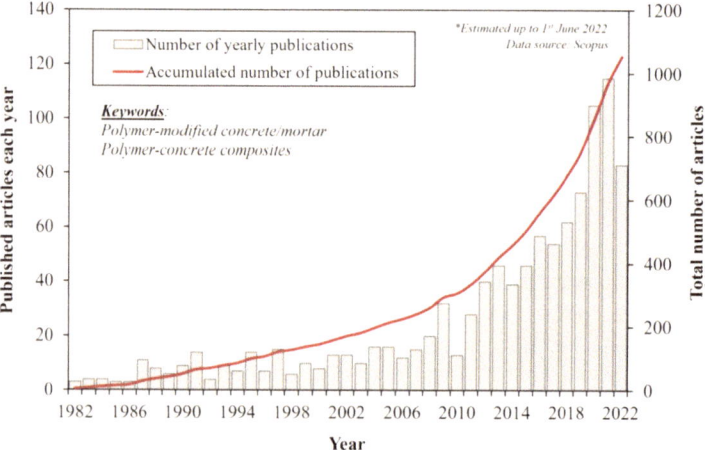

Fig. 4. The growth trajectory of publications with topics on polymer-modified cementitious composites over the past 40 years (1982–2022) [Data source: SCOPUS – Search dated: 1 June 2022] [8].

2 From Application of Polymers in a Mineral Matrix to Understanding Interactions Between Organic and Mineral Phases

The combination of polymers with cement concrete dates back to the beginning of the twentieth century because the origin of cement concrete itself only dates back to the second half of the nineteenth century. Natural polymers, however, were already used in ancient times to enhance the properties and durability of plasters, mortars and concretes [9]. Today, analysis of the composition of ancient binders and mortars with proven durability reveals the hardening activation methods, and these methods serve as guidance to improve the hydration mechanisms of pozzolans and industrial by-products to develop more sustainable binders for the construction industry [10].

Although it was soon recognised that the combination of hydraulic cement and polymer in concrete creates opportunities for beneficial synergies that result from the intended interactions between cement and polymer particles in the fresh mix, between cement hydration and polymer hardening systems during curing, and from the interactions between hardened cement hydrates and hardened polymer structures, it took several decades and successive modelling steps to understand and master the microstructure formation in C-PC [11].

Several milestones can be observed on the road to building the PCC microstructure formation model, which run parallel with the history of the ICPIC Congresses, see Table 3 and [12].

Table 3. Milestones in the development of C-PC, marked by International Congresses on Polymers in Concrete, ICPIC [6]

Congress No	Year	Main theme
I	1975	Innovation – progress in concrete technology: PIC, PCC, PC
II	1978	Applications – trials and errors
III	1981	Usability testing
IV	1984	Material model
V	1987	Control of properties
VI	1990	Effectiveness of polymer utilisation
VII	1992	Evaluation, simulation, optimisation
VIII	1995	Modelling: durability and synergy
IX	1998	Micro–macro-structure relations (nanotechnology, application of plastic recycles)
X	2001	Sustainable C-PC
XI	2004	Integrated PCC model; synergy. Water-soluble polymers as modifiers

(continued)

Table 3. (*continued*)

Congress No	Year	Main theme
XII	2007	Sustainability; nanotechnology as the driving force
XIII	2010	C-PC of high usability
XIV	2013	Modelling the processes of binding and curing. Synergy between the components. Pursuing new development directions
XV	2015	Great expectations. Potentially "mature" C-PC
XVI	2018	Does polymer still create progress in concrete technology?

At the first ICPIC Congress in 1975, H.R. Sasse [13] already presented the interaction between polymer admixtures and cement hydrates. He assumed that the polymers formed extremely thin resin films or net-like structures on the hydrate surfaces. During hydration, these films are penetrated and swallowed up by newly formed hydration products, thus losing their effectiveness. That assumption, however, is only valid for the low-ratio polymer admixtures in his study. At first, the models only envisaged the interaction of polymer, cement paste and aggregates in the hardened state [14, 15]. The original three-step model proposed by Y. Ohama [16] took into account the hardening process of the polymer phase. Subsequently, numerous specifications and modifications to this model have been presented [12]. Beeldens et al. [17] proposed an integrated model, in which the interaction between cement hydration and polymer hardening is integrated. Dimmig-Osburg [18] included the adsorption of polymer on cement particles, whereas Ye [19] considered possible flocculation effects of the polymer particles, leading to discontinuous distribution of polymer throughout the microstructure. Enhancement of SEM resolution and magnification capabilities also enabled to study [20] the effect of very low amounts of polymer on the microstructure at the nano-scale, e.g. to study the positioning and influence of water-soluble polymer in between hexagonal Portlandite plates [21]. The difference between micro- (Fig. 5a) and nano-interaction (Fig. 5b) is clearly represented in Fig. 5. It is obvious that the above-presented Polymer Cement Concrete technology developments and microstructure models only and solely involve physical mechanisms and physical interactions by which the synergy phenomena are obtained. Compared to physical interaction, chemical interaction is not considered in many cases.

3 Lessons from the Past

In searching for new developments, looking at the past is frequently a sparkling source of inspiration. As a sector, the construction industry accounts for more than 10% of global GDP (in developed countries, construction comprises 6–9% of GDP), and employs around 7% of the total employed workforce around the globe (accounting for over 273 million full- and part-time jobs in 2014). Construction is a major source of employment and a driving force of the economy in most countries [22].

a)

b)

Fig. 5. (a) Polymer bridging between crack-edges of concrete on micro-level [49]; (b) Polymer bridging between Portlandite plates on nano-level [11]

The construction sector has an important impact on the environment: as much as 50% of all materials extracted from the earth's crust are transformed into construction materials and products. In relation to the total energy consumed globally, energy used in the construction sector in the process of erecting and operating buildings accounts for 40% of this consumption. Moreover, these same materials when they enter the waste stream, account for some 35% of all waste generated prior to recovery [23]. Greenhouse gas emissions from material extraction, manufacturing of construction products, as well as the construction and renovation of buildings are estimated at 5–12% of the total national GHG emissions. Greater material efficiency could save up to 75% of these emissions by 2060, compared to the 2017 level [24]. These numbers put the construction industry "in the picture" for sustainability efforts and sustainable consumption and production action plans.

Cement production is a major energy and primary materials consumer, as well as a major greenhouse gas emitter. Therefore, if we only consider cement consumption as an indicator for the contribution of construction to ecological improvement, the numbers are not flattering (Fig. 6). Worldwide consumption of cement is still increasing, only the economic recession kept the level of worldwide consumption of cement nearly constant between 2013 and 2018 but, since then, consumption is steadily increasing again.

Material resource efficiency can be applied across a construction project's life cycle, with the greatest benefits at the early stages, where more opportunities arise to design out waste and investigate material choices. Greater material resource efficiency requires the various parts of the construction supply chain to work together for a common goal, as a decision by one part could adversely affect another. There is an increasing awareness that improved material resource efficiency would produce benefits across the industry, such as cost savings, reduced environmental impact, and an enhanced reputation. Exploiting the synergies between construction materials in composites, e.g. C-PC, is a way towards efficiency enhancement. In all of the above development areas, the combination of mineral and organic binders is still a virgin research field where the same benefits as in C-PC will be possible.

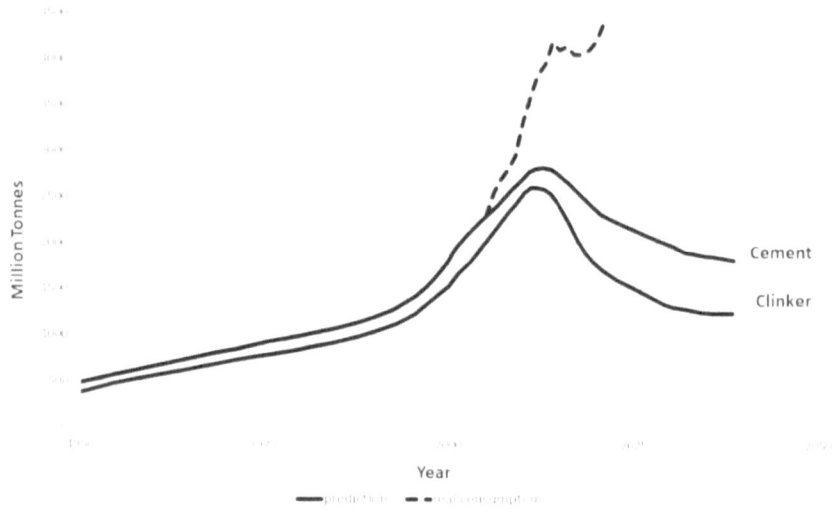

Fig. 6. Estimated evolution of cement and clinker consumption in sustainable cement industry. Data source [25]

4 Searching for Novelties

4.1 Assumptions of the Search

The question arises: "where should we seek novelty?" The most promising areas of research for the future are:

- new material solutions;
- new technological methods;
- new applications;
- novelties in unexpected areas.

It should be assumed that this novelty will be found in the frame of sustainability requirements; sustainability remains the cogent commandment [26, 27]. It also seems significant that, in general terms, the circular economy will define new paths of development and new challenges [28, 29]. The present state of knowledge can already bring some suggestions. According to the above categorisation, it could be:

- the mineral-organic reaction as the gateway to the new generation of C-PC;
- nano-modification as the new thrust in C-PC technology;
- PCC as waste polymer product storage – paradoxically a new application area;
- "novelties in unexpected areas" means, by definition, a collection of innovations of which we are not yet aware.

4.2 The Mineral-Organic Interaction in C-PC Properties Control

The physical and chemical interactions between polymers and Portland cement components and mineral filler define the technical properties of concrete. In comparison to physical interactions, there are more controversies and even lack of knowledge in regard to chemical interactions. Some of the reactions explain the failure mode of the materials, such as the hydrolysis of ester groups. Others explain the strengthening mechanisms, such as the formation of chelates.

Chemical bonding (ionic, covalent or metallic bonds), as an aspect of bonding of polymer materials to concrete at the molecular scale, has already been considered by Sasse and Fiebrich in 1983 [30], but they attributed bonding primarily to van der Waals forces (internal dipoles originating from dispersion, induction, orientation effects) and to micromechanical interlocking mechanisms. Recent studies, however, show evidence of chemical interactions between polymers and hydrating Portland cement. Chemical interaction may result in the formation of complex structures, as well as in changes in the morphology, composition, and quantities of hydrated cement phases [31]. Further research on mastering and exploitation of chemical interactions between mineral and organic phases is needed.

Chemical and physical interactions between cement and polymers are two sides of the same coin. Only when the chemical interactions between polymers and cement are clearly understood one can better explain the micro- and macro-structure relationship in concrete-polymer composites, which in turn serves the purpose of developing higher-performance materials. A clear picture of the chemical interaction that takes place between cement and polymer will be a good supplement to theories on physical interactions. Today, we are just a step away from this potential becoming a reality. Looking at the problem from another perspective will help to understand the application performance of these polymers in cement-based material modification [32].

Further progress will be made with an organic-inorganic composite, in which some components are chemically bonded, parallel to the physical interactions. If the two phases, polymer and Portland cement paste, are additionally partially linked together through strong chemical covalent or iono-covalent bonds, this gives extra cohesion to the whole structure and enhances the technical properties.

Existing studies have demonstrated ample evidence of chemical interaction between polymers and cement components in concrete-polymer composites through various analytical methods, including IR spectroscopy, thermal analysis, NMR microscopy etc. Recently, Molecular Dynamics simulations have proven to be the most effective method to study the interactions between inorganic-organic composites at this stage [33].

At the nanoscale, the interaction between polymers and cement hydration products contains several aspects, i.e. chemical bonding, van der Waals forces, hydrogen bonding etc. Different polymers may have different types of interactions with cement hydration products:

- changing the molecular structure of hydration products and forming interactions with the hydration products, including chemical bonding and intermolecular force;
- polymers affect the hydration reaction of cement and the molecular structure of hydration products;
- in some cases, polymers significantly retard early hydration reactions.

All those interactions between the polymer and cement-based materials affect the end properties of concrete. Understanding these interactions is important to elucidate the relationship between the microstructure and macroscopic properties of polymer-modified cement-based materials. There is need for a scientific tool, which reveals the impact of polymer on concrete in macro performance and which will be able to control concrete performance in a reliable way.

4.3 Nanomodified Polymers for Resilient C-PC Infrastructure

In the past three decades, nanomodified polymers were introduced as a new alternative to standard polymers. Nanomodified polymers are synthesised by dispersing nanoparticles into the polymer resin. Changes in the polymer characteristics included but were not limited to low viscosity, high adhesion with other materials (e.g. aggregate and hardened cement), high tensile strength, high or low modulus of elasticity, low creep compliance, high thermal and electrical conductivity, as well as improved durability. Examples of nanoparticles used to modify polymers included carbon nanotubes (CNTs), graphene nanoplatelets (GNPs), alumina nanoparticles (ANPs) and nano clay (NC) particles, to name but a few. The change in polymer characteristics with the addition of nanoparticles is attributed to a dual effect of the nanoparticles that can induce physical and chemical changes in the polymer matrix. The first effect is that the nanoparticles act as a reinforcing particle/fibre in the polymer matrix, creating a particulate or fibre composite. On the other hand, the submicron scale of the nanoparticles allows them to interfere with and alter the polymerisation reaction and produce a new polymer with the desired properties. Researchers showed that nanoparticle content of less than 1.0–2.0% by weight of the polymer resin is sufficient to alter the polymer characteristics. The relatively low concentration of the nanomaterials necessary to induce significant changes in the polymer matrix is attributed to the very large surface area of the nanomaterials. The ability of nanomaterials to alter the properties of a polymer is mainly dependent on the efficiency of the dispersion technique [34]. Numerous dispersion techniques, including ultrasonication, centrifuging and magnetic stirring, have been reported in the literature with different levels of success. The level of dispersion success is a function of polymer rheology, the nanoparticle geometry, and its contents. Researchers showed that adding surface functional groups might improve the dispersion and enable the nanoparticles to interfere with the polymerisation process further [35].

Researchers have shown the ability to produce polymer concretes using nanomodified polymers [36, 37]. Polymer concretes with attractive characteristics, such as improved impact strength, high ductility, fracture toughness, and superior electrical conductivity were reported [36, 38]. Researchers demonstrated that Styrene-Butadiene Rubber (SBR) polymer-modified concrete (PMC) incorporating functionalised CNTs has much improved failure strain (Fig. 7) by up to 400% [38].

This improvement in PMC failure strain was attributed to the ability of CNTs to alter SBR polymerisation. Furthermore, scanning electron micrographs (SEM) of an SBR film and SBR PMC incorporating functionalised CNTs are shown in Fig. 8 (a) and (b) respectively. The SEM micrographs demonstrate that CNTs also act as microfibres bridging microcracks, thus improving PMC failure strain [39].

Furthermore, researchers also showed the ability of a mix of pristine and functionalised CNTs to improve the mechanical strength, ductility, and fracture characteristics of epoxy polymer concrete (PC) [40]. Stress-strain diagrams with Epoxy PC incorporating a mix of pristine and functionalised CNTs demonstrate an increase in the material

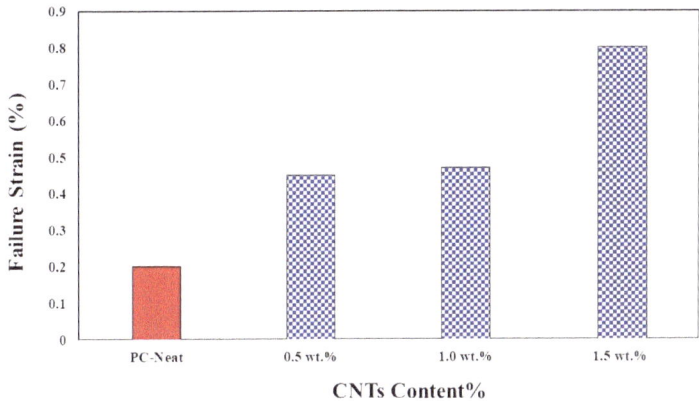

Fig. 7. Failure strain for SBR PMC incorporating functionalised CNTs at seven days of age [39]

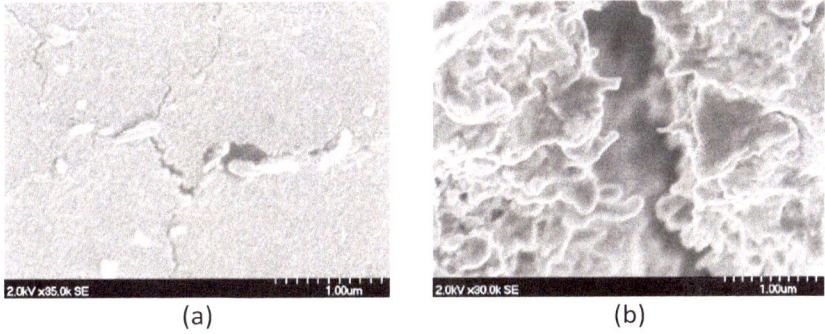

Fig. 8. SEM micrographs showing the interwoven functionalised CNTs affecting the polymer matrix and bridging its microcracks [a] SBR film [b] SBR PMC [39]

failure strain by up to 74%, while maintaining an appreciable tensile strength of 10 MPa (Fig. 9).

Fracture toughness measurements of Epoxy PC incorporating CNTs showed an increase of about 100% in the non-linear fracture toughness represented by the total critical J-Integral [41]. The improvement of the ductility and fracture toughness of PC incorporating this mix of CNTs was attributed to the ability of the CNT mix to increase the tensile failure strain and the shear transfer between the epoxy matrix layers, as demonstrated by the schematic model (Fig. 10).

Finally, it was recently also shown that non-functionalised CNTs can significantly improve the electrical conductivity of PC [38]. The improved electrical conductivity of PC incorporating CNTs is attributed to the ability of the CNTs to percolate the polymer matrix with a CNT content close to 2.0%. Similar observations were reported for glass fibre-reinforced polymer composites incorporating CNTs [42]. The nanomodified PC with improved electrical conductivity can be used as a smart material due to its self-sensing capability.

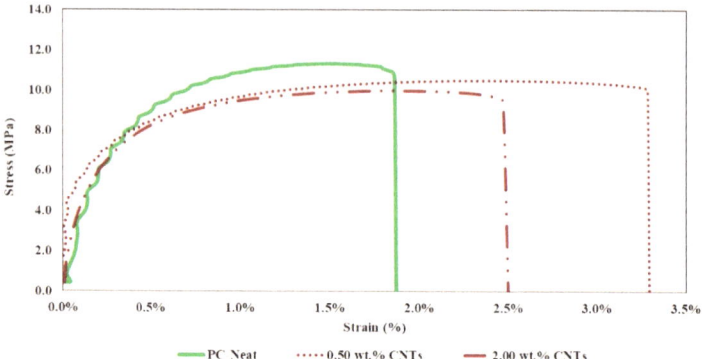

Fig. 9. Stress-strain curves of Epoxy PC incorporating a mix of pristine and functionalised CNTs compared with neat epoxy PC [39]

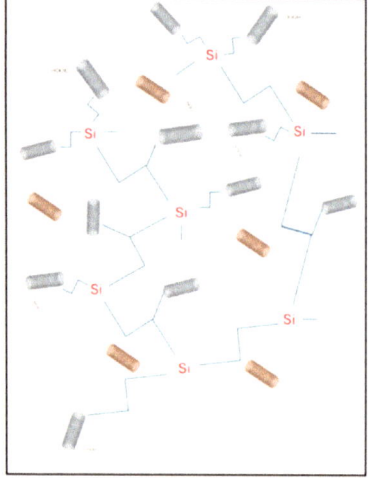

Fig. 10. Schematic representation showing the effect of a mix of functionalised and pristine CNTs on the polymer matrix in Epoxy PMC. The mixed CNTs allowed improving the failure strain and shear transfer in the Epoxy PC

Such PC with superior ductility, improved fracture toughness, and self-sensing capability can be used as a structural material in resilient structures observing significant seismic activities.

4.4   PCC as Waste Polymer Product Storage

In general, in the categories we have stated ahead of the problem (see Sect. 4.1), 'even barely acceptable concrete from waste components', will be a necessity dictated by the civilization. If yes, the request claimed in this sub-chapter is justified. Shortly after the 2018 ICPIC, I (L.Cz) considered the question, "Will recycled plastics be a driving force in concrete technology?" [43]. This is a dramatic forecast, to link together concrete – the absolute premium construction material with which the works of our civilisation are created – with plastic waste, but it is a social responsibility, and accented recycling is a way of survival.

What is more, I have outlined that it should be a paradigm setting the direction for a new development cycle, but there are several particularities to this [6]:

- the objective is not to develop better concrete owing to polymer introduction but concrete with non-deteriorated characteristics despite the use of plastic waste;
- the modifier is not the original liquid polymer used as a binder or co-binder but solid plastic waste as partial filler;
- the changes should be attributed not to the pursuit of concrete refining and its provision with specific functionalities, but to "taking the load off" from plastic landfills, which now contain 70 billion tonnes of plastic and are still growing, estimated to last for 450–600 years. During this period, they will contaminate water, destroying organisms living in natural water, and damage the environment and its appearance;
- the action does not result from the fact that concrete needs more polymer but from the environment not being able to take on more plastic waste, that is a higher amount of used plastic products. In this context, the slogan "good concrete is sustainable concrete" gains new meaning.

A few numbers to estimate how serious the problem is, are:

- plastic production is 300 million tons annually;
- seven billion tons of plastics are already in the landfill.

Roughly speaking, it will be possible to replace around 8% of aggregate by mass with plastics. This means that plastics in the landfill will at least no longer increase in volume. More information can be found in the breakthrough monography (492 pages): F. Pacheco-Torgal et al. (eds.): Use of recycled plastics in eco-efficient concrete. Elsevier 2018 [43].

4.5   Novelties in Unexpected Areas – Selected Examples

"Novelties in unexpected areas" represent a collection, which is the result of a task-oriented study. Such innovations advocate going beyond what is currently possible, and this call captures the public imagination. In civil engineering, however, we should play it safe according to the basic requirements of construction works. This social responsibility does not hamper building innovation but, instead, makes it more sophisticated. Some examples are presented, which at a given moment of disclosure could be treated as unexpected. "Unexpected" means that they are the result of a research programme, but also that it is not easy to forecast them. Retrospectively considering the research work of a given scholar, an achievement is always the result of a chain of values: ideas – research – innovation – validation/verification – implementation. At some moment during this change, a discovery loses its "unexpected" value. Nevertheless, the selected examples originate mainly in the scientific activity of Y. Ohama and D. Van Gemert and may be able to serve as inspiration.

Hardener-free epoxy as modifier is a promising new concept [44, 45, 50]. Conventional epoxy-modified mortars and concretes have inferior applicability due to the two-component mixing of the epoxy resin and the hardener, the toxicity of some hardeners like polyamine or polyamide, and the obstruction of cement hydration by the polymer. Even without hardeners, however, epoxy resin can harden in the presence of the alkalis or hydroxide ions produced by the hydration of cement in the epoxy-modified mortars. Such new epoxy-hydraulic cement systems provide an increase in flexural strength and a marked improvement in carbonation or chloride ion penetration resistance.

In hardener-free epoxy-modified mortars with polymer-cement ratios of 20% or less, the hardening degree of the epoxy resin is 50% to 90%, and unhardened epoxy remains. It is considered that the unhardened epoxy resin may be sealed during the hardened epoxy resin phase in the epoxy-modified mortars. In that way, the epoxy resin phase forms self-capsuled epoxy resin droplets. The self-capsuled epoxy resin can be broken at the cracking of the epoxy-modified mortar under loading. The unhardened resin in the self-capsuled epoxy phase may fill microcracks, thus providing a self-healing capacity to the mortar. The hydroxide ions, set free at hydration, act as hardener elements for the epoxy resin. The mechanism is presented in Fig. 11 and Fig. 12.

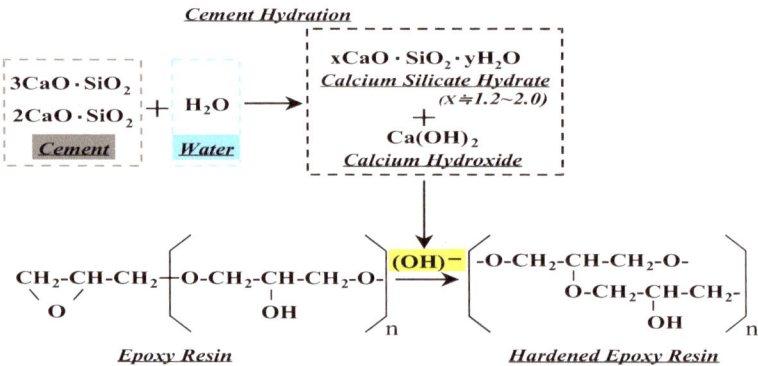

Fig. 11. Hardening mechanism of epoxy resin in hardener-free epoxy-cement system [50]

Nitrite-type hydrocalumite as corrosion inhibitor of reinforced steel. Nitrite-type hydrocalumite [$3CaO.Al_2O_2.Ca(NO_2)_2.nH_2O$] is a corrosion-inhibiting admixture or anticorrosive admixture which can adsorb the chloride ions (Cl^-) that cause corrosion in reinforcing bars, and it liberates the nitrite ions (NO_2^-) that inhibit the corrosion. This is expressed by the formula in Fig. 13. It provides excellent corrosion-inhibiting properties to the reinforcing bars in concrete.

Consequently, polymer-modified mortar with superior corrosion-inhibiting properties and durability is expected when combining the use of nitrite-type hydrocalumite and hardener-free epoxy-resin. It is used as an effective repair material for deteriorated reinforced concrete structures [7, 40].

Water-soluble polymer (WSP) as an effective nano modificatory. If we change the polymer position from a micro-area into a nano-area, it will be bridging not between the crack-edges but between the hexagonal plates of Portlandite. The situation and the

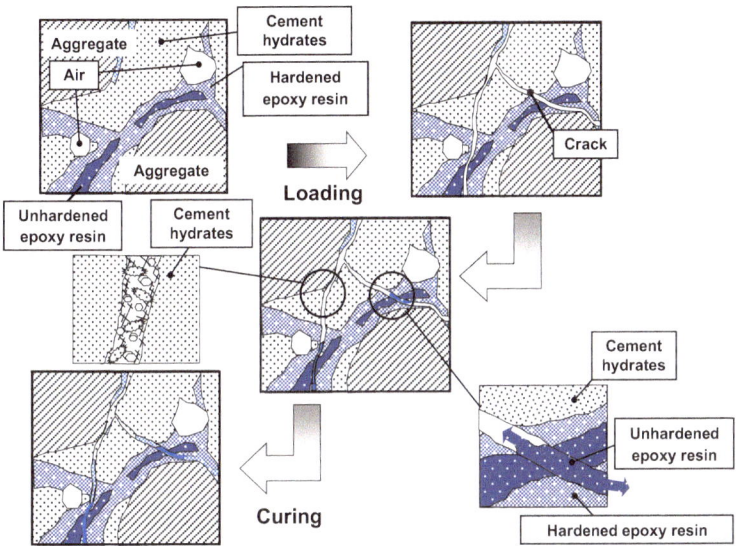

Fig. 12. Cracks caused by mechanical loading liberate unhardened epoxy from capsules. Epoxy hardens in contact with cement hydrate [50]

$$3CaO \cdot Al_2O_3 \cdot Ca(NO_2)_2 \cdot nH_2O + 2Cl^-$$

(Nitrite-type hydrocalumite)

$$\rightarrow 3CaO \cdot Al_2O_3 \cdot CaCl_2 \cdot nH_2O + 2NO_2^-$$

(adsorb)   (liberate)

Fig. 13. Chloride ions adsorption mechanism of nitrite-type hydrocalumite

result will change drastically. On a micro-level, we will use 10% of polymer and receive 10 MPa of tensile strength. On the nano-level, we will use only 1% of polymer (ten times less) and receive 15 MPa of tensile strength. On a conceptual level, however, *changing the polymer position from a micro-area to a nano-area* is easier said than done. Ideas create innovation, but how is this technologically implemented? This has been possible due to the breakthrough achievement of D. Van Gemert and E. Knapen [46, 47]. If we use water-soluble polymer, WSP, instead of liquid polymer, the WSP, due to the thermodynamic conditions, will be placed in the "nano-area". Understanding the nature of polymer-modified materials and various practical reasons, as well as the *logic of concrete technology*, shows that the water-soluble polymers could be a very promising modification of concrete. Only very few (literally) publications [13] addressed the particular microstructure of water-soluble polymer cement mortars and a fair amount of those microstructure–technical properties have been published before D. Van Gemert and E. Knapen's discovery. This research field should be developed further. Researchers are focused on self-cleaning, self-repairing, high-adhesive, active products, such as air pollution reduction and nano-porous insulation products [48]. Progress has also been made with nano-coating on concrete. This is still at the stage of expectations, however, with a very promising outlook.

5   Conclusions

According to the authors' statement, there is need for a new paradigm of C-PC development; it seems to be both the proper time and the proper reasons. A new paradigm should be found within the framework of sustainable development. The circular economy will define the paths of development and new challenges. The following have been considered as novelties:

- new material solutions;
- new technological methods;
- new applications.

It turns out that it is difficult to formulate one "development paradigm" on a general level, but one that is substantive enough to be of practical meaning. There are three directions that seem to be the most promising:

- the mineral organic as the gateway to the new generation of C-PC;
- nanomodification as the new thrust in C-PC technology;
- PCC as waste polymer product storage – paradoxically a new application area.

"Novelties in unexpected areas" are not included. The list may be short but it is not exhaustive!

References

1. Cresson, L.: Latex-hydraulic cement-system. British Patent 19/474 (1923)
2. Kukacka, L.E. et al.: Introductory course on concrete-polymer materials. Brookhaven National Laboratory, Upton, MA (1974)
3. Hop, T., Miodyński, I.: Polymer-mineral mixtures as new building materials. Build. Sci. **2**, 147–163 (1967)
4. Czarnecki L.: Polymers in concrete on the edge of the millennium. In: Proceedings of 10th International Congress on Polymers in Concrete, International Center Aggregate Research (ICAR), Universality of Texas at Austin (2001)
5. Czarnecki, L., Reda Taha, M.M., Wang, R.: Are polymers still driving forces in concrete technology? In: Reda Taha, M.M. (ed.) International Congress on Polymers in Concrete (ICPIC 2018). Polymers for Resilient and Sustainable Concrete Infrastructure, pp. 219–225. Springer, Cham (2018)
6. Czarnecki, L.: My pursuit of truth in building materials engineering. ACE **66**(3), 3–35 (2020)
7. Reda Taha, M.M. (ed.): International Congress on Polymers in Concrete (ICPIC 2018). Polymers for Resilient and Sustainable Concrete Infrastructure. Springer, Cham (2018)
8. Tran, N.P., Nguyen, T.N., Ngo, T.D.: The role of organic polymer modifiers in cementitious systems towards durable and resilient infrastructures: a systematic review. Constr. Build. Mater. **360**, 129562 (2022). https://doi.org/10.1016/j.conbuildmat.2022.129562
9. Chandra, S., Ohama, Y.: Polymers in Concrete. CRC Press, Boca Raton, FL (1994)

10. Van Gemert, D., Czarnecki, L., Wang, R., Cizer Ö.: Contribution of concrete-polymer composites and ancient mortar technology to sustainable construction. In: Reda Taha, M.M. (ed.) International Congress on Polymers in Concrete (ICPIC 2018). Polymers for Resilient and Sustainable Concrete Infrastructure, pp. 299–305. Springer, Cham (2018)

11. Van Gemert, D.: Modeling structural built up in polymer concrete and in polymer-cement concrete. In: Van Gemert, D., Kabashi, N., Nushi, V., Baha, R. (eds.) Proceedings of 1st Kosovo Seminar on Polymers in Concrete, pp. 39–54. Hasan Prishtina, Kosovo (2013)

12. Van Gemert, D., Beeldens, A.: Evolution in modeling microstructure formation in polymer-cement concrete. Restor. Build. Monum. **19**(2–3), 97–108 (2013)

13. Sasse, H.R.: Water-soluble plastics as concrete admixtures. In: Proceedings of First International Congress on Polymer Concretes, pp. 168–173. The Construction Press, Lancaster (1975)

14. Bareš, R.: A conception of a structural theory of composite materials. In Brandt, A., Marshall, I. (eds.) Brittle Matrix Composites I, pp. 25–48 (1985)

15. Van Gemert, D., Czarnecki, L., Bareš, R.: Basis for selection of PC and PCC for concrete repair. Int. J. Cem. Compos. Light. Concr. **10**, 121–123 (1988)

16. Ohama, Y.: Principle of latex modification and some typical properties of latex modified mortars and concretes. ACI Mater. J., 511–518 (1987)

17. Beeldens, A., Van Gemert, D., Schorn, H., Ohama, Y., Czarnecki, L.: From microstructure to macrostructure: an integrated model of structure formation in polymer modified concrete. RILEM Mater. Struct. **38**(280), 601–607 (2005)

18. Dimmig-Osburg, A.: Microstructure of PCC—effects of polymer components and additives. In: Proceedings of 12th ICPIC, pp. 239–248. Chuncheon, Korea (2007)

19. Tian, Y., Li, Z., Ma, H., Jin, N.: An investigation on the microstructure formation of polymer modified mortars in the presence of polyacrylate latex. In: Leung, C., Wan, K. (eds.) Proceedings of International RILEM Conference on Advances in Construction Materials Through Science and Engineering, pp. 71–77. Hong Kong (2011)

20. Czarnecki, L., Schorn, H.: Nanomonitoring of polymer cement concrete microstructure. Restor. Build. Monum. **13**(3), 141–152 (2007)

21. Knapen, E., Van Gemert, D.: Cement hydration and microstructure formation in the presence of water-soluble polymers. Cem. Concr. Res. **39**(1), 6–13 (2009)

22. Construction - Wikipedia (14/02/2021)

23. Buildings and construction | Internal Market, Industry, Entrepreneurship and SMEs (https://europa.eu) (2021)

24. International Energy Agency Publications, Material efficiency in clean energy transitions, p. 36 (March 2019)

25. Cement production global 2021 | Statista (access 24/08/2022)

26. Czarnecki, L., Van Gemert, D.: Innovation in construction materials engineering versus sustainable development. BPASTS **65**(6), 765–771 (2017). https://doi.org/10.1515/bpasts-2017-0083

27. Czarnecki, L., Justnes, H.: Sustainable & durable concrete. Cem. Lime Concr. **17**(79, No. 6), 341–362 (2012)

28. Tomaszewska, J.: Construction versus circular economy. ICPIC (2023)

29. Falaciński, P., Machowska, A.: Conditions of circular economy in Poland. ICPIC (2023)

30. Sasse, H.R., Fiebrich, M.: Bonding of polymer materials to concrete. RILEM Mater. Struct. **16**, 293–301 (1983)

31. Wang, R., Wang, G., Wang, P.: Status of research and application of C-PC in China. In: Wong Sook Fun, Tan Kang Hai, Ong Khim Chye Gary (eds.) Proceedings of 15th International Congress on Polymers in Concrete (ICPIC 2015), Advanced Materials Research, Vol. 1129, pp. 59–68 (2015)

32. Wang, R., Li, J., Zhang, T., Czarnecki, L.: Chemical interaction between polymer and cement in polymer-cement concrete. Bull. Pol. Acad. Sciences. Tech. Sci. **64**, 785–792 (2016)
33. Wang, R.: Interaction between polymer and cement: A review. ICPIC (2023)
34. Yu, N., Zhang, Z.H., He, S.Y.: Fracture toughness and fatigue life of MWCNT/epoxy composites. Mater. Sci. Eng.: A **494**, 380–384 (2008)
35. Theodore, M., Hosur, M., Thomas, J., Jeelani, S.: Influence of functionalization on properties of MWCNT–epoxy nanocomposites. Fracture toughness and fatigue life of MWCNT/epoxy composites. Mater. Sci. Eng.: A **528**(3), 1192–1200 (2011)
36. Daghash, S.M., Soliman, E., Kandil, U.F., Reda Taha, M.M.: Improving impact resistance of polymer concrete using CNTs. Int. J. Concr. Struct. Mater. **10**(4), 539–553 (2016). https://doi.org/10.1007/s40069-016-0165-4
37. Emiroglu, M., Douba, A. E., Tarefder, R., Kandil, U.F., Reda Taha, M.M.: New polymer concrete with superior ductility and fracture toughness using alumina nanoparticle. J. Mater. Civ. Eng. **29**(8) (2017). https://doi.org/10.1061/(ASCE)MT.1943-5533.0001894
38. Reda Taha, M.M., Douba, A.E., Emiroglu, M., Kandil, U.F.: Electrically and thermally conductive polymer concretes. U.S. Patent Number 10,494,299 B2, issued Dec. 3, 2019 (2019)
39. Soliman, E., Kandil, U.F., Reda Taha, M.M.: The significance of carbon nanotubes on styrene butadiene rubber (SBR) and SBR modified mortar. Mater. Struct. **45**(6), 803–816 (2012). https://doi.org/10.1617/s11527-011-9799-5
40. Douba, A.E., Emiroglu, M., Kandil, U.F., Reda Taha, M.M.: Very ductile polymer concrete using carbon nanotubes. Constr. Build. Mater. **196**(30), 468–477 (2018). https://doi.org/10.1016/j.conbuildmat.2018.11.021
41. Douba, A.E., Emiroglu, M., Tarefder, R., Kandil, U.F., Reda Taha, M.M.: Use of carbon nanotubes to improve fracture toughness of polymer concrete. Transp. Res. Rec.: J. Transp. Res. Board **2612**(1), 96–103 (2017)
42. El-Sabagh, A., Taha, E., Kandil, U.F., Nasr, G.M., Reda Taha, M.M.: Monitoring damage propagation in glass fiber composites using carbon nano-fibers. Nanomaterials **6**(169) (2016). https://doi.org/10.3390/nano6090169
43. Pacheco-Torgal, F., et al. (eds.): Use of Recycled Plastics in Eco-efficient Concrete. Elsevier (2018)
44. Ota, M., Ohama, Y., Tatematsu, H.: Properties of polymer-modified mortars using hardener-free epoxy resin with nitrite-type hydrocalumite. In: Proceedings of ISPIC 2006, Guimarães, pp. 49–59 (2006)
45. Ohama, Y., Ota, M.: Recent trends in research and development activities of polymer-modified paste, mortar and concrete in Japan. In: Advanced Materials Research, vol. 687, Proceedings ICPIC XIV, Shanghai, pp. 26–34 (2013)
46. Knapen, E., Beeldens, A., Van Gemert, D.: Water soluble polymeric modifiers for cement mortars and concrete. In: Proc. Con. Mat. 2005, Vancouver, BC (2005)
47. Knapen, E., Van Gemert, D.: Cement hydration and microstructure formation in the presence of water-soluble polymers. Cem. Concr. Res. **1**, 6–13 (2007)
48. Broekhuizen, F., Broekhuizen, P.: Nano-products in the European construction industry—State of the Art 2009 Report commissioned by EFBWW and FIEC, Amsterdam (2009)
49. Van Gemert, D., Czarnecki, L., Beeldens, A., Łukowski, P. et al.: Cement concrete and concrete-polymer composites: two merging worlds. A report from 11th ICPIC Congress in Berlin, 2004. In: Proceedings of the 11th International Congress on Polymers in Concrete – ICPIC, Berlin (2004)
50. Ohama, Y., Van Gemert, D., Ota, M.: Introducing process technology and applications of polymer-modified mortar and concrete in construction. Restor. Build. Monum. **19**(6), 369–392 (2013)

Construction Versus Circular Economy

Justyna Tomaszewska-Krygicz[✉]

Deloitte Advisory Sp. z .o.o. Sp. k., Warsaw, Poland
jtkrygicz@deloittece.com

Abstract. Closing the loop of materials circulation is certainly the right way to decrease the pressure humans place on the environment. Although many efforts have been made toward effective mitigation of anthropogenic impacts, mostly on the policy dimension, there is still much more to do. The transformation affects every phase of the building's lifecycle, which therefore requires the engagement of all value chain actors and suppliers, often assigning them new roles and responsibilities. The experiences gained from over a decade of the EU's journey towards CE clearly indicate that, regardless of how good the legal regulations are and how effective the educational efforts are, achieving the goals of maintaining resources in the economy is not possible without the implementation of innovation and proper business models. Considering that materials and resources marketplaces are among the most common areas of Contech investment, it is worth considering what role polymer concrete composites (CPC) may play in the sector's quest for circularity. This article will try to find an answer to this dilemma by discussing the meaning of CE for the sector, its main drivers, and implications for the supply chain.

Keywords: Circular Economy (CE) · Construction sector · Supply chain

1 What Does CE Mean for the Construction Sector?

The scale and specificity of products delivered by the construction sector means that it can be described using three key attributes, such as resource—and energy-intensive and contributing to the creation of significant amounts of waste. Due to the observed increase in awareness in recent years regarding the consequences of excessive use of fossil fuels, including the deteriorating quality of inhaled air that we all experience on a daily basis, the high-emissions label has been added to this set. The dynamic development of increasingly advanced ConTech and PropTech technologies dedicated to construction, including modular construction [1] digital twin [2, 3], and BIM [4, 5], is a necessary condition in striving to improve the sector's characteristics and broadly understood savings. Without the implementation of innovation, achieving these formulated goals is rather unlikely. However, today we already know that strengthening the technological area is not sufficient to radically change the growth trajectory and additionally requires coordinated legislative and educational stimuli. The European transformation towards a closed-loop economy, initiated in 2014 with the publication of the EC communication "Towards a

L. Czarnecki et al. (Eds.): ICPIC 2023, 61, pp. 22–34, 2025.
https://doi.org/10.1007/978-3-031-72955-3_2

circular economy: a zero waste programme for Europe (COM(2014) 398 final)" and then in 2015 with the "An EU action plan for the Circular Economy (COM(2015) 614)", is a challenge facing the well-established construction sector, rooted in its traditions and practices. However, this transformation is increasingly seen as a unique opportunity for the entire sector to reconsider its approach and build a new concept of construction that co-creates and even lives in symbiosis with the environment. This is the idea behind the CE model, initiated at the design stage, taking a life-cycle perspective, and based on three pillars: maximizing the potential accumulated in materials, eliminating waste and pollution, and mitigating negative changes to the environment. Operating in the CE model requires the involvement and collaboration of the entire supply chain and transparency in actions taken. The ultimate reward is a situation in which each party benefits irrevocably. CE aims to maintain a balance between environmental and economic benefits, which in light of the paralyzing amounts of requirements and regulations and the lack of appropriate educational, economic and technological support, can be nearly intangible. Nevertheless, without unnecessary discussion or reference to the opinions of prominent experts, it can be clearly stated that the actions currently being taken at various levels of social life to protect nature and mitigate the effects of previous destructive human activity are a right direction and, thankfully, increasingly advocated by younger generations. The bitter fact that the main motivation of to undertake all these actions is the fear of severe consequences arising from the way man interacts with the environment and experienced in the form of increasingly frequent disasters.

2   Life Cycle Perspective

2.1   Status Quo

Despite the fact that the Construction Project Regulation (CPR 305/2011), which lists among its basic requirements the seventh stating that the construction works to be designed, built, and demolished in a way that ensures durability, reuse, and recycling of resources, was introduced more than a decade ago, the actual development of sustainable building is only taking place now as a result of political, economic, technological, and social movements. This is observed in the form of an increasing number of buildings undergoing voluntary multi-parameter evaluations verifying compliance with often stringent requirements for sustainable building, including both the construction process and future operational parameters such as BREAM, LEED, DGNB or WELL. According to the latest analyses by the Polish Association of Ecological Building (PLGBC) [6], the number of certified buildings in Poland has already reached almost 1400, with a total usable area of 28.6 million square meters, giving a 24% increase over the year. Currently, three main trends are observed in the domestic market: a dynamic increase in certified warehouse space by 4 million square meters per year, thus downgrading certified office space, which dominated all multi-criteria certifications in Poland from the very beginning; a decrease in new certifications among commercial properties, most likely due to the COVID-19 pandemic, and an increasing number of WELL awarded, with 43 analyzed during the annual period, compared to seven the previous year. It is also worth noting that currently 45% of all certified buildings in the Central and Eastern Europe region are in Poland [7].

2.2 Environmental Impacts Through Life Cycle

The concept of life cycle approach involves an evaluation of the environmental impacts occurring throughout the entire lifecycle, including the upstream and downstream stages. As can be seen, the life cycle perspective is becoming an integral part of the European policy framework despite having been recognized as valuable in the construction sector for some time [8–10]. The objective of reducing greenhouse gas emissions (GHG), also referred to as a carbon footprint, is to be distinguished at various stages of a product's life cycle, including embodied carbon associated with the construction phase [11], is becoming the overarching goal of introduced policies aiming to improve buildings energy performance [12, 13]. GHG emissions are also one of the most frequently analyzed environmental indicators by businesses. However, it is important to remember that the life cycle analysis method allows for the study and quantification of a significantly larger number of environmental impacts than just GHG emissions. The ISO 14040–14044 series of standards set by the International Standardization Committee defines the principles, structure, and methodology for conducting a life cycle assessment (LCA) of environmental impacts. In Europe, the EN 15804 standard is used for the assessment of construction products and systems, while the EN 15798 standard is used for buildings and the Level(s) [14] assessment system is gaining increasing interest. These standards distinguish four fundamental stages in the life cycle of a product and building, including the product stage, construction stage, operation stage, and end of life stage. The benefits and burdens that occur outside the boundaries of the system, resulting directly from the possibility of recovery, reuse, or recycling of the resources involved, are also considered. The assessment of environmental impacts of construction products is becoming an increasingly common practice, although it is still voluntary [15]. For over a decade, the results of such assessments have been published by manufacturers in the form of Type III Environmental Product Declarations (EPD) that comply with the ISO 14025 guidelines [16]. In Europe, a non-profit organization called EcoPlatform [17], which brings together EPD Program Operators and LCA practitioners, has been in existence since June 2013. In Poland, the Institute of Building Technology has so far been the entity issuing EPDs and is a member of EcoPlatform. Since 2022, EPDs published by members of EcoPlatform are available in digital form [18]. There are indications that EPDs will in the future become a part of the mandatory technical assessment of products in accordance with CPR [19]. The extension of product life cycles and the change in their disposal methods at the final stage of existence are the elements that fundamentally differentiate the CE from the linear model, based on the principle of "take-produce-dispose". Maintaining resources in the economic cycle requires focusing special attention on the final and initial stages of the life cycle. The way in which products that have been withdrawn from the operation stage are handled should allow for the maximum recovery of the resources accumulated in the product, so that they can be successfully used in the next economic cycle, in accordance with the European waste hierarchy. Adaptation abandoned buildings by providing them with new functions is an excellent demonstration of how to implement the principles of a circular economy in our daily lives. Figure 2 presents a soviet sewing factory of Tbilisi that has been revived and transformed into a multi-functional urban space (Fig. 1).

Fig. 1. 'Fabrica' in Tibilis as an example of successful building adaptation providing a multi-functional urban space with a hostel, caffe shops, bars, artist studios, shops, educational institutions, and co-working space. (photo by author)

With regards to building products, the first requirements for the use of minimum recycled content are already appearing. An example that heated up the situation for Polish manufacturers importing building materials to Italy, among others, was the CAM (Criteri Ambientali Minimi) [20] requirement, which expects the use of at least 10% recycled material in polystyrene products used for insulation. It is expected that more and more such requirements will appear in line with the announced actions of the European Commission in promoting the use of recycled materials.

2.3   Energy Performance

It is estimated that buildings are responsible for about 40% of energy consumption and 36% of related GHG emissions in the EU. Of this, 80% of energy is used for cooling, heating, and hot water. The statistics that nearly 75% of buildings in the EU are energy inefficient are additionally alarming [21]. Considering the above, acting towards improving the energy efficiency of buildings, including existing ones, plays a key role in the sector's strive towards sustainability and circularity, which requires involvement from multiple supply chain actors. A milestone in improving the energy efficiency of the sector was supposed to be the 2010 Energy Performance of Buildings Directive (2010/31/EU), which later underwent changes, introducing several ideas for improving building efficiency, such as minimum energy performance standards (MEPs), energy performance certificates (EPCs), concepts of nearly-zero-energy buildings (nZEB), and deep renovation, as well as announcing the establishment of the Long-Term Renovation Strategy (LTRS) and the Smart Readiness Indicator (SRI). However, in practice, despite the establishment of additional support programs enabling EU member countries to cooperate, exchange experiences and best practices such as CA EPBD, in most cases, the provisions of the directive were not reflected in national regulations or only to a limited extent. EC is not giving up and despite previous experiences, has decided to tighten the regulations on building energy efficiency. Changes in the ongoing revision of the EPBD directive (COM/2021/802 final), which is part of the "Fit for 55" package [22] are very ambitious and will certainly strongly impact the further development path of the building sector. Among the basic assumptions are, among others, that from 2030 all new buildings should be zero-emission buildings, and by 2050 existing buildings should be transformed into zero-emission buildings. After long and turbulent discussions, the member states agreed to introduce minimum standards for the energy performance of existing buildings that would correspond to the maximum amount of primary energy that buildings can consume in per square meter per year. The aim is to initiate a wave of thermomodernization and lead to a gradual phasing out of buildings with the lowest parameters. Exceptions are to be made for historical buildings, places of worship, and buildings used for defense purposes. Improving the energy efficiency of building operational systems, including heating, cooling, lighting, and transitioning to clean energy sources, is crucial in reducing the energy demand of buildings and minimizing their environmental impact, and will therefore play a significant role in the shift towards a circular economy in the sector. The International Energy Agency (IEA) [23] reports that since 2020 the rate of investment in clean energy has increased to 12% totaling around USD 260 billion in 2021 in Europe.

Fig. 2. Bosco Verticale in Milano. A well-known example of sustainable and energy-efficient residential multi-storey building. (Photo by author)

3   Implications for Supply Chain

3.1   A Circular Economy Action Plan

Construction and building are among the seven key areas of the value chain listed in the revised European Commission's new Circular Economy Action Plan for a cleaner and more competitive Europe, published in 2020 (COM(2020) 98 final) [24]. This plan references the International Resource Panel (IRP) [25] suggestions that the implementation of material efficiency strategies across the entire building sector could lead to a substantial decrease in greenhouse gas emissions, with the potential for a reduction of up to 80%. The path towards this improvement is through the "Strategy for a Sustainable Built Environment," which will ensure uniformity in important policy areas such as climate, energy efficiency (including the 'Renovation Wave' initiative [36]), resource management, waste management from construction and demolition, accessibility, digitization, and skills. The strategy will promote the whole life cycle perspective of buildings through:

- Sustainability performance of construction products, including the requirement of recycled content,
- Increasing the longevity and adaptability of built structures and developing digital logbooks for buildings,
- Integrating a life cycle assessment perspective into public procurement and the EU sustainable finance framework using the Level(s) scheme,
- Re-evaluating the material recovery targets set in EU legislation for construction and demolition waste,
- Reducing soil sealing, revitalizing abandoned or contaminated brownfields, and promoting the safe, sustainable, and circular use of excavated soils.

The implementation of these postulations will undoubtedly have an impact on all actors in the supply chain, especially on construction product manufacturers who will be forced to invest in the reformulation of their products. This, in turn, will lead to the need to incur further expenditures for laboratory research necessary for these products to be admitted to the building materials market.

3.2   Ecodesign for Sustainable Construction Products

The legislative demand for more environmentally sustainable and circular products is also evident in the proposed new Ecodesign for Sustainable Products Regulation, which was published in March 2022 (COM(2022)142). This proposal builds upon the existing Ecodesign Directive (2009/125/EC), which only covers energy-related products, however, it is estimated that in just 2021 alone, it resulted in a savings of 120 billion euros in energy costs for EU consumers and a reduction of 10% in the annual energy consumption of the products under its jurisdiction [27]. The revised regulation proposal emphasizes the need for products that align with a climate-neutral, resource-efficient, and circular economy, reducing waste and making it the norm for products to perform well in terms of sustainability. As regards construction products, the proposal assumes that requirements will only be established if the implementation of the revised CPR regulation

(COM(2022) 144 final) does not achieve the environmental sustainability objectives set forth under this regulation. The regulation will continue to apply to energy-related products, as before.

3.3 Sustainable Investment

The idea of supporting and promoting sustainable investments is established in the 2020 classification system, introduced by the regulation, commonly known as the Taxonomy (2020/852/EU). It indicates six key environmental goals against which investments are to be evaluated, including: (1) Climate change mitigation, (2) Climate change adaptation, (3) The sustainable use and protection of water and marine resources, (4) The transition to a circular economy, (5) Pollution prevention and control and (6) The protection and restoration of biodiversity and ecosystems. The aim of this initiative is to differentiate between investments that cause environmental harm and those that are neutral or environmentally friendly and will contribute to achieving climate neutrality in the long term. This method of classification is intended to support financial institutions in the decision-making process and allow safe redirecting of capital without incurring additional reputational risk. The lack of clearly specified criteria for determining which investments are environmentally sustainable has led to the spread of greenwashing. The establishment of harmonized classification rules is meant to be a solution to this problem. It is worth noting that the Taxonomy does not ban investment in activities harmful to the environment but grants additional preferences to ecological solutions. Technical classification criteria regarding the first two environmental goals and not causing significant harm to any of the other environmental goals, were established in the first delegated act (2021/2139/EU) to the Regulation (2020/852/EU). In the case of infrastructure construction, all investments intended for the transport or broadly defined storage of fossil fuels have been categorically excluded. In the context of new buildings, the Taxonomy refers to the EU Level(s) [14] assessment system that supposed to be a common language for assessing and reporting on the sustainability performance of buildings. In February of this year Platform on Sustainable Finance [28] has been reactivated. The platform serves as an advisory body established under Article 20 of the Taxonomy Regulation and operates under the Commission's horizontal rules for expert groups.

3.4 Non-Financial Reporting—Environment, Society & Governance (ESG)

Until recently, investment decision-making was based on the analysis of a company's financial results relative to the results of entities operating in similar macroeconomic conditions. The issue of sustainable financing and investing is gaining increasing importance as investors, especially in mature capital markets, increasingly recognize that evaluating and valuing specific categories of business risks associated with companies that more transparently communicate non-financial data, specifically Environmental, Social, and Governance (ESG) data, is easier and more precise [29]. This makes these companies a safer potential investment target.

In a study conducted by Deloitte in 2022, involving over 2,000 representatives of management (CxO, C-level) from 24 countries worldwide, representing the most important sectors of the economy, nearly all respondents stated that their companies experienced the consequences of climate changes in the past year. According to the Deloitte study, environmental issues are often mentioned by survey participants as one of the three most influential factors on companies, only slightly trailing economic perspectives, and clearly ahead of areas such as innovation, talent search, or supply chain challenges. Nearly all respondents admitted that their companies experienced the consequences of climate changes in the past year, and 61% of them expect that in the next three years, they will have a significant or very significant impact on their businesses' strategies or operational activities. 36% indicate that the impact will be moderate, and only 3% that it will be negligible or none. As a result, over 75% of CxO assess that their organizations increased spending on sustainable investment in the past year, and almost 20% indicate that it was done significantly [30]. Investing in ESG assets is so profitable that more and more companies are starting to label themselves as such, even though they have nothing to do with sustainable development. The desire to raise the ESG ratio causes companies to increasingly advertise their products as sustainable and environmentally friendly, even if it is not true. Mandatory non-financial reporting is intended to prevent this. In January 2023, the Corporate Sustainability Reporting Directive (CSRD) (2022/2464/EU) became effective, fortifying the regulations surrounding the social and environmental reporting that companies must provide. A wider range of large corporations, as well as publicly traded small and medium-sized enterprises, will now have to disclose sustainability information, affecting approximately 50,000 companies in total [31]. Companies subject to the CSRD will have to report according to European Sustainability Reporting Standards (ESRS) which are to be issued by the EC in first half of 2023. The standards encompass 84 mandatory disclosures accounting to 1144 data related to environment (E), society (S) and governance (G) (Table 1).

Table 1. Phases of CSRD reporting obligation for EU entities.

Phase	Reporting year	Reporting period	Scope	Type of entity
1st phase	2025	2024	full ESRS	public interest entities and large listed companies that are currently subjected to the NFRD directive
2nd phase	2026	2025	full ESRS	all large entities and large capital groups
3rd phase	2027	2026	choice to apply full or simplified ESRS	Small and medium-sized listed companies and specific types of entities

The new reporting requirement will provide investors and other stakeholders the information they need to assess investment risks arising from climate change and other sustainability issues.

4 Technology as a Circularity Driver

4.1 Siginificance of ConTech and PropTech

The development of construction technologies (ConTech), including solutions that provide an access to green energy and IT solutions finding wide application in digitization and management of building resources (PropTech), are a decisive asset to enable transforming the construction industry. Providing a constant and accessible supply of safe and green energy from renewable sources would significantly bring forward the timeline for change. The traditionally conservative and resistant-to-change construction sector is beginning to recognize dozens of benefits resulting from using ConTech and PropTech technologies what paradoxically was triggered by the COVID-19 pandemic.

The flexibility of these technologies means that they are constantly adapted to the challenges facing the construction industry. An example is the increasingly widespread integration of BIM technology with LCA for efficient quantification and monitoring of a building's environmental impacts [5, 6], also in combination with modular construction, which allows work to be conducted off-site [1]. There are also examples of successful use of locally available materials, native clays and soils, as well as supplementary cementitious materials (SCM) and geopolymers for 3D printing, which is projected to become an industrial reality in the near future [32]. Meanwhile, the increasing use of digital twin technology for project management or lean management is no longer surprising anyone and is gradually becoming part of our daily lives [2, 3]. Among the most common areas of Contech investment in 2022 were project designing including planning, scheduling, specification and budgeting (40.8%), off-site and modular construction (8.3%), materials and resources marketplaces (7.9%) as well as sustainable materials (4.4%). Despite years of record investment reaching 5.38 billion USD (31% of all investments took place in Europe), ConTech saw no growth and only a slight decrease of less than 1% compared to 2021—5.4 billion USD—due to unfavorable macroeconomic conditions, which are predicted to persist throughout the current year [33].

4.2 A Role of CPC in CE Model

The ability to shape the properties of polymer concrete (CPC), which constitute a large group of composite materials hidden under the acronyms PMC, PCC, PIC, PC, PFiC, causes them to have a great potential for use in construction [34]. Considering that materials and resources marketplaces as well as sustainable materials are among the most common areas of Contech investment and the fact that construction sector is becoming open to innovations like never before, it is worth considering what role polymers "in concrete" and "on concrete" may play in the sector's quest for circularity. It is worth to mention that the production of CPC in the 1950-1970s were a manifestation of "progress" and "modernity" [35, 36]. As has been emphasized many times, one of the fundamental assumptions of CE is to extend the life span through proper design, including allowing repairs in the life cycle. An excellent example of a material that meets these requirements is concrete, which reveals its potential for self-healing and crack-healing as a result of its synergistic interaction with polymers (CPC), thus enhancing the durability and longevity of concrete structures [37–39]. Another possibility of shaping the properties of concrete polymer composites (CPC) is the use of various modifications, including increasingly popular nanomodifications in which the use of nanometric polymer particles, characterized by a high ratio of surface area to volume, leads to improved dispersion and improved adhesion to the cement matrix [40]. The advent of lightweight polymer concrete came in response to frequent complaints about the weight of concrete, which at times leads to its rejection as a building material. The use of lightweight polymers, such as foam and expanded polystyrene, reduces the weight of concrete structures while simultaneously improving its energy efficiency [41]. Recently, there has been growing interest in using recycled plastics in CPC to reduce the amount of waste, which has fortunately received various bystander reactions. Nevertheless, researchers are exploring the use of recycled polymers as partial replacements for traditional concrete components, such as aggregates and binders [42, 43].

Given the incredibly attractive functional properties of CPC, it is worth considering what the disposal of this type of material would look like after the end of the use stage.

Is the statement that CPC using recycled materials is an innovative approach for promoting and mitigating environmental impacts with regards to the extraction of new materials and waste disposal actually true? Despite the absence of negative impact reported in the life cycle of CPC [43], it is worth answering this question by analyzing the life cycle perspective that is so promoted. Therefore, after the long phase of use resulting from the outstanding performance of CPC, including self-healing and self-cracking properties, would it be possible to effectively recover the raw material potential accumulated in these materials? In the author's opinion, it is rather doubtful at the moment, but we can hope that the rapidly developing market for recycling-dedicated technologies and the long-life cycle of CPC structures will provide conditions for developing a solution leading to effective recovery of material resources engaged and efficient maintenance of them in the economic cycle.

5   Conclusions

The transformation of the construction sector towards Circular Economy (CE) affects every phase of the building's lifecycle, starting from the design phase, through construction, operation, and utilization to the end of the lifecycle, which therefore requires the engagement of all value chain actors and suppliers, often assigning them new roles and responsibilities. The experiences gained from over a decade of the EU's journey towards CE clearly indicate that, regardless of how good the legal regulations are and how effective the educational efforts are, achieving the goals of maintaining resources in the economy is not possible without the implementation of innovation and proper business models. Driving innovation, simplifying processes, embracing digitization, and adopting more sustainable practices are decisive assets to enable the path towards change. Designing with a whole-life perspective must be evident in relation to all building sector products, indicating repair, reuse, adaptation, and recycling as a natural method of proceeding. The goal of efficient resource utilization and effective reduction of emissions should inspire every action.

Only this approach will allow for an effective change in the sector's characteristics and the objects used to describe it. This is particularly important in the face of the current geopolitical situation and the clear prospect of economic slowdown. The results of the recent market sentiment survey carried out by Deloitte [44] are not encouraging. Over half of the respondents from Central Europe predict a slowdown in overall market activity, affecting both investors, developers, and market advisors alike. This is confirmed by the fact that only 15% of surveyed investors plan to develop new projects soon, while the rest plan to focus on their existing portfolio and closely monitor the market situation. Developers are also not enthusiastic. While the shortage of investment land was the biggest problem for these group four years ago, today the greatest concerns are related to financing investments.

References

1. Ansah, M.K., Chen, X., Yang, H., Lu, L., Lam, P.T.I.: Developing an automated BIM-based life cycle assessment approach for modularly designed high-rise buildings. Environ. Impact Assess. Rev. **90**, 106618 (2021)
2. Çetin, S., Gruis, F., Straub, A.: Digitalization for a circular economy in the building industry: multiple-case study of Dutch social housing organizations. Resour. Conserv. Recycling Adv. **15**, 100200 (2022)
3. Hickey, B., Gachon, C., Cosgrove, J.: Digital twin—a tool for project management in manufacturing. Procedia Comput. Sci. **217**, 720–727 (2023)
4. Soust-Verdaguer, B., Llatas, C., García-Martínez, A.: Critical review of BIM-based LCA method to buildings. Energy Build. **136**, 110–120 (2017)
5. Khan, M.: Integrating BIM with ERP systems towards an integrated multi-user interactive database: reverse-BIM approach. In: Ranadive, M.S., Das, B.B., Mehta, Y.A., Gupta, R. (eds.) Recent trends in construction technology and management. Lecture Notes in Civil Engineering, vol. 260, pp. 99–110. Springer, Singapore (2023)
6. PLGBC Homepage, https://plgbc.org.pl/, last accessed 2023, February 13
7. PLGBC Zrównoważone certyfikowane Budynki. Raport 2022
8. Bekker, P.C.F.: A life-cycle approach in building. Build. Environ. **17**(1), 55–61 (1982)
9. Visentin, C., Trentin, A.W.S., Braun, A.B., Thomé, A.: Life cycle sustainability assessment: A systematic literature review through the application perspective, indicators, and methodologies. J. Cleaner Prod. **275**, 122509 (2020)
10. Zuo, J., et al.: Green building evaluation from a life-cycle perspective in Australia: a critical review. Renew. Sustainable Energy Rev. **70**, 358–368 (2017). https://doi.org/10.1016/j.rser.2016.11.251
11. Cabeza, L.F., Boquera, L., Chàfer, M., Vérez, D.: Embodied energy and embodied carbon of structural building materials: Worldwide progress and barriers through literature map analysis. Energy Build. **231**, 110612 (2021)
12. Russell-Smith, S.V., Lepech, M.D., Fruchter, R., Meyer, Y.B.: Sustainable target value design: integrating life cycle assessment and target value design to improve building energy and environmental performance. J. Clean. Prod. **88**, 43–51 (2015)
13. Huang, J., Wang, L., Siddik, A.B., Abdul-Samad, Z., Bhardwaj, A., Singh, B.: Forecasting GHG emissions for environmental protection with energy consumption reduction from renewable sources: a sustainable environmental system. Ecol. Modelling **475**, 110181 (2023)

14. Level(s) Homepage, https://environment.ec.europa.eu/topics/circular-economy/levels_en, last accessed 2023, February 13
15. Michalak, J., Michalowski, B.: Understanding sustainability of construction products: answers from investors, contractors, and sellers of building materials. Sustainability **14**(5), 3042 (2022)
16. Marsh, E., Allen, S., Hattam, L.: Tackling uncertainty in life cycle assessments for the built environment: a review. Build. Environ. **231**, 109941 (2023)
17. EcoPlatform Homepage, https://www.eco-platform.org/home.html. last accessed 2023, February 13
18. EcoPortal Homepage, https://www.eco-platform.org/epd-data.html. last accessed 2023, February 13
19. Michalak, J.: Standards and assessment of construction products: Casestudy of ceramic tile adhesives. Standards **2**(2), 184–193 (2022)
20. CAM—Criteri Ambientali Minimi Homepage. https://gpp.mite.gov.it/Home/Cam. Last accessed 2023, February 13
21. EC energy efficiency in buildings homepage. https://energy.ec.europa.eu/topics/energy-efficiency/energy-efficient-buildings_en, last accessed 2023, February 13
22. Fit for 55 Homepage, https://www.consilium.europa.eu/en/policies/green-deal/fit-for-55-the-eu-plan-for-a-green-transition/, last accessed 2023, February 13
23. International Energy Agency (IEA). Energy investment 2022. Report (2022)
24. Hughes, R.: The EU circular economy package—life cycle thinking to life cycle law? Procedia CIRP **61**, 10–16 (2017)
25. The International Resource Panel (IRP): Resource efficiency and climate change: material efficiency strategies for a low-carbon future. Journal **2**(5), 99–110 (2016)
26. Renovation Wave Homepage. https://energy.ec.europa.eu/topics/energy-efficiency/energy-efficient-buildings/renovation-wave_en, last accessed 2023, February 13
27. Ecodesign for sustainable products Homepage. https://commission.europa.eu/energy-climate-change-environment/standards-tools-and-labels/products-labelling-rules-and-requirements/sustainable-products/ecodesign-sustainable-products_en, last accessed 2023, February 13
28. Platform on Sustainable Finance Homepage. https://finance.ec.europa.eu/sustainable-finance/overview-sustainable-finance/platform-sustainable-finance_en, last accessed 2023, February 13
29. Green is good, Miasto2077. Rola i znaczenie ESG w strategii spółek giełdowych oraz w ich komunikacji z rynkiem finansowym. Report (2022)
30. Deloitte: CxO Sustainability Report. Accelerating the Green Transition. Report (2023a)
31. Corporate Sustainability Reporting Directive (CSRD) Homepage, https://finance.ec.europa.eu/capital-markets-union-and-financial-markets/company-reporting-and-auditing/company-reporting/corporate-sustainability-reporting_en, last accessed 2023, February 13
32. Robayo-Salazar, R., Mejía de Gutiérrez, R., Villaquirán-Caicedo, M.A., Delvasto Arjona, S.: 3D printing with cementitious materials: challenges and opportunities for the construction sector. Automation Constr. **146**, 104693 (2023)
33. Cemex. Cemex top 50 construction technology startups. Report (2023)
34. Czarnecki, L., Reda Taha, M., Wang, R. Are polymers still driving forces in concrete technology? ICPIC 2023 (2023)
35. Czarnecki, L.: Polymer concretes. Cement, Lime, Concrete **2**, 63–85 (2010)
36. Czarnecki, L., Łukowski, P.: Polymer-cement concretes. Cement, Lime, Concrete **2**, 243–258 (2010)
37. Czarnecki, L.: Polymer-Concrete composites for the repair of concrete structures. MATEC Web of Conferences **199**, 01006 (2018)
38. Chindasiriphan, P., Yokota, H., Pimpakan, P.: Effect of fly ash and superabsorbent polymer on concrete self-healing ability. Constr. Build. Mater. **233**, 116975 (2020)

39. Ayeleru, O.O., Olubambi, P.A.: Concept of self-healing in polymeric materials. Mater. Today: Proceedings **62**(1) (2022)
40. Monteiro, H., Moura, B., Soares, N.: Advancements in nano-enabled cement and concrete: Innovative properties and environmental implications. J. Build. Eng. **56**, 104736 (2022)
41. Shafigh, P., Che Muda, Z., Beddu, S., Zakaria, A., Almkahal, Z.: Thermo-mechanical efficiency of fibre-reinforced structural lightweight aggregate concrete. J. Build. Eng. **60**, 105–111 (2022)
42. Taurino, R., Bondioli, F., Messori, M.: Use of different kinds of waste in the construction of new polymer composites: review. Mater. Today Sustainability **21**, 100298 (2023)
43. Eskander, S.B., Saleh, H.M., Tawfik, M.E., Bayoumi, T.A.: Towards potential applications of cement-polymer composites based on recycled polystyrene foam wastes on construction fields: impact of exposure to water ecologies. Case Studies in Constr. Mater. **15**, e00664 (2021)
44. Deloitte. Deloitte Real Estate Confidence Survey for Central Europe 2023. Survey (2023)

Condition of Circular Economy in Poland

Paweł Falaciński and Agnieszka Machowska$^{(\boxtimes)}$

Faculty of Building Services, Hydro and Environmental Engineering, Warsaw University of Technology, Warszawa, Poland
agnieszka.machowska@pw.edu.pl

Abstract. The article covers the state of circular economy implementation in Poland. The consumption of raw materials is presented, as well as indicators of monitoring the transformation of the CE in EU and Poland acc. to the COM/EC, OECD and oto-GOZ project. Poland's priorities within the circular economy are also presented. They include innovations, markets for secondary raw materials as well as ensuring their high quality and service. The main financial, organizational, social and technological barriers defined. There are a few examples presenting technical solutions implementing the CE in Poland (e.g. building materials, hydraulic binders). The public authorities are recognized as leaders in the implementation of the CE also in the construction sector. Introducing the circular business models is also necessary.

Keywords: circular economy · raw materials · consumption

1 Introduction

Global urbanization and growing population makes the construction industry one of the fastest growing industries in the world. Unfortunately circular economy is not implemented in construction sector in a high extent. The collected data reveals that the industry annually generates around 1.0 billion tons of solid waste globally (according to the World Bank, building materials account for half of the solid waste generated globally each year). It results in the emission of significant amounts of greenhouse gases. 39% of the world's total CO_2 emissions occur during the production of building materials, building construction and its use [1, 2].

The traditional economic model used in the construction industry must be replaced with a circular model which focuses on using materials and products of the highest quality and limiting the amount of waste to the maximum. The aim is to extend the life of buildings, as well as parts and materials, therefore maintaining their maximum value in the economic cycle.

Thus it is of great importance to implement the principle of estimating the life cycle assessment of materials and products, also involving service of products and flow of information regarding buildings, parts and materials used in them. Closing the loop by re-using the materials and products in structures results in saving the money invested in

© The Author(s) 2025
L. Czarnecki et al. (Eds.): ICPIC 2023, 61, pp. 35–46, 2025.
https://doi.org/10.1007/978-3-031-72955-3_3

buildings and structures and access to complete information on the parts and materials enables owners to optimize maintenance and subsequent capital investments [1].

To implement the circular economy the whole construction process requires indispensable modifications, e.g. methods of waste collection, logistics, production and changing roles in the construction market - producers will also become suppliers.

2   The State of Knowledge About Circular Economy in Poland

Poles try to live ecologically. However, they are not aware of the principles of the circular economy. 75% of respondents have never heard of it, according to the study "The state of knowledge of Poles on the circular economy (CE)" [3]. Meanwhile, it is currently the most realistic concept that will not only protect us against ecosystem degradation, but above all will redefine the perception of needs in a world oversaturated with products. According to the study by Stena Recycling [3], basic pro-ecological activities are very popular among Poles. Most people segregate waste (74%) and save energy at home or at work (71%). Most of us also save water (68%) and pay attention not to waste food when shopping carefully (67%). More than half of Poles minimize the amount of waste by using reusable products (54%) or avoiding the use of disposable bags when shopping (52%). However, some pro-ecological behaviors, important due to the implementation of circular economy in Poland, are still undertaken much less frequently. Only 28% of respondents choose products which are environmentally friendly or in recycled packaging. The fewest people (18%) consciously choose the services or products of socially responsible companies, i.e. guided not only by economic benefits, but also by taking actions for the benefit of ecology or society.

The study showed to what extent the pro-ecological attitudes of Poles translate into awareness of the implementation of the circular economy principles in Poland. Almost three out of four respondents have not come across the concept of the circular economy – over 40% have never heard of it, and as many as 30% do not know whether they know it. Among Poles who have heard of the circular economy (29%), most people associate this term with environmental issues, i.e. the possibility of reducing the number of landfills and waste (57%) and the general improvement of the natural environment (49%). The fewest respondents mentioned circular economy in the context of new EU regulations (24%), ecological product design (28%) and CSR (29%). The activities that make up the ecological portrait of Poles and their level of knowledge about the circular economy are closely related. Poles have great potential and are ready to act for the benefit of the environment. However, Poles still too rarely pay attention to the social responsibility of companies, as well as whether the packaging of the products they buy comes from recycling. Also, too few people have heard of the circular economy in the context of ecological design. Meanwhile, the implementation of the circular economy requires high consumer awareness in these areas. The cooperation with entrepreneurs and motivating each other for positive changes is also important.

3   Consumption of Raw Materials

More than half of CO_2 emissions come from coal mining and the production of building materials. Cement production in Poland in 2022 amounted to 18.8 million tons and was 2.4% lower than in the previous year, while cement sales in 2022 amounted to 18 million tons. Ready-mix concrete production in 2022 amounted to 40.1 million m^3, while sales amounted to 38.8 million m^3 and were lower by 2.8% than in the previous year. The use of aggregate for the production of concrete was 38 million tons [4].

A decrease in production and sales, compared to 2021, was also recorded for other binders, such as lime and gypsum: 1.56 million tons of lime and 1.71 million tons of gypsum binder were produced, and 1.32 million tons and 0.92 million tons were sold, respectively [4].

In the case of wall masonry materials, there was also a decrease in both production and sales, e.g. aerated concrete produced 5.5 million m^3 and 5.1 million m^3 sold, building ceramics 3.6 million m^3 produced and 3.1 million m^3 sold, and silicates produced 1.4 million m^3, and 1.365 million m^3 were sold. The production of mortars in 2022 amounted to 4.3 million tons, and sales to 3.8 million tons [4].

The polish construction sector utilizes tremendous amounts of raw materials, energy and water – 228.6 million tons per year. The increase factor of resources excluded from the economic cycle equals 35.2%. As the polish construction sector expands, the need for resources increases [5].

4   Indicators for Monitoring the Circular Economy Transformation in the EU and Poland

The new circular economic model needs new economic indicators, both domestic and business. They have been created in recent years and are starting to be used to monitor progress and plan activities. The indicators used to measure the circularity of the economy at the global, European Union and Polish level, as well as for a single company are presented.

Monitoring the circular economy is a challenge due to the complexity of the circular economy idea. Thus, the most commonly used set of indicators examines various aspects of circular economy, such as the use of recycled materials or the production of waste. The material flow is also studied, often visualized on the Sankey diagram (Fig. 1) [6]. It presents the flow of raw materials in the economy, from their acquisition to becoming waste - which should be closed in circular economy.

Eurostat also calculates the Union's circularity index, also known as the material circularity index. It determines the level of materials recovered and returned to the economy (in other words, it determines what percentage of materials used in the EU economy come from recycling). Currently, the circularity ratio is 11.9%. Its tendency is growing – it is 3 percentage points more than in 2014. In 2023, the framework for monitoring the circular economy in the EU is to be updated to include the tasks included in the new Circular Economy Action Plan adopted in 2020. Indicators on resource use, including consumption and material footprints, are to be developed.

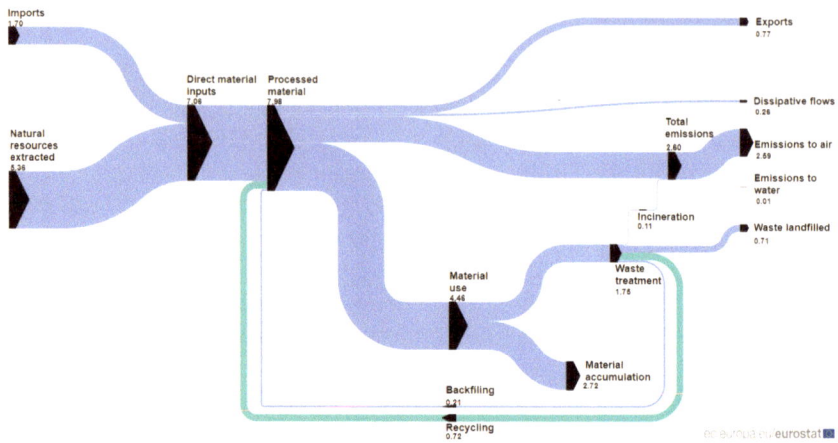

Fig. 1. Sankey diagram presenting material flow in the EU [6].

Poland has not yet developed its own circular economy monitoring framework. This task is included in the government road map [7]. The first recommendations have already been made under the "Oto-GOZ" Gospostrateg project [8]//. In order to track the circularity of the Polish economy, the indicators of sustainable development calculated by the Central Statistical Office can be used. The Central Statistical Office is responsible for transferring data from Poland to Eurostat as part of the EU circular economy monitoring. Data on Poland is available on the Eurostat website. Poland's circularity index (circular use of materials) is 9.8%. It is lower than the EU average and has been decreasing since 2010.

4.1 According to COM EC

The indicators defined by the European Commission are based on four groups:

- Production and consumption - EU self-sufficiency in raw materials, GPP, waste generation, food waste,
- Waste management - total recycling rate, recycling rates for individual waste streams,
- Secondary raw materials - the impact of recycled materials on the demand for raw materials, trade in raw materials subject to the recycling process,
- Competitiveness and innovation - private sector investment, jobs and gross value added, patents.

Selected indicators according to COM EC [6] are shown in Fig. 2.

4.2 According to the OECD

Between 2018 and 2020 the OECD Inventory of Circular Economy Indicators collected more than 400 circular-economy-related indicators, which are classified into five main categories:

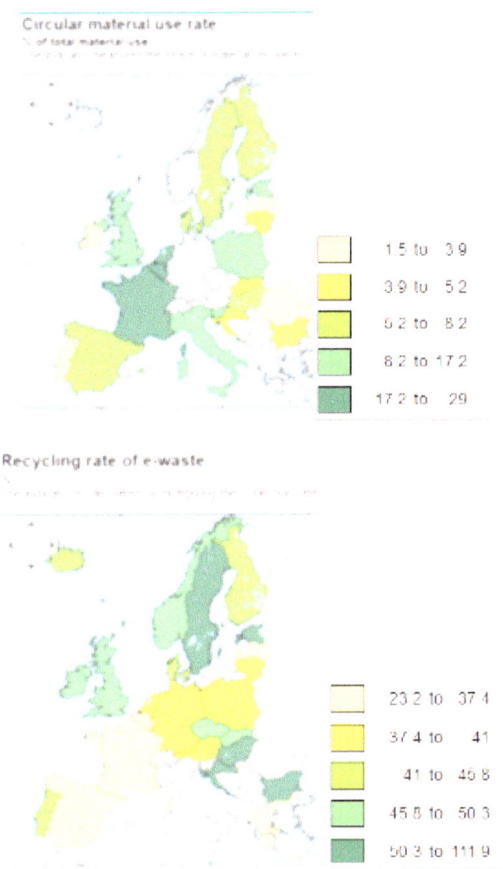

Fig. 2.   Selected indicators according to COM EC [6]

- Environment (39%): indicators such as emissions, output material process and production and consumption, which have a direct impact on the ecosystem,
- Governance (34%): indicators related to education, capacity building and regulation,
- Economic and business (14%): indicators expressed in monetary units such as the value-added of the circular economy and the public investment in circular economy projects, as well as those indicators specifically focusing on activities performed by and within companies,
- Infrastructure and technology (8%): Covers all the indicators that aim to measure the existence of tools, technologies and spaces that boost the circular economy.

The fifth category concerns jobs (5%), which consists of indicators associated with employment and human resources [9].

In addition, sectoral indices, which are presented in Fig. 3, should be mentioned.

Analyzing the Fig. 3 it can be observed that a total of 7% of indicators are devoted to the built environment sector. All five categories are present in this sector: 44% of

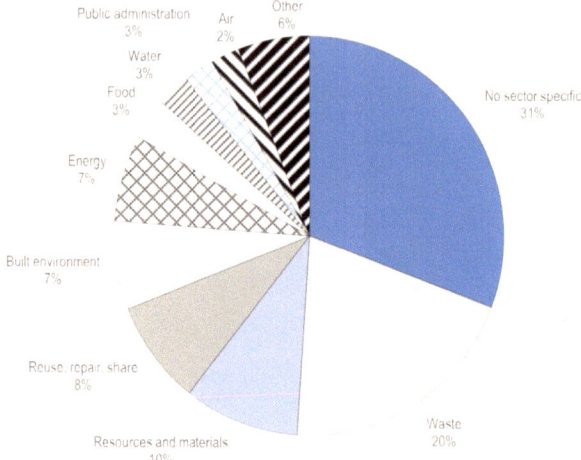

Fig. 3. Indicators by sector [9]

it is the governance category, 32% - environment, 12% - infrastructure and technology, 9% - economy and business and 3% - social category. The whole life cycle of buildings, beginning with the design (e.g. construction works with circular design and projects incorporating smart design), to the end of life (e.g. recovery rate of construction and demolition waste) is covered in the indicators. Also topics such as the use and consumption of materials, including construction and demolition waste usage rate, the recovery rate of construction and demolition waste, the recovery rate of construction waste as material, are covered. Awareness of the construction sector's environmental impact results in circular-economy-related certifications for buildings. Such indicators include the number of companies with certification based on life cycle or eco-design, percentage of construction projects applying to certification programs and the inclusion of eco-designed products [9].

4.3   According to the *OTO-GOZ* Project

The aim of the OTO-GOZ project was to elaborate the set of measurement indicators which can be used to evaluate the progress of transformation towards circular economy, also in the construction sector. This includes the possible economic, ecological and social benefits as well as an increase in competition between the companies, also the influence of the CE on social and economic development, within the regions as well as the whole country. As a result of the OTO-GOZ project the eight CE indicators were established to measure the rate of transformation of polish companies and institutions towards the circular economy and impact of transformation on social and economic development. Four indicators were used on a local level (city and company), two indicators were implemented on a regional level (Małopolskie Voivodeship) and the last two indicators were established for national range and integrated with a project of productivity strategy. These actions, which are of great necessity in the context of sustainable development and climate changes, allow for development of CE monitoring system in Poland [10].

The report presented key CE indicators for construction sector. They are:

- the amount of raw material consumption/income,
- the amount of recycled material consumption/income,
- the amount of critical resource consumption/income,
- the amount of recycled waste,
- the amount of waste submitted to other recovery operations,
- the amount of waste disposed,
- environmental footprint LCA,
- the amount of money invested in CE projects,
- the number of investments adjusted to CE.

It is stated in the report [11] that the construction sector marginally reports circular indicators, only due to image and marketing purposes. It is expected that within five years there will be an increase in the significance of circular indicators due to the impact of CO_2 emissions on production volume and the share of renewable energy in total energy consumption. The research presented in the report [11] that each city or entity should develop its own individual indicators based on identified materials flow and supply chains.

5   Action Plan - Road Map of Circular Economy in Poland

In 2019 the Council of Ministers approved the "Road map of the circular economy" prepared by the Ministry of Development and Technology [7]. Poland's priorities within the circular economy include:

- Innovativeness, strengthening cooperation between the industry and the science sector, and as a result implementing innovative solutions in the economy,
- Creating a European market for secondary raw materials, where their flow would be easier,
- Ensuring high-quality secondary raw materials resulting from sustainable production and consumption,
- Development of the service sector.

The definition of the circular economy presented in the CE Polish Road map is a model of economic development in which the added value of raw materials/resources and products is maximized or the amount of generated waste is minimized and the resulting waste is managed in accordance with the waste management hierarchy (waste prevention, preparation for reuse, recycling, other recovery methods, disposal). The efficiency condition is maintained.

The CE Polish Road map emphasises activities at all stages of the life cycle, starting with product design, through raw material acquisition, processing, production, consumption, waste collection to its management as an alternative to "take - make - use - throw away". The document is divided into 5 chapters covering the topics of sustainable industrial production, sustainable consumption, bioeconomy, new business models and implementation, monitoring and financing of circular economy.

In the construction sector the producer is responsible for collecting more waste and recycling as much as it is possible within the construction process. This is to be implemented already within the design process by incorporating technological and material solutions which limit the amount of waste to the minimum. Recommendations also apply to LCA (Life Cycle Assessment) product specifications and its life cycle.

The CE Polish Road map indicates circular bioeconomy, biological cycle of renewable resources and wastes from food production as one of the greatest potentials for the circular economy in the historical and constructional context. The scale of building products in this context varies from facades, structures, modular construction to furniture and fabrics. There is potential for interdisciplinary and cross-sectoral innovations [12].

6 Status of Circular Economy Implementation in Poland

According to the report of the Supreme Audit Office on actions to reduce the generation of plastic wastes and its proper management [13] the actions of the Minister of Climate and Environment, as the body responsible for shaping the policy related to waste management in the country, as well as other public administration bodies subject to inspection, as well as the advancement of conceptual and legislative work by both the Minister of Climate and Environment and the Minister of Development, were insufficient. It is a significant obstacle in implementation of the CE model in Poland.

The basic barriers on the road to circular construction in Poland are:

- financial,

  - lack of economies of scale,
  - unfavourable investment financing model,

- organizational,

  - regulatory support,
  - hasty regulations,
  - lack of tracking of waste streams,

- social,

  - perception of reuse of building materials and parts,
  - declaration vs. practice,
  - lack of knowledge and competence of investors,

- technological,

  - limited possibilities of recycling materials,
  - cost and pace of construction vs. circular properties.

Nowadays new activities have been undertaken to promote the CE in the polish construction sector. The Minister of Climate and Environment with support of the National Fund for Environmental Protection and Water Management signed the contract to finance the CIRCON project – the circular economy in the construction sector: eco-design of circular buildings. The aims of the project are to strengthen the implementation of CE in construction sector, to prepare a handbook on design of the buildings according to the rules of circular economy and to disseminate knowledge on circular buildings' design among the architects, engineers, academic teachers, students and production, investment and execution companies [14].

As the ecological awareness of the polish society is constantly rising [3] the opportunities to implement the circular economy in companies and institutions become more real.

7   Examples of Organizational and Technical Solutions Promoting Circular Economy in Poland

The implementation of the circular economy in Poland encounters numerous difficulties and obstacles, both legislative and technical. However, it should be emphasized that Polish companies see the need for change and the needs of the market and adapt their activities to new challenges.

In the construction sector, advanced solutions in accordance with the assumptions of the circular economy are introduced both in prefabrication and in the production of building materials. Innovative actions extend the life cycle of products by increasing their durability and maintaining the expected properties at a very high level. An example of such is the technology of insulating permanent formwork made of high-density expanded polystyrene modified with the addition of graphite, which ensures high thermal insulation parameters of the building, thanks to which it achieves low energy demand, low GHG emissions and a long service life (100–150 years). The technology of making this material allows the use of recycled waste, as well as the subsequent recovery of raw materials and thermal insulation components [15].

Another example is the formwork made of polyisocyanurate (PIR) polymer plates used for the construction and thermal insulation of foundations, basements and external walls of buildings. The slabs are connected in parallel with ladders made of reinforcing steel. The advantage of this type of formwork is the reduction of waste and CO_2 emissions associated with formwork transport [15].

The assumptions of the circular economy are also introduced in the production technology of hydraulic binders, which include ash from coal combustion. This by-product of combustion, after appropriate transformation, is widely used in geotechnical applications, e.g. as the base of roads, car parks, for elements of road surface structures, embankments and as cubature filling for land levelling [15].

The producers' offer also includes 100% recycled aggregates used for the production of concrete and mixes for unbound foundations - road stabilization or land hardening. The lower bulk density of this type of aggregate allows material savings of 25% compared to the consumption of natural aggregate in the same implementation [15].

A major problem at present is the management of ashes from the incineration of municipal waste, which, due to their properties, are currently only stored. New technologies for solidifying dust from municipal waste incineration have appeared, thanks to which a durable product is created, with high strength and reduced leachability of harmful substances, which can be used for road foundations, soil stabilization, but also safely stored in landfills without negative effects on the environment [15].

Research was also conducted on the possibility of making clinker-free binders, which would be used in slurries, mortars and hydrotechnical mass concrete. These binders include only a by-product of fluidized bed combustion of brown coal (fly ash) and a product of the metallurgical industry in the form of granulated blast furnace slag. The clinker-free slag-fly ash binder does not require chemical activators for the binding reaction, and the composites mixed with it present properties characteristic to cement composites [16, 17].

An alkali-activated slag binder was also developed, which can be used for the production of mortars and concretes with high resistance to chemical corrosion [8].

Similarly, low-emission concretes are produced, the production of which reduces CO_2 emissions by up to 60% compared to the standard technology [18].

Research is also conducted on the use of ashes from municipal waste incineration as an additive to hardening slurries during the construction of cut-off walls in hydro-engineering structures [19].

All the examples cited above allow for a significant reduction in CO_2 emissions to the atmosphere, both by using waste for their production instead of natural resources, but also thanks to good technical parameters enabling significant savings in energy consumption.

8 Summary and Conclusions

The Polish construction industry significantly operates in a traditional linear model of economy, although new technologies and management methods are implemented. Transition to a circular economy model is necessary for ecological and economic reasons as well as social good. The main aim is to significantly limit the negative impact of the construction sector on the environment. Pursuing this goal may achieve many advantages. Transforming to the circular economy model in construction sector means utilizing the materials and products to the maximum degree and limiting the amount of waste, thus lengthening the life of buildings and materials. This approach will bring measurable benefits such as maintaining the highest values of buildings and materials in an economic circulation.

The identified barriers in the financial, organizational, social and technological areas, especially slow legislative processes are necessary to overcome. Public authorities should assume the role of a leader in the implementation of the circular economy, also in the construction sector. Without changing the regulations only a small percentage of companies would transform their activities towards the CE. For this purpose, it is necessary to use public procurement as a stimulus for innovation and growth of the market for circular products. The next step is to introduce the circular business models and technologies which are to be implemented and which play a fundamental role in the building design process and should be applied at the same beginning [20].

Nowadays, public awareness of circular economy increases and the difficult times (climate changes, high inflation, war in Ukraine) may turn out to be a driving force for changes towards the CE also in the construction sector.

References

1. Global status report for buildings and construction (2021)
2. LNCS Homepage, http://www.circularhotspot.pl/pl/budownictwo
3. Ekobarometer. On the way to a green society. Research report—measurement IV. SW RESEARCH, November 2022
4. LNCS Homepage, https://www.locja.pl/raport-rynkowy/produkcja-materialow-budowl anych-2022,247
5. The circularity gap—Poland by Innowo et al.
6. LNCS Homepage: https://ec.europa.eu/eurostat. Last accessed 20 Feb 2023
7. Annex to the Regulation of the Council of Ministers of 2019. ROAD MAP. Transformation towards a circular economy
8. Project "Oto-GOZ" Gospostrateg. LNCS Homepage, http://www.circularhotspot.pl/pl/ oto-goz, as of February 19, 2023
9. LNCS Homepage, https://oecd-ilibrary.org. Last accessed 20 Feb 2023
10. LNCS Homepage, https://www.gov.pl/web/rozwoj-technologia/projekt-oto-goz
11. Kulczycka, J., Nowaczek, A., Kopyciński, P., Głowacki, J.: Opracowanie systemu wskaźników pomiarowych umożliwiających ocenę postępu w transformacji w kierunku gospodarki o obiegu zamkniętych oraz wpływu gospodarki o obiegu zamkniętym na rozwój społeczno-gospodarczy na poziomie mezoekonomicznym (regionów) i makroeko-nomicznym (gospodarki narodowej). Raport końcowy. Kraków (2021)
12. Road Map towards the transition to circular economy (2019)
13. LNCS Homepage, https://www.nik.gov.pl/aktualnosci/odpady-z-tworzyw-sztucznych.html
14. LNCS Homepage, https://www.gov.pl/web/klimat/circon---gospodarka-o-obiegu-zamkni etym-w-budownictwie-ekoprojektowanie-budynkow-cyrkularnych. Last accessed 20 May 2023
15. Eco-solutions in the construction sector. Polish products for the transformation towards a circular economy. IGSMiE PAN, Kraków (2021)
16. Kledyński, Z., Machowska, A., Wilińska, I., Pacewska, B.: Investigation of hydration products of fly ash-slag pastes. J. Therm. Anal. Calorim. **130**, 351–363 (2017)
17. Machowska, A.: Properties of mass concrete with CFBC fly ash-slag binder. Arch. Civ. Eng. **4**, 471–484 (2020)
18. LNCS Homepage: https://www.gorazdze.pl/pl/ecocrete. Last accessed 27 Feb 2023
19. Falaciński, P., Szarek, Ł: Potential use of municipal waste incineration ash as a hardening slurry ingredient. Minerals **12**(5), 1–12 (2022)
20. Kulczycka, J.: Gospodarka o obiegu zamkniętym w polityce i badaniach naukowych. In: Kulczycka, J. (ed.). IGSMiE PAN, Kraków (2019)

Emerging Materials and Technologies for Next-Generation Sustainable and Resilient Polymer Concrete

Daniel Heras Murcia and Mahmoud Reda Taha$^{(\boxtimes)}$

Department of Civil, Construction & Environmental Engineering, University of New Mexico, Albuquerque, NM, USA
`mrtaha@unm.edu`

Abstract. Emerging materials and technologies (EMTs) are introduced to improve the sustainability and resilience of infrastructure. Polymer concrete (PC) has been used for the last 80 years in infrastructure applications where extreme environmental conditions and exposures are dominant. We suggest that EMTs can enable the development of next-generation PC that can contribute to infrastructure resilience and sustainability. This paper presents an overview of the latest developments in using innovative PC by incorporating a myriad of EMTs to improve infrastructure resilience and sustainability. These EMTs include nanotechnology, bio-based polymers, 3D printing, and textile reinforcement. Using nanotechnology, we demonstrate the possible production of a PC with superior ductility and self-sensing capabilities. We also show that a bio-based polyurethane PC with appreciable compressive strength of 20 MPa can be produced. We demonstrate rheological testing of polymer concrete leading to innovative 3D printed polymer concrete structures. We finally show the ability to produce superior flexural load capacity and textile-reinforced PC (TRPC) ductility compared with cementitious textile-reinforced concrete (TRC). We conclude by demonstrating the potential production of 3D printed TRPC. We suggest that the EMTs will enable a quantum leap in using PC to produce sustainable and resilient infrastructure.

Keywords: Nanotechnology · Biopolymer · 3D Printing · Textile Reinforcement · Resilience · Sustainability

1 Introduction

The importance of infrastructure resilience cannot be overstated, as it ensures that critical services and systems can function during and after disruptive events such as natural disasters or cyberattacks. Resilient infrastructure can minimize downtime, protect public safety, reduce economic losses, and enhance community well-being [1–4]. Infrastructure resilience is critical in ensuring essential services and systems continuity during and after disruptive events such as earthquakes or tornadoes [5, 6]. The significance of resilient infrastructure lies in its ability to minimize downtime, protect public safety, reduce economic losses, and enhance community well-being [5]. These benefits underline the

© The Author(s) 2025
L. Czarnecki et al. (Eds.): ICPIC 2023, 61, pp. 47–58, 2025.
https://doi.org/10.1007/978-3-031-72955-3_4

importance of infrastructure resilience in maintaining community safety, security, and stability [1–3]. Emerging materials and technologies (EM&Ts) can be defined as the materials and technologies whose development and/or field application is still in a state of growth. Rotolo et al. [7] characterized emerging technologies on the basis of five attributes—(i) radical novelty, (ii) relatively fast growth, (iii) coherence, (iv) prominent impact, and (v) uncertainty and ambiguity. EM&Ts have been identified as a potential solution to enhance the resilience of civil infrastructure, providing improved absorptive, adaptive, and restorative capabilities [6]. EM&Ts are increasingly becoming an integral part of the engineering community's work and will significantly impact the future of civil infrastructure.

Using EM&Ts in civil infrastructure is expected to improve infrastructure resilience capacities, enabling infrastructure to withstand or efficiently recover from disruptive events. Three disruptive technologies affecting infrastructure resilience have been identified in the literature [6]: innovative materials, advanced construction technology, and advanced sensing technology. These EM&Ts will enhance the four characteristic elements of infrastructure resilience: redundancy, robustness, rapidity, and resourcefulness. For instance, the significance of using advanced construction technology, such as 3D printing, and advanced materials, such as self-sensing polymer concrete, can improve infrastructure resilience in the case of an extreme event, as depicted in Fig. 1. A framework has been proposed to provide common ground for stakeholders such as ET companies, infrastructural professionals, and infrastructure owners to evaluate, position, and communicate ET's performative roles in establishing or enhancing civil system resilience [8, 9]. This framework emphasizes five properties: resourcefulness, robustness, redundancy, responsiveness, and rapidity.

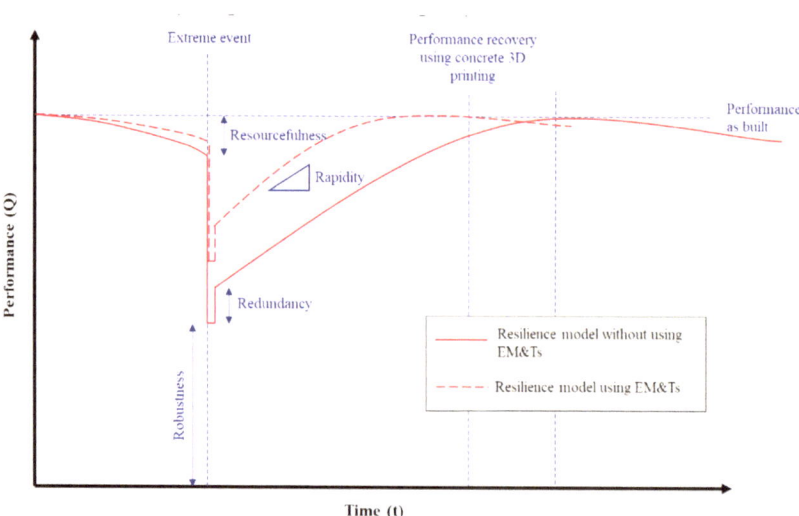

Fig. 1. Resilience model with and without the use of 3D printing technology. Adapted from [9].

Polymer concrete (PC) has the potential to enhance infrastructure resilience due to its ability to withstand harsh conditions and maintain structural integrity over time [10]. PC is highly durable and resistant to environmental degradation factors such as extreme temperatures, chemicals, and moisture. Its good impact resistance makes it an ideal choice for infrastructure exposed to high traffic or heavy loads [11–13]. Moreover, PC has a faster setting time than traditional concrete [10], which means that infrastructure projects can be completed more quickly and with less disruption to the surrounding community. The main applications of PC in infrastructure include bridge decks, industrial overlays, waste-water pipes and containers, manholes, underground communication and transmission line boxes, building façade panels, machine foundations, and other elements subjected to dynamic and cyclic loads [12]. The reduced maintenance needed for PC due to its long-term durability and resistance to performance drops due to extreme events (e.g., earthquakes) would limit damage and help infrastructure resilience. Polymers have also been used as a partial substitute of the cementitious material for concrete applications that require meeting high structural and durability demands [14]. Details regarding characterization of polymer-cement concretes incorporating polymers as a partial replacement of cement can be found elsewhere [15, 16].

However, it is essential to note that polymers, like other industrially processed materials, produce greenhouse gases (GHG), contributing to climate change [17, 18]. Thus, reducing the carbon footprint of PC is necessary to limit the adverse effects of GHG on the environment and public health [17–19]. Many countries have introduced regulations to reduce greenhouse gas emissions, including those related to the production and use of polymers [20]. Promoting sustainable PC can be achieved by decreasing our reliance on non-renewable polymers and examining alternative polymers from natural resources.

In this context, this paper provides an overview of the latest advancements in EM&Ts that promote resilient and sustainable PC. Through exploring EM&Ts, we propose solutions that can help mitigate the environmental impact of PC while enhancing infrastructure resilience.

2   Nanotechnology for Resilient Polymer Concrete

In the past decades, nanotechnology has revolutionized a vast majority of industries. The use of nanotechnology in PC has the potential to enhance the material's properties at many levels. Using nanotechnology in PC has been shown to improve its properties, including fracture toughness, self-sensing capabilities, impact resistance, and electrical and thermal conductivity, thus improving infrastructure resilience.

Carbon nanotubes (CNTs) have been widely used as nanofillers to improve the properties of polymer composites. Daghash et al. proved that the impact resistance of PC was improved by adding CNTs. The results showed that the addition of CNTs significantly enhanced the toughness and impact resistance of the PC [21]. This is supported by other studies in which the addition of CNTs significantly improved the fracture toughness of PC [22]. Moreover, Soliman et al. demonstrated the importance of CNTs in enhancing the mechanical properties of polymer-modified concrete incorporating styrene butadiene rubber [23]. Alumina nanoparticles have also been investigated for their potential to improve PC properties. In the study by Emiroglu et al., a new PC with superior ductility and fracture toughness was developed using alumina nanoparticles. The study showed that adding alumina nanoparticles led to an increase in the fracture toughness of the PC [24]. Furthermore, Douba et al. developed a very ductile PC using CNTs, as depicted in Fig. 2. This PC also exhibited improved compressive and tensile strength [25]. This was further supported by the study of Douba et al. [22], where PC's fracture toughness was improved by adding a combination of CNTs and alumina nanoparticles. Reda Taha et al. developed electrically and thermally conductive PCs using CNTs. The study demonstrated the potential use of CNTs to improve PC's electrical and thermal conductivity, making them suitable for self-sensing smart infrastructure [26]. In smart infrastructure with self-sensing capability, damage can be detected by observing the change in the electrical conductivity of the PC. This is demonstrated in Fig. 3, where the change in the electrical resistance of an electrically conductive PC can be used for damage detection and quantification [27].

Fig. 2. PC with very high levels of ductility achieved by Douba et al. [25] by incorporating CNTs in the polymer mixtures prior to producing PC. Adapted with permission from the authors.

It is evident that implementing nanotechnology in PC has enhanced its properties, including fracture toughness, impact resistance, and electrical and thermal conductivity. CNTs and alumina nanoparticles have been identified as promising nano modifications for PC, potentially improving its resilience in infrastructure applications. These findings suggest that further research is needed to fully explore the potential of nanotechnology in PC and develop new and innovative applications for this material. The use of nanotechnology has also been applied to other aspects of resilience infrastructure such as fiber-reinforced polymer composites for reinforcement bars as a way to improve the mechanical performance of concrete structures [28].

3   Bio-Based Polymers for Sustainable Polymer Concrete

The need to reduce GHG emissions has prompted researchers to investigate innovative solutions to investigate materials with limited carbon footprint [19]. One potential solution is to use sustainable bio-based polymers in infrastructure applications [17, 18].

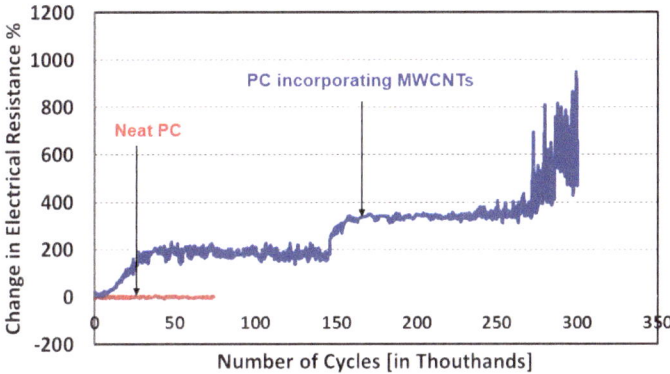

Fig. 3. Comparison of the change in electrical resistance of neat PC and PC incorporating MWC-NTs under cyclic loading showing the significant difference in the electrical resistance of PC incorporating MWCNTs under cyclic stress, which would enable self-sending PC for damage detection in infrastructure applications [27].

Bio-based polymers, derived from renewable resources, have attracted increasing interest due to their limited carbon footprint of construction materials. These polymers can be used in various applications, such as adhesives, coatings, and composites, and can also be used as binders in concrete production. For instance, researchers explored using lignin-based polymers for CO_2 sequestration [29–31]. Lignin, a paper industry byproduct, is a complex polymer that can substitute synthetic polymers in various applications [32]. Using sustainable bio-based polymers for CO_2 sequestration can potentially reduce the total carbon footprint of the construction industry. Recently, researchers have investigated the use of bio-based polyurethane as a binder in the production of PC [18]. The new PC, known as bio-PC (BPC), is produced by replacing cement with bio-based polyurethane. The main challenges in using Bio-based polyurethane include its very fast setting and foaming due to the reaction of polyurethane with moisture, as shown in Fig. 4.

Fig. 4. PC with unmodified bio-based polyurethane with fast setting and foaming

It was suggested that the concrete set and foaming could be controlled by nano-modifying the bio-based polyurethane using carboxyl-functionalized multi-walled carbon nanotubes (MWCNTs) dispersed in benzoic acid before mixing it with the aggregate [18]. Figure 5a shows PC samples produced using bio-based polyurethane with 30 min setting time and no foaming. Figure 5b shows the compressive strength of sustainable BPC incorporating bio-based polyurethane with and without heat curing. A PC with 20–30 MPa compressive strength is attainable using bio-based polyurethane. Researchers also reported that this BPC has a low carbon footprint, with a 50% reduction compared with ordinary Portland cement concrete, as depicted in Fig. 6. The BPC also displayed excellent durability characteristics, making it a promising alternative for infrastructure projects [18].

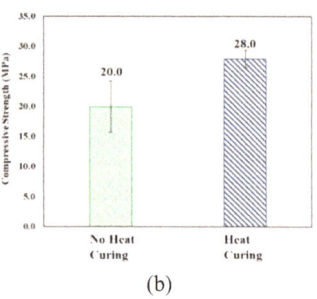

(a)   (b)

Fig. 5. [a] Samples of nano-modified bio-based polyurethane BPC with 30 min setting time and no foaming [b] Compressive strength of BPC with and without heat treatment

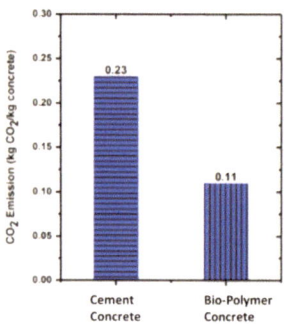

Fig. 6. Carbon footprint analysis for cement-based concrete vs. BPC concrete [18].

4 3D Printing for Resilient Polymer Concrete

Concrete 3D printing commonly relies on the deposition of layered filaments of the material. Cement-based materials have been used for this application due to their ability to stay as pseudo-solid before hardening when their flow properties are properly designed1. Numerous PC applications are directed toward precast elements, making PC amenable to 3D printing technology. Researchers [10] reported 3D printing of PC incorporating Novolac epoxy resin mixed with 44% Benzyl alcohol hardener fumed silica and medium-graded silica sand with a nominal maximum size of 2.36 mm. The maximum aggregate size was selected to ensure proper printing resolution for the 40 mm nozzle diameter. Class-F fly ash and silica fume were also used as fillers to increase the packing fraction and, consequently, the mechanical performance of a 3D printed PC. The fly ash to silica fume weight ratio was kept constant at 2:1 since it has been proven to yield a good compressive strength for PC [33]. Different proportions of rheology modifiers (fumed silica), fillers (silica fume and fly ash), and aggregate (sand) were studied to investigate their effect on the rheological properties of PC. The PC mixes were changed to keep the polymer (resin and hardener) volume fraction of all the mixes around 40%. Rheology measurements were performed using a Brookfield RST soft solid tester rotational rheometer. For the PC, a shear vane spindle with a diameter of 20 mm, a length of 40 mm, and a shear stress range of 5.2 to 3400 Pa was used. The test was performed using the hysteresis technique widely used for thixotropic particulate suspensions. The shear rate was ramped up from 0 s^{-1} to 100 s^{-1} in 60 s and then down from 100 s^{-1} to 0 s^{-1} in 60 s, as shown in Fig. 7. The rheological behavior of the polymer (mixed resin and hardener) with no fillers were studied following the same procedure but using a coaxial spindle (40 mm diameter). The rheological testing allowed quantifying PC thixotropy and thus reaching a printable PC mix.

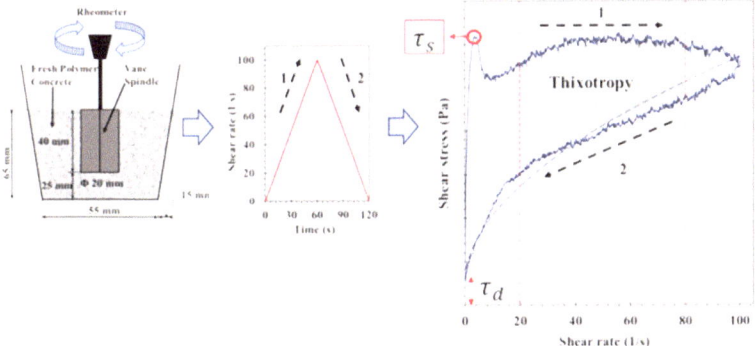

Fig. 7. Rheology test. From left to right, testing configuration, shear rate input over time, and typical material response for PC. Arrow 1 indicates an increase of shear rate while arrow 2 indicates a reduction in shear rate. τ_d is dynamic yield stress, and τ_s is static yield stress.

An optimal PC mix was 3D printed using a gantry robot 3D printer with 3 degrees of freedom and a printable area of 2.0 x 2.0 x 2.0 m. Cube specimens of 45 x 45 x 45 mm were also printed by using a 40 mm diameter nozzle, 10 mm layer height, and 50 mm/s linear printing speed. The extrusion rate represents the rotational speed of the helicoidal extruder that pushes the material through the nozzle. The extrusion rate was kept constant at 19.2 L/min. A build-up rate (H) of 0.5 mm/s was used for the geometry printed and the printing settings reported herein. The 3D printing of PC is shown in Fig. 8. PC was air-cured for 7 days at controlled ambient conditions of 23 ± 2 °C. The compressive strength of the 3D printed concrete cubes was tested in the vertical (Z) direction. The 3D printed PC showed a compressive strength of 30 MPa and a thixotropy growing from 10,000 to 60,000 Pa/s over 45 min. Both measurements proved the optimal PC mix able to produce good 3D printable material.

Fig. 8. 3D printing of the PC. Inset shows 3D printed PC filament and PC buildability.

5   Textile Reinforcement for Resilient Polymer Concrete

Textile Reinforced Concrete (TRC) is a composite material that was introduced to civil infrastructure as a new emerging technology to repair and strengthen deteriorated structures. TRC benefits included design flexibility, lightweight, and high durability[34–37]. TRC typically consists of a matrix of cement binder and small-size aggregates reinforced with high-performance textile reinforcements, either in 2D or 3D [38, 39]. Several types of textile reinforcement materials have been used in TRC, such as glass, polypropylene, carbon, and basalt [40]. However, the cementitious matrix's improper impregnation of the fiber fabrics has led to premature debonding between the textile yarns and the cement matrix, potentially affecting TRC's load-carrying capacity and ductility [39, 41, 42]. Researchers have proposed using organic coats to mitigate this problem to enhance the bond behavior between the fibers and mortar.

A new type of TRC has been proposed to overcome the above performance limitations by replacing the cement binder with a polymer resin to produce textile-reinforced polymer concrete (TRPC) [42]. The new TRPC is created using a fine-graded aggregate, a polymer resin, and a fiber textile fabric. Researchers showed that a methyl methacrylate polymer and basalt fibers could be used to produce a good TRPC. Four different specimen configurations were manufactured by embedding 0, 1, 2, and 3 textile layers in concrete.

Through 3-point bending tests, flexural performance was analyzed and compared with reference TRC specimens with similar compressive strength and reinforcement configurations. TRPC showed significantly improved flexural capacity, superior ductility, and substantial plasticity compared with TRC. TRPC represents an excellent new material for civil infrastructure applications. Figure 9 compares the performance of TRPC incorporating different layers of textile reinforcement with cementitious TRC. It is evident that TRPC overperforms TRC and can achieve a much higher structural capacity and ductility than conventional TRC. Figure 10a shows a light-microscopy picture at a sagittal cut (through the textile-matrix interface) for TRC (top) and TRPC (bottom) and demonstrates the significant ability of the polymer to impregnate the textile fibers compared with the cement binder and thus leading to the improved performance reported above. Finally, Fig. 10b shows the potential production of TRPC using 3D printing technology.

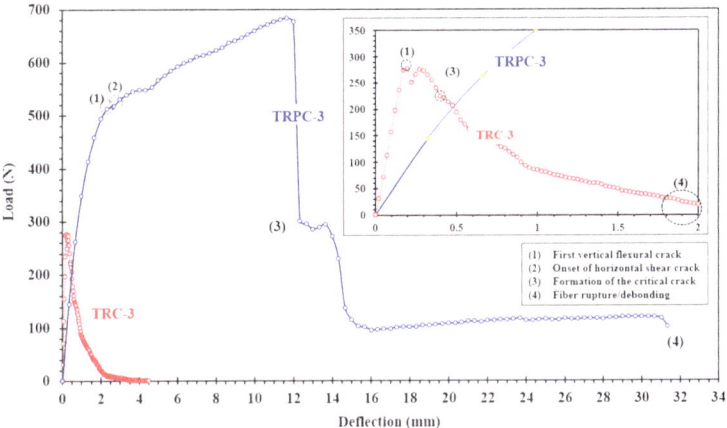

Fig. 9. Comparison of flexural load capacity and ductility (displacement at failure) of TRPC with TRC with 3 intermediate basalt fiber textile layers.

6   Conclusions

Infrastructure resilience ensures critical services and systems function during and after disruptive events such as natural disasters and cyberattacks. PC is a durable material that can withstand harsh conditions and maintain structural integrity over time. However, the production of polymers contributes to climate change, and reducing the carbon footprint of polymers is necessary to mitigate its impact on the environment and public health. We suggest that EMTs can help alter PC contributions and enable producing a sustainable and resilient PC.

Nanotechnology has been shown to enhance PC properties, including superior ductility and improved electrical conductivity. Such developments can lead to smart PC with self-sensing capabilities limiting damage propagation and improving PC ability to respond to loading in extreme events, thus improving infrastructure resilience. We also

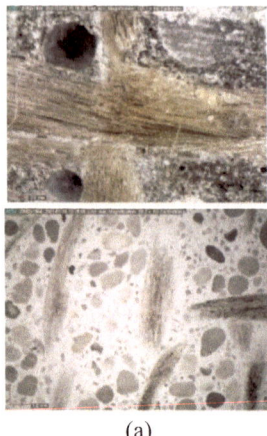

(a)   (b)

Fig. 10. (a) Light-microscopy pictures at a sagittal cut (through the textile-matrix interface) for TRC (top) and TRPC (bottom). (b) 3D printed TRPC panels showing printing over textile reinforcement. Adapted from [42] with permission from the authors.

demonstrated that sustainable bio-based polymers derived from renewable oil could be used to produce a PC with appreciable strength and excellent durability. We showed the possible production of 3D printed polymer concrete, which can enable fast recovery in extreme events. We also showed that such polymer concrete could have superior load capacity and ductility by incorporating textile reinforcement during the 3D printing process. EMTs have the potential to facilitate a resilient and sustainable PC for future infrastructure applications.

References

1. America's Infrastructure Report Card 2017 | GPA: D+ n.d. https://2017.infrastructurereportc ard.org/. Accessed 19 Feb 2023
2. Reduction UNO for DR.: Global Assessment Report on Disaster Risk Reduction 2022: Our World at Risk: Transforming Governance for a Resilient Future. UN (1901)
3. McAllister, T.P.: Community resilience planning guide for buildings and infrastructure systems. NIST **I**
4. Ayyub, B.M.: Systems resilience for multihazard environments: definition, metrics, and valuation for decision making. Risk Anal. **34**, 340–355 (2014). https://doi.org/10.1111/risa. 12093
5. Bruneau, M., Chang, S.E., Eguchi, R.T., Lee, G.C., O'Rourke, T.D., Reinhorn, A.M., et al.: A framework to quantitatively assess and enhance the seismic resilience of communities. Earthq. Spectra **19**, 733–752 (2003). https://doi.org/10.1193/1.1623497
6. Reda Taha, M., Ayyub, B.M., Soga, K., Daghash, S., Heras Murcia, D., Moreu, F., et al.: Emerging technologies for resilient infrastructure: conspectus and roadmap. ASCE-ASME J Risk Uncertain Eng Syst Part Civ Eng **7**, 03121002 (2021). https://doi.org/10.1061/AJRUA6. 0001134
7. Rotolo, D., Hicks, D., Martin, B.R.: What is an emerging technology? Res. Policy **44**, 1827–1843 (2015). https://doi.org/10.1016/j.respol.2015.06.006

8. Soga, K., Hubbard, P.G., Reda Taha, M., Chen, Z., Heras Murcia, D., Tang, P., et al.: Evaluation of emerging technologies for system resilience contributions: case studies. Rev n.d

9. Hubbard, P.G., Chen, Z., Soga, K., Reda Taha, M., Heras Murcia, D., Tang, P., et al.: A framework for evaluating emerging technologies' contributions to system resilience. Rev n.d

10. Murcia, D.H., Abdellatef, M., Genedy, M., Taha, M.M.R.: Rheological characterization of three-dimensional-printed polymer concrete. ACI Mater. J. **118**, 189–201. https://doi.org/10.14359/51733123

11. Committee 548 ACI.: Guide for the use of polymers in concrete. J. Proc. **83**, 798–829 (1986). https://doi.org/10.14359/10674

12. Fowler, D.W.: Polymers in concrete: a vision for the 21st century. Cem. Concr. Compos. **21**, 449–452 (1999). https://doi.org/10.1016/S0958-9465(99)00032-3

13. Fowler, D.W.: Polymers in concrete: where have we been and where are we going? ACI Symp. Publ. **214** (2003). https://doi.org/10.14359/12765

14. Łukowski, P.: Polymer-Cement Composites and Their Use in Construction. In: Horszczaruk, E., Brzozowski, P. (eds.), pp. 77–83. West Pomerianian University of Technology (2021)

15. Wang, R., Li, J., Zhang, T., Czarnecki, L.: Chemical interaction between polymer and cement in polymer-cement concrete. Bull. Pol. Acad. Sci. Tech. Sci. (2016)

16. Czarnecki, L., Lukowski, P.: Polymer-cement concretes. Cem. Wapno Beton **5**, 243–258 (2010)

17. Correa, J.P., Montalvo-Navarrete, J.M., Hidalgo-Salazar, M.A.: Carbon footprint considerations for biocomposite materials for sustainable products: a review. J. Clean. Prod. **208**, 785–794 (2019). https://doi.org/10.1016/j.jclepro.2018.10.099

18. Murcia, D.H., Al Shanti, S., Hamidi, F., Rimsza, J., Yoon, H., Gunawan, B., et al.: Development and characterization of a sustainable bio-polymer concrete with a low carbon footprint. Polymers **15**, 628 (2023). https://doi.org/10.3390/polym15030628

19. Sivakrishna, A., Adesina, A., Awoyera, P.O., Rajesh, K.K.: Green concrete: a review of recent developments. Mater Today Proc **27**, 54–58 (2020). https://doi.org/10.1016/j.matpr.2019.08.202

20. Nations U. Net Zero Coalition. U N n.d.: https://www.un.org/en/climatechange/net-zero-coalition. Accessed 19 Feb 2023

21. Daghash, S.M., Soliman, E.M., Kandil, U.F., Taha, M.M.R.: Improving impact resistance of polymer concrete using CNTs. Int. J. Concr. Struct. Mater. **10**, 539–553 (2016). https://doi.org/10.1007/s40069-016-0165-4

22. Douba, A., Emiroglu, M., Kandil, U.F., Reda Taha, M.M.: Very ductile polymer concrete using carbon nanotubes. Constr. Build. Mater. **196**, 468–477 (2019). https://doi.org/10.1016/j.conbuildmat.2018.11.021

23. Soliman, E.M., Kandil, U.F., Taha, M.M.R.: The significance of carbon nanotubes on styrene butadiene rubber (SBR) and SBR modified mortar. Mater. Struct. **803** (2012)

24. Emiroglu, M., Douba, A.E., Tarefder, R.A., Kandil, U.F., Taha, M.R.: New polymer concrete with superior ductility and fracture toughness using alumina nanoparticles. J. Mater. Civ. Eng. **29**, 04017069 (2017). https://doi.org/10.1061/(ASCE)MT.1943-5533.0001894

25. Douba, A.E., Genedy, M., Tarefder, R., Reda Taha, M.: Improving fracture toughness of polymer concrete using MWCNTs. In: Proceeding of the 9th International Conference on Fracture Mechanics of Concrete and Concrete Structures, IA-FraMCoS (2016). https://doi.org/10.21012/FC9.184

26. US10494299B2—Electrically and thermally conductive polymer concrete—Google Patents n.d.: https://patents.google.com/patent/US10494299B2/en. Accessed 19 Feb 2023

27. Douba, A.E., Genedy, M., Reda, Taha, M.: Self-sensing polymer concrete incorporating carbon nanotubes. Mater. Struct. n.d.; In Review

28. Ogrodowska, K., Łuszcz, K., Garbacz, A.: Nanomodification, hybridization and temperature impact on shear strength of basalt fiber-reinforced polymer bars. Polymers **13**, 2585 (2021). https://doi.org/10.3390/polym13162585
29. Liu, H., Xu, T., Liu, K., Zhang, M., Liu, W., Li, H., et al.: Lignin-based electrodes for energy storage application. Ind. Crops Prod. **165**, 113425 (2021). https://doi.org/10.1016/j.indcrop.2021.113425
30. Klapiszewski, Ł., Klapiszewska, I., Ślosarczyk, A., Jesionowski, T.: Lignin-based hybrid admixtures and their role in cement composite fabrication. Molecules **24**, 3544 (2019). https://doi.org/10.3390/molecules24193544
31. Stojanovska, E., Pampal, E.S., Kilic, A., Quddus, M., Candan, Z.: Developing and characterization of lignin-based fibrous nanocarbon electrodes for energy storage devices. Compos. Part B Eng. **158**, 239–248 (2019). https://doi.org/10.1016/j.compositesb.2018.09.072
32. Bajwa, D.S., Pourhashem, G., Ullah, A.H., Bajwa, S.G.: A concise review of current lignin production, applications, products and their environmental impact. Ind. Crops Prod. **139**, 111526 (2019). https://doi.org/10.1016/j.indcrop.2019.111526
33. Bărbuţă, M., Harja, M., Baran, I.: Comparison of mechanical properties for polymer concrete with different types of filler. J. Mater. Civ. Eng. **22**, 696–701 (2010)
34. Kulas, C.: Actual applications and potential of textile-reinforced concrete. Reinf. Concr. n.d.:11
35. Lepenies, I.G., Richter, M., Zastrau, B.W.: A multi-scale analysis of textile reinforced concrete structures. PAMM **8**, 10553–10554 (2008). https://doi.org/10.1002/pamm.200810553
36. Volkova, A., Paykov, A., Semenov, S., Stolyarov, O., Melnikov, B.: Flexural behavior of textile-reinforced concrete. MATEC Web Conf. **53**, 01016 (2016). https://doi.org/10.1051/matecconf/20165301016
37. Williams Portal, N., Lundgren, K., Wallbaum, H., Malaga, K.: Sustainable potential of textile-reinforced concrete. J. Mater. Civ. Eng. **27**, 04014207 (2015)
38. Triantafillou, T.: Textile Fibre Composites in Civil Engineering. Woodhead Publishing (2016)
39. Peled, A., Bentur, A., Mobasher, B.: Textile Reinforced Concrete. CRC Press (2017)
40. Wu, G., Wang, X., Wu, Z., Dong, Z., Zhang, G.: Durability of basalt fibers and composites in corrosive environments. J. Compos. Mater. **49**, 873–887 (2015). https://doi.org/10.1177/0021998314526628
41. Hartig, J., Häußler-Combe, U.: A model for the uniaxial tensile behaviour of textile reinforced concrete with a stochastic description of the concrete material properties. In: Bicanic, N., Borst, R., Mang, H. (eds.), pp. 153–162 (2010)
42. Murcia, D.H., Çomak, B., Soliman, E., Reda Taha, M.M.: Flexural behavior of a novel textile-reinforced polymer concrete. Polymers **14**, 176 (2022). https://doi.org/10.3390/polym14010176

Interaction Between Polymer and Cement: A Review

Ru Wang[(✉)] and Shiwei Zhang

Key Laboratory of Advanced Civil Engineering Materials of Ministry of Education, School of Materials Science and Engineering, Tongji University, Shanghai, China
ruwang@tongji.edu.cn

Abstract. Polymer-modified cement-based materials are commonly used in engineering applications and have achieved good results. The interactions between polymer and cement have received extensive attention. In this paper, the interaction between them is discussed and summarized by reviewing the existing technologies. Traditional experimental methods do not provide a comprehensive picture of the interaction between polymers and cement-based materials, molecular dynamics (MD) simulations were used recently in the study of inorganic-organic phase interactions. People almost reach a consensus on the modification mechanism of polymers on concrete at micro-scale. But at nano-scale, the interaction between polymers and cement is an ongoing work, researches show that it contains several aspects, i.e., chemical bonding, hydrogen bonding, van der Waals forces, etc. Different polymers may have different types of interactions with cement. Understanding these interactions is important to elucidate the relationship between the microstructure and macroscopic properties of polymer-modified cement-based materials. Molecular dynamics simulation has proved to be an effective method to study the interactions between inorganic-organic composites at this stage but has some limitations.

Keywords: Polymer · Cement-based materials · Interactions · Micro-scale · Nano-scale

1 Introduction

Cement-based materials are most widely used in construction material in the twentieth century, which is closely related to the daily life of human beings. However, concrete belongs to heterogeneous, brittle composite cementitious material, has a low tensile strength (1/10 ~ 1/20 of compressive strength) [1], that easily cracks under tensile loading [2]. The cracks will weaken the bearing capacity and stability of the structure, and will also accelerate the failure of concrete.

To improve the crack resistance and durability of concrete, one of the approaches is to add modifiers. Polymers are popular as modifiers for concrete. It has been used in many projects and has achieved good results [3–11]. There are many types of polymers. Among them, the most commonly used polymers are styrene-butadiene rubber (SB), styrene-acrylic ester (SA), and ethylene-co-vinyl acetate (EVA). The polymer can not

© The Author(s) 2025
L. Czarnecki et al. (Eds.): ICPIC 2023, 61, pp. 59–72, 2025.
https://doi.org/10.1007/978-3-031-72955-3_5

only improve the fluidity of cement paste [12–14] and water retention [15, 16] but also improve tensile strength, toughness [17], crack resistance, impermeability [18], durability [19], and bonding performance of concrete [20–23].

Until now, some researchers have studied the mechanism of polymer-modified cement-based materials through a range of physical and chemical methods, to reveal the impact of polymer on concrete in macro performance. Cement-based materials have multi-scale characteristics, and the mechanisms of each scale are also different. This article mainly reviews the mechanism at micro-scale ($10^{-3} \sim 10^{-7}$ m) and especially nano-scale ($<10^{-7}$ m).

2   Mechanism at Micro-scale

In recent years, researchers have mainly explored the modification mechanism of polymers on a micro-scale. They believed that the decentralization of polymer particles and forming a polymer film are the main causes of modification. Two famous models, Ohama's and Konietzko's models were proposed based on the characteristics of polymer forming a thin film inside concrete [24, 25]. The Konietzko's model assumed that the polymer film and cement paste penetrated each other to form a mesh structure, while Ohama's model assumed that the cement paste was encapsulated in a polymer film. Fichet et al. observed by SEM that polymer particles were distributed in the middle of cement hydration products and on the surface of unhydrated cement particles, which reduced the hydration rate of cement [26]. With the hydration of cement, polymer emulsion encapsulates the surface of cement particles and hydrated products, filling the pores. The porosity of concrete and the material properties could be improved by this physical effect [27]. The polymer also forms a bridging effect inside cement-based materials [28], consuming the energy generated internally by the external force and increasing the tensile strength of the cementitious material. Su et al. found that polymers also adhere to the inner walls of cement-based material pore channels to form thin films [29], closing the internal pores and improving the durability of cement-based materials. Figure 1 shows the evolution of polymer (SB) morphology and structure in polymer/cement composites cementation materials. Optical microscope and environmental scanning electrical microscope were used to investigate the dispersion and absorption of polymers in mono-dispersed cement system [30]. The mono-dispersed cement system can be achieved at the water to cement ratio of 10:1. Proper amount of SA and SB is beneficial to the dispersion of the cement. SA and SB can be dispersed on the surface of mono-dispersed cement particles as well as the solution in the system proportionally. The absorption density of polymers on the cement particle grows with the increasing polymer to cement ratio (mp/mc). Through sedimentation test, it can be found that the absorption amount of both polymers in the mono-dispersed cement grows with mp/mc but decreases with water to cement ratio (mw/mc). The absorption rate of both polymers in the mono-dispersed cement declines with mp/mc and mw/mc. The polymer coverage on the surface of the mono-dispersed cement particle increases with mp/mc but decreases with mw/mc. Both of the coverage is less than 100%, indicating the absorption of these two polymers on the surface of the cement particle is single layered.

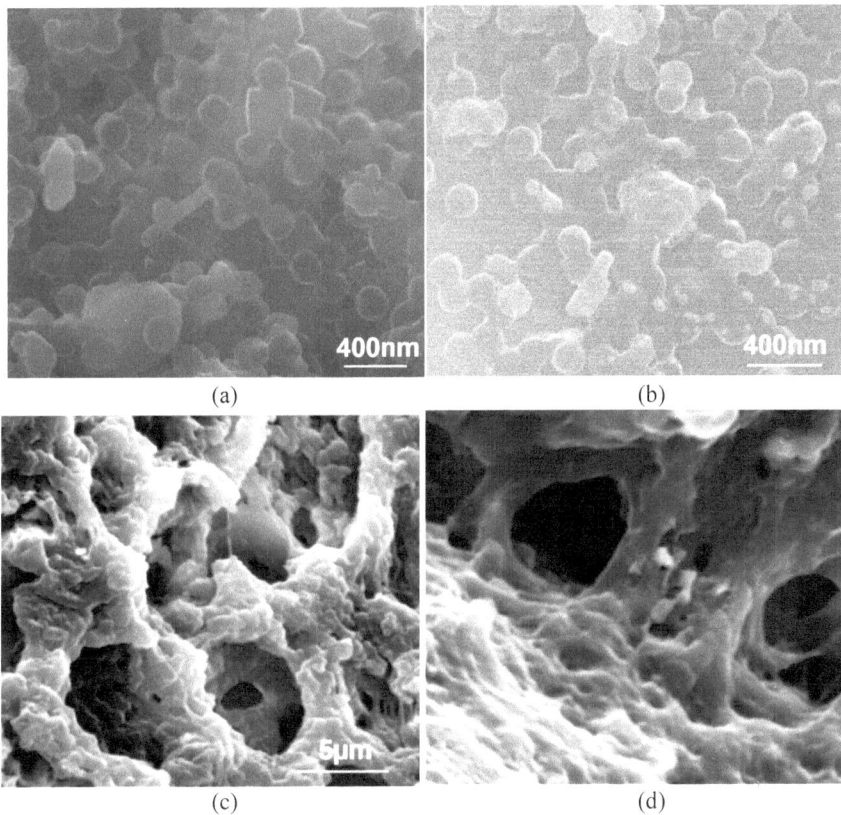

Fig. 1. Evolution of polymer (SB) morphology and structure in polymer/cement composite cementitious materials: (a) Polymer particles dispersed on the surface of cement granules after mixing for 10 min, (b) Polymer particles deformed and fused together after mixing for 3 h, (c) Network polymer film formed at 28 days, (d) Polymer bridges at the interface of cement paste and aggregate at 28 days

3 Mechanism at Nano-Scale

Although the modification mechanism of polymers has been revealed at the micro-scale, there are still some issues to be solved. How do the polymers interact with the cement-based materials? Researchers are needed to delve further into the nano-scale to explore the intrinsic mechanism. The influence of polymers at the nano-scale consists of two main types, i.e., changing the molecular structure of hydration products and forming interactions with the hydration products, including chemical bonding and intermolecular forces.

3.1 Changing the Molecular Structure of Hydration Products

Polymers affect the hydration reaction of cement and the molecular structure of hydration products. Some studies have shown that polymers significantly retard early hydration reactions. Zhang et al. found that SA significantly retarded the hydration of cement-based materials [31]. Wang et al. investigated the effect of SB and SA on hydration products, and the results showed that the polymers promoted the formation of calcium aluminate, but reduced the content of calcium hydroxide [32, 33]. The hydration and hardening processes of polymer-modified concrete are mainly influenced by both cement hydration and polymer film formation processes [34]. Polymers also alter the molecular structure of some hydration products (e.g., C-S-H). Wang et al. used NMR to demonstrate the effect of the incorporation of SB on the polymerization of $[SiO_4]^{4-}$ tetrahedron in C–S–H gel and the results showed that the degree of polymerization was depressed significantly (Fig. 2) [35]. Peng et al. found that polymer increased the Ca/Si ratio of C-S-H (Fig. 3) [36]. In addition to the degree of polymerization and Ca/Si ratio, polymers also change the layer spacing of C-S-H. Matsuyama first reported that polymers can be intercalated, the used polymers were nonionic poly(vinyl alcohol) (PVA), anionic poly(acrylic acid) (PAA) and cationic poly(diallyl dimethylammonium chloride) (PDC), as indicated by the shift of the (002) basal reflections to smaller diffraction angles (Fig. 4), and later found that the addition of PAA increased the Q^2 intensity by ^{29}Si NMR spectra, indicating lengthening of the silicon chain [37]. Subsequently, some scholars also analyzed the shift in the (002) peak position through XRD and found that polyethylene glycol (PEG) caused the expansion of the C-S-H layer spacing [38]. The overall results of the literatures suggested that the success of intercalation depended on the Ca/Si ratio of C-S-H, the nature of the polymer and its molecular weight, and the synthesis process [39].

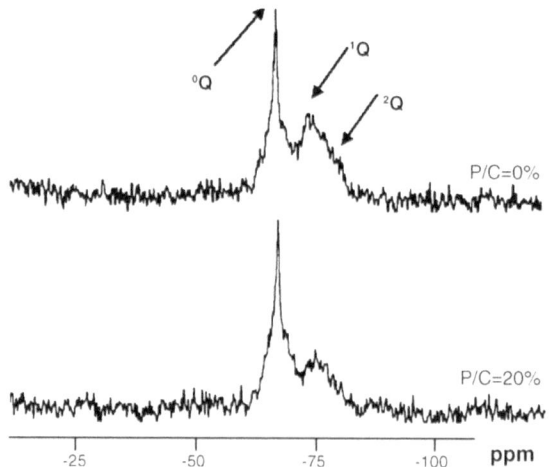

Fig. 2. ^{29}Si NMR spectra of SB-modified cement pastes hydrated for 28 days [35].

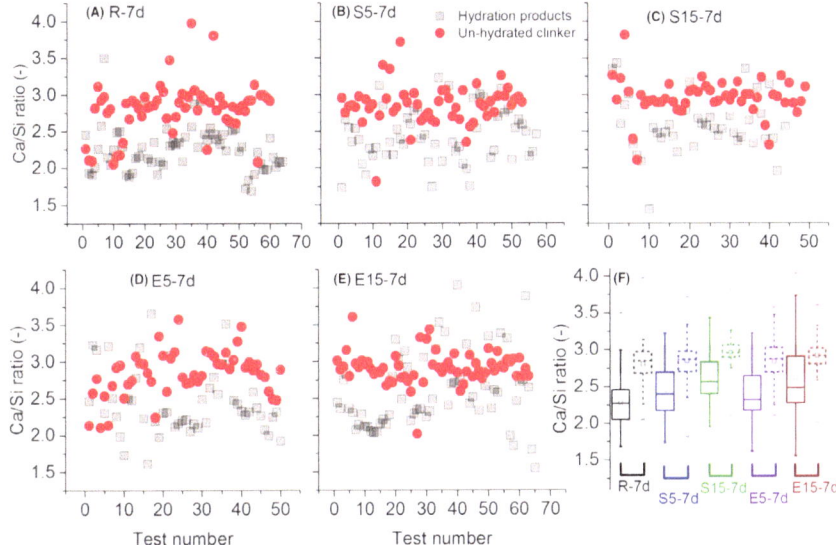

Fig. 3. Ca/Si ratios of unhydrated cement particles and hydration products for all cement pastes cured for 7 d: (A) Reference, (B) SB-5%, (C)SB-15%, (D) EVA-5%, and (E) EVA-15%. The filled circle represents the data of the unhydrated cement particles and the open square represents these of the hydration products. All data are statistically compared in the panel (F) [36].

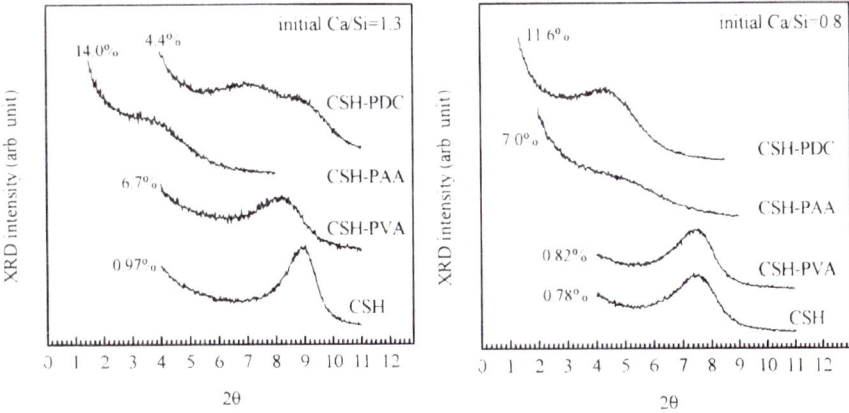

Fig. 4. Changes in (002) basal spacing of C-S-H precipitated with different polymers. Percentages are weight percent of carbon in the complexes [37].

3.2 Interaction Between Polymer and Cement-Based Materials

3.2.1 Research Using Microstructure Characterization

Researchers have conducted several experiments to investigate the chemical reaction between polymers and cement-based materials. Silva et al. found the presence of calcium acetate ($Ca(CH_3COO)_2 \cdot H_2O$) inside the EVA-modified cement paste by Fourier-transform infrared spectrometer (FTIR) analysis, where the carboxyl group is thought to originate from the hydrolysis of the acetate group in the EVA molecule in a high alkali solution environment [40]. Wang et al. investigated the chemical reaction of SA in cement paste and found that the absorption peak which occurred between 1729 and 1735 cm^{-1} corresponds to $C = O$ in the COO^- group of SA powder modified cement pastes. The same group in pure SA powder appears at 1728 cm^{-1}. Figure 5 shows that the absorption peak of $C = O$ begins to shift to a higher wave number. The possible reason is that the lone pair of electrons belonging to -O- in the COO^- group is transferred to the Ca^{2+} ion in the system, this reduces the electronic cloud density around –O– and also around $C = O$ which is connected to it. The spectrum changes indicate that $O \rightarrow Ca^{2+}$ coordination bonds exist [33, 41]. Wang et al. studied the chemical interaction between polyacrylate emulsions and cement paste that the carboxyl groups of polyacrylate chains reacted with Ca^{2+} from $Ca(OH)_2$ to produce $Ca(HCOO)_2$ (as shown in Fig. 6) [42].

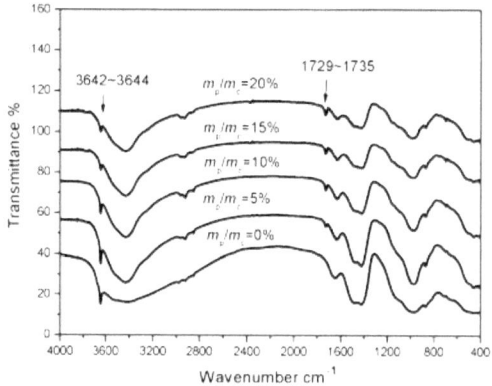

Fig. 5. FTIR patterns of cement pastes with and without SAE powder cured for 28 days [33, 41].

3.2.2 Research Using Molecular Dynamics Simulation

While the above studies have demonstrated some chemical interactions between polymers and cement using methods such as FTIR and XPS, the interactions between atoms have not yet been visualized. Furthermore, there are no feasible experimental approaches to explore the van der Waals force between molecules as well as the hydrogen bonding. Molecular dynamics (MD) simulations are based on force fields and can go down to the

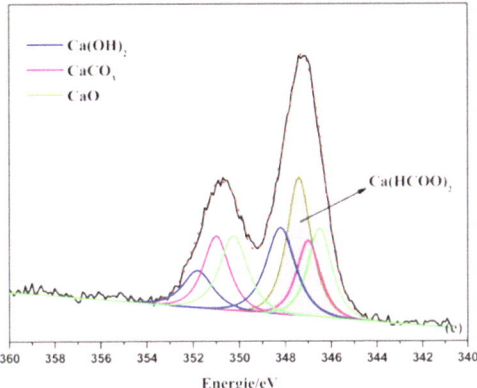

Fig. 6. XPS survey spectra of Ca 2p [42].

molecular/atomic scale to characterize the interactions between polymers and cement-based materials. MD was used to study the mechanism of the interaction between polymers and cement hydration products, which makes up for the shortage of existing experimental conditions. The results of molecular dynamics simulations are closely related to the selection of force field types, however, unfortunately, there are no rigorous guidelines for selecting an appropriate force field. There are two major categories of force field types in cement-based materials research, (i) ReaxFF reactive force field, which can simulate the breaking and formation of chemical bonds; (ii) Classical force fields (ClayFF, CSH-FF and COMPASS), which cannot simulate the breaking and formation of chemical bonds.

3.2.3   Interaction at Molecular-Level

A reactive MD simulation based on the ReaxFF force field was employed to study the strengthening mechanisms of C–S–H incorporating polyethylene glycol (PEG), polyvinyl alcohol (PVA) and polyacrylic acid (PAA) polymers. It was found that all three polymers can be inserted into the interlayer of C-S-H and new Si-O-C bonds are formed after breaking the C-C bonds in the PVA, PAA and PEG polymer chains (Fig. 7(a, b, c)) [43], which has been verified by first-nature principle calculations in some similar organic/inorganic system [44, 45]. Additionally, it was reported that the PEG/C–S–H composite displayed a stronger bond due to the breaking of the C–O bond in polymer by the Ca ions presented in the gel and form a new C-Ca bond (as shown in Fig. 7(d)) [46], which has also been proven by the first-nature principle studies, which reduced the system energy and made the system more stable [47]. Although the utilization of ReaxFF force fields can reflect the breaking and formation of chemical bonds, allowing researchers to better understand the interaction between polymers and C-S-H, the parameters in the ReaxFF force field have a huge impact on the results [48].

Researchers have also tried to analyze the interaction between polymers and concrete using classical force fields. Hou et al. investigated the interfacial interaction of three polymers, PEG, PVA, and PAA, with C-S-H gels, which exhibited brittle fracture in the

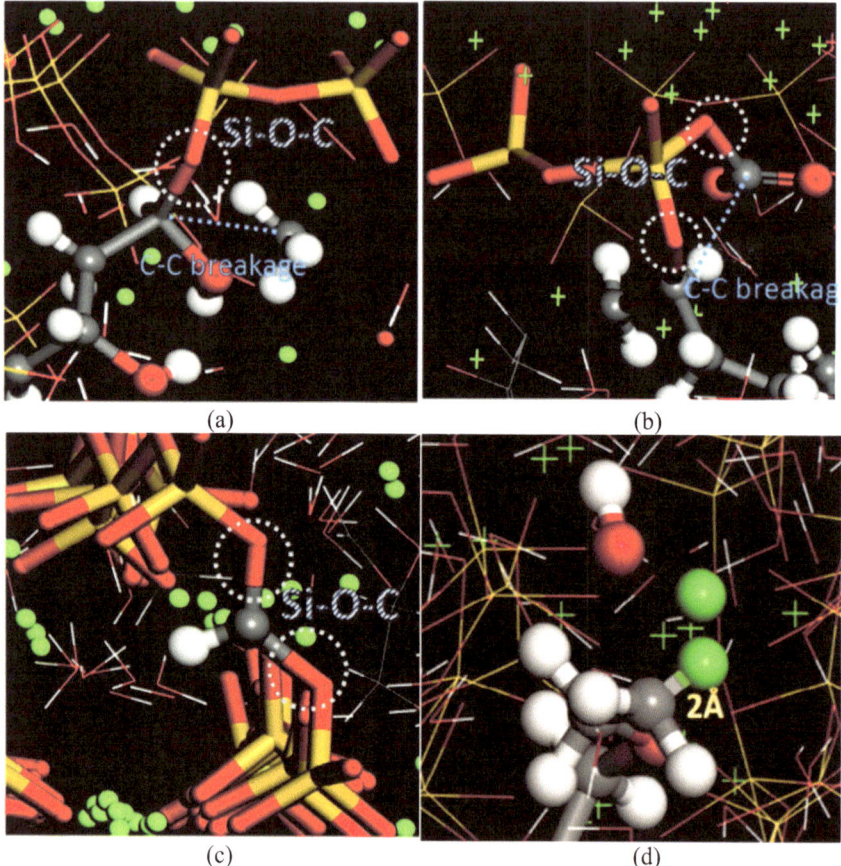

Fig. 7. Local snapshots after structural rearrangements: (a) PVA, (b) PAA and (c)(d) PEG (red and yellow sticks indicate silicate tetrahedra, white and red lines indicate water molecules, and the balls of other colors indicate the polymer atoms: grey for carbon, white for hydrogen, red for oxygen, and green for calcium) [43, 46].

z-direction due to the presence of weak hydrogen bond (H-bond) interactions. After the addition of polymers, the stress rebounded at a later stage and the brittle properties were improved. The strongest interaction was found in C-S-H with PAA which was mainly due to two types of strong and stable connections. One was the coordination of double-bonded oxygen atoms from the carboxyl groups around Ca ions from the C–S–H surface (Fig. 8(a)) and the other was the H-bonds formed via high-reactivity nonbridging oxygen atoms from silicate tetrahedra of C–S–H accepting hydrogen atoms from carboxyl groups of polymers (Fig. 8(b)) [49]. Similarly, there were also coordination and H-bond interactions between hydroxyl groups of PVA and C–S–H surface (as shown in Fig. 8(c, d)), and thus PVA was intermediate in terms of affinities, greater than PEG (as shown in Fig. 8(e, f)), in which no obvious connections were found [50]. The interaction strength between the polymer and C-S-H affects the mechanical properties (Young's modulus and tensile strength) of the composite materials, and the higher the interaction strength the better the mechanical properties.

In addition to the study of polymer-C-S-H interface interactions, MD simulations have been applied to study the effect of polymers on cement hydration. The adsorption of PCE on the cement particles, represented as C_2S, was investigated. The results revealed that the presence of PCE perturbed the dense water layer above the C_2S surface and lowered the water density, which affected the hydration reaction of cement [51]. Chaudhari et al. studied the effect of carboxylic and hydroxycarboxylic acids on cement hydration. It suggested that the chelate complex of the hydroxycarboxylic acid retards cement hydration by adsorbing onto the reactant cement mineral phases particularly hydroxylated C_3S [52].

3.2.4 Effect of Service Environment on the Interaction

It is known that the interaction between different materials plays a crucial role in the structural integrity and durability of concrete under extreme environmental conditions. In fact, it is difficult to probe the effect of the service environment on polymer-cementitious material interactions at the molecular level by existing technical means. MD can bridge this gap very well.

Many macro-scale experimental studies have explored the adverse effects of external environmental exposures on the epoxy/concrete interface. However, there is a lack of sufficient intrinsic understanding of how the bond evolution and stress transfer at the epoxy/concrete interface are disrupted or altered by aggressive environments. Hou and Wang et al. used MD to reveal the mechanism of debonding between epoxy and C-S-H under sulfate environment conditions, they found that water molecules and sulfate ions weakened the interaction between C-S-H and epoxy resin by breaking the Ca-O and H-bonds [53]. Yu et al. also demonstrated that water molecules disrupt the bonding interaction between the epoxy resin and C-S-H, leading to a reduction in bond strength [54]. Zhang conducted a study on the effect of temperature on the bonding properties between tannic acid and C-S-H and showed that the bonding properties were improved after high temperatures due to the increase in contact area caused by the larger radius of gyration of tannic acid [55].

4 Summary

The application of polymers can improve the mechanical properties and durability of cement-based materials. The modification mechanism of polymers at micro-scale and nano-scale have been studied. It is recognized that at the micro-scale the film-forming action of polymers inside cement-based materials can bridge the microcracks to improve the mechanical properties of the materials, and also close the pores to improve the durability. However, it is difficult to reveal clearly the basic mechanism of the action of polymers with cement-based materials at the micro-scale.

The action of polymers on cement-based materials at the nano-scale consists of two parts. Firstly, it changes the number and molecular structure of the hydration products, such as the Ca/Si ratio, the polymerization of $[SiO_4]^{4-}$ tetrahedron and the layer spacing

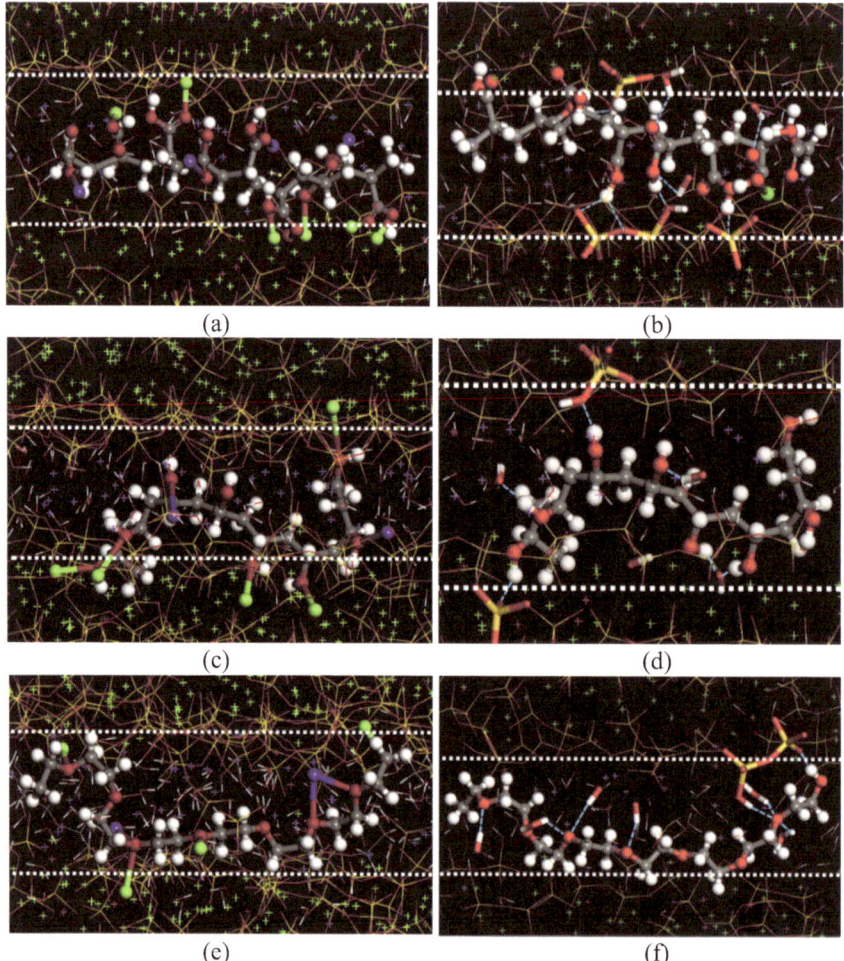

Fig. 8. Overall view of calcium silicate hydrates with (a) (b) PEG (c) (d) PVA (e) (f) PAA inter-calated and enlarged snapshots of the Os-Hp connection. (Red and yellow sticks represents the silicate tetrahedra, green and purple balls correspond to the intralayer calcium atoms and the inter-layer calcium atoms respectively, white and red lines represent water molecules and hydroxy, balls of other colors are for the polymer atoms: grey for carbon, white for hydrogen, red for oxygen, and blue dotted line represents H-bonds) [49].

of C-S-H gel. Secondly, it is the molecular interaction between polymers and cement-based materials. Currently, the chemical interaction between polymers and cementitious materials have been demonstrated by XPS and FTIR techniques, and these chemical interactions are generated mainly between the polymer with polar functional groups and cement-based materials. However, some commonly used polymers such as SB that does not contain polar functional groups but still shows good performance. The interaction between polymers and cement-based materials is still controversial.

The application of the first-nature principle and molecular dynamics can also solve the problems of interaction between polymers and cement-based materials, and some progress has been made in this area. However, there are still some shortcomings in molecular dynamics simulation studies that need further investigation: (1) Limited by the bulk volume and time length of the molecular dynamics simulation itself, its simulation results may differ from the actual experimental results, and it is an important research task to relate the simulation results to the experimentally observed macroscopic behavior; (2) Most of the current studies represent cement-based materials with individual calcium silicate hydrates, without considering the effect of other compositions and porosity within the material on its performance; (3) Although force fields such as ClayFF and CVFF have been used to describe the interaction between polymers and cement-based materials and are considered reasonable, such classical force fields cannot describe the chemical reactions that exist between the polymer and the cementitious material, and therefore more suitable force fields need to be developed for polymer-cement systems.

Understanding of the nature of materials may bring new solutions for materials. Elucidation of the modification mechanism of polymers on cement-based materials at nano-scale is helpful to understanding the nature of the material, which may open the gate to create a new generation of polymer-cement based composite.

Acknowledgement. The authors acknowledge the financial support by the National Natural Science Foundation of China (Grant No. 51872203 and 51572196) and the Top Discipline Plan of Shanghai Universities-Class I (2022-3-YB-17).

References

1. Chen, X.Q., Yuan, J.W., Dong, Q., Zhao, X.K.: Meso-scale cracking behavior of cement treated base material. Constr. Build. Mater. **239**, 117823 (2020)
2. Van Tittelboom, K., De Belie, N.: Self-healing in cementitious materials—A review. Materials **6**(6), 2182–2217 (2013)
3. Jones, S.: Polymer-modified asphalt for the paving industry. Asphalt **15**(1), (2000)
4. Wu, H.L., Bi, S., Zhao, S.L., Lu, L.S.: Application of polymer mortar in the treatment of cracks in the corridor of Huilong Power Station. Yellow River **07**, 55–56 (2006). (in Chinese)
5. Pacheco-Torgal, F., Jalali, S.: Sulphuric acid resistance of plain, polymer modified, and fly ash cement concretes. Constr. Build. Mater. **23**(12), 3485–3491 (2009)
6. Zhang, X.W., Liu, W.D., Zhong, H.R., Zhou, W.B.: Application of spraying insulation polymer-modified mortar for energy-saving and reinforcement to external wall. Eng. Mech. **28**(S1), 167–171 (2011). (in Chinese)
7. Lee, H.X.D., Wong, H.S., Buenfeld, N.R.: Self-sealing of cracks in concrete using superabsorbent polymers. Cem. Concr. Res. **79**, 194–208 (2016)
8. Wang, R., Ma, D.X., Wang, P.M., Wang, G.Y.: Study on waterproof mechanism of polymer-modified cement mortar. Mag. Concr. Res. **67**(18), 972–979 (2015)
9. Wang, R., Zhang, L.: Mechanism and durability of repair systems in polymer-modified cement mortars. Adv. Mater. Sci. Eng. **2015**, 594672 (2015)
10. Li, B., Wang, F., Fang, H.Y., Yang, K., Zhang, X., Ji, Y.: Experimental and numerical study on polymer grouting pretreatment technology in void and corroded concrete pipes. Tunn. Undergr. Space Technol. **113**, 103842 (2021)

11. Li, L., Liu, K., Chen, B., Wang, R.: Effect of cyclic curing conditions on the tensile bond strength between the polymer modified mortar and the tile. Case Stud. Constr. Mat. **17**, e01531 (2022)
12. Felekoğlu, K.T., Felekoğlu, B., Yalçınkaya, Ç., Baradan, B.: Influence of styrene acrylate and styrene butadiene rubber on fresh and mechanical properties of cement paste and mortars. In: 7th Asian Symposium on Polymers in Concrete, pp. 145–155. İstanbul (2012)
13. Ukrainczyk, N., Rogina, A.: Styrene–butadiene latex modified calcium aluminate cement mortar. Cement Concr. Compos. **41**, 16–23 (2013)
14. Wang, R., Li, L.: Experimental study on the rheology and setting behavior of calcium sulfoaluminate cement paste modified with styrene-butadiene copolymer dispersion. J. Mater. Civ. Eng. **34**(4), 04022015 (2022)
15. Han, Y., Zhao, W.: Study on properties of PB-g-PS latex-modified cement mortars. In: 2011 International Conference on Consumer Electronics, pp. 3600–3603. IEEE (2011)
16. Wan, Q., Wang, Z.J., Huang, T.Y., Wang, R.: Water retention mechanism of cellulose ethers in calcium sulfoaluminate cement-based materials. Constr. Build. Mater. **301**, 124118 (2021)
17. Czarnecki, L., Sokołowska, J.J.: Optimization of polymer-cement coating composition using material model. In: Key Engineering Materials, pp. 191–199. Trans Tech Publications Ltd (2011)
18. Ma, D.X., Liu, Y., Lai, Y.: The influence of pore structure on the waterproof performance of polymer modified mortar. In: Applied Mechanics and Materials, pp. 1130–1134. Trans Tech Publications Ltd (2014)
19. Wang, R., Fan, Y.S., Wang, Z.J., Huang, T.Y., Zhang, T.: Performance development of styrene-butadiene copolymer-modified calcium sulfoaluminate cement mortar under different curing conditions. J. Zhejiang Univ. Sci. A (Appl. Phys. Eng.) **22**(12), 1005–1026 (2021)
20. Barluenga, G., Hernández-Olivares, F.: SBR latex modified mortar rheology and mechanical behaviour. Cem. Concr. Res. **34**(3), 527–535 (2004)
21. Yang, Z.X., Shi, X.M., Creighton, A.T., Peterson, M.M.: Effect of styrene–butadiene rubber latex on the chloride permeability and microstructure of Portland cement mortar. Constr. Build. Mater. **23**(6), 2283–2290 (2009)
22. Zhang, Y.R., Kong, X.M.: Influences of superplasticizer, polymer latexes and asphalt emulsions on the pore structure and impermeability of hardened cementitious materials. Constr. Build. Mater. **53**, 392–402 (2014)
23. Guo, S.Y., et al.: Mechanical and interface bonding properties of epoxy resin reinforced Portland cement repairing mortar. Constr. Build. Mater. **264**, 120715 (2020)
24. Omaha, Y.: Polymer-modified mortars and concretes, Concrete Admixtures Hand Book: Properties, Science and Technology. Noyes Oublications, Park Ridge. NJ, USA (1984)
25. Konietzko, A.: Polymerspezifische auswerkungen auf das tragverhalten modifizierter zement-gebundenen beton (PCC). Braunschweig (1988)
26. Ollitrault-Fichet, R., Gauthier, C., Clamen, G., Boch, P.: Microstructural aspects in a polymer-modified cement. Cem. Concr. Res. **28**(12), 1687–1693 (1998)
27. Aggarwal, L.K., Thapliyal, P.C., Karade, S.R.: Properties of polymer-modified mortars using epoxy and acrylic emulsions. Constr. Build. Mater. **21**(2), 379–383 (2007)
28. Van Gemert, D., Beeldens, A.: Evolution in modeling microstructure formation in polymer-cement concrete. In: 7th Asian Symposium on Polymers in Concrete, pp. 59–73. İstanbul (2012)
29. Su, Z., Sujata, K., Bijen, J.M.J.M., Jennings, H.M., Fraaij, A.L.A.: The evolution of the microstructure in styrene acrylate polymer-modified cement pastes at the early stage of cement hydration. Adv. Cem. Based Mater. **3**(4), 87–93 (1996)
30. Shi, X.X., Wang, R., Wang, P.M.: Dispersion and absorption of SBR latex in the system of mono-dispersed cement particles in water. Adv. Mat. Res. **687**, 347–353 (2013)

31. Zhang, H.B., Wang, C., Li, D.L.: Study on performances and mechanism of high-strength cement-based materials modified by styrene-acrylic-emulsion. Bull. Chin. Ceram. Soc. **33**(1), 164–169 (2014). (in Chinese)

32. Wang, R., Wang, P.M.: Formation of hydrates of calcium aluminates in cement pastes with different dosages of SBR powder. Constr. Build. Mater. **25**(2), 736–741 (2011)

33. Wang, R., Yao, L.J., Wang, P.M.: Mechanism analysis and effect of styrene-acrylate copolymer powder on cement hydrates. Constr. Build. Mater. **41**, 538–544 (2013)

34. Rossignolo, J.A., Agnesini, M.V.C.: Mechanical properties of polymer-modified lightweight aggregate concrete. Cem. Concr. Res. **32**(3), 329–334 (2002)

35. Wang, R., Li, X.G., Wang, P.M.: Influence of polymer on cement hydration in SBR-modified cement pastes. Cem. Concr. Res. **36**(9), 1744–1751 (2006)

36. Peng, Y., Zeng, Q., Xu, S.L., Zhao, G.R.: BSE-IA reveals retardation mechanisms of polymer powders on cement hydration. J. Am. Ceram. Soc. **103**(5), 3373–3389 (2020)

37. Matsuyama, H., Young, J.F.: Intercalation of polymers in calcium silicate hydrate: a new synthetic approach to biocomposites? Chem. Mater. **11**(1), 16–19 (1999)

38. Beaudoin, J.J., Dramé, H., Raki, L., Alizadeh, R.: Formation and properties of C-S-H–PEG nano-structures. Mater. Struct. **42**(7), 1003–1014 (2009)

39. Pelisser, F., Gleize, P.J.P., Mikowski, A.: Structure and micro-nanomechanical characterization of synthetic calcium–silicate–hydrate with Poly (Vinyl Alcohol). Cement Concr. Compos. **48**, 1–8 (2014)

40. Silva, D.A., Monteiro, P.J.M.: Hydration evolution of C_3S–EVA composites analyzed by soft X-ray microscopy. Cem. Concr. Res. **35**(2), 351–357 (2005)

41. Wang, R., Li, J., Zhang, T., Czarnecki, L.: Chemical interaction between polymer and cement in polymer-cement concrete. Bull. Polish Acad. Sci. Tech. Sci. **64**, 785–792 (2016)

42. Wang, M., Wang, R.M., Zheng, S.R., Farhan, S., Yao, H., Jiang, H.: Research on the chemical mechanism in the polyacrylate latex modified cement system. Cem. Concr. Res. **76**, 62–69 (2015)

43. Zhou, Y., Hou, D.S., Geng, G.Q., Feng, P., Yu, J., Jiang, J.Y.: Insights into the interfacial strengthening mechanisms of calcium-silicate-hydrate/polymer nanocomposites. Phys. Chem. Chem. Phys. **20**(12), 8247–8266 (2018)

44. Tsetseris, L., Pantelides, S.T.: Encapsulation of floating carbon nanotubes in SiO_2. Phys. Rev. Lett. **97**, 26 (2006)

45. Wang, S., et al.: Bonding at the SiC–SiO_2 interface and the effects of nitrogen and hydrogen. Rev. Lett. **98**, 26101 (2007)

46. Zhou, Y., Hou, D.S., Jiang, J.Y., She, W., Yu, J.: Reactive molecular simulation on the calcium silicate hydrates/polyethylene glycol composites. Chem. Phys. Lett. **687**, 184–187 (2017)

47. Liu, L., Jin, J., Lin, Y., Hou, F., Li, S.: The effect of calcium on nitric oxide heterogeneous adsorption on carbon: A first-principles study. Energy **106**, 212–220 (2016)

48. Duque-Redondo, E., Bonnaud, P.A., Manzano, H.: A comprehensive review of C-S-H empirical and computational models, their applications, and practical aspects. Cem. Concr. Res. **156**, 106784 (2022)

49. Hou, D.S., Yu, J., Wang, P.: Molecular dynamics modeling of the structure, dynamics, energetics and mechanical properties of cement-polymer nanocomposite. Compos. B Eng. **162**, 433–444 (2019)

50. Zhou, Y., et al.: Interfacial connection mechanisms in calcium–silicate–hydrates/polymer nanocomposites: a molecular dynamics study. ACS Appl. Mater. Interfaces. **9**(46), 41014–41025 (2017)

51. Zhao, H.X., Wang, Y.W., Yang, Y., Shu, X., Yan, H., Ran, Q.P.: Effect of hydrophobic groups on the adsorption conformation of modified polycarboxylate superplasticizer investigated by molecular dynamics simulation. Appl. Surf. Sci. **207**(15), 8–15 (2017)

52. Chaudhari, O., Biernacki, J.J., Northrup, S.: Effect of carboxylic and hydroxycarboxylic acids on cement hydration: experimental and molecular modeling study. J. Mater. Sci. **52**, 13719–13735 (2017)
53. Hou, D.S., et al.: Unraveling disadhesion mechanism of epoxy/C-S-H interface under aggressive conditions. Cem. Concr. Res. **146**, 106489 (2021)
54. Yu, Z.H., Zhou, A., Ning, W.Y., Tam, T.: Molecular insights into the weakening effect of water on cement/epoxy interface. Appl. Surf. Sci. **553**, 149493 (2021)
55. Zang, Y., et al.: Molecular dynamics simulation of calcium silicate hydrate/tannic acid interfacial interactions at different temperatures: configuration, structure and dynamic. Constr. Build. Mater. **326**, 126820 (2022)

Soft Means of Concrete Modification – Curing Conditions

Piotr Woyciechowski[✉], Wioletta Jackiewicz-Rek, and Beata Jaworska

Department of Building Materials Engineering, Warsaw University of Technology, Warsaw,
Poland
piotr.woyciechowski@pw.edu.pl

Abstract. Concrete curing is an important activity from the point of view of shaping all the properties of concrete, and the way it is carried out depends primarily on the type of binder used. The optimal care effect is a kind of soft method of positive modification. Choosing such an optimal method is not an easy task, especially if the composite contains a complex binder whose components have different care requirements. The article review considerations on the optimal method of polymer-cement concrete curing, as well as the possibility of using various forms of polymers in the curing process of cement concretes.

Keywords: concrete curing · polymer-cement concrete · superabsorbent polymers

1 Concrete Curing - Functions and Methods

The term "concrete curing" is not clearly defined in European standards. It appears many times both in PN-EN 206 [1] and, above all, in PN-EN 13670 [2], which are two basic documents regulating the principles of concrete technology and concrete works. However, it is not defined, but only described by giving methods, rules and requirements. Generally accepted definitions in concrete technology formulated e.g. in [3, 4], refer to the definition taken from the terminology dictionary of the American Concrete Institute (successive versions of ACI 116 [5] and currently subsequent versions of the ACI Concrete Terminology dictionary (current ACI CT-16 of 2016 [6]), which states: "curing - action taken to maintain moisture and temperature conditions in a freshly placed cementitious mixture to allow hydraulic cement hydration and (if applicable) pozzolanic reactions to occur so that the potential properties of the mixture may develop." Other definitions often include the phrase "…actions taken from the moment of placing and compacting the concrete mix…" ([3, 4]), which actually limits the concept of maintenance to activities on the outer surface of the element after placing and compaction, which includes maintaining the appropriate temperature and humidity of concrete and its protection against atmospheric factors.

The classic definitions refer to the curing of cement concrete and do not take into account the different conditions for shaping the structure of concrete containing polymeric co-binder, which are the main subject of this study. In addition, for the purposes

© The Author(s) 2025
L. Czarnecki et al. (Eds.): ICPIC 2023, 61, pp. 73–86, 2025.
https://doi.org/10.1007/978-3-031-72955-3_6

of this study, the concept of care has been extended to include issues related to activities undertaken at other stages of the technological process or related to other than surface impact on concrete, which are necessary and can have a significant impact on the formation of the concrete microstructure and the obtained properties of the composite. This made it possible to include issues of internal curing, issues of preparation of formwork related to curing (formwork inserts, anti-adhesive agents) or issues of active thermal curing during the initial hardening period. Proper curing, taking into account the specificity of the used binder, co-binders and modifiers, is a prerequisite for ensuring the durability of concrete, next to the correct selection of materials, construction design and technology of work.

Uncured or improperly cured cement concrete primarily shows weakening of the surface layer. This is due to physical and chemical phenomena related to two main factors: the flow of moisture and the flow of heat, while the scale of threats resulting from each of these factors is related to climatic conditions. At an ambient temperature above $+10\,°C$, the phenomena related to the lack of moisture are dominant, and the possible negative effects of errors are the more dangerous the higher the ambient temperature in which the concrete works are carried out and the higher the wind force. In low temperature conditions, ensuring proper humidity is equally important, but thermal curing is also becoming more important. In extreme cases, the effects of curing errors can be noticed in a short time, but often the defects are hidden in the concrete structure and reveal themselves during its use, causing deterioration of durability. One of the main defects of this type is the increase in the porosity of the concrete surface layer (i.e. the reinforcement cover), which deteriorates its protective properties and tightness [7–12]. The proper scenario of PCC curing is also crucial for bond development to composites and tiles [13].

As a result, the mechanical properties of the near-surface layer of concrete in the structure deteriorate, as well as the durability characteristics, such as water tightness, frost resistance, chemical resistance, including resistance to chloride penetration and carbonation. This impact is significantly greater in the case of concretes containing mineral additives, used as a substitute for part of the cement. The level of the w/c ratio is also important: with medium and low w/c values, the role of curing is particularly important due to the optimal structure of the surface layer. At high w/c - when concrete is assumed to be weak (porous), the effect of curing decreases. The classification of cement concrete curing methods takes into account the method of its conduct (surface and internal), the type of surface (formed, unformed) and the method of impact (humidity, heat). Other forms of impact were also taken into account, including treatment in an environment with increased carbon dioxide concentration, which is the direction of sustainable development of concrete technology. In the above approach - regarding cement concretes as a base - it is difficult to take into account the specificity of polymer-modified concrete and polymer concrete curing. The so-called low-polymer cement-concretes, with a polymer content of less than 5% of the cement mass used as a modifying admixture, are subject to the curing methods and rules indicated above.

In case of cement-free polymer concretes (or PC) the curing conditions are different than in the case of ordinary concretes or PCC as there is no need to provide water to the internal structure needed for the hydration of cement. Moreover, it is not recommended or sometimes even not allowed to produce and cure the polymer concrete specimens in the atmosphere of increased humidity. In many cases the increased humidity may interfere with the binding process of the polymer binder (here specifically: resins), and in extreme cases, even prevent setting and hardening [14, 15]. Therefore, it is recommended to cure polymer concretes in a dry environment. What is more, in order to avoid moisture in the polymer concrete mix and the negative impact of the presence of water on setting, in many cases dried fillers are used for the production of polymer concrete. This approach is recommended and used in their research, among others by Sokołowska et al. [16–18]. The ambient temperature during application and maintenance cannot be too low or too high, because it affects the dynamics of resin setting (delaying or accelerating setting, respectively), although the setting time can be adjusted by using appropriate amounts of chemical regulators dosed into the resin together with the curing system [18]. Many of the above mentioned methods do not apply to cement concretes with a higher polymer content (PCC), and those that do apply require a different approach related to different bonding conditions of both binders.

Concrete curing process effects in positive changes in concrete microstructure and properties, so in this context it could be treated as a soft mean of concrete modification. The positive effect of this modification is related to the optimal course of curing. The issue of the role of polymers in concrete curing processes can be considered in many aspects, including: the influence of the resin co-binder on the optimal course of concrete curing (Sect. 2), the role of superabsorbent polymers as an internal curing agent for cement concrete (Sect. 3), spray polymeric film-forming agents as a form of surface treatment of cement concrete (Sect. 4), plastic films as a material for the care of cement concrete, treatment of polymer-modified concretes in an environment with increased concentration of carbon dioxide (Sect. 5). Selected issues will be discussed in this article.

2 Polymer-Cement Concrete Curing

2.1 Recommendation from Literature

Polymer-cement composites (PCC) contain two types of co-binder: hydraulic and organic. Optimum conditions for the setting of both of them are different. Cement hydration and hardening of the organic binder. The wet curing of the cement concrete fosters cement hydration while the organic resin hardening generally prefers dry conditions. Thus, the optimum conditions of PCC curing are the compromise between initial period of wet curing and subsequent period of dry hardening. Wet curing period recommended for cement composites varies from 1 to 21 days, depending on the cement type [19–24]. In the case of PCC extended time of maintaining high humidity may significantly worsen the effect of polymer hardening [25]. Hardening of PCC leads to the development of thin polymer film on the cement grains [26], which decreases the hydration degree. If the water access to the material in the first period of the process is not limited (due to the water curing) this film could be redissolved and redispersed in pores [27], giving opportunity to further cement hydration. After wet curing, in dry conditions polymer films are finally formed by coagulation on the cement and hydrates grains as an effect of water evaporation. The type of polymer also is important, as the intensity of hardening disturbance by the high moisture environment varies for different types of organic binders.

This effect is related to the different diffusivity of water through the continuous polymer phase. Standards, guidelines and recommendations show different requirements for the proper PCC curing regime. The literature references can be found for PCC on site curing [28, 29] as well as for curing of PCC specimens in the laboratory [19, 20, 25–27] and [30–39] (Fig. 1). Recommended duration of wet curing of PCC is usually from one to seven days, with some guidelines recommending the higher humidity even after that time. It is hard to find in these documents any information about the quantitative impact of the course of curing on the technical characteristics of PCC [40, 41]. Also, the guidelines for the handling of samples prepared in the laboratory and for concrete in the structure differ significantly. In the latter case, the key is technical simplicity of on-site processes which leads even to the complete withdrawal of the humid conditions according to guidelines [25]. In the laboratory conditions, it is essential to unify the procedures in different laboratories and optimize conditions for the co-binder in order to take advantage of the potential of both binders. Technical complexity of the curing process is less important here, as the size of the specimens is usually small.

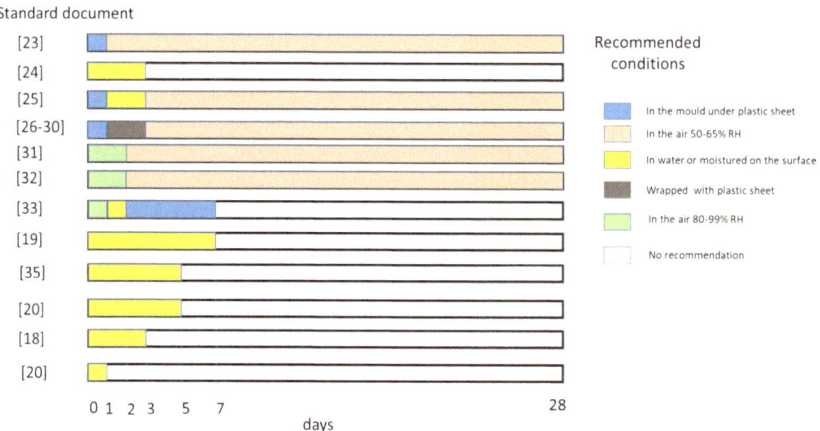

Fig. 1. Time of PCC curing in humid conditions by various recommendations

Research on optimizing the time of wet and dry curing of PCC, depending on the type of polymer and its content, was carried out by the authors of this publication many times, taking into account the impact on various properties of the composite. Exemplary results (Fig. 2) refer to research program for polymer-cement concretes modified with pre-mix and post-mix polymer binders, with polymer content 7% and 15% of the cement mass [42, 43]. Portland cement CEM I 42.5R was used as mineral binder. Natural river sand and gravel were used as aggregates. As the pre-mix modifier aqueous dispersion of polyacrylic esters (PAE) was used.

Six scenarios of curing in constant temperature were adopted (Fig. 3), including 1-day in the form under the plastic sheet, and then - after demolding - in water at 20 °C temperature (according to the recommendations of EN 12390-2, concerning samples of cement concrete) or in air conditions (RH approximately 60%, temperature approx.

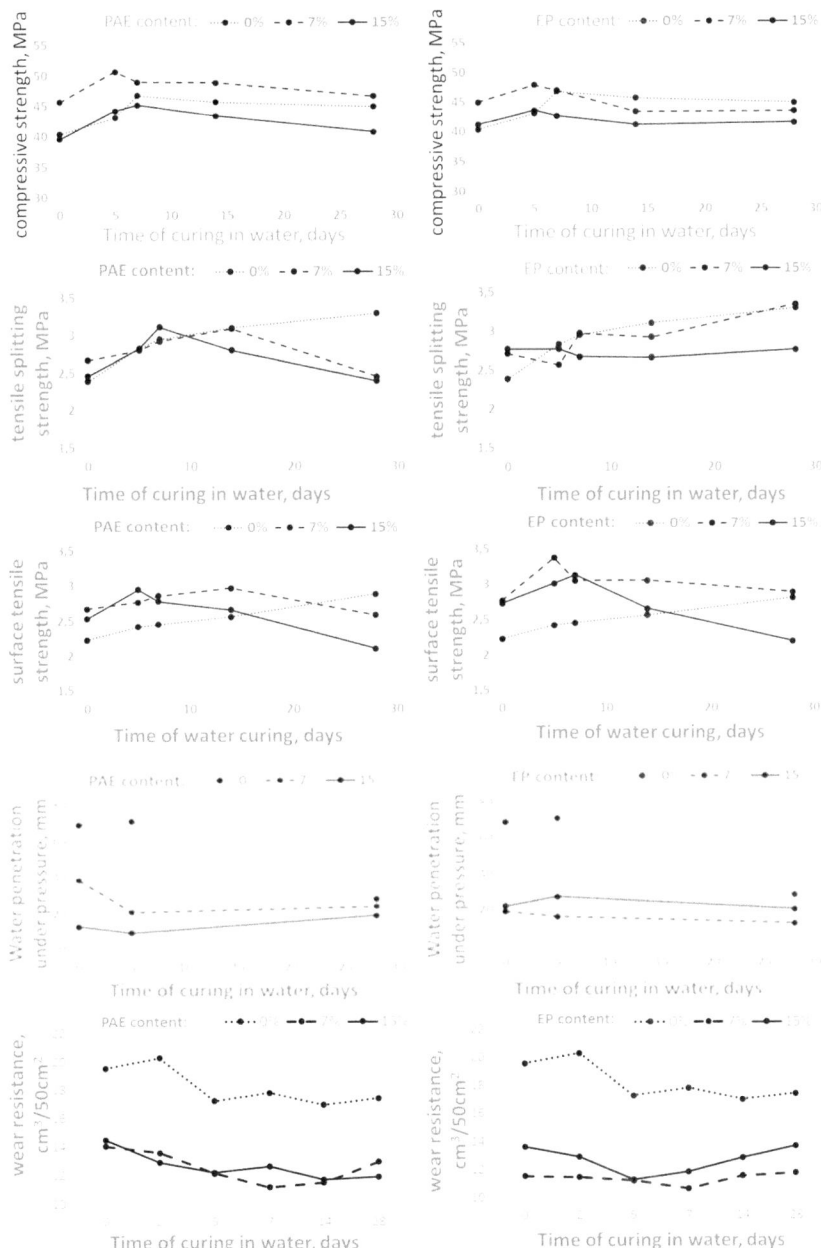

Fig. 2. Effect of water curing time on properties of PCC with PAE (a) and EP (b). Test performed for 28-days old concrete

20 °C - in line with European Standards for materials with a polymer-cement binder for repairs). As extreme variants (1 and 6 in Fig. 3), the curing conditions preferred for

only one of the co-binder components, i.e. 27 days in water after demolding as optimal conditions for cement (scenario 1) or 27 days in air after demolding as optimal conditions for polymer (scenario 6) are chosen. The intermediate variants covered the alternating conditions, i.e. the initial curing in water (for 2, 5, 7 or 14 days) and then in air-dry conditions until 28 day, i.e. the time of testing of the concrete properties.

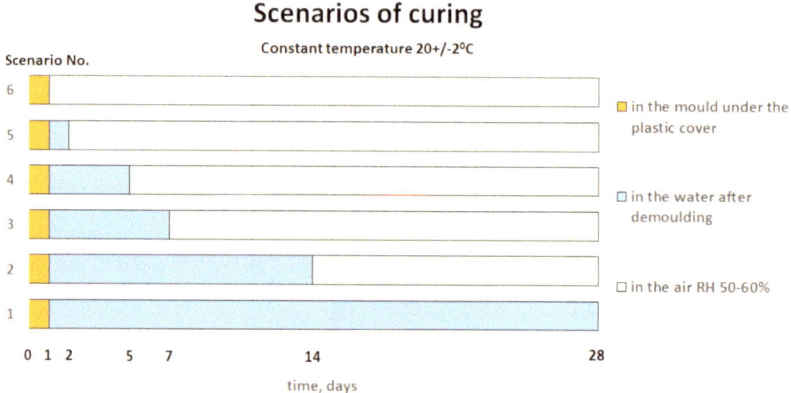

Fig. 3. Scenarios of PCC curing adopted in the research program

The conducted research allows to formulate the following conclusions:

- the optimum time of water curing of the investigated PCC ranged from 5 to 14 days, which is longer than recommended by most of the literature sources;
- the optimum time of curing depends on the polymer type and content;
- all tested PCC properties are sensitive to the way of curing, with the largest impact of this process on the results of water penetration under pressure test;
- each of the considered properties can be used as a measure for PCC curing effectiveness;
- the results confirmed the negative impact of the extension of wet curing of PCC beyond 14th day on its properties, which is probably due to the disturbance of polymer hardening process in water;
- the progress of polymer-cement concrete carbonation in time could be described by hyperbolic model similar to the models developed earlier by author for cement concrete.

The tests were carried out for two substantially different polymer binders (pre-mix, aqueous dispersion of acrylic polymer vs. post-mix, liquid epoxy resin), with low and medium content of the polymer in the binder. The results of evaluation of curing effectiveness are in all cases similar. The results shown in the paper do not allow for establishing of the very precise rules of PCC curing. The optimum time of water curing exceeds 5 days but the accurate value depends on type and content of polymer in the binder. Further studies, particularly on the intermediate curing scenarios, but also for the higher contents of polymers and other kinds of the polymer modifiers are necessary. Such comprehensive approach should give the bases for general guidelines for the selection of the optimum time of water curing for PCC, taking into account both material aspects and conditions of the works on site.

3   Superabsorbent Polymers for Internal Curing of Concrete

The curing of cement concrete with the use of superabsorbent polymers (SAP) is based on changes in water transport process in the pore network of hardening cement composites caused by this modification. Superabsorbent polymers absorb water as a result of high osmotic pressure present in the polymer structure - water absorption is electrochemical in nature. The ability to absorb significant amounts of water by superabsorbent polymers is related to high osmotic pressure caused by the accumulation of ions (e.g. sodium, potassium) in the polymer structure. The process of water absorption causes the polymer to swell, thereby moving the ions apart, which reduces the osmotic pressure. Assuming such a model of SAP operation, their ability to absorb is limited not only by reducing the osmotic pressure as a result of water absorption, but also by the influence of external pressures resulting from the change in polymer volume. This property of SAP determined their use in concrete technology, because in case of loss of equilibrium between the osmotic pressure of the water-saturated polymer and the internal stresses of concrete, SAP is able to reduce its volume, i.e. to release water. This is synonymous with the ability of these polymers to internally cure the composite. With the introduction of SAP to the concrete mix, the polymer grains absorb part of the mixing water. As a result of absorption, their physical properties change - from the form of dry granules in a water-unsaturated state, SAP goes into the form of a hydrogel, with a polymer structure stretched as far as it is allowed by chemical structure and the properties of the water absorption environment (e.g. alkali content in the environment).

From a chemical point of view, the group of superabsorbent polymers includes cross-linked polyelectrolytes (e.g. acrylic polyesters with acrylic acid) that swell when they come into contact with water. Many materials, both naturally occurring and synthetic, fit the definition of hydrogels. The classification of hydrogels is extensive and includes a number of factors. Hydrogels can be of natural origin (e.g. proteins such as collagen or gelatin) or synthetic, resulting from the polymerization process. Synthetic hydrogels are obtained by polymerizing one type of monomer (homopolymer hydrogels) or two or more (copolymer hydrogels). An example of copolymer hydrogels are IPN hydrogels (multipolymer interpenetrating polymeric hydrogel), made of two cross-linked polymer chains [44]. Currently, the most commonly used hydrogels are hydrogels of petrochemical origin, produced from acrylic monomers. Acrylic acid (AA) and its sodium or potassium salts and acrylamide (AD) are most often used in the production of hydrogels [45]. SAP is most commonly available as hard, dry, granular powders with particle sizes ranging from 100 to 1000 μm. They are produced by block polymerization (gel polymerization) or suspension polymerization [46, 47].

The SAP particles can be added into the concrete mix both in dry form and in the form of a hydrogel (i.e., SAP premixed with a portion of the mixing water). The effect of introducing SAP polymers in a dry form to the concrete mix is the absorption of part of the mixing water, which significantly exceeds the weight and volume of added polymer. In the case of a polymer previously mixed with mixing water (dosing in the form of a hydrogel), the mass ratio of saturated polymer to the mass of cement changes about a hundredfold. This is due to the different absorptivity of SAP in environments with different pH (at pH = 7 and pH = 13, the absorptivity of polyacrylate SAPs differs about 10 times). By introducing SAP into the cement composite, in a dry state or in the form of a hydrogel, after mixing with the other components of the composite, an additional phase of the material is created - in the form of quasi-pores filled with hydrogel, which over time, after fulfilling the function of ensuring continuity of hydration and after desorption of all water from the SAP structure, passes into the pore phase [44].

Table 1. Influence of method of SAP adding to the concrete mix on properties of concrete (↑ improved, ↑↑ intensively improved, ↓ deteriorated, ↓↓ intensively deteriorated, - lack of effect)

Properties	SAP adding form	
	SAP in dry form	SAP in hydrogel form
Consistency of concrete mix	↓	↓
Air content in concrete mix	↑	↑
Homogeneity of SAP distribution	↓↓	
Compressive strength	↓/↓↓	↑/↑↑
Tensile strength	↓/↓↓	-/↑
Frost resistance	-/↑	↑/↑↑
Autogenous shrinkage	↑↑	↑↑
Absorbability	-/decrease	decrease
Water tightness	-	↑
Chloride ions diffusion coeff	decrease	decrease

The effectiveness of internal curing with the use of SAP is based on its effect on the kinetics of water migration during the hydration of the binder in cement composites. Internal stresses in the cement matrix resulting from hydration and self-drying contribute to the release of previously absorbed water from the SAP structure. Taking into account a number of properties of cement composites that change as a result of the use of SAP - mechanical properties, degree of binder hydration, autogenous shrinkage, pore network distribution and others affecting the durability of the material - it is crucial to collect as much information as possible regarding both the course of absorption and desorption water from the SAP structure. The two mentioned methods have a different impact both on the issue of the homogeneity of the water absorption process and the characteristics of its course, and above all on the moment of the appearance of the stage of water desorption from the polymer structure over time. As a result of these differences, the impact of SAP introduced in different ways to the concrete mix is also different (Table 1).

The use of superabsorbent polymers is an interesting example of modification of cement composites by changing the rules governing the transport of water in the pore network of the material. By introducing an additional phase into the material which, depending on the circumstances, exhibits both solid and liquid characteristics, it is possible to exercise better control over the internal moisture of the pore network, and thus effectively influence a number of properties of the internally cured material. The introduction of SAP into the composite in the form of a hydrogel is the latest approach to the issue. The purpose of such a modification of the SAP dosing method is to eliminate the negative phenomena associated with the method of dosing SAP in a water-unsaturated state, including increasing the homogeneity of the distribution of SAP particles in the cement matrix, ensuring a positive impact of SAP on both material strength and durability-related properties. The authors, while emphasizing the advantages of internal curing, would like to strongly emphasize that internal curing cannot fully replace the activities that make up external curing - both forms of curing complement each other, using different mechanisms of influencing the properties and characteristics of the microstructure of the cement matrix and affecting different areas of concrete element.

4 Polymers for Surface Curing of Concrete

Surface curing of cement concrete in the traditional approach is carried out "wet", i.e. with the use of additional water, introduced to the surface of the treated concrete element. An alternative is the so-called coating curing, i.e. preventing the evaporation of the mixing water from the concrete by introducing a tight barrier material on the surface of the fresh concrete. Polymer materials are ideal for this role. For a long time, sheet materials have been used in this role, such as polyethylene foil, which can be spread on the surface several hours after concreting, so that it does not stick to fresh concrete. This delay in the start of surface protection is a disadvantage of the method, because curing in the first hours after concreting are the key to obtaining optimal effects. An earlier start of coating curing is possible with the use of film-forming polymer agents that can be applied to fresh concrete immediately after concreting, i.e. after a time sufficient to absorb the cement laitance from the surface. Film-forming agents are liquid substances used to cover the surface of fresh concrete in order to obtain a coating that protects against water evaporation. They can be divided into three groups: solutions of macromolecular substances in organic solvents, water emulsions that are dispersions of organic substances and low viscosity resins. Commercially available agents are in the form of resin solutions and in the form of emulsions. Currently, solutions of the following resins are most often used: acrylic, vinyl, styrene butadiene. These resins are diluted with highly volatile solvents. Film-forming agents in the form of emulsions are made with the use of wax or paraffin. In order to ensure uniform coverage of the concrete surface, a white dye is added to the film-forming preparations, which disappears in time. The requirements for these agents are currently not standardized within European standards. Old national guidelines [48] and American guidelines [49] are in use. Based on the research conducted by the authors, an estimated comparison of several care solutions is shown (Fig. 4). The effectiveness of each variant was assessed using a method based on the European standard draft CEN/TS 14754-1:2007 [50], which includes a unified method for assessing the effectiveness of film-forming preparations by assessing the reduction evaporation of water from the protected concrete surfaces. Water loss from the unprotected surface was assumed as 100%, for individual combinations of materials the loss was expressed as a % loss from the unprotected surface.

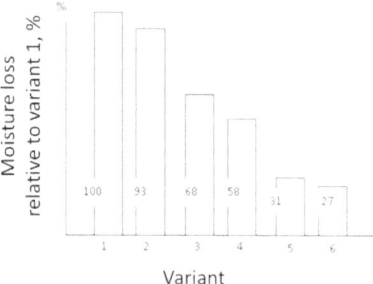

Fig. 4. Comparison of moisture loss from the concrete surface with its protection: 1- without protection (comparative); 2 - transparent film-forming agent based on acrylic resin in one layer; 3 - like 2 but with a double layer; 4 - like 3, but the agent has additionally white pigment; 5 - dark polyethylene foil; 6 - white polyethylene foil

5   Polymer-Modified Concretes Cured with CO_2

Nowadays we are struggling with the problem of excessive carbon dioxide emissions into the atmosphere. A tremendous contribution to CO_2 production comes from the construction industry which emits about 39% of CO_2 and the manufacture of construction materials which alone takes up as much as 11% of global carbon dioxide emissions [51]. A certain amount of carbon dioxide is captured during the service-life, the end-of-life and secondary usage stages of construction, but nevertheless there is crucial need to reduce these emissions and potential capture and avoidance of emitting CO_2.

Accelerated carbonation curing holds the key to capturing and storing CO_2 emissions from the cement industry for the production of value-added concrete products. Carbon sequestration is the process of separating and capturing carbon dioxide from exhaust gases in order to reduce its emissions to the atmosphere. Sequestration methods are categorized into direct, indirect and advanced.

Mineral carbonation is used in the construction industry. In favor of using it as a method of reducing carbon dioxide emissions into the atmosphere is supported primarily by the fact that it is a natural process, occurring in nature and the products resulting from this process are inert for the environment [52]. Mineral carbonation as a method of carbon dioxide sequestration was proposed by Seifritz (1990), and in 2005 was finally defined in the IPCC Special Report on Carbon Dioxide Capture and Storage, part 7. Mineral Carbonation and Industrial uses of Carbon Dioxide [53]. The idea of using CO_2 to cure building materials was proposed as early as the 1970s. However, the method was reluctantly considered because of the costly production of pure CO_2 and the possible negative effects of atmospheric carbonation. However, the need to reduce greenhouse gas emissions has led to renewed interest in the topic. Curing by CO_2 sequestration has a twofold effect on the engineering properties of products: in the early term, the rapid reaction of cement accelerates the development of concrete strength; in the longer term, durability is enhanced as a result of the change in the chemical composition and the microstructure caused by the precipitation of $CaCO_3$ in the cement paste. $CaCO_3$ precipitation impermeable microstructure which affects the reduction of total and capillary porosity, which ultimately modifies the transport properties of the paste, reducing the absorption and permeability of the concrete [54]. In the ECC (fiber reinforced concrete) with fly ash, after CO_2 curing, the early term tensile and compressive strength were accelerated by 57% and 41%, respectively. The matrix and fiber/matrix interface were found densified which increased the ultimate tensile strength by 22% compared to non-carbonated reference. Also dense fiber/matrix interface improved the chemical and frictional bonds leading to tighter crack width which reduced the ECC's water permeability in loaded condition, despite a more permeable matrix due to a larger pore size associated with the lowered pozzolanic reaction [55]. The nature of the carbonation phenomena is similar in cement and polymer-cement concrete [42]. For PCC, besides the proper curing procedure, the duration of component exposition to CO_2, or the composite age, the important factors that affect the reaction with CO_2 are the polymer modifier type and its quantity. The lower carbonation was reported for polymer-modified cement because of the presence of the polymer in the pore which partially buffer the carbonation reaction despite the pore concentration abundance [56]. Referenced research establishes the viability of applying carbonation curing to polymer-modified concrete, with technical merits. Beyond the remarkable CO_2 sequestration capacity at the manufacturing stage, concrete after carbonation curing is anticipated to lower the lifecycle emissions as an example of beneficially utilizing CO_2 for durable precast construction products.

6 Conclusions

The scale of modification of concrete properties with the use of appropriately selected, effective curing methods can be significant, as shown in the research on polymer-cement composites presented in the article. This is particularly important in the context of durability, but also in the context of the cost of ensuring the expected utility of concrete. In this sense, research on the optimization of the concrete curing process is a part of the paradigm of sustainable development of concrete technology. This problem is of particular importance in relation to polymer-cement composites, in which the binder co-components need different conditions to optimally shape their structure - making mistakes in this case can be particularly expensive. The article also draws attention to other aspects of the use of polymers in the curing process of cement and polymer-cement composites. The possibility of curing PCC concretes with the use of carbon dioxide was also initially considered, pointing to this issue as an undiscovered research area.

References

1. PN-EN 206: 2014: Beton. Wymagania, właściwości, produkcja i zgodność
2. PN-EN 13670-1: 2011: Wykonywanie konstrukcji betonowych
3. Jamroży, Z.: Beton i jego technologie. PWN, Warszawa - Kraków (2000)
4. Neville, A.M.: Właściwości betonu. Polski Cement. Kraków (2000)
5. ACI 116R – 13 Cement and Concrete Terminology
6. ACI CT-16: ACI Concrete Terminology - An ACI Standard
7. Jackiewicz-Rek, W., Woyciechowski, P.: Pielęgnacja – klucz do zapewnienia trwałości betonu w konstrukcji. BTA **3**(59), 54–58 (2012)
8. Bajorek, G.: Pielęgnacja betonu w okresie dojrzewania, SPC, Kraków (2016)

9. Kurdowski, W.: Chemia cementu i betonu, SPC/PWN Kraków/Warszawa (2010)
10. Piotrowicz, M., Romanowski, P., Woyciechowski, P.: Klasy pielęgnacji betonu według PN-EN 13670: 2011–kryteria wyboru i wpływ na kształtowanie właściwości betonu. Prace Instytutu Ceramiki i Materiałów Budowlanych **6**, 14/2013, 27–40
11. Woyciechowski, P., Jackiewicz-Rek, W.: Rola pielęgnacji w kształtowaniu trwałości betonu. Materiały Budowlane **5**, 44–48 (2012)
12. Woyciechowski, P., Piotrowicz, M.: Ocena wpływu klasy pielęgnacji na wybrane właściwości eksploatacyjne betonu - Materiały Budowlane, 2/2014 (498), 14–16
13. Li, L., Liu, K., Chen, B., Wang, R.: Effect of cyclic curing conditions on the tensile bond strength between the polymer modified mortar and the tile. Case Stud. Constr. Mater. **17**, e01531 (2022)
14. Czarnecki, L.: Betony żywiczne, Arkady (1982)
15. Czarnecki, L.: Polymer concretes. Cem. Lime Concr. **15**, 63–85 (2010)
16. Sokołowska, J.J., Woyciechowski, P.P.: Chemical resistance of vinyl-ester concrete with waste mineral dust remaining after preparation of aggregate for asphalt mixture. In: Taha M. (eds.) International Congress on Polymers in Concrete (ICPIC 2018), pp. 491–497. ICPIC 2018. Springer, Cham (2018). https://doi.org/10.1007/978-3-319-78175-4_63
17. Sokołowska, J.J., Woyciechowski, P.P., Łukowski, P., Kida, K.: Effect of perlite waste powder on chemical resistance of polymer concrete composites. Adv. Mater. Res. **1129**, 516–522 (B-7) (2015)
18. Sokołowska, J.J.: Long-term compressive strength of polymer concrete-like composites with various fillers. Materials **13**(5), 1207 (2020). https://doi.org/10.3390/ma13051207
19. ACI (American Concrete Institute) (2009) Report No 548.3 R-09: Report on Polymer-Modified Concrete, American Concrete Institute, Farmington Hill, MI, US
20. Evbuomwan NFO: Flexural behavior of reinforced polymer modified mortar under wet conditions. In: Proceedings of 3rd Southern African Conference on Polymers in Concrete, Johannesburg, RSA, pp. 117–124 (1997)
21. Wang, R., Fan, Y., Wang, Z., Huang, T., Zhang, T.: Performance development of styrene-butadiene copolymer-modified calcium sulfoaluminate cement mortar under different curing conditions. J. Zhejiang Univ. Sci. A (Appl. Phys. Eng.) **22**(12), 1005–1026 (2021)
22. Lin Li, R., Wang, S.Z.: Effect of curing temperature and relative humidity on the hydrates and porosity of calcium sulfoaluminate cement. Constr. Build. Mater. **213**, 627–636 (2019)
23. Wang, R., Li, L., Yundong, X.: Influence of curing regimes on the mechanical properties, water capillary adsorption, and microstructure of CSA cement mortar modified with styrene-butadiene copolymer dispersion. J. Mater. Civ. Eng. **31**(1), 04018344 (2019)
24. Li, N., Linglin, X., Wang, R., Li, L., Wang, P.: Experimental study of calcium sulfoaluminate cement-based self-leveling compound exposed to various temperatures and moisture conditions: hydration mechanism and mortar properties. Cem. Concr. Res. **108**, 103–115 (2018)
25. German Federal Ministry for Transport: Technical guidelines for concrete repair systems made of cement mortar/concrete with a polymer additive TP BE-PCC (1990) and German Federal Ministry for Transport: Additional Technical Contract Conditions and Guidelines for the Protection and Repair of Concrete Construction Components, ZTV-SIB (1991)
26. Knapen, E., van Gemert, D.: Effect of underwater storage on bridge formation by water-soluble polymers in cement mortars. Constr. Build. Mat. **23**(11), 3420–3425 (2009)
27. Kwan, W.H., Ramli, M., Cheah, C.B.: Accelerated curing regimes for polymer-modified cement. Mag. Concr. Res. **67**(23), 1233–1241 (2015)
28. ASTM C 1439:2013 Standard Test Methods for Evaluating Latex and Powder Polymer Modifiers for use in Hydraulic Cement Concrete and Mortar
29. Specification and guidelines for polymer-modified cementitious flooring as wearing surfaces for industrial and commercial use (2001), EFNARC

30. NT BUILD 428:1994, Nordtest Method. Concrete and mortar, polymer cement (PCC): moulded test specimens – curing. [19]
31. PN EN 1542:2000 Products and systems for the protection and repair of concrete structures. Test methods. Measurement of bond strength by pull-off
32. PN EN 12190:2000 Products and systems for the protection and repair of concrete structures. Test methods. Determination of compressive strength of repair mortar
33. PN EN 13295:2005 Products and systems for the protection and repair of concrete structures. Test methods. Determination of resistance to carbonation
34. PN EN 13396:2005 Products and systems for the protection and repair of concrete structures. Test methods. Measurement of chloride ion ingress
35. PN EN 13412:2008 Products and systems for the protection and repair of concrete structures. Test methods. Determination of modulus of elasticity in compression
36. Makhtar, A.M.: Properties and performance of polymer modified concrete (1997) Ph.D. Thesis, The University of Leeds, Leeds
37. Łukowski, P.: The role of polymers in formation of properties of polymer-cement binders and composites, Warsaw University of Technology Ed, Warsaw, (in Polish) (2008)
38. RILEM Technical Committee 105 C-PC Concrete - Polymer Composites, State-of-the-Art Report (1996), Warszawa-Praha, RILEM
39. Kasai, Y., Matsui, I., Fukushima, Y.: Physical properties of polymer-modified mortars (1981), In: 3rd International Congress on Polymers in Concrete, Koriyama, Japan, pp. 178–192
40. Bhutta, M.A., Ohama, Y.: Recent status of research and development of concrete-polymer composites in Japan. Concr. Res. Lett. **1**(4), 125–130 (2010)
41. Sokołowska, J., Woyciechowski, P., Adamczewski, G.: Influence of acidic environments on cement and polymer-cement concretes degradation. Adv. Mater. Res. **687**, 144–149 (2013)
42. Woyciechowski, P.: Effect of curing regime on polymer-cement concrete properties. Arch. Civil Eng. **66**(1) (2020)
43. Łukowski, P., Woyciechowski, P., Adamczewski, G., Rudko, M., Filipek, K.: Curing of polymer-cement concrete–search for a compromise, Adv. Mater. Res. **1129**, 222–229 (2015)
44. Kalinowski, M., Woyciechowski, P.: Pielęgnacja wewnętrzna betonu, Go Green Concrete Sympozjum – Reologia w Technologii Betonów Niskoemisyjnych, s.1–11. (2022)
45. Zohuriaan-Mehr, M.J., Kabiri, K.: Superabsorbent polymer materials: a review. Iran. Polym. J. **17**, 451–477 (2008)
46. Ahmed, E.M.: Hydrogel: Preparation, characterization, and applications: A review. J. Adv. Res. **6**, 105–112 (2015). https://doi.org/10.1016/j.jare.2013.07.006
47. Mechtcherine, V., Reinhardt, H.W.: Application of superabsorbent polymers (SAP) in concrete construction: State of the art report prepared by technical committee 225-SAP. Springer, Netherlands, Dordrecht (2012)
48. Wstępne wytyczne oceny przydatności preparatów powłokowych oraz powłok do pielęgnacji nawierzchni z betonu cementowego. IBDiM Ośrodek informacji Naukowej, Technicznej i Ekonomicznej Drogownictwa. Zeszyt 16
49. ASTM C309-02: Standard specification for Liquide Membrane - Forming Compounds for Curing Concrete. ACI Committee 2002
50. PKN-CEN/TS 14754-1:2007 Curing compounds. Test methods. Part 1: Determination of water retention efficiency of common curing compounds
51. Uliasz-Bocheńczyk, A., Mokrzycki, E.: Przegląd możliwości utylizacji ditlenku węgla, Wiertnictwo Nafta Gaz, Tom 22/1 (2005)
52. Huijgen, W.J.J., Comans, R.N.J.: Carbon dioxide sequestration by mineral carbonation, ECN, ECN-C-03-016 (2003)
53. Mazzotii, M.: IPCC Special Report on Carbon Dioxide Capture and Storage, part 7. Mineral Carbonation and Industrial uses of Carbon Dioxide. Coordinating Lead (2005)

54. Jaworska, B., Łukowski, P.: Stosowanie CO2 w procesie wiązania i twardnienia betonu, Go Green Concrete Sympozjum – Reologia w Technologii Betonów Niskoemisyjnych [referaty prelegentów], s.1–11 (2022)
55. Zhang, D., Ellis, B.R., Jaworska, B., Hu, W.-H., Li, V.C.: Carbonation curing for precast Engineered Cementitious Composites. Const. Build. Mat. **313**, 125502 (2021)
56. Elbakhshwan, M.S., et al.: Structural and chemical changes from CO2 exposure to self-healing polymer cement composites for geothermal wellbores. Geothermics **89**, 101932 (2021)

Innovative Building Materials Containing Post-Consumer Plastics: A Rewarding Example of Circular Economy in Construction

Mariaenrica Frigione[1]([⊠]) and José Luís Barroso de Aguiar[2]

[1] Department of Innovation Engineering, University of Salento, Lecce, Italy
mariaenrica.frigione@unisalento.it
[2] Department of Civil Engineering, University of Minho, Braga, Portugal
aguiar@civil.uminho.pt

Abstract. Circular Economy, which it is among the priorities of the European Commission, is defined as an economy in which the value of products, materials and resources is maintained for as long as possible and the production of waste is reduced to minimum. Keeping in mind the impact on the environment caused on the one hand by post-consumer plastic waste and on the other hand by production processes of concrete, it is possible to find a solution able, at least partly, to mitigate these two issues. Following the principles of the circular economy, in fact, it is possible to reuse post-consumer plastic waste as fine aggregates in concrete: in this way, post-consumer plastic from waste becomes a resource; at the same time, the use of other natural resources is limited, such as the minerals traditionally used as aggregates in concrete. However, this virtuous solution still presents some problems to study and solve: this work aims to illustrate some of these issues, and provides indications on the aspects to be analyzed and solved.

Keywords: Circular economy · Concrete · Post-consumer plastics · Recycling · Sustainable building materials

1 Introduction: Circular Economy Principles

Circular Economy (CE) is defined as an economy in which the value of products, materials and resources is maintained for as long as possible; this can be achieved by reusing, repairing, reconditioning and recycling existing materials, extending their life cycle and minimizing the production of waste. It is among the priorities of the European Commission to support sustainable growth and job creation [1]. The demand for raw materials is, in fact, continuously increasing while the natural resources are going to be depleted, resources that are essential for an economy struggling with a constantly growing world population. Furthermore, the procurement of raw materials (not available internally) leads to dependence on other Countries, and we have recently learned how much this can damage a Country's economy. Actions in this direction give a strong boost to product and process innovation, with an increase in competitiveness and economic growth.

© The Author(s) 2025
L. Czarnecki et al. (Eds.): ICPIC 2023, 61, pp. 87–97, 2025.
https://doi.org/10.1007/978-3-031-72955-3_7

The adoption of the principles of the CE is also an effective response to current environmental issues, as the processes of extraction and use of raw materials produce a great impact on the environment and increase energy consumption and CO_2 emissions. Furthermore, according to CE waste can become a profitable resource, no longer a problem to be managed: it is possible to generate value from waste. The economic model long pursued thanks to a large availability of materials and energy, which involved an enormous exploitation of resources with massive production of waste, must be replaced by a more sustainable model in which waste must be reused/recycled, thus returning economic value to an asset, at the same time reducing its disposal costs. To this regard, Circular Economy strategies can offer new opportunities also to plastic waste which can be recycled instead of being dumped into the environment causing the well-known serious environmental problems. CE represents, therefore, a model for a closed system which promotes the reuse of plastic products in a logic of conservation of resources, generates value from waste producing new eco-sustainable products, and avoids sending a material that is still recoverable to landfill.

2 Post-Consumer Plastics: Disposal and Recycle

In 2020, the world plastic production reached 367 Mtons while the European plastic production was 55 Mtons [2]. After a significant decline in the first half of 2020 due to the pandemic, the production of plastics has recovered since the second half of the same year. The recovery was conditioned by the impact of the pandemic which increased the demand for plastics by major industries, for the production of disposable personal protective equipment and to cope with the massive demand for packaging for shipping of products during the global lockdown. At the same time, many companies have faced supply chain disruptions, raw materials shortages and rising energy prices.

In 2020, the demand of plastic materials in Europe reached 49.1 Mtons, the packaging and construction sectors representing by far the largest end users of such materials. The polymers mainly used in these industries are polyolefins (i.e. polyethylene, PE, and polypropylene, PP), polystyrene (PS), polyvinyl chloride (PVC) and polyethylene terephthalate (PET). At the end of their useful life (probably very short in packaging applications), these polymers are not completely biodegradable, their biodegradation being extremely slow and occurring to a limited extent only in appropriate environmental conditions. On the other hand, if released in the environment, these plastic materials can represent a serious threat to the environment and ecosystems. Therefore, their recovery would avoid these problems allowing to decrease materials extracted from non-renewable resources. In 2020, more than 29 Mtons of post-consumer plastic waste were collected in Europe [2]. Of these, more than a third was recycled and over 40% was recovered as energy. However, over 23% of collected plastic waste were still sent to landfill.

If from the one hand plastic waste represents a valuable resource that can be employed to produce new polymeric materials to manufacture plastic parts and products, on the other hand the recycle of such materials presents many challenges: it is not always practicable and economically convenient. In fact, post-consumer plastic flows are only partially exploited [3], and when, in most cases, their recycling is not possible, they are burned to generate energy, this solution implying a waste of non-renewable natural resources. Therefore, alternative solutions are needed to fully exploit plastic waste to produce new materials and products.

3 Valorization of Post-Consumer Plastic Waste in Construction Sector: Issues

An example of possible route to recover and valorize the post-consumer plastic waste is represented by their use in construction, and in particular to produce concrete.

Considering that concrete is the most commonly used construction material [4], if post-consumer plastics, for instance packaging and disposable items, were systematically employed on an industrial level as raw materials for the production of concrete, this would benefit the environment in several ways. This use as raw materials could reduce the use of minerals traditionally used as aggregates. Post-consumer plastic, dispersed in small particles can, in fact, satisfactorily replace the fine aggregates, enhancing some of the final properties of concrete. Plastic wastes have less weight per unit volume than concrete aggregates, therefore their use as aggregates will reduce the unit mass of concrete structures. The insulation characteristic of post-consumer plastic-concrete is also going to increase.

Research is very active in this field: different experimental studies describe, for instance, the use of post-consumer PET bottles in concrete [5, 6]. The concretes containing PET particles as aggregate are generally reported to be very resistant in both compression and flexure compared to conventional Portland cement concrete. The tensile strength has been found generally increased due to the bridging action of plastic fibers in concrete. Referring to workability of fresh concrete, some literature reported an increase in workability with the addition of small percentages of waste PET while others reported an opposite influence, probably due to the different shape, size, mechanical properties and origin of the waste plastics. It is generally recognized, however, that if a proper mix design is identified, improvements of mechanical and physical properties can be achieved in PET-modified concrete. Additionally, the waste PET does not require any particular treatment before the addition in concrete, i.e. neither cleaning nor removal of colors.

On the other hand, there are many aspects still to be analyzed and investigated. First of all, the cost of production of the post-consumer plastic-concrete is high, still not competitive to conventional concrete. It is therefore necessary to identify applications in which these new materials can be conveniently used, i.e. where they are competitive on the market also in relation to their performance.

Post-consumer plastic waste typically contains different types of polymeric materials in different percentages and with different shapes. The streams of post-consumer plastics are, in fact, composed of different polymers, largely polyolefins (high density and low density polyethylene, i.e. HDPE and LDPE, respectively, and polypropylene), polystyrene, polyethylene terephthalate (Fig. 1), while the international scientific literature mainly analyzes the use of individual polymers (PET only, PE only, etc.) in concrete [7]. On the other hand, the separation of the different polymers composing post-consumer plastic waste is a quite expensive process, not always easy to implement (different polymers have very similar densities). Therefore, if the separation of different types of polymers would be necessary before their addition into the concrete, this would lead to a further increase in production costs. It is therefore necessary to evaluate if by using different plastic waste as aggregates in concrete, without their prior separation, it is possible to obtain concretes with properties and performances comparable to traditional concretes.

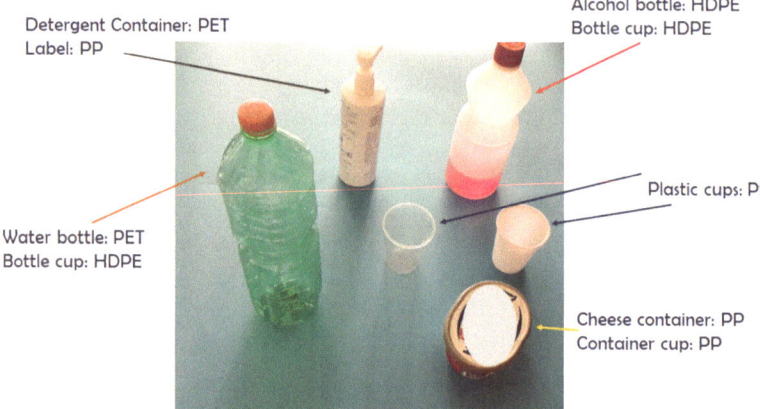

Fig. 1. Some plastic single use containers/items that can be found in post-consumer plastic streams. (PET: polyethylene terephthalate; PP: polypropylene; HDPE: high density polyethylene; PS: polystyrene)

4 Effect of Replacing Fine Aggregate with Plastic Waste of Different Chemical Nature on the Properties of Concrete

Multiple studies in literature illustrate the effects on the slump of concrete due to the introduction of plastic waste in replacement of aggregates. Ismail and AL-Hashmi [8] noticed that the values of slump of concrete mixes containing plastic waste (consisting of about 80% PE and 20% PS) tended to decrease with increasing the waste content. This reduction was attributed to the different shapes of the waste particles: some were angular while others had non-uniform shapes. Similar results were found also by Rai et al. [9], as illustrated in Fig. 2. The study stated that the sand in concrete was partially replaced by waste plastic flakes, without specifying the chemical nature of the polymer.

The same study [9] reported that fresh density of concrete decreased by 5.0%, 8.7% and 10.7% by replacing sand with 5%, 10% and 15% of plastic waste, respectively. This behavior can be attributed to the density of plastic waste that is much lower (by about 70%) than that of sand, resulting in a reduction of the fresh density of concrete. These results are in line with those reported by Ismail and EL-Hashmi [8]. Figure 3 illustrates the effect of an increase in the amount of plastic waste on the fresh density of concrete.

Referring to the effects on mechanical strength of concrete, Ismail and AL-Hashmi [8] found that the compressive strength decreased with increasing the quantity of plastic waste, irrespective to the curing age of concrete. This result, aligned with findings

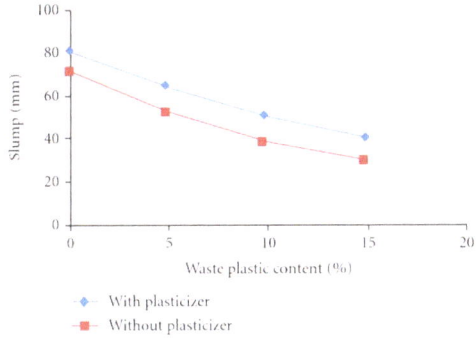

Fig. 2. Slump of concrete mixes with different % of plastic waste [reprinted with permission from reference 9].

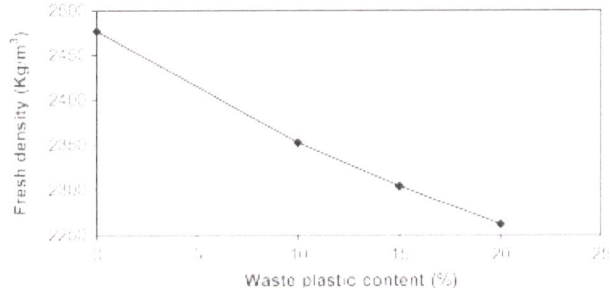

Fig. 3. Effect of plastic waste content on the fresh density of concrete [reprinted with permission from reference 8].

reported by other Authors [9], is largely attributed to the low adhesion strength developed between the plastic (PE and PS) particles and the cement paste. Similarly, Mustafa and co-workers [10] found that the compressive strength of the plastic-concrete decreased as the plastic content increased. Polycarbonate (PC) particles from industrial waste were used in this study. The decrease in strength was again mostly attributed to a low adhesion between the cement paste and polycarbonate aggregate, partly to the lower resistance and stiffness that characterize the plastic material. Figure 4 gives an example of how the compressive strength of concrete is influenced by an increasing addition of plastic waste to replace inorganic aggregates.

Passing to analyze the tensile strength properties, Rahmani et al. [11] studied the influence of the addition of PET particles, in quantities up to 15%, on the tensile strength of concretes based on two water-cement ratios, i.e. 0.42 and 0.54, respectively. The PET particles were obtained by grinding post-consumer bottles. A decrease in tensile strength was observed, again attributed to the limited adhesive strength between the plastic aggregate and the cement paste. Figure 5 presents the tensile strength values as a function of the percentage of plastic waste added to the concrete, as found in [11].

Albano and co-workers [12] analyzed the replacement of sand with PET, up to 20% by volume, in concrete with two values of the water-cement ratio, namely 0.50 and 0.60.

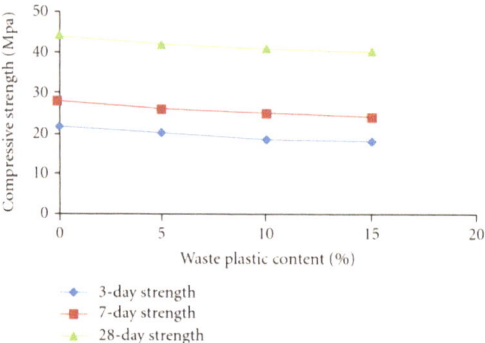

Fig. 4. Compressive strength of concrete as a function of the % of plastic waste added and the age of curing [reprinted with permission from reference 9].

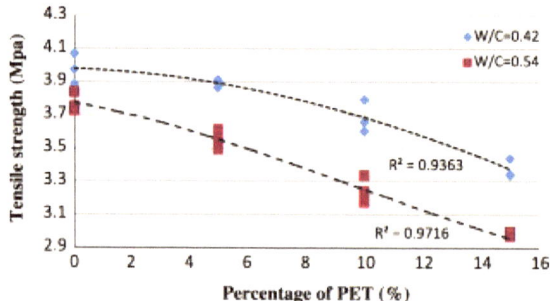

Fig. 5. Tensile strength at 28 days vs. the percentage of plastic waste, at two values of water/cement ratio [reprinted with permission from reference 11].

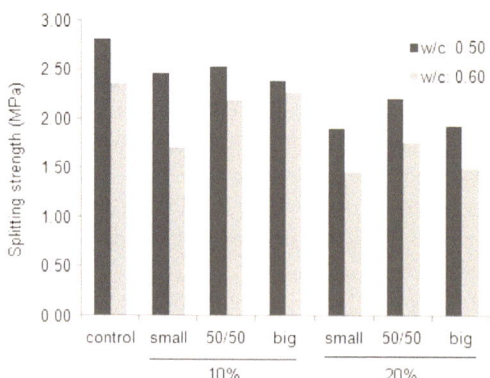

Fig. 6. Splitting tensile strength of Concrete-PET blends at different water/cement ratios [reprinted with permission from reference 12].

The average dimensions of the PET particles were 0.26 (small granules) and 1.14 cm (large granules); PET came from recycled-waste bottles. The Authors of the study found that a decrease in the splitting tensile strength at a water-cement ratio equal to 0.50 compared to the reference concrete, regardless of the size of the PET particles. The decrease in strength was more significant when the amount of ground PET reached 20%, attributed to the high porosity of the concrete at this high amount of PET. A very similar trend was also observed for a water-cement ratio of 0.60: the tensile strength values decreased compared to the control concrete, even more significant. Figure 6 presents the splitting tensile strength values as a function of the percentage of plastic waste added to the concrete, as found in [12].

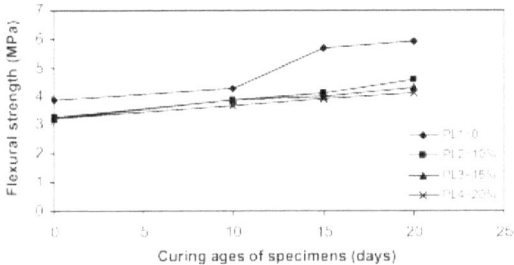

Fig. 7. Effect of waste plastic (content up to 20%) on flexural strength of concrete as a function of the age of curing [reprinted with permission from reference 8].

Finally, the effect of the addition of plastic waste on the flexural strength of concrete was evaluated by Ismail and AL-Hashmi [8]. These Authors found that the flexural strength of plastic waste-concrete decreased with increasing the content of polyethylene/polystyrene particles. These results are substantially in line with those found by Rai et al. [9]. Harini and Ramana [13] found that the flexural strength decreased by increasing plastic granules (PET, less than 4.75 mm in size) in place of fine aggregate, irrespective to the curing age. This result was again attributed to the low bond strength between the surface of the plastic particles and the cement paste. The effect of plastic waste on the flexural strength of concrete is summarized in Fig. 7.

In all the examples just illustrated, the replacement of the fine aggregate in concrete with post-consumer plastic particles, based on different polymers, led to a general decrease in the mechanical properties of concrete, except for the effect of fibrous plastic particles on flexural strength. This finding was attributed to the poor adhesion that develops at the interface between an organic material, i.e. plastic particles, and an inorganic one, i.e. cement paste. To solve this problem, several solutions have been proposed. A first possibility involves the use of plastic particles with specific shape, size and aspect ratio. Typically, plastic fibers act as a physical bridge capable of limiting the propagation of fractures in concrete. In [14], the Authors revealed that the geometry of the PET fibers can play a significant role in achieving good mechanical properties of concrete. For instance, fibers with variable cross sections produced a substantial improvement in compressive strength over straight fibers.

A second possibility involves the modification of the surface properties of the plastic particles so that they exert an adequate adhesive strength to the cement paste. This can be achieved through an appropriate functionalization of the surfaces of the plastic material, in order to increase the chemical interactions with the cement paste: this solution is more effective but also has higher costs. An example is illustrated in the work by Akçaözoğlu and co-workers [15]. The Authors investigated the possibility to modify the surface of waste PET exposing this polymer to selected bacterial strains, in order to improve its chemical affinity and adherence with cement paste. These experiments were successful: they found that the concrete produced with the waste PET exposed to a bacterial strain achieved greater compressive and flexural strength values with respect to concrete containing un-treated PET.

5   Recent Attempts to Reuse Disposable COVID-19 Masks in Concrete

The Covid 19 pandemic that spread across the globe in the last two and a half years (2020–2022) resulted in a huge entry into the environment of disposable personal protective equipment (PPE), especially face masks. The polymeric material mainly employed in the fabrication of disposable face masks is polypropylene, the most of masks being realized in non-woven fabric PP. Several researchers have proposed different solutions to reuse/recycle the billions of single-use masks continuously released into the environment; some of them suggested to reuse them as aggregate in concrete in the wake of what has been experienced in the last years with post-consumer plastics.

In this case, the issues that arise are even greater than those already analyzed. The face masks consist of different layers: they contain a metallic part (to tighten the mask on the nose) and rubber ear loops. The different materials must be separated by disassembling the mask. The second important issue is represented by the possible need to sanitize the used masks before using them in concrete. Both procedures, that is the separation of the various components and their sanitization, involve an increase in production costs, justified only by a considerable increase in the performance of the concrete. Scientific research is, however, progressing fast in the search for a solution to this alarming environmental problem.

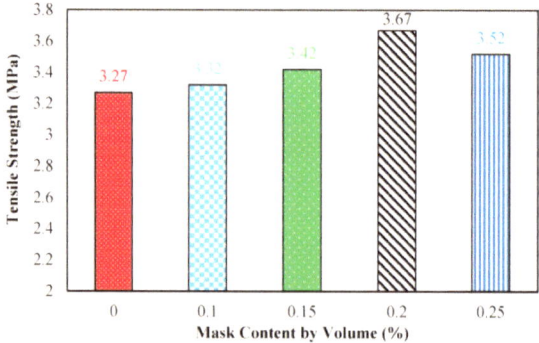

Fig. 8. Tensile strength at 28 days of concrete containing shredded face masks [reprinted with permission from reference 16].

The Authors of the study reported in [16] analyzed the effect of the addition of small amounts of shredded face masks on the mechanical properties of concrete. They found advantages in terms of improved compressive strength, indirect tensile strength (as illustrated in Fig. 8) and modulus of elasticity by including in concrete very small contents of masks, i.e. up to 0.2% by volume.

Encouraging results were recently found also by Ajam and co-workers [17], i.e. an increase in both compressive and flexural strength upon addition in cement mortars of up to 5% in volume of pieces of single use surgical masks.

6 Conclusions: Areas Where Further Research is Needed

The above overview has shown that post-consumer plastics can be effectively used as an aggregate in concrete, with multiple benefits for the environment [18, 19]. However, there are still many aspects to be clarified or deepened.

The presence of a random mix of different polymers in the plastic waste can lead to opposite behaviors in the properties of concrete. Future research should focus on identifying compositions able to achieve adequate concrete properties, regardless of the type of polymers contained in the waste, possibly playing on the shape and aspect ratio of plastic fibers.

New, low-cost and feasible methods must be developed to increase the adhesion at the interface between the plastic particles and the cement paste.

The high costs of these new building materials represent a limit to their wider use: it will be necessary to identify applications where high costs are justified by specific performances.

The literature is very lacking as regards the durability of plastic waste-concrete: this gap must be filled because, in addition to the mechanical performance, the durability of the concrete is an essential feature for their actual application.

Finally, there is an urgent need of standard codes and updated guidelines to reliably use polymer waste in concrete.

References

1. https://www.europarl.europa.eu/news/it/headlines/priorities/economiacircolare. Accessed 18 Jul. 2022
2. https://plasticseurope.org/wp-content/uploads/2021/12/Plastics-the-Facts-2021-web-final.pdf. Accessed 18 Jul. 2022
3. Adelodun, A.A.: Plastic recovery and utilization: from ocean pollution to green economy. Front. Environ. Sci. 9(683403), 1–12 (2021). https://doi.org/10.3389/fenvs.2021.683403
4. Donadkar, M.U., Solanke, S.S.: Review of e-waste material used in making of concrete. Int. J. Sci. Technol. & Eng. 2, 66–69 (2016)
5. Jethy, B., Paul, S., Das, S.K., Adesina, A., Mustakim, S.M.: Critical review on the evolution, properties, and utilization of plastic wastes for construction applications. J. Mater. Cycles Waste Manage. 24, 435–451 (2022). https://doi.org/10.1007/s10163-022-01362-4
6. Eyni Kangavar, M., Lokuge, W., Manalo, A., Karunasena, W., Frigione, M.: Investigation on the properties of concrete with recycled polyethylene terephthalate (PET) granules as fine aggregate replacement. Case Stud. Constr. Mater. 16(e00934), 1–14 (2022). https://doi.org/10.1016/j.cscm.2022.e00934

7. Bhagat, G.V., Savoikar, P.P.: Durability related properties of cement composites containing thermoplastic aggregates – a review. J. Build. Eng. **53**(104565), 1–33 (2022). https://doi.org/10.1016/j.jobe.2022.104565
8. Ismail, Z.Z., AL-Hashmi, E.A.: Use of waste plastic in concrete mixture as aggregate replacement. Waste Manage. **28**(11), 2041–2047 (2008). https://doi.org/10.1016/j.wasman.2007.08.023
9. Rai, B., Rushad, S.T., Kr, B., Duggal, S.K.: Study of waste plastic mix concrete with plasticizer. Int. Sch. Res. Netw. Civ. Eng. **2012**(469272), 1–5 (2012). https://doi.org/10.5402/2012/469272
10. Mustafa, M.A.T., Hanafi, I., Mahmoud, R., Tayeh, B.A.: Effect of partial replacement of sand by plastic waste on impact resistance of concrete: experiment and simulation. Structures **20**, 519–526 (2019). https://doi.org/10.1016/j.istruc.2019.06.008
11. Rahmani, E., Dehestani, M., Beygi, M.H.A., Allahyari, H., Nikbin, I.M.: On the mechanical properties of concrete containing waste PET particles. Constr. Build. Mater. **47**, 1302–1308 (2013). https://doi.org/10.1016/j.conbuildmat.2013.06.041
12. Albano, C., Camacho, N., Hernández, M., Matheus, A., Gutiérrez, A.: Influence of content and particle size of waste pet bottles on concrete behavior at different w/c ratios. Waste Manage. **29**(10), 2707–2716 (2009). https://doi.org/10.1016/j.wasman.2009.05.007
13. Harini, B., Ramana, K.V.: Use of recycled plastic waste as partial replacement for fine aggregate in concrete. Int. J. Innov. Res. Sci., Eng. Technol. **4**(9), 8596–8603 (2015). https://doi.org/10.15680/IJIRSET.2015.0409106
14. Marthong, C., Sarma, D.K.: Influence of PET fiber geometry on the mechanical properties of concrete: an experimental investigation. Eur. J. Environ. Civ. Eng. **20**(7), 771–784 (2016). https://doi.org/10.1080/19648189.2015.1072112
15. Akçaözoğlu, S., Adigüzel, A.O., Akçaözoğlu, K., Deveci, E.Ü., Gönen, Ç.: Investigation of the bacterial modified waste PET aggregate VIA *Bacillus safensis* to enhance the strength properties of mortars. Constr. Build. Mater. **270**(121828), 1–11 (2021). https://doi.org/10.1016/j.conbuildmat.2020.121828
16. Kilmartin-Lynch, S., Saberian, M., Li, J., Roychand, R., Zhang, G.: Preliminary evaluation of the feasibility of using polypropylene fibres from COVID-19 single-use face masks to improve the mechanical properties of concrete. J. Clean. Prod. **296**(126460), 1–8 (2021). https://doi.org/10.1016/j.jclepro.2021.126460
17. Ajam, L., Trabelsi, A., Kammoun, Z.: Valorisation of face mask waste in mortar. Innov. Infrastruct. Solut. **7**(130), 1 (2022). https://doi.org/10.1007/s41062-021-00729-0
18. Pacheco-Torgal, F., Khatib, J., Colangelo, F., Tuladhar, R. (eds.): Use of recycled plastics in eco-efficient concrete. Woodhead Publishing, UK (2019)
19. Czarnecki, L.: Would recycled plastics be a driving force in concrete technology. J. Zhejiang Univ. Sci. A (Applied Physics & Engineering), **5**(7), 384–388 (2019)

Application and Challenges for C-PC in Circular Economy

Current Status of Resin Concrete in Japan

Nobuhiro Kai[1]([✉]), Makoto Kawakami[2], Masahisa Kido[1], Kei Ishitsuka[1],
and Yuki Kuwahara[1]

[1] Hinode Ltd, Iwasaki Harukoga Miyaki-Cho, Miyaki-Gun, Japan
n-kai@hinodesuido.co.jp
[2] Akita University, Akita, Japan

Abstract. Resin concrete, which has high strength, high early strength development and excellent chemical resistance, has been widely applied for repair and reinforcement of concrete structures, and precast products since the 1950's. The research and development of resin concrete in Japan have evolved in tandem with various regions and international activities. It is a common and indispensable material for infrastructures in the world at present. On the other hand, there are so many different Acts of God, such as earthquakes, typhoons, and torrential rains, that occur in Japan, and disaster measures are an important issue. In this paper, the current status and practical application of resin concrete mainly applied for sewage products in Japan were investigated and discussed. Based on the regional characteristics of Japan, a resin concrete manhole of high quality to compensate for the defects of cement concrete was developed. Furthermore, the future trends in the research and development of resin concrete including environmental issues such as carbon dioxide emissions reduction, application of recycled materials and bioplastics to replace natural aggregates and petroleum-derived resin for concrete are proposed and discussed. A homogenization analysis method is introduced particularly to bring out the ability of resin concrete as composite materials and to carry out material development efficiently.

Keywords: resin concrete · manhole · environmental measures · homogenization analysis

1 Introduction

Research and development of resin concrete started in the USSR, USA, West Germany, UK and Japan in the 1950's [1]. In Japan, it was successfully commercialized in1971, and is widely used for precast products such as block manholes and pipes, as well as cast-in-place mortar and concrete. Along with growing use and expanding applications of resin concrete, test methods and quality of resin concrete applied for structural members were established as a Japanese Industrial Standard in 1978 [2–7]. As a high-strength and lightweight precast concrete, resin concrete is widely used for boxes for protecting valves of water pipes, manholes for sewage facilities, and sewage pipes. Unsaturated polyester resin is mainly used as a binder for resin concrete.

© The Author(s) 2025
L. Czarnecki et al. (Eds.): ICPIC 2023, 61, pp. 101–110, 2025.
https://doi.org/10.1007/978-3-031-72955-3_8

Japan is located in the Circum-Pacific Mobile Belt where seismic and volcanic activities occur constantly. Although the land area of Japan covers only 0.25% of that on the planet, the number of earthquakes and active volcanoes is quite high. Three-fourths of Japan's land area is mountainous and hilly, and flat land for people to live in is limited. As a result, the population is concentrated on the plains, and there are areas where the roads are too narrow that large vehicles cannot pass. Thus, infrastructures in Japan should have high seismic resistance and corrosion resistance in hot spring areas associated with volcanic activities, and it is necessary to reduce the size of facilities to accommodate narrow land.

In addition, because of geographical and meteorological conditions, the country is subject to frequent natural disasters such as typhoons, torrential rains and heavy snowfalls, as well as earthquakes and tsunamis. As Japan is an island country surrounded by the sea on all sides, there are concerns about salt damage to coastal structures. The environment surrounding Japan, issues and measures for sewage pipeline facilities as one of the infrastructures are summarized in Table 1.

Table 1. Environment surrounding Japan, issues and measures for sewage pipeline facilities.

Environment		Influence on sewerage	Measures for pipeline facilities
Many disasters	Earthquake	Destruction of facilities	Higher strength and improved impact resistance
	Torrential rains	Flood damage	Efficient pipeline augmentation
Many corrosive environments	Volcanic gas/hot spring area	Early deterioration due to corrosion	Measures against hydrogen sulfide
	Salt damage		Measures against salt damage
Topographic features	Narrow road	Work in places where heavy machinery cannot enter	Miniaturization
	Little flat area	Corrosion deterioration due to hydrogen sulfide at manhole pump discharge destinations and densely populated areas	Measures against hydrogen sulfide
Aging population/declining working population		Decrease in sewage fee income and workers	Efficient maintenance and long life
Aging facilities		Frequent occurrence of road subsidence accidents	

In this paper, the application of resin concrete manholes based on the superior material characteristic and development of the small manhole suited for severe conditions in Japan were investigated. Furthermore, a homogenization analysis method to plan the material optimization of resin concrete was studied and the obtained results were confirmed by the experiments.

2 Development of Small Resin Concrete Manhole

A small resin concrete manhole [8] suited to the Japanese environment was developed as shown in Fig. 1. The development concept of this product is summarized in Table 2. The standard manhole is a combination of cover and body. The inner diameters of the cover and body are 600 mm and 900 mm, respectively. On the other hand, for small manholes, the diameters are both 300 mm. Since people cannot enter inside small manholes, inspection and cleaning inside manholes is done by machines instead of humans. In order to enhance production efficiency and workability, and to reduce material cost and construction cost, small manholes are used in Japan. Reinforced Concrete Small Assembled Manhole and Small Manhole Made of Hard Vinyl Chloride for Sewerage had been established as Japan Sewage Works Association Standard [9, 10]. The standardization played an important role in the popularization of small resin concrete manholes. The developed resin concrete manholes are superior to the former in workability and durability, and to the latter in strength, workability and cost balance, respectively. Furthermore, the inside of the resin concrete manhole was coated by FRP to increase the flexural strength and impact resistance. The improvement in strength made it possible to reduce the thickness and weight of the product. As a result, it is possible to improve the efficiency of transportation and save labor in on-site construction. In addition to the above, the flow function of this resin concrete manhole can be maintained even in the event of a large-scale earthquake such Level 2 earthquake ground motion. The function was confirmed by pull-out/push-in test, bending test and shearing test as shown in Fig. 2. There was no occurrence of lateral slip and cracks in the pull-out/push-in test, bending test and shearing test.

2.1 Durability

Resin concrete has almost no water absorption and no penetration of salt, and thus the products have excellent durability such as resistance to freezing and thawing in cold regions and resistance to permeability of salt in coastal areas. Furthermore, it is characterized by excellent corrosion resistance in hot spring areas and places with high hydrogen sulfide gas concentrations.

Excellent corrosion resistance of resin concrete against chemicals and hydrogen sulfide gas was clarified by the immersion test at the hot spring site and laboratory as described below. Specimens were prepared with FRP attached to the top and bottom of the resin concrete. The sandwiched specimen was thickness 16.5 mm, width 30 mm and length 300 mm and the thickness of resin concrete was 8.0 mm, and that of FRP plates pasted was 3 mm at the top and 5.5 mm at the bottom, respectively. The specimen was immersed in 10% sulfuric acid at 23 °C for up to one year. The flexural strength decreased by 10% in the first month, but remained constant thereafter as shown in Fig. 3.

Table 2. Development concept of a small resin concrete manhole.

Element	Concept
Workability	• Weight reduction by using high-strength esin concrete • Manpower construction in narrow spaces is possible without using heavy machinery • Short construction time by assembly
Durability	• Improved lateral flexural strength and impact resistance by combining with FRP • Applicable to hot spring areas where acid resistance is required and cold areas with severe freezing and thawing • Flow capacity against level 2 earthquake motion
Cost	• Since heavy machinery is not required, transportation and construction costs are low • Excellent durability in harsh environments and low life cycle costs

Fig. 1. Appearance of a small resin concrete manhole.

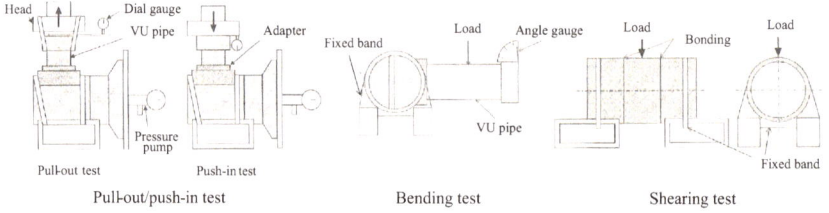

Fig. 2. Durability test against level 2 earthquake ground motions.

Disk-shaped resin concrete specimens of diameter 50 mm and thickness 10 mm were immersed in an acidic hydrogen sulfide spring of pH 1.8 and 55–70 °C for 3 months. Figure 4 shows the appearance of the specimens after 3 months of immersion. The surface was slightly whitened, but there was no change in mass. Cement mortar specimens for comparison became brittle throughout, and its mass was decreased by about 50%. The results of these experiments show that resin concrete can be used under severe corrosive environments such as hot springs and sewage treatment facilities.

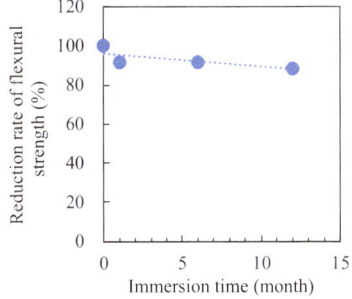

Fig. 3. Changes in flexural strength due to 10% sulfuric acid immersion test.

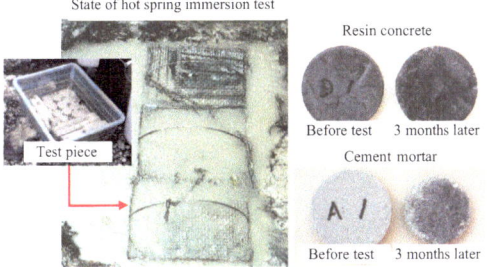

Fig. 4. Result of hot spring immersion test (pH 1.8, 55 ~ 70 C).

2.2 Workability

Reducing the size of manholes provides great benefits. A comparison of construction cost and product weight of portland cement concrete and resin concrete is shown in Fig. 5. Replacing a standard cement concrete manhole at a depth of 2 m with a small resin concrete one, the product weight can be reduced by about 80% and construction cost can be decreased by 30% or more. Due to the miniaturization, construction can be done manually in narrow spaces and construction work is easy even on narrow roads.

3   Future Prospects of Resin Concrete

3.1   Environmental Adaptability

A comparison of CO_2 emissions related to construction, transportation and manufacture of portland cement concrete and resin concrete manhole is shown in Fig. 6 when the effective inner diameter is fixed [11]. The amount of CO_2 emitted per manhole of resin

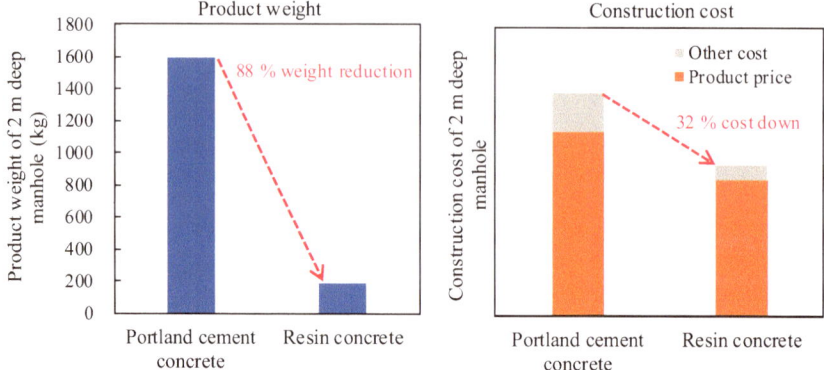

Fig. 5. Comparison of construction cost and product weight of portland cement concrete and resin concrete (2 m deep).

concrete is about 30% less than that of cement concrete. Due to the high performance of resin concrete's high strength and excellent durability, the weight of products is decreased and CO_2 emissions are significantly reduced. Additionally, resin concrete is no elution of harmful substances and environmentally friendly material and then the products can be expected to decrease the amount of materials and long-term service.

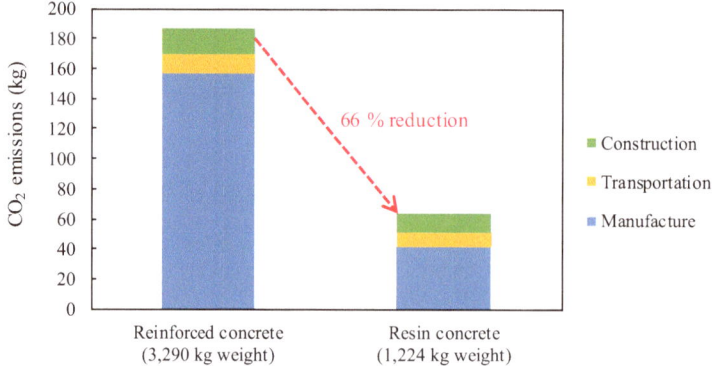

Fig. 6. Comparison of CO_2 emissions of reinforce concrete and resin concrete manholes (3 m depth).

In order to reduce further carbon footprint, three major research and development were tried as follows;

1) Application of an unsaturated polyester resin from biomass resources. Instead of conventional petroleum-based materials, biomass resin was applied for the resin concrete products.

2) The recycle of resin concrete products. The recycled aggregates from waste resin concrete products were examined. The material was able to be applied to coarse

aggregate by crushing it to a size of about 5 mm, and considering the reduction of flowability and strength, material placement of coarse aggregate was possible up to 10 %. These crushed aggregates can be also used as recycled base course material.

3) Utilization of blast furnace slag and coal gasification grinded slag. Application of blast furnace slag produced as a by-product in the steel manufacturing process, and coal gasification slag generated at Integrated coal Gasification Combined Cycle abbreviated to IGCC, as recycled aggregate was investigated. Fig. 7 shows the appearance of the recycled aggregates. Since the slag reduced the fluidity at the time of material filling, the particle size was adjusted in advance.

The strengths of the new resin concrete containing biomass resin and recycled aggregate were almost the same as those of conventional ones as shown in Fig. 8. Indeed, the strength of the resin concretes using slag is slightly low, but that can satisfactorily meet the required values. Although it is necessary to investigate CO_2 emissions during transportation and procurement costs, it may be used as an alternative material in the future.

Fig. 7. Appearance of recycled aggregates.

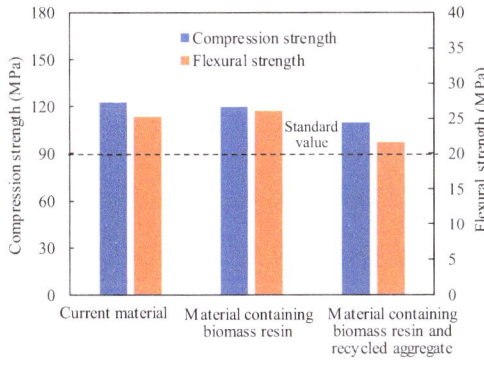

Fig. 8. Influence of strength on use of biomass resin and recycled aggregate.

3.2 Application of Homogenization Analysis to Material Development

In the mixture design of resin concrete, according to product requirement, molding conditions and manufacturing method, the material type and mixture proportion are determined. Determining the optimum combination from a myriad of options in a short time leads to efficiency in material development. Recently, homogenization analysis applied in the field of FRP has been investigated [12]. Generally, composite materials are discretized by finite elements, and macroscopic material properties are estimated by homogenization analysis for microscopic models considering periodicity.

Examples of creating 3D micro models for analysis in which coarse aggregate and fine aggregate are filled with a resin paste (filler + resin) are shown in Fig. 9. Young's modulus and Poisson's ratio corresponding to aggregate ratios are obtained from homogenization analysis as shown in Fig. 10. These analyzed results suggest the approximate mixture proportion of resin concrete to satisfy the specified physical quantities.

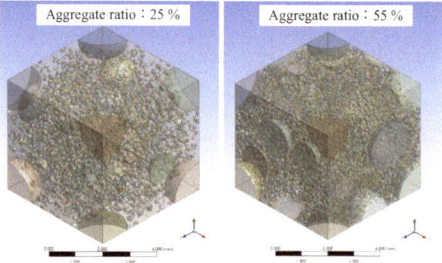

Fig. 9. Example of 3D micro model used for homogenization analysis.

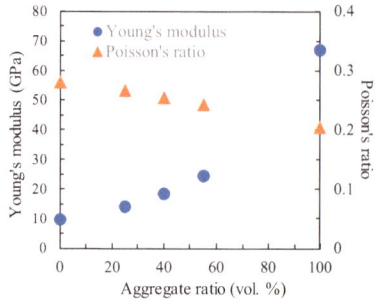

Fig. 10. Result of homogenization analysis.

After test specimens with 55% aggregate content were actually fabricated, strength tests were performed. The test results were compared with the homogenization analysis results. The dumbbell and cylindrical specimens were used for tests. The sizes of test pieces were 25.4-mm diameter with 25.4 mm height at the central parallel part and total 105.4-mm height including the end for tensile strength tests and 75-mm diameter with 150-mm height for compressive tests, respectively. Figure 11 shows the comparison of the results of homogenization analysis and experiment. The strain and strength gradient in the elastic region that are important for product design are in good agreement with the experimental results. Therefore, the apparent physical property values obtained from this analytical method are expected to be used to improve the efficiency of material development.

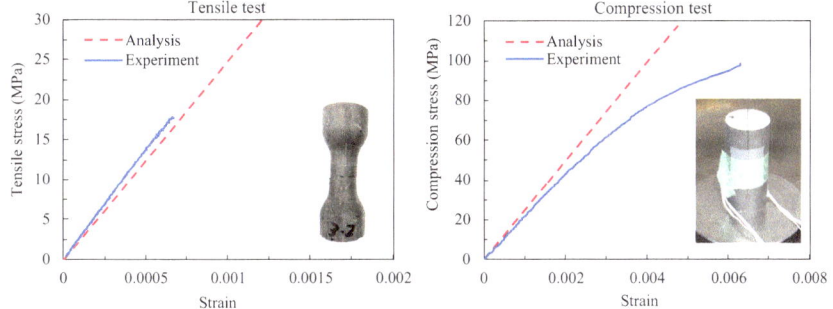

Fig. 11. Comparison of results of analysis and experiment.

4 Conclusions

Development and practical application of resin concrete applied for sewage products in Japan were investigated and discussed. The main conclusions obtained from the results are summarized in the following:

- In Japan, resin concrete with unsaturated polyester resin has been developed mainly in the field of civil engineering. The standardization played an important role in the popularization of resin concrete.
- Considering severe conditions of sewer infrastructures such as a narrow working environment, long-term resistance to corrosive environment and cost rationalization, small size resin concrete manholes were developed and haves been widely used.
- Application of collected used products, plant-derived biomass resin, blast furnace slag and coal gasification slag as recycled aggregate for environmental-friendly resin concrete was tried and proposed.
- The homogenization analysis method including 3D micro models is effective in estimating the material properties to permit efficient material formulation design to become possible.

References

1. Ohama, Y., Yamaguchi, S.: Polymer concrete as a construction material. J. Soc. Mater. Sci., Jpn. **54**(9), 971–978 (2005)
2. JIS A 1181: Method of making polyester resin concrete specimens, Japanese Industrial Standards, Japan (1978)

3. JIS A 1182: Method of test for compressive strength of polyester resin concrete, Japanese Industrial Standards, Japan (1978)
4. JIS A 1183: Method of test for compressive strength of polyester resin concrete using portions of beams broken in flexure, Japanese Industrial Standards, Japan (1978)
5. JIS A 1184: Method of test for flexural strength of polyester resin concrete, Japanese Industrial Standards, Japan (1978)
6. JIS A 1185: Method of test for splitting tensile strength of polyester resin concrete, Japanese Industrial Standards, Japan (1978)
7. JIS A 1186: Measuring methods for working life of polyester resin concrete, Japanese Industrial Standards, Japan (1978)
8. JSWAS K-10: Resin concrete manholes for sewerage, Japan Sewage Works Association Standard, Japan (1997)
9. JSWAS A-10: Reinforced concrete small assembled manhole for sewerage, Japan Sewage Works Association Standard, Japan (1997)
10. JSWAS K-9: Small manhole made of Hard Vinyl Chloride for sewerage, Japan Sewage Works Association Standard, Japan (1996)
11. Japan Resinconcrete Products Association Homepage, https://www.jrpa.gr.jp/, last accessed 2023/2/28
12. Ishitsuka, K., Kuwahara, Y., Shinohara, N., Kai, N.: Calculation of Young's modulus and applicability to material development of resin concrete using homogenization analysis, Proceedings of the 8th Annual Scientific Lecture, 65–66 (2021)

Recycled Mixed Plastic Fine Aggregate in Cement Concrete

Kevin Jia Lee[1][(✉)] and Sook Fun Wong[1,2]

[1] Centre for Urban Sustainability, School of Applied Science, Temasek Polytechnic, Singapore, Singapore
kevin_lee@tp.edu.sg
[2] American Concrete Institute-Singapore Chapter, Singapore, Singapore

Abstract. The literature extensively examines the utilization of sorted single-type plastic waste from post-consumer waste streams as a sustainable substitute for natural sand in cement concrete. However, severe heterogeneity of plastic waste in municipal solid waste streams, including variations in polymer types, grades, shapes, sizes, and cross-contamination with other commingled waste materials, poses a significant challenge in adopting findings from prior research that necessitates high-purity single-type plastic waste for concrete applications. This paper reports the characterization of cement concrete incorporated with mixed plastic fine aggregate (rMPFA) containing an optimized blend of plastic types produced using a proprietary mixed plastic recycling process. Five concrete mixtures containing 0% (M0), 10% (M10), 20% (M20), 30% (M30), and 40% (M40) rMPFA by volume of natural sand were investigated in this study. The laboratory results show that concrete mixture M20 had comparable compressive strength and water penetration test results when compared to control mixture M0. Additionally, toxicity characterization of concrete mixture M20 demonstrated a reduction of heavy metals in the leachate solution when compared to control mixture M0. Furthermore, microplastic detection analysis results of concrete mixtures M0 and M20 were comparable and stable.

Keywords: Mixed plastics · Cement concrete · Fine aggregate replacement · Leachate · Microplastic

1 Introduction

Plastic waste constitutes a significant portion of municipal solid waste (MSW), making up approximately 10% to 15% of the total amount, with less than 10% currently being recycled [1]. Domestic and industrial entities generate plastic waste, primarily in the form of food packaging, trash bags, and plastic bottles. In Singapore, the annual plastic waste generation has risen sharply over the past two decades, from 546,537 tonnes in 2001 to 982,000 tonnes in 2021 [2]. Improper disposal of non-biodegradable plastic waste has become a pressing societal and environmental issue. Thus, converting plastic waste into valuable products can contribute to establishing a sustainable circular economy and achieving carbon neutrality.

© The Author(s) 2025
L. Czarnecki et al. (Eds.): ICPIC 2023, 61, pp. 111–118, 2025.
https://doi.org/10.1007/978-3-031-72955-3_9

One promising strategy for addressing this issue is recycling plastic waste as value-added materials for building and infrastructure, such as an alternative to natural sand. However, plastic waste in MSW streams is highly heterogeneous in terms of polymer types, grades, dimensions, and other commingled materials, including metal, paper, and cardboard. The non-homogeneity and contamination of mixed plastics are significant obstacles to recycling them as feedstock for building materials. Hence, past studies reported on the use of sorted single-type plastic waste such as, polyethylene (PE), polypropylene (PP), polyethylene terephthalate (PET), polystyrene (PS), polycarbonate (PC), acrylonitrile butadiene styrene (ABS) in cement concrete as natural sand replacement [3] is cost-ineffective and difficult for implementation in practice.

In response to these challenges, Khalil et al. [4] investigated the use of irregular mixed plastic waste as coarse aggregate in concrete materials, ranging from 15 to 45 vol%. Their study revealed that samples containing 45 vol% mixed plastic coarse aggregate can be classified as structural lightweight concrete, along with significant improvement in thermal conductivity. Similarly, Lee et al. [5] explored the use of 10, 15 and 20 vol% mixed plastic fine aggregates, made up of an optimized blend containing different plastic types, as a substitute for natural sand in cement concrete. They concluded that the effects of mixed plastic aggregate on the mechanical and durability properties are comparable to those of recycled plastic fine aggregate made of sorted single-type plastic waste. This suggests that with proper recycling techniques, the need for sorting mixed plastic waste to achieve high-purity homogeneous plastic recycling streams for use in cement concrete can be minimized, thus increasing the recovery efficiency of mixed plastic waste. Additionally, Lee et al. [6] studied the use of mixed plastic fiber as reinforcement, with similar polymer composition to plastic waste in MSW streams for fiber-reinforced concrete.

Presently, existing studies on the use of heterogeneous waste materials in building and infrastructural applications have largely been limited to the fresh, durability and mechanical properties. In addition to technical performance, the objective of this study is also to examine the effects of recycled mixed plastic fine aggregate (rMPFA) on the environmental properties of cement concrete designed for general concreting application at various natural sand (NS) replacement levels.

2   Materials and Methods

2.1   Materials

In this study, concrete mixtures were prepared with constituent materials conforming to SS EN 206 [7]. The raw ingredients include ordinary Portland cement (OPC, CEM Type I 42.5 N, specific gravity (SG) 3.15), NS (4 mm nominal size, SG 2.60) as fine aggregate, rMPFA made of an optimized concoction of multiple plastic types manufactured via a proprietary mixed plastic recycling facility (4 mm nominal size, SG 0.95) as fine aggregate replacement, natural gravel (NG, 20 mm nominal size, SG 2.65) as coarse aggregate, tap water and commercial chemical admixtures, i.e., retarder and water reducing superplasticizer.

Five concrete mixtures containing 0% (M0), 10% (M10), 20% (M20), 30% (M30), and 40% (M40) rMPFA by volume of NS content were designed in this study. The mix design of the concrete mixtures is shown in Table 1. The water/binder ratio, aggregate/binder ratio and dosages of chemical admixtures were kept constant to capture the effects of rMPFA content on the properties of all concrete mixtures investigated.

Table 1. Mix design of concrete mixtures.

Mix No	OPC	Water	NG	NS	rMPFA
M0	1	0.45	2.5	1	0
M10	1	0.45	2.5	0.9	0.1
M20	1	0.45	2.5	0.8	0.2
M30	1	0.45	2.5	0.7	0.3
M40	1	0.45	2.5	0.6	0.4

2.2 Sample Preparation

The dry ingredients namely, OPC, NS, NG and rMPFA were first mixed in a rotary mixer for 3 min. Tap water dosed with chemical admixtures was then added to the dry mixture and blended for an additional 3 min. The freshly mixed concrete mixtures were transferred and compacted into 150 mm cube formworks. After 24 h, the concrete samples were removed from the formworks and stored in a sheltered open space to air cure till its designated test dates.

2.3 Laboratory Tests

The compressive strength of the concrete samples was evaluated in accordance with BS EN 12390-3 [8] using 150 mm cube samples after 1, 7 and 28 days of curing. To evaluate water penetration, 150 mm cubic concrete specimens were tested according to the procedures outlined in BS EN 12390-8:2019 [9]. After 28 days of curing, one surface of each specimen was subjected to 500 kPa water pressure for 72 ± 2 h. The specimens were then split into two to measure the depth of water penetration.

The leachate solutions of concrete mixes were prepared using the standard batch leaching method for granular waste materials, as specified in BS EN 12457-1 [10]. The amount of heavy metal content in the solutions was determined in accordance with the methods outlined by the American Public Health Association [11, 12]. Microplastic detection analysis was performed by filtering the leachate solutions obtained from the standard batch leaching procedure through a 0.45 μm cellulose nitrate membrane. The residual particles on the filter membrane were collected with a surgical blade and then transferred to a diamond cell for analysis using microscopic Fourier-Transform Infrared Spectroscopy (FTIR). The collected particles were matched against a library for identification.

3 Results and Discussion

The compressive strength tests of different concrete mixtures were conducted at various curing ages, and the results are illustrated in Fig. 1. The findings indicated that the compressive strength trends of the five concrete mixtures remained relatively consistent over time. Concrete mixtures M0 and M40 demonstrated the highest and the lowest compressive strengths, respectively. This observation is consistent with past literatures reported on the effects of increasing recycled plastic content over the mechanical properties of concrete mixtures [5]. Additionally, the results showed that the compressive strength of concrete mixture M20, which contained 20 vol% rMPFA, decreased only slightly in comparison to control mixture M0.

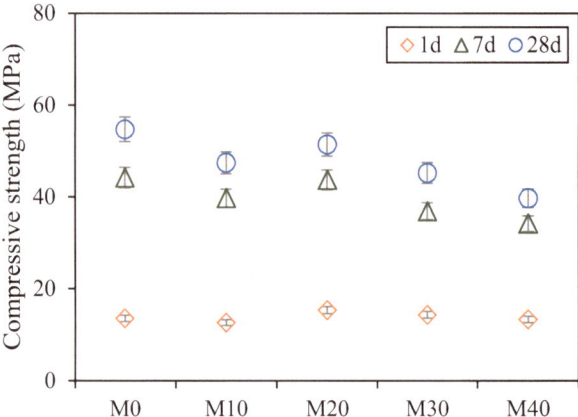

Fig. 1. Compressive strength of designed concrete mixtures.

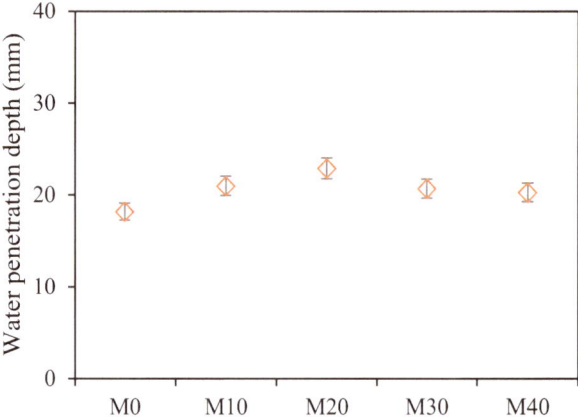

Fig. 2. Water penetration depth of designed concrete mixtures.

The water penetration tests of the designed concrete mixtures were also conducted after 28 days of curing, and the results are displayed in Fig. 2. It is evident that concrete mixtures containing rMPFA demonstrated a rise in water penetration depth in contrast to the control mixture M0. This increase in water penetration depth could be linked to the poor adhesion bonds between rMPFA and the cementitious matrix, leading to an increase in porosity which allows for easier medium transport within the microstructure of the concrete mixtures.

From the preliminary investigation, concrete mixture M20 was selected for further examination of its environmental properties, compared to control mixture M0. The limit of reporting (LOR) and leaching test results are presented in Table 2. The computation of leachate concentration for various parameters is expressed as milligram equivalent mass of a parameter per kilogram of solid sample, with test results below the LOR indicated as <LOR value.

Most of the test parameters for concrete mixtures M0 and M20 showed similar concentrations below the LOR values, with the exception of Ba, F, Pb, Hg, and phenolic compounds. Concrete mixture M20 demonstrated lower concentrations of Ba, Pb, Hg, and phenolic compounds than control mixture M0. Further investigation revealed that the slight increase in F content was due to the presence of fluoride in natural aggregates rather than the addition of rMPFA.

Table 2. Leaching test results (mg parameter/kg solid sample).

Parameters	LOR	M0	M20
Arsenic (As)	0.01	< 0.01	< 0.01
Barium (Ba)	1.00	3.53	3.23
Cadmium (Cd)	0.001	< 0.001	< 0.001
Chromium (Cr)	0.01	< 0.01	< 0.01
Copper (Cu)	0.01	< 0.01	< 0.01
Fluoride (F)	1.00	2.90	2.94
Lead (Pb)	0.01	0.02	0.01
Manganese (Mn)	0.10	< 0.01	< 0.01
Mercury (Hg)	0.001	0.002	< 0.001
Nickel (Ni)	0.01	< 0.01	< 0.01
Phenol	1.00	0.46	0.31
Selenium (Se)	0.01	< 0.01	< 0.01
Silver (Ag)	0.01	< 0.01	< 0.01
Zinc (Zn)	0.10	< 0.1	< 0.1

Figure 3 displays the FTIR spectra of residual particles found in the leachate solutions of the designed concrete mixtures. Through visual examination and matching with a FTIR library, a significant portion of the particles was identified as ground calcium carbonate, which is consistent with the hydration products formed within the cementitious matrix. Only a small amount of polymeric particles, such as kaolin, cellulose, PP and PET, were detected on the filter membrane. These particles are likely to have been introduced from external sources, such as materials storage and handling equipment, during the sample preparation process involving material mixing and transfer.

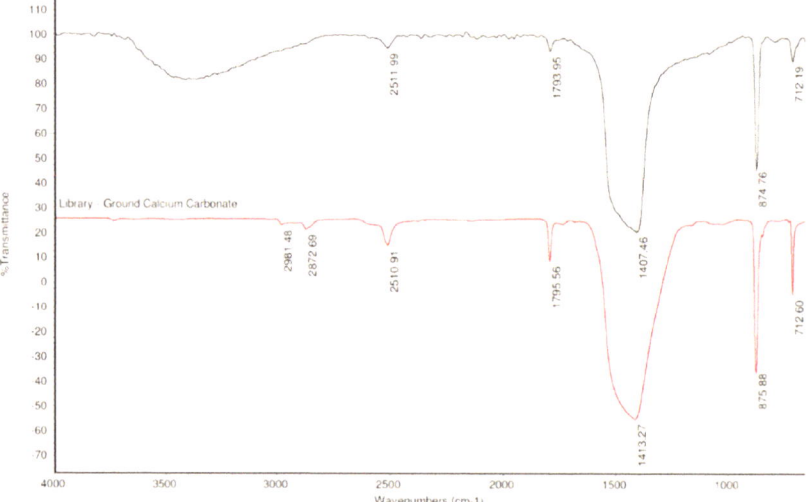

Fig. 3. FTIR spectra of residual particles after filtration of leachate solutions derived from concrete mixtures.

4   Conclusions

This paper presents a study on the technical performances and environmental properties of cement concrete incorporated with rMPFA as partial NS replacement, produced from a proprietary mixed plastic recycling process. Five different concrete mixtures containing rMPFA content ranged from 0 to 40 vol% of NS were characterized and the key findings are:

- A higher rMPFA content in concrete mixtures led to lower compressive strength across all curing ages, well-aligned with findings reported in the literature.
- Concrete mixtures containing rMPFA exhibited a slight increase in water penetration depth when benchmarked with the control mixture M0, possibly due to the increase in porosity caused by reduced adhesion interfacial bonds between rMPFA and the cementitious matrix.
- Upon further analysis, concrete mixture M20 revealed improvements in leachate content with lower or comparable leachate concentration as control mixture M0. Despite an increase in porosity that eased transport within the microstructure of concrete mixtures, the addition of rMPFA had little influence on the mobility of the heavy metals.

- The microplastic detection test results showed that the concrete mixture M20 with 20 vol% rMPFA exhibited little or no formation of microplastic, comparable to the control mixture M0.

This study primarily focused on the laboratory characterization of concrete mixtures containing rMPFA. Field investigation on concrete mixtures containing rMPFA should be carried out to characterize the short- and long-term durability performances, mechanical properties, and environmental behaviors under natural weathering conditions in actual operative environments. Further studies could also be carried out to produce recycle mixed plastic coarse aggregate of suitable particle size to substitute NG in concrete mixtures to further increase the recycled mixed plastic content. Finally, life cycle assessment could be performed to quantify the effects of recycled mixed plastic ingredients on the embodied carbon of concrete mixtures, which can provide valuable information for future development of greener concrete mixtures.

Acknowledgements. This research is supported by the National Research Foundation, Singapore, and National Environment Agency, Singapore, under its Closing the Waste Loop Funding Initiative (Award No. USS-IF-2019-2). The authors are grateful to Mr. Lim Yin Yen and the technical staff from the Centre for Urban Sustainability at the School of Applied Science, Temasek Polytechnic. The authors would also like to express their gratitude to Mr. Lim Guang Jie Jonathan, Dr. Wang Su, and the technical staff from Pan-United Corporation Pte. Ltd. for their tremendous support.

References

1. OECD, Global Plastics Outlook: Economic drivers, environmental impacts and policy options. OECD (2022). https://doi.org/10.1787/de747aef-en
2. National Environment Agency.: Waste statistics and overall recycling. https://www.nea.gov.sg/our-services/waste-management/waste-statistics-and-overall-recycling (2021). Accessed 17 Aug 2021
3. Babafemi, A., Šavija, B., Paul, S., Anggraini, V.: Engineering properties of concrete with waste recycled plastic: a review. Sustainability **10** (11), 3875 (Oct 2018). https://doi.org/10.3390/su10113875
4. Khalil, W.I., Mahdi, H.M.: Some properties of sustainable concrete with mixed plastic waste aggregate. IOP Conf. Ser.: Mater. Sci. Eng. **737** (1), 012073 (Feb. 2020). https://doi.org/10.1088/1757-899X/737/1/012073
5. Lee, K.J.L., Wong, S.F.: Multi-objective taguchi optimization of cement concrete incorporating recycled mixed plastic fine aggregate using modified fuller's equation. Buildings **13**(4), 893 (Mar.2023). https://doi.org/10.3390/buildings13040893
6. Lee, K.J.L., Wong, S.F.: Optimization of fiber-reinforced concrete composite with recycled aggregate and fiber produced with mixed plastic waste. In: Materials Today: Proceedings, p. S2214785323009264, Mar. 2023, https://doi.org/10.1016/j.matpr.2023.02.362
7. Concrete: specification, performance, production and conformity, First revision. Singapore: SPRING Singapore (2014)
8. BS EN 12390-3: Testing hardened concrete. Part 3: Compressive strength of test specimens. The British Standards Institution (2019)
9. BS EN 12350-8: Testing hardened concrete. Part 8: depth of penetration of water under pressure. The British Standards Institution (2019)

10. BS EN 12457-1, Characterisation of waste - Leaching - Compliance test for leaching of granular waste materials and sludges. Part 1, One stage batch test at a liquid to solid ratio of 2 1/kg for materials with high solid content and with particle size below 4 mm (without or with size reduction). The British Standards Institution (2002)
11. APHA 3120: APHA method 3120: standard methods for the examination of water and wastewater. American Public Health Association (1992)
12. APHA 3125: APHA method 3125: metals in water by ICP/MS. American Public Health Association (2011)

Cement Mortars with Incorporation of Foundry Industry Wastes: Physical, Mechanical and Durability Behavior

Sandra Cunha(✉) ⓘ, Raphael Silva, and José Aguiar ⓘ

Centre for Territory, Environment and Construction (CTAC), University of Minho, Campus de Azurém, 4800-058 Guimarães, Portugal
sandracunha@civil.uminho.pt

Abstract. Planet Earth is facing real challenges that require urgent and significant measures. It is necessary to give a new direction to the construction sector, making it essential to change the way that raw material is selected, giving preference to industrial by-products. The utilization of industrial wastes allows minimize the high consumption of natural raw materials, energy consumption and waste deposition in landfills. It is important to note that the use of waste in the construction industry is a great opportunity, however, the heterogeneity of these materials and sometimes their contamination can compromise the durability. The lost-wax process in foundry industry is currently an expanding area, so more and more manufacturing industries have serious problems related to their waste management. During its production process, wastes of ceramic mold shells and paraffinic wax are generated and until now any practical application is known. The main objective of this study was the correlation between the physical, mechanical behavior and durability of cement mortars with incorporation of paraffin wax and ceramic mold shells. The main results revealed a decrease in water absorption, flexural strength, and compressive strength of the mortars, along with a slight increase in degradation during freeze-thaw cycles. Additionally, a correlation was observed between the physical, mechanical performance, and durability of the mortars. This included factors such as water absorption through immersion and capillarity, as well as the relationship between compressive strength and the mass loss suffered during freeze-thaw tests.

Keywords: Foundry wastes · Freeze-thaw tests · Mechanical properties

1 Introduction

Our planet is facing serious environmental challenges due to the excessive consumption of natural raw materials and energy resources during their extraction and processing, as well in maintaining the energy needs associated with the different activity sectors. Earth's Overshoot Day is occurring earlier each year, currently falling around mid-year (August 2 in 2023), indicating that nowadays the humanity requires approximately 1.7 planets to sustain its needs.

© The Author(s) 2025
L. Czarnecki et al. (Eds.): ICPIC 2023, 61, pp. 119–127, 2025.
https://doi.org/10.1007/978-3-031-72955-3_10

The retreat of Overshoot Day is also linked to the consumption of natural resources, which has been exacerbated by the significant levels of urbanization we are currently experiencing. It is known that replacing traditional concrete by a concrete made from recycled aggregate could immediately delay Earth Overshoot Day by 2.4 days [1]. Therefore, the construction industry needs to invest in reusing industrial by-products from its own sector, like construction or demolition waste, as well as by-products from other industries, such as the foundry industry, to create new and sustainable construction materials.

The foundry industry works with various types of metals, producing metal pieces for the automotive, domestic, military and agro-industrial industry all over the world. Consequently, its waste production is enormous. In 2020, approximately 105. 5 million metric tons of solid foundry waste were generated globally [2–4]. Precision foundry industry also known as lost-wax foundry industry, produces two main types of waste: ceramic mold shells and paraffinic wax. Initially, paraffinic wax is used to create a replica of the metal piece to be molded. This replica is then coated with multiple layers of ceramic material to create the mold into which the liquid metal will be poured.

Next, the set will be placed in an oven to create a ceramic mold with increased resistance and to facilitate the removal of the paraffinic wax, which liquefies due to the temperature. This process results in the extraction of the first residue, the paraffin wax waste. Finally, the molten metal is poured into the ceramic mold cavity, where it cools down and solidifies, resulting in a flawless final piece. During the demolding of the metal pieces, it is necessary to break the molds, obtaining in this way the ceramic mold shells waste [5]. It is important to note that these wastes cannot be reincorporated in any stage of a new production process. Until now, very little is known about the use of these wastes in construction materials. Only few studies conducted by this research team have explored the behavior of these wastes in cement mixtures [6–8].

However, it was necessary to establish a treatment process for the ceramic mold shells, to eliminate the alkali-aggregate reaction caused by the presence of sodium, potassium, calcium and magnesium in their chemical composition. The results of this work [7] revealed that washing process was a practical, simple, cheap and effective method for using this waste as a substitute for natural aggregate in mortars. Thus, the main objective of this work was the correlation between the physical, mechanical behavior and durability of cement mortars with incorporation of paraffin wax and ceramic mold shells.

2 Experimental Program

2.1 Materials

The materials used considered other research works carried out by this research team [6–8]. It was selected a Portland cement CEM I 42.5 R, a natural sand, a ceramic mold shell waste, a paraffin wax waste and a superplasticizer. The raw materials densities are present in Table 1.

The waste used as a substitute for natural aggregate underwent a treatment process, to enable its incorporation into cement mortars. In this way, the ceramic mold shells underwent a crushing and washing process, while the paraffin wax waste only underwent a crushing process. Table 2 presents the average particle size and water absorption capacity of the different aggregates used in this study.

Table 1. Materials densities.

Materials	Density (kg/m^3)
Cement	3184
Natural Sand	2569
Ceramic mold shells	2630
Paraffin wax	1013
Superplasticizer	1041

Table 2. Particle average size and water absorption of the aggregates.

Materials	Water absorption (%)	Particle average size (mm)
Natural Sand	1.2	0.68
Ceramic mold shells	6.6	1.25
Paraffin wax	1.6	1.8

2.2 Compositions

Six different compositions were developed with different contents of ceramic mold shells and paraffin wax. The development of these mortars was carried out in a previous study [8], developed by this research team. Table 3 presents the mixture of aggregates used to produce the different mortars, along with their water/cement ratio. A cement dosage of 750 kg/m^3 and a superplasticizer dosage of 7. 5 kg/m^3 were used.

Table 3. Mortars aggregates mix and water-cement ratio.

Composition	Aggregates mixture	Water/Cement
REF	100% natural sand	0.36
CMS100	100% ceramic mold shells waste	0.42
CMS80PW20	80% ceramic mold shells waste and 20% of paraffin wax waste	0.41
CMS60PW40	60% ceramic mold shells waste and 40% of paraffin wax waste	0.40
CMS40PW60	40% ceramic mold shells waste and 60% of paraffin wax waste	0.38
CMS20PW80	20% ceramic mold shells waste and 80% of paraffin wax waste	0.37

2.3   Test Procedures

The developed mortars were tested in order to evaluate their physical, mechanical and freeze-thaw behaviors. Thus, several tests were performed based in European and national standards.

The behavior in the fresh state was evaluated according to consistence determination tests, by flow table method, in accordance with the specification EN 1015–3 [9]. Having established an average spreading diameter of $200 \pm 5mm$ to fix the water cement ratio (Table 3).

The behavior in the hardened state was evaluated according to tests of water absorption by capillarity based in the specification EN 1015–18 [10], water absorption by immersion in accordance with the Portuguese specification LNEC E 394 [11], flexural and compression strengths in accordance with the specification EN 1015–11 [12] and the evaluation of the freeze-thaw behavior according to the specification CEN/TS 12390–9 [13].

3   Results and Discussion

3.1   Physical Properties

The physical properties studied were water absorption by capillarity and immersion (Table 4).

Table 4. Water absorption by immersion and water absorption by capillarity coefficient of the mortars.

Composition	Water absorption by immersion (%)	Water absorption by capillarity coefficient $(kg/(m^2.min^{0.5}))$
REF	14.49	0.0014
CMS100	20.88	0.002
CMS80PW20	13.24	0.0012
CMS60PW40	8.26	0.0008
CMS40PW60	5.84	0.0005
CMS20PW80	3.62	0.0003

The water absorption by immersion and the water absorption by capillarity coefficient show the same tendency (Fig. 1). Mortars with a lower water absorption capacity by immersion are also those with a lower water absorption coefficient by capillarity. In this way, it is possible to verify that there is a relationship between the macroporosity and microposity of mortars. Mortars produced from a mixture of 100% aggregate from ceramic mold shells exhibit a higher water absorption capacity compared to mortar with 100% natural aggregate and mortars with a combination of ceramic mold shells and paraffin wax. This behavior is influenced by the nature of the aggregate, leading to a higher water-cement ratio as shown in Table 3. Mortars containing a mixture of aggregates (ceramic mold shells and paraffin wax) exhibit a lower water absorption capacity. This is attributed to the hydrophobic nature of paraffin wax and lower water-cement ratio in the mortars.

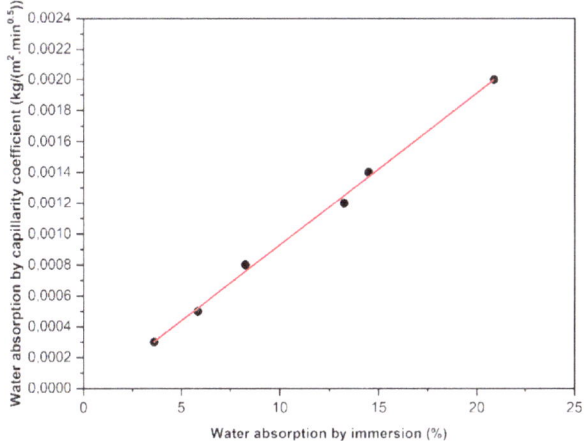

Fig. 1. Relationship between water absorption by immersion and water absorption by capillarity coefficient of mortars.

3.2 Mechanical Properties

The mechanical properties studied were the flexural and compressive behavior (Table 5).

Table 5. Flexural and compressive strength of the mortars.

Composition	Flexural strength (MPa)	Compressive strength (MPa)
REF	7.9	41.6
CMS100	7.3	34.7
CMS80PW20	5.5	21.2
CMS60PW40	4.6	16.7
CMS40PW60	3.9	13.4
CMS20PW80	3.9	11.9

Mortars produced with 100% of natural sand exhibit superior mechanical properties, showing increased flexural and compressive strengths. These properties are influenced by the shape and composition of the aggregate, as well as the lower water-cement ratio. Mortars developed with wastes, especially mortars incorporating aggregate made up entirely of ceramic mold shells, show a decrease in their mechanical performance, due to the higher water cement ratio and porosity (Table 4). Finally, mortars incorporating ceramic mold shells and paraffinic waxes show the greatest losses in mechanical performance, especially in mortars with a higher paraffinic wax content, even presented lower water cement ratios. This behavior is justified by the lower adhesion of the paraffin wax particles to the mortar matrix, as well as by the presence of a higher water-cement ratio.

It should be noted that the incorporation of ceramic mold shells and paraffin wax, although reducing the compressive strength of the developed mortars, does not prevent them from exhibiting mechanical behavior suitable for the construction industry. All developed mortars meet the maximum resistance class according to the European specification NP EN 998–1 [14], based on their compressive strength.

3.3 Durability

The durability of the mortars was evaluated based on their resistance to freeze-thaw cycles, as outlined in the European standard NP EN 998–1 [14]. The freeze-thaw resistance was directly connected with the mass loss of the specimens during the freeze and thaw cycles. Table 6 presents the mass loss values of the specimens due to degradation from 56 freeze-thaw cycles conducted on various mortars.

Table 6. Mass loss suffered during mortar freeze-thaw tests.

Composition	Mass loss (MPa)
REF	0.43
CMS100	0.70
CMS80PW20	0.77
CMS60PW40	1.72
CMS40PW60	1.70
CMS20PW80	0.80

Mortars incorporating waste as a substitute for natural aggregate show greater degradation compared to mortar with 100% aggregate of natural origin. This situation is aggravated for mortars incorporating paraffin wax, which can be justified by their lower mechanical performance (flexural and compressive strengths). Figure 2 illustrates a correlation between mass loss in freeze-thaw tests and the compressive strength of the mortars, indicating that those with lower compressive strength experience more significant degradation. However, it is important to note that all developed mortars exhibit very high resistance to freeze-thaw action, without a total degradation or significant superficial degradation of the specimens.

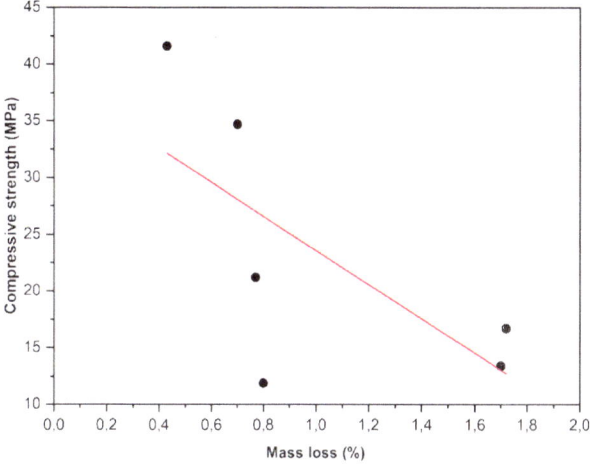

Fig. 2. Relationship between compressive strength and mass loss of mortars.

4 Conclusions

This study made it possible to evaluate the possibility of successfully incorporating waste from the foundry industry (ceramic molds shells and paraffin wax) into mortars for buildings applications, as well establishing correlations between the performance of several of their properties.

Based on the physical properties of the mortars, it is possible to determine a close relationship between the macroporosity and microporosity of the mortars. Additionally, it can be observed that the presence of ceramic mold shells increases the water absorption capacity of the mortars. This increase is largely due to the greater absorption capacity of the recycled aggregate, as well as the higher water-cement ratio. However, the presence of paraffin wax resulted in a decrease in the porosity of the mortars, due to the hydrophobic nature of the paraffin wax, as well as the lower water-cement ratio of the mortars.

Regarding mechanical performance, it was found that the presence of paraffin wax weakens the mortars, due to the low adhesion of the paraffin wax particles to the cement paste.

Concerning the durability of mortars, particularly in freeze-thaw tests, a correlation was observed between the degradation of specimens and their compressive strength. Mortars with higher resistance resulted in less degradation of specimens and, therefore, less mass loss.

The reuse of these wastes remains an area with significant research needs. However, the mortars studied exhibit adequate mechanical behavior and durability suitable for the construction industry. This approach can also be viewed as a potential contribution to reducing the consumption of raw materials and energy associated with the extraction of natural resources.

Acknowledgements. This work was supported by FCT/MCTES through national funds (PID-DAC) under the R&D Unit Centre for Territory, Environment and Construction (CTAC) under reference UIDB/04047/2020.

References

1. Earth Overshoot day, https://www.overshootday.org/portfolio/construction-waste-recycler/. Last accessed 2023/04/04
2. Paiva, F., et al.: Effect of phenolic resin content in waste foundry sand on mechanical properties of cement mortars and leaching of phenols behaviour. Sustainable Chemistry and Pharmacy **31**, e100955 (2023)
3. Qasrawi, H.: Fresh properties of green SCC made with recycled steel slag coarse aggregate under normal and hot weather. J. Clean. Prod. **204**, 980–991 (2018)
4. Torres, A., Bartlett, L., Pilgrim, C.: Effect of foundry waste on the mechanical properties of Portland Cement Concrete. Constr. Build. Mater. **135**, 674–681 (2017)
5. Wang, J., Sama, S., Lynch, P., Manogharan, G.: Design and topology optimization of 3D-Printed wax patterns for rapid investment casting. Procedia Manufacturing **34**, 683–694 (2019)
6. Cunha, S., Tavares, A., Aguiar, J., Castro, F.: Cement mortars with ceramic molds shells and paraffin waxes wastes: Physical and mechanical behavior. Construction and Building Materials 342 Part B, e127949 (2022)
7. Cunha, S., Costa, D., Aguiar, J., Castro, F.: Mortars with the incorporation of treated ceramic mold shells wastes. Constr. Build. Mater. **365**, e130074 (2023)
8. Cunha, S., Silva, R., Aguiar, J., Castro, F.: Performance of eco-friendly cement mortars incorporating ceramic molds shells and paraffin wax. Materials **16**(17), e5765 (2023)
9. European Committee for Standardization (CEN). EN 1015–3. Methods of test for mortar for masonry—Part 3: determination of consistence of fresh mortar (by flow table) (1999)
10. European Committee for Standardization (CEN). EN 1015–18. Methods of test for masonry—Part 18: Determination of water absorption coefficient due to capillary action of hardened mortar (2002)
11. National Laboratory for Civil Engineering (LNEC). Specification E 394. Concrete—Determination of water absorption by immersion (1993). (in Portuguese)
12. European Committee for Standardization (CEN). EN 1015–11. Methods of Test for Mortar for Masonry—Part 11: Determination of Flexural and Compressive Strength of Hardened Mortar (1999)
13. European Committee for Standardization (CEN). CEN/TS 12390–9. Testing hardened concrete—Part 9: Freeze-thaw resistance (2006)
14. Portuguese Institute for Quality (IPQ). NP EN 998–1. Specification for Masonry Mortars—Part 1: Plastering Mortars for Interior and Exterior (2013). (in Portuguese)

Eco-cement Cobblestones with Polyurethane Wastes

Raquel Arroyo$^{(\boxtimes)}$, Sara González-Moreno, Lourdes Alameda Cuenca-Romero, and Verónica Calderón

Departamento de Construcciones Arquitectónicas E I.C.T., Universidad de Burgos, Burgos, Spain
rasanz@ubu.es

Abstract. With the aim of implementing a circular economy in all manufacturing processes, by reducing the use of raw materials, while minimizing the use of natural resources and valuing several industrial wastes, efforts are focused on the development of new techniques that allow the use of wastes in the construction sector, in order to turn them into raw materials for the development of new materials, thus achieving all the processes towards a sustainable environment.

With this purpose, different precast cement cobble has been manufactured using recovered polyurethane waste from complete vehicle roofs generated in the automotive industry, turning them into a raw material. Depending on the amount of waste used, with progressive substitutions of sand of 20%, 40% and 60% of polymer waste, the final properties are achieved according to the application requirements, both in the fresh and hardened state, reaching the values required by current standards.

As well as water properties and microstructure, accelerated aging tests in freeze-thaw cycles and crystallization of salts have been tested, stablishing the compressive strength before and after, guaranteeing the properties in outdoor environments in which these materials can be placed.

The characterization has been completed with tests on the generation of volatile organic compounds (VOCs), fire resistance and Life Cycle Assessment (LCA).

In this way, the development of innovating solutions is achieved, valorizing waste that is generated in significant quantities, being able to be used in prefabricated products to be used in the building sector.

Keywords: Cement mortar cobblestones · lightweight prefabricated products · polyurethane waste

1 Introduction

The construction sector is one of the most polluting sectors today, as it generates CO_2 emissions of about 10 GtCO$_2$, increasing by 5% since 2020 and 2% more than the maximum obtained in 2019 [1]. In this sector, which uses a large amount of non-renewable raw materials, one of the main materials used for more than a century has been cement,

L. Czarnecki et al. (Eds.): ICPIC 2023, 61, pp. 128–136, 2025.
https://doi.org/10.1007/978-3-031-72955-3_11

for the manufacture of concrete and mortar [2]. The impact that the manufacture of these materials has on the environment is very high, and that is why the European Union has established the objective of trying to reduce these impacts and get closer to a green economy through several action plans [3]. The impact of the manufacture of these mortars and concretes could be reduced by using other raw materials, replacing either part of the cement or the aggregates used. Several studies have analyzed the use of some industrial sub products such as complementary cementitious materials (SCM) [4], volcanic dust and recycled concrete aggregates [5], recycled concrete aggregates and fly ash [6], polyurethane foam wastes [7] or lime sludge, evaluating the impacts that these substitutions have with respect to the conventional products used [8]. For all these reasons, and with the aim of trying to obtain new construction products in order to reduce the dependence on non-renewable raw materials, this research tries to use waste from the construction sector in mortars to manufacture prefabricated products. Thus, new ecological products are obtained using industrial waste, avoiding landfill and its consequent costs, and the amount of aggregate used in mortars, which is a non-renewable natural material, is reduced, trying to get closer to a more sustainable model in a sector as polluting as the construction sector.

2 Experimental Design

The research that has been carried out is the obtaining of lightened prefabricated products in the shape of cobblestones, in which industrial waste from the automotive sector is incorporated. The waste comes from the inside of complete vehicle roofs made of polyurethane and other materials such as adhesives or cardboard. They are incorporated crushed into prefabricated cobblestones, replacing part of the aggregate used in the manufacture of these cement mortars, with the aim of reducing waste treatment costs and preventing them from being deposited in landfills. They are designed to be used in both building and civil engineering.

2.1 Raw Materials

The prefabricated cobblestones are made with cement CEM I 52.5 R, washed river sand of rounded and slightly angular shapes, typical of natural granular aggregates and a non-ionic additive with a high hydrophobic/hydrophilic composition, that improves their properties by reducing the amount of water needed for mixing. The cement/conglomerate dosage is 1/3, the conglomerate being the addition of the aggregate and the waste in the cases in which this is part of the prefabricated cobblestones. A first reference dosage is manufactured and subsequently the amount of aggregate is substituted by residue, in volume, of 20%, 40% and 60%, incorporating the additive that provides a better compaction of the matrix due to the lower water demand of the mortar, thus improving its mechanical properties.

2.2 Manufacturing Process

For its preparation, the materials are mixed, on the one hand, the complete roof waste with the cement and aggregate, and on the other hand, the water and the additive. They are kneaded in the concrete mixer and then placed into the molds where they are vibrated and compacted. The manufactured cobblestones have dimensions of $200 \times 100 \times 60$ mm^3, complying market requirements.

The prefabricated cobblestones have been tested at 28 days, complying with the indications established by European standards. Samples have been manufactured in accordance with the regulations for each of the tests carried out, so that the required conformity criteria are always satisfied in each one of them.

After a visual examination of the cobblestones, which must not present defects such as cracks, exfoliations or efflorescence, the behavior of the cobblestones has been studied with tests carried out to determine their possible application in building and civil engineering.

3 Results

The results obtained when testing the prefabricated cobblestones with the different dosages replacing aggregate with polyurethane waste show how. in general, they satisfy the current standard on cobblestones EN 1338:2004 "Concrete paving blocks. Specifications and test methods" (Table 1).

Table 1. Properties in fresh and hardened state.

Dosages. Substitution sand-waste	Bulk density (kg/m^3)	Hardness Shore (C)	Mechanical strength (MPa)	Abrasive wear resistance (mm)	Slip resistance (USRV)	Water absorption (%)	Reaction to fire
Reference	2317	93.1	4.8	14	62	4.4	A1
20	2179	90.3	5.1	14	56	5.1	A1
40	2152	85.7	4.3	16	59	5.9	A1
60	2053	84.6	3.8	17	60	7.6	A1

3.1 Density and Hardness Shore C

The tests were carried out on all the dosages described in accordance with the standard for cobblestones. The results obtained show how the density decreases progressively as the amount of aggregate replaced by residue increases, due to the lower density of the waste compared to the aggregate it replaces (Fig. 1). These enhanced results also allows a better workability, the handling, transport and the installation on site. Shore C hardness also decreases as the amount of aggregate replaced by waste increases, which allows a lower resistance to external penetration, although the values are still high, so this decrease is not very significant (Fig. 2).

3.2 Mechanical Strength

All the values obtained of the mechanical strength are above the minimum value established in the standard of 3.6 MPa, including those of greater substitution, such as those

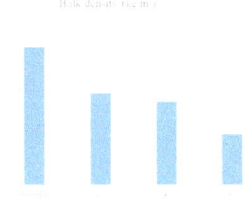

Fig. 1. Bulk density

Fig. 2. Hardness Shore C

of 60% replacement of aggregate by waste, whose value of 28-day fracture is 3.8 MPa (Fig. 3).

In the case of the 20% substitution of aggregate for waste, the compressive strength result is even higher than the reference value, improved thanks to the effect of the additive, which means that less water is required for mixing, improving the compaction of the matrix, with the consequent improvement of its mechanical properties.

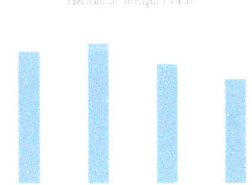

Fig. 3. Mechanical strength

3.3 Abrasion Resistance and Slip/skid Resistance

The results obtained in the abrasion resistance test show in Fig. 4. All the dosages achieve the requirements of the standard, as they are less than 20 mm, so they can be classified as Class 4, Marking I. In all cases, the results are very close to the reference values without residue, being very satisfactory results. The results obtained in the abrasion resistance test are important because their use in floor pavements exposes them continuously to external friction, which can lead to accelerated aging.

Regarding the results obtained in the Slip/Skid Resistance (USRV) test, the results obtained show values close to the value obtained in the reference dosages, and always higher than 45 USRV, so that adequate values are always obtained.

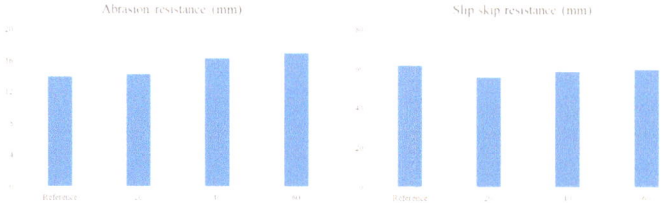

Fig. 4. Abrasion resistance (mm) and Slip/skip resistance (mm).

3.4   Water Absorption

The values obtained after carrying out the water absorption tests show how the values increase as a greater amount of waste is incorporated into the cobblestones. (Fig. 5) Almost all the values accepted by the standard except those of replacing 60% of the aggregate with residue. In this sample, an absorption of 7.6% is obtained while maximum value accepted by the standard is 6%. This result is due to the great amount and characteristics of the waste, which has polyurethane and other components such as cardboard, a compound capable to captivate more water. Therefore, all dosages except this one can be classified as Class 2, B marking.

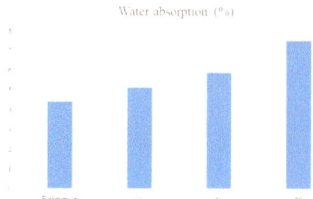

Fig. 5. Water absorption (%).

3.5   Fire Performance

Prefabricated cobblestones, according to the EN 1338:2004 standard "Concrete paving blocks. Specifications and test methods" belong to class A1 of the reaction to fire test, without the need for testing, since they meet the requirements for external fire performance.

3.6   Accelerated Aging Tests in Freeze-Thaw and Salt Crystallization Cycles

In order to evaluate the behaviour of the cobblestones when placed in outdoor environments, accelerated aging tests have been carried out in freeze-thaw cycles and salt crystallization, establishing the compressive strength before and after, and consequently it can be evaluated the pertinence to use them in these conditions. (Table 2).

Table 2. Mechanical resistance after accelerated aging tests.

Dosages. Substitution sand-waste	Mechanical strength after frost/thaw cycles (MPa)	Mechanical strength after crystallization of salts (MPa)
Reference	6.5	5.3
20	6.1	5.6
40	6.2	7.7
60	4.4	5.1

The freeze-thaw test was carried out and subsequently, the cobblestones subjected to these cycles were tested in compressive strength. The results after the test give higher values than the initial ones, improved in all the dosages, and always comply with standard since it is never lower than 3.6 MPa (Fig. 6).

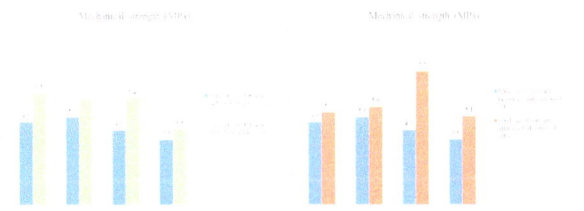

Fig. 6. Mechanical resistance after accelerated aging tests (MPa).

The salt crystallization test has also been carried out, and in the compressive strength results obtained after this test it is possible to see how, as in the previous case, the mechanical strengths are higher than those initially obtained. The result obtained in the dosage of 40% replacement, increases up to 7.7 MPa. This result is probably due to the effect of the additive, which, when combined with the mortar with waste, compacts the matrix in such a way that its resistance value improves notably, with very satisfactory results obtained.

3.7   Life Cycle Assessment (LCA)

It has been evaluated the environmental impact of the manufacture of these cobblestones, to quantify the impact of replacing part of the natural aggregate used in cement mortars by a waste product from the automotive industry, comparing it with the reference ones. For this purpose, a Life Cycle Analysis (LCA) of the different dosages of the prefabricated cobblestones has been carried out. The functional unit used is 1 m^2 of cobblestone. It was calculated using the SIMAPRO program, with the CML-IA baseline V3.06/EU25 method.

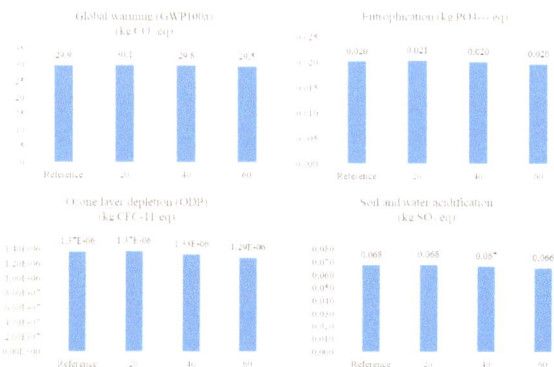

Fig. 7. Environmental impacts of the studied precast cobblestones.

As can be seen in Fig. 7, the results show certain variations in the studied impacts of the cobblestones with waste with respect to the reference ones. In the case of Global Warming (kg CO_2 eq), the 20% substitution increases slightly with respect to the reference value, possibly due to the additive, which means that, although less aggregate is used, it has to be considered in the overall estimate of impacts. However, for the others substitutions, the results obtained are lower than the reference values, improving this parameter by including the waste in our cobblestones. Therefore, we consider these results really positive, since the use of an industrial waste replacing a natural resource, such as aggregates, even when using the additive, contributes less to global warming than the reference cobblestones currently used in the market.

Regarding eutrophication, an important parameter in our environment, the values obtained in the dosages with waste are very similar to the values of the reference cobblestones, so it is a parameter that is not significantly influenced by the use of waste or additives.

The ozone layer depletion has also been studied, which is reduced when we use waste replacing the reference aggregate.

Finally, soil and water adification has also been studied, a parameter that benefits when waste is used. As the amount of aggregate replaced by waste increases, this value is slightly reduced, so we obtain positive results by reducing this parameter.

It can be concluded that, by incorporating waste to replace part of the aggregate in the cement mortars, the environmental impacts generated are reduced. For all these reasons, we consider that the environmental results are slightly better than those of the cobblestones currently on the market.

4   Conclusions

The research carried out involves the study of prefabricated cobblestones manufactured with cement mortar made up of cement, aggregate, water, an additive and a waste of industrial origin with which the aggregate of the reference mortar is substituted in different percentages, in volume. With these substitutions, products lighter than the reference mortar are obtained, with the consequent savings in the base structure and the improved workability that this provides. Several tests have been carried out to study its behaviour, in terms of mechanical resistance, resistance to abrasion resistance and slip/skip resistance, water absorption, fire resistance, as well as accelerated aging tests in freeze-thaw cycles and crystallization of salts. In all of them, except in some specific cases with a 60% substitution of waste for aggregate in the absorption test, the cobblestones comply with the requirements of the standard. Finally, the environmental impact of the inclusion of these wastes in the precast products has been evaluated by carrying out a Life Cycle Analysis (LCA) of all the dosages tested. The parameters analysed show how the incorporation of these wastes into the precast products improves their environmental performance, so that the use of these ecological cobblestones could be considered, making the construction sector more circular and greener, by using wastes as raw materials, avoiding their deposit in landfills, reducing the use of natural materials, and thus reducing both energy and environmental costs.

Acknowledgements. Authors gratefully acknowledge the Regional Government of Castilla y León (Junta de Castilla y León) and by the Ministry of Science and Innovation MICIN and the European Union NextGenerationEU / PRTR.

References

1. United Nations Environment Programme: Global Status Report for Buildings and Construction: Towards a Zero-Emission, Efficient and Resilient Buildings and Construction Sector. UN Environment Programme (2022)
2. Simonnet, C.: Concrete: History of a material. Ed. NEREA EDITRIAL, S.A. (2009)
3. https://environment.ec.europa.eu/strategy/circular-economy-action-plan_en
4. Hossain, M.U., Dong, Y., Thomas, S.: Influence of supplementary cementitious materials in sustainability performance of concrete industry: a case study in Hong Kong. Case Studies in Construction Materials, Vol. 15 (2021)
5. Letelier, V., Ortega, J.M., Tarela, E., Muñoz, P., Henríquez-Jara, B.I., Moriconi, G.: Mechanical performance of eco-friendly concretes with volcanic powder and recycled concrete aggregates. Sustainability **10**(9), 3036 (2018)
6. Kurda, R., Silvestre, J. D., de Brito, J.: Life cycle assessment of concrete made with high volume of recycled concrete aggregates and fly ash. Resour. Conserv. Recyc. **139** (2018)
7. Calderón, V., Gutierrez-Gonzalez, S., Gadea, J., Rodríguez, Junco, C.: Construction applications of polyurethane foam wastes. Recycling of Polyurethane Foams. Plastics Design Library, 125 (2018)
8. Madrid, M., García Frómeta, Y., Cuadrado, J., Blanco, J. M.: Life cycle analysis in concrete blocks: comparison of the impact produced between traditional blocks and blocks made with by-products. Informes de la Construcción **74**:566 (2022)

Properties of Eco-Cement Blocks Made with Polymer Wastes and Graphene

Verónica Calderón[1]([⊠]), Raquel Arroyo[2], Cristina Alía[2], Lucía Garijo[2], and Sara González-Moreno[1]

[1] Departamento de Construcciones Arquitectónicas, Escuela Politécnica Superior, Universidad de Burgos, Burgos, Spain
vcalderon@ubu.es
[2] Departamento de Ingeniería Mecánica, Diseño Industrial, ETSID, Universidad Politécnica de Madrid, Madrid, Química, Spain

Abstract. The inclusion of polyurethane wastes as recycled and reusable materials to replace variable amounts of aggregates is interesting in the production of new construction materials due to their final properties. In this research, the effects of waste polymer replace by sand (25%) and graphene oxide on mortars (0.5, 1, 1.5, 2, 2.5 y 3% with respect to the cement) have been investigated. To maintain and even improve the final properties, graphene oxide modify aspect as thermal conductivity and electrical properties, water behavior, mechanical properties and final contribution to fire.

Keywords: graphene · mortar · physical and mechanical properties

1 Introduction

The incorporation of graphene into mortars is still a relatively emerging field, and research is ongoing to explore its full potential and optimize its usage. The dosage, dispersion, and other factors related to graphene incorporation in mortars can affect the overall performance, and careful consideration should be given to ensure appropriate usage and compatibility with other components of the mortar mix [1]. Consultation with materials experts, testing, and adherence to relevant standards and guidelines are recommended when using graphene in mortars or any other construction materials.

Nanotechnology applied to polymer mortars such as graphene oxide make available the novelty in cement-based materials by adding an innovative vision to building materials.

The introduction of graphene oxide into cement can provide several significant benefits as decrease the porosity and consequently increase the durability of structures reducing long-term maintenance costs, enhanced thermal and electrical conductivity with better heat distribution [2]. The addition of graphene oxide to cement can also help decrease the formation of cracks and fissures due to its ability to improve material cohesion and toughness. This results in increased structural integrity and reduced likelihood of premature failure. These are just some of the potential benefits of introducing

L. Czarnecki et al. (Eds.): ICPIC 2023, 61, pp. 137–141, 2025.
https://doi.org/10.1007/978-3-031-72955-3_12

graphene oxide into cement. However, it's important to note that the effectiveness and outcomes may vary depending on the dosage and specific application conditions [3].

The economic cost of introducing graphene into concrete can vary depending on several factors, including the amount of graphene used, the production process, and the availability of graphene in the market. In this way, it is important to consider the balance between cost and potential benefits when introducing graphene into concrete. On the other hand, as technology and graphene production methods continue to advance, it is possible that the costs associated with its use may gradually decrease.

To develop the final products in this research, the microstructure (obtained by SEM and optical microscopy), thermal and electrical conductivity, spectrophotometry and density with suitable outcomes is determinate. To complete the study, additional destructive test as compressive strength, non-combustibility test, TGA and water absorption has been also carried out.

2 Samples and Procedure

The mortars have been fabricated using CEM I type Portland, replacing a 25% of aggregates by polyurethane grinded waste. Relation of cement aggregate is 1/3. The graphene oxide has been added on mortars with 0.5, 1, 1.5, 2, 2.5 y 3% with respect to the cement weight. The amount of water is the required to achieve enough workability according EN 1015-19. The size and dimensions of samples are normalized with $160 \times 40 \times 40$ mm^3.

3 Results and Discussion

To maintain and even improve the final properties, graphene oxide modify aspect densities, mechanical properties and water behavior.

3.1 Bulk Density

Bulk density test for mortar is a procedure used to determine the mass per unit volume of a mortar sample. It is a measure of the density or compactness of the mortar, and is an important parameter in assessing the quality and performance of mortar used in construction.

The bulk density of mortar is determined by dividing the mass of the mortar sample by its corresponding volume. The mass of the mortar sample is calculated using a scale, and the volume is determinate using a container of known volume. Results obtained for all the samples can be seen in Fig. 1.

Graphene is a two-dimensional material with a high aspect ratio (length-to-width ratio), which means it can form a network-like structure when incorporated into mortars. This can increase the packing density of the mortar, leading to higher bulk density when the amount of graphene increase in the sample. Graphene also can potentially influence the hydration of cement, which is a chemical reaction that occurs when cementitious materials react with water to form a solid matrix, indirectly affecting the bulk density of mortars.

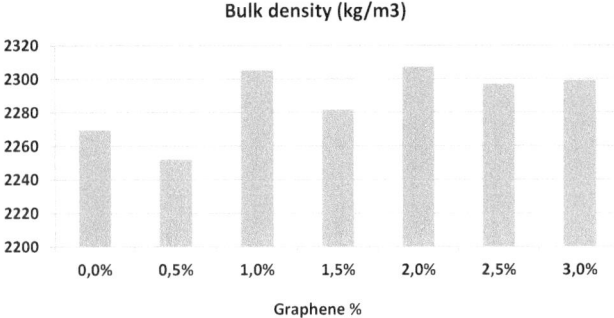

Fig. 1. Bulk density.

3.2 Water Absorption

Water absorption in mortars refers to the ability of mortar to absorb and retain water. The water absorption characteristics of mortars can have important implications for their performance and durability in various applications. Several factors can affect water absorption in mortars, including the type and proportions of the constituents used in the mortar mix, curing conditions, and environmental factors such as temperature and humidity.

To calculate the water absorption of mortar, you would need to determine the weight of the dry mortar and the weight of the saturated mortar, and then use the following formula:

$$\text{Water Absorption (\%)} = ((\text{Weight of Saturated Mortar} - \text{Weight of Dry Mortar})/\text{Weight of Dry Mortar}) \times 100.$$

The results obtained for this measure are all in the range of 5.4%-6.2%, very similar for all the samples as can be seen in Fig. 2, the addition of graphene not seems decisive for this property.

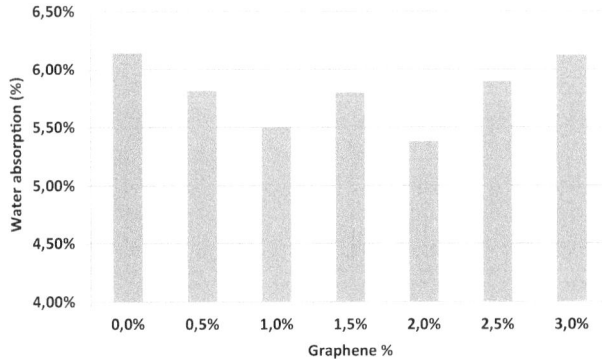

Fig. 2. Water absorption.

3.3 Compressive Strength

Compressive strength is an important mechanical property of mortars, which refers to the maximum amount of load that a mortar specimen can support before failing in compression and is experimentally through testing EN 1015-11.

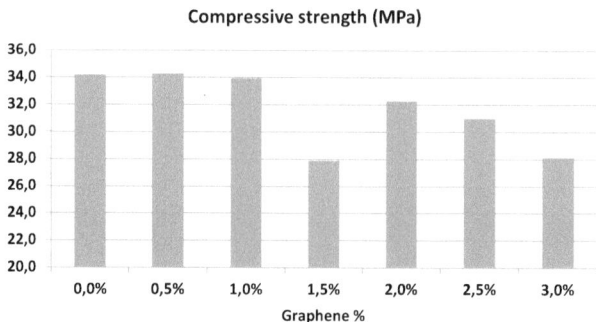

Fig. 3. Compressive strength.

The optimal dosage of graphene in mortars can vary depending on factors such as the type of graphene used, the mix design of the mortar, and the curing conditions. Our research has shown that an optimal dosage of graphene is achieved until a dosage of 1.0%. Exceeding this dosage, it is observed in Fig. 3 with a diminishing returns or even detrimental effects on the properties of the mortar. To avoid the tendency of graphene to agglomerate, nowadays we work adding surfactant-assisted dispersion, to ensure that graphene is evenly distributed throughout the mortar mix.

3.4 Thermal Conductivity

Graphene has an exceptionally high thermal conductivity, which refers to its ability to conduct heat. When incorporated into mortars, graphene can significantly enhance their thermal conductivity, allowing for more efficient heat transfer within the material. This can be advantageous in applications where good thermal management is important, such as in construction materials for buildings, where enhanced thermal conductivity can improve insulation properties and reduce energy consumption.

The thermal conductivity of reference mortars without graphene typically ranges from about 1.6 W/m·K. The temperature of test at isothermal conditions is also an interesting manner to determinate the final thermal conductivity testing involves measuring the ability of a material to conduct heat. The equipment used to determine the thermal conductivity is the FOX 50 according standards ASTM C518 e ISO 8301 to ensure accurate and reliable results. (Fig. 4). By means of heat flow meter method, a temperature gradient pass from side to side the samples measuring the heat flow through it. The addition of graphene to this mortars increase the thermal conductivity with reported values ranging around 1.5 times, depending on the composition, with slight influence of graphene concentration.

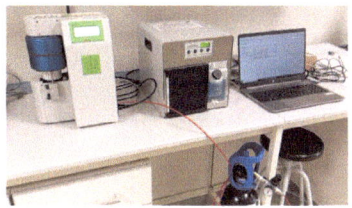

Fig. 4. Equipment FOX 50 and size of samples used.

4 Conclusions

The successful incorporation of graphene into mortars requires careful consideration of factors such as graphene dispersion, concentration, and compatibility with other components of the mortar mix as polymer wastes. Additionally, dosage, cost and scalability of graphene production are important aspects to consider for practical applications of graphene in mortars. Further research and development are ongoing to fully understand the potential of graphene in mortars and other cement-based materials.

Acknowledgements. This work is supported by the Regional Government of Castilla y León (Junta de Castilla y León) and by the Ministry of Science and Innovation MICIN and the European Union NextGenerationEU/PRTR.

References

1. Chintalapudi, K., Mohan, R., Pannem, R.: An intense review on the performance of graphene oxide and reduced graphene oxide in an admixed cement system. Constr. Build. Mater. **259**, 598–618 (2020)
2. Li, W., Qu, F., Dong, W., Mishra, G., Shah, S.P.: A comprehensive review on self-sensing graphene/cementitious composites: a pathway toward next-generation smart concrete. Constr. Build. Mater. **331**, 127284 (2022)
3. Lin, Y., Du, H.: Graphene reinforced cement composites: a review. Constr. Build. Mater. **265**, 120312 (2020)

Optimization of Tire Rubber-Concrete Core Materials for Application in New Sandwich-Structured Cementitious Composites

Matteo Sambucci[1,2]($\boxtimes$), Giulia Gullo[1], and Marco Valente[1,2]

[1] Department of Chemical Engineering, Materials, Environment, Sapienza University of Rome, 00184 Rome, Italy
matteo.sambucci@uniroma1.it

[2] INSTM Reference Laboratory for Engineering of Surface Treatments, UdR Rome, Sapienza University of Rome, Rome, Italy

Abstract. Implementing tire rubber-concrete mixtures to produce sandwich-structured cementitious composites can represent an attractive route in the perspective of lightweight design, energy efficiency, and sustainability for the building and construction industry. This work deals with a DOE multi-response optimization study on rubber-concrete mixes designed with different proportions of fine and coarse rubber aggregates to achieve the best formulation to be applied in the manufacturing of cementitious sandwich composites. The "sand-free" concrete mixture made up of 70% of rubber powder and 30% of rubber granules was optimal in terms of mechanical properties, physical characteristics, and thermo-acoustic insulation behavior. Sandwich-structured composite incorporating the optimum mix as a core layer showed significant improvement in terms of flexural performance over the monolithic rubberized materials and strength value in the range of RILEM "class II" lightweight construction materials.

Keywords: Rubber-concrete · Sandwich-structured composites · DOE optimization

1 Introduction

Along with rapid urbanization, the century is observing the biggest increase in the built environment through the construction of buildings, road networks, dams, pavements, etc., leading to an increase in the consumption and demand of natural raw materials. Hence, alternative sources of construction materials are required to reduce the demand for virgin resources and to preserve the environment. The use of recycled rubber from end-of-life tires (*ELTs*) as aggregate in concrete has great potential to positively affect the engineering and environmental performance of cement-based materials for a wide spectrum of civil and architectural applications where the use of ordinary mineral aggregates is not needed (lightweight concrete, non-bearing concrete brick walls, noise barriers, pavement, improved thermal insulation for flooring in buildings, railway track beds). It was claimed that rubberized concrete has abilities to absorb a large amount of plastic

© The Author(s) 2025
L. Czarnecki et al. (Eds.): ICPIC 2023, 61, pp. 142–151, 2025.
https://doi.org/10.1007/978-3-031-72955-3_13

energy under compressive and tensile loads, improve shock wave absorption, provide resistance to cracking, lower heat conductivity, and improve the acoustical environment which is advantageous in the applications mentioned above [1]. However, the use of tire rubber in concrete faces a problem of low mechanical strength performance which is the main barrier to full scalability in the construction industry. On this side, the challenge lies in investigating solutions that aim to enhance the strength behavior of rubberized mixtures, investigating both material optimization and advanced design solutions. Sandwich-structured composites (*SSCs*), consisting of a thick lightweight core and thin stiff skins, are well-established in lightweight design in many applications including automotive, aerospace, and building. *SSCs* offer high strength and stiffness while maintaining reduced weight, low heat and acoustic signature, and enhanced impact energy absorption characteristics. Core material properties govern many of the functionality of *SSC* including thermo-acoustic insulation, toughness, and lightweight [2]. Within a context of "green" design of building applications, energy efficiency, mechanical performance, and material optimization the tire rubber-concrete mixtures can represent an attractive solution for manufacturing cement-based *SSCs* in construction. A preliminary investigation on rubberized *SSCs* was conducted by the authors in previous research [3], demonstrating that the synergistic effect between a lightweight rubber-concrete core and stiff cementitious skins resulted in better mechanical properties (both static and impact response) over the monolithic rubberized materials and satisfactory acoustic insulation performance for paving unit applications. Comprehensive know-how of this novel approach requires addressing many aspects including the influence of production parameters (e.g., layering time) on *SSC* performance, the sandwich design (e.g., skin-core-skin thickness ratio), and the optimization of the constituent materials (core and skins). The present work faced one of these research gaps. Specifically, it presents the results of an optimization investigation of rubber-concrete mixtures that could be scaled as effective core layers for cementitious SSCs. The rubberized concrete samples were produced with two types of ground tire rubber particles, 0–1 mm rubber powder (*RP*) 1–3 mm rubber granules (*RG*) in different proportion ratios, as a total aggregate fraction of the mix design. A multimethodological experimental analysis, including static and dynamic mechanical testing, porosity and water absorption evaluation, acoustic and thermal insulation analysis of all the samples was performed, demonstrating how the size of rubber fractions influences the properties of the final concrete. The optimal mixture was achieved by MINITAB software using the design of experiment (*DOE*) "mixture design" approach that predicted the best parameters by investigating the combined effect of different factors simultaneously. The goal was to reach the "best" combination of fine and coarse rubber aggregates that maximizes the insulation properties, and, at the same time, maintains the physical mechanical properties at a suitable value. The optimized formulation was then scaled up to the fabrication and first characterization of the rubberized SSC aiming at verifying the truthfulness of the results.

2   Materials and Methods

2.1   Materials

The materials involved in the experimental work were cement, fine river sand, ground tire rubber (*GTR*), and tap water. CEM IV/A-type pozzolanic cement (strength grade of 42.5 R) was the cementitious binder used in this research. Fine-grained river sand (maximum nominal size of 1 mm) was procured from a local supplier. The *GTR* was supplied by the European Tyre Recycling Association (ETRA, Brussels, Belgium). Two rubber aggregate fractions, obtained by ambient grinding processing of *ELTs*, were used in the experimental work: 0–1 mm *RP* and 1–3 mm *RG*. The specific gravity of *GTR*, measured by the ethyl alcohol pycnometer method, was 1.40.

2.2   Mix Proportion and Preparation of Test Specimens

The main purpose of the study was to determine the optimal rubber-concrete formulation to implement like a core layer in the manufacturing of SSC. The control (*CTR*) mix design, containing no rubber, was produced with the selected water-to-cement (*W/C*) ratio of 0.42 and slump index below 0.1. It consisted of 720 g of cement, 300 g of water, and 1200 g of fine sand, all per liter (*L*) of mix. Four rubberized mix designs (Table 1) were studied by total volumetric replacement of sand with the two rubber aggregates, following different combinations of *RP* and *RG*. The amount of cement was held constant while the water content was adjusted during the operation to meet the slump properties and workability of *CTR* mix. As reported in Table 1, the fine *GTR* was always preserved in all designed mixtures. In the best authors' experience, fine rubber fraction is crucial to enhance the compactness and static stability of the compound.

Table 1. Mix proportions of rubberized concrete formulations.

Sample ID	Cement [g/L]	Water [g/L]	W/C ratio	RP [g/L]	RG [g/L]
RP100	720	338	0.47	550	-
RP75RG25	720	327	0.45	412	138
RP50RG50	720	300	0.42	275	275
RP25RG75	720	248	0.34	138	412

All required materials to produce the samples were firstly weighted, then the cement and the aggregates (sand and *GTR*) were drill-mixed until homogenization. Then the water was added, and the wet mixture was further mixed to form the fresh compound. Fresh mixes were then molded in prism molds with the standard dimensions of 40 × 40 × 160 mm and vibrated to reduce the amount of entrained air. After 24 h, the concrete prisms were demolded and placed in small containers for water curing until the time of testing (28 days). Figure 1 displayed the *GTR* distribution within the concrete composites investigated.

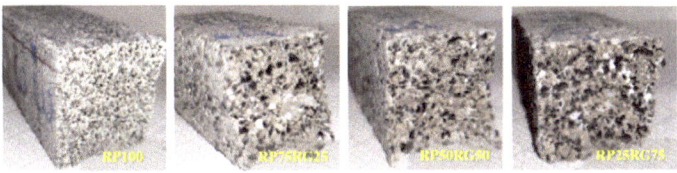

Fig. 1. *GTR* distribution in rubberized concrete samples.

2.3   Materials Testing

An experimental program related to mechanical (flexural strength, flexural modulus, compression strength, and impact energy absorption) physical (porosity and water absorption), acoustic (sound reduction index), and thermal (thermal conductivity) characteristics of the materials was planned in a multi-response optimization scenario. Flexural test was conducted on three beams for each mix. The test was performed on a Zwick-Roell Z10 universal testing machine using a three-point configuration with 100 mm span length (ASTM C 348-02, 2008). Compression test was conducted on 40 mm-side cube specimens (six samples for each formulation) extracted from broken beams after the bending test. According to ASTM C 109/109M standard method, the test was performed by using a Zwick-Roell Z150 testing system, with preload equal to 1 MPa and a test speed to 2 mm/min. Puncture impact test was performed on an Instron 9400 drop weight impact testing machine on $40 \times 40 \times 20$ mm samples (three tested specimens for each mix). The impact energy was set at 10 J. Permeable porosity and water absorption were determined using the vacuum saturation method, conforming to ASTM C1202 standard procedure. Four specimens for each mix were investigated. Thermal conductivity of the samples was studied by a C-Therm Thermal Conductivity Analyzer. The test was performed following the standard method ASTM D7984 on three specimens for each formulation. Sound reduction index was considered as an indicator to assess the acoustic insulation behavior of the materials. Acoustic insulation tests were made on two $50 \times 60 \times 80$ mm test samples for mix by means of the impedance tube method that is explained in detail in Ref. [3].

2.4   DOE Optimization

The mixture design method (*MDM*) in Minitab software was used to design the optimum rubberized concrete mixture to be scale as a core layer in SSC manufacturing. The proposed methodology consisted of several phases. First, the upper and lower limits of the component variables (*RP* and *RG*) and the process variable (*W/C* ratio) had been decided. The w/c ratio was divided into two categories: w/c ratio over 0.42 (named *1*) and w/c ratio lower than 0.42 (named *0*). Then, mechanical, physical, and thermo-acoustic properties of the investigated samples (responses), from the experimental characterization described above, were inserted into the analysis. Finally, the tool allows getting the optimum performance frame by maximizing or minimizing the desirable and undesirable properties respectively. Mechanical and acoustic insulation response parameters selected for maximization were flexural strength (*Fmax fle*), flexural modulus (*Emod*), compression strength (*Fmax com*), puncture energy (*Puncture*), and sound reduction index

(*Acoustic*). Thermal conductivity (*Thermal*), permeable porosity (*Porosity*), and water absorption (*WA*) were instead minimized. Minitab Response Optimizer tool identifies the combination of variable settings that jointly optimize a set of responses implemented in the model.

2.5 SSC Manufacturing and First Characterization

The rubberized *SSC* was based on three different layers: two skin layers (10 mm-thick), made of *CTR* mix, separated by a rubberized concrete core (20 mm-thick). The mixture used for the core was the one resulting from *DOE* optimization. The sandwich-based sample was produced by a 'three-steps' casting method, using 40 × 40 × 160 mm prismatic plastic molds (Fig. 2). A layering time of 1 h was selected for manufacturing. After the same curing procedure described in Sect. 2.2, the optimized rubber-concrete core and the related sandwich samples were mechanically tested in flexural. The mechanical characterization aimed to provide the improvement effect of sandwich configuration with respect to the monolithic rubberized core material. In addition, flexural strength results were compared with the performance of "no-optimized" *SSC* investigated by the authors in their previous research [3] to verify the reliability of *DOE* optimization.

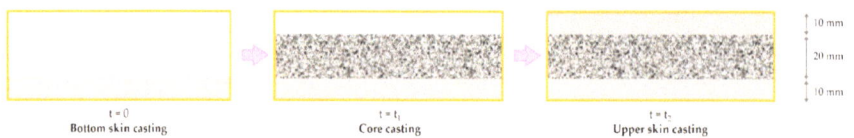

Fig. 2. SSC preparation procedure and specimen configuration.

3 Results

3.1 Materials Testing Results

Table 2 reports the mean value results of the experimental characterization conducted on *CTR* mix and the potential rubberized concrete "candidates" for the *SSC* design. The best performance detected in the *GTR*-based mixtures for each property investigated in the characterization is highlighted in bold.

Concrete mixtures including *GTR* of any type and combination proved to reduce the static mechanical strength (flexural and compression) and stiffness compared to *CTR* mixture. Moreover, the addition of rubber slightly increased the water permeability of the material. The advantage of using *GTR* as an aggregate fraction involves improvements in terms of impact energy absorption, noise reduction, and thermal resistivity. This finding matches those obtained by other researchers [4–6]. By detailing the influence of the rubber particle size on the performance of the rubberized composites, favoring the fine fraction (*RP*) implied better static mechanical performance. Small-sized rubber particles tend to increase the concrete compactness of concrete, reduce the stress singularity at internal voids and hence reduce the likelihood of fracture [4]. Because of the smaller specific surface area, coarse rubber aggregates (*RG*) are more likely to develop interface defects with the cement matrix, worsening the material's strength. Furthermore, the higher the *RP* content, the lower the thermal conductivity of the material and therefore better heat insulation performance.

Table 2. Overview of results related to the materials characterization.

	CTR	RP100	RP75RG25	RP50RG50	RP25RG75
Mechanical properties					
Flexural strength [MPa]	5.28	1.15	**1.29**	1.13	1.19
Flexural modulus [GPa]	2.51	0.40	**0.51**	0.31	0.33
Compression strength [MPa]	37.60	**3.29**	2.34	3.06	1.81
Puncture energy [J]	1.67	5.85	7.01	7.66	**8.00**
Physical properties					
Permeable porosity [%]	21.39	27.37	24.68	23.91	**22.57**
Water absorption [%]	16.55	22.53	19.44	18.36	**16.88**
Thermo-acoustic properties					
Sound reduction index [dB]	14.39	14.82	15.94	**15.99**	13.66
Thermal conductivity [W/m × K]	2.29	**0.42**	0.43	0.64	0.58

Indeed, the best heat insulation behavior was found in the RP100 mix with a drop in thermal conductivity of about 82% over the CTR sample. *GRT* had non-polar rough surfaces, permitting the entrainment of air (porosity) in concrete. Consequently, the smaller the particle size the greater the surface area for the same mass of rubber and thus, the greater the opportunity to entrain air [5]. This would verify the lowest thermal conductivity and highest permeability detected in the rubberized composites loaded with the high volume of *RP*. The coarse rubber fraction induced positive contributions in terms of mechanical-dynamic properties and acoustic dissipation. Compared to *CTR* mix, the maximum improvements occurred in *RP25RG75* mix (about 380% increase in puncture energy absorption) and *RP50RG50* mix (11% increase in sound reduction index), respectively. During impact load application, the failing rubber-concrete specimen will be capable of absorbing significant plastic energy and withstanding large deformations without full disintegration. The increased ability of the material to absorb and dissipate energy is consequently related to its sound insulation capabilities [7]. The dissipative behavior of rubberized concrete is more effective when coarse rubber particles are used in replacement of mineral aggregates than fine rubber [6]. However, by focusing on the noise abatement behavior, higher *RG* content (*RP25RG75* sample) aggravated the insulating characteristics of the material due to the adverse effect of rubber-matrix interface gaps on its sound dissipation ability.

3.2 DOE Optimization

Targeting the maximum performance in mechanical behavior (static and dynamic) and acoustic insulation and the minimum value of thermal conductivity, porosity, and water absorption the optimal percentage of *RP* and *RG* in the concrete mixture was achieved by the Minitab Response Optimizer tool (Fig. 3). The optimal solution, possessing a higher composite desirability value (*D*), was 70% of *RP* and 30% of *RG* (*RP70RG30* mix). *D*-value, expressed as a combination of the maximum or minimum of individual response desirability, is found 0.7005 which is close to 1, pointing out that settings are favorable for all responses. The optimal rubber-concrete formulation was therefore implemented as a core layer in the manufacturing of *SSC* samples (named *SSC-RP70RG30*).

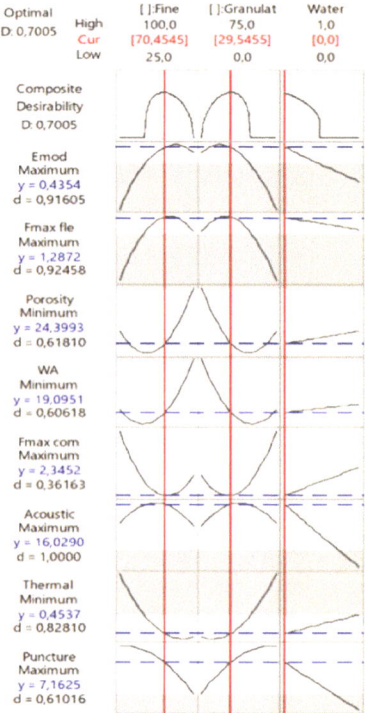

Fig. 3. Response optimizer for mixture design using Minitab software.

3.3 Mechanical Characterization of the Optimized SSC

Three-point flexural testing was conducted on *SSC-RP70RG30* samples and the respective core mixture to preliminarily assess the effect induced by the sandwich design on the mechanical behavior. As can be seen from Fig. 4, *SSCs* exhibit a flexure failure pattern typical of monolithic materials and no detachments emerged between the skins and the rubberized core. This demonstrates adequate compatibility and synergy between the layers resulting from a proper selection of the mixtures' rheology and the layering time. Figure 5 displayed the results of mechanical testing in terms of flexural strength properties. The average flexural strength of the "optimum" *RP70RG30* mix (1.32 MPa) is close to that predicted by Minitab (1.29 MPa), indicating a good match between the statistical analysis and the experiment. *SSC-RP70RG30* samples showed a clear improvement in flexural properties over the core material of about 200%. Moreover, the optimized sandwich design performed better mechanically than the *SSC* samples investigated by the authors in a previous research work [3] where two not-optimized rubberized cores with different proportions of *RP* and *RG* were studied. Compared to *CTR* mix, while the rubberized core material suffered a strong mechanical decrease (75% drop in strength), the *SSC* sample experienced a milder loss of strength (25% drop in strength), providing satisfactory mechanical characteristics for use as lightweight construction material "class II", as stipulated in the RILEM classification.

Fig. 4. Flexural testing on *SSC-RP70RG30* sample and failure pattern.

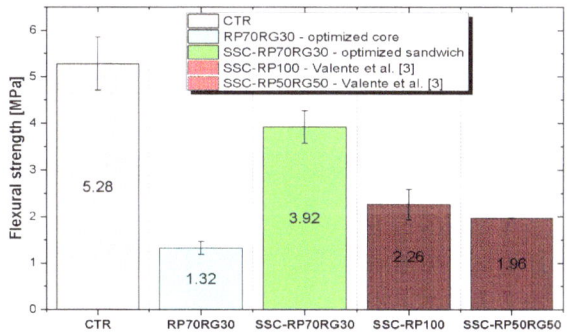

Fig. 5. Flexural strength test results.

4 Conclusions

In this study, an optimization analysis of *GTR*-cement mixtures to be implemented in the production of cementitious *SSCs* was conducted. Following a multi-methodological experimental characterization, the formulation with the optimal proportion of *RP* and *RG* was determined through *DOE*, targeting the maximization of the mechanical, durability and thermo-acoustic insulation characteristics. Then the optimum mix was employed in the development of *SSC* specimens and a preliminary mechanical characterization was performed. The main results are listed below:

- In the investigation of rubberized concrete mixtures, it was ascertained that the size of the rubber aggregate had a significant influence on the characteristics of the material. A high content of *RP* preserved better static mechanical performance and heat insulation. The addition of coarse aggregate improved the dynamic mechanical and acoustic behaviors of the material.
- The optimum mix from *DOE* analysis, by using Minitab Response Optimizer tool, was *RP70RG30* (70% of *RP* and 30% of *RG* in total substitution of sand). The accuracy of the statistical analysis was demonstrated by the good match between the real strength properties (experimental) of *RP70RG30* and those predicted by the software.
- *SSC* produced with the optimum core mix (*SSC-RP70RG30*) showed significant improvements in flexural strength (+ 200%) over the monolithic rubberized mixtures (non-sandwich configuration) and over not-optimized rubberized sandwich composites investigated by the authors in past research.
- *SSC-RP70RG30* met the technical requirement of lightweight construction materials "class II" (RILEM classification).

In addition to providing a more complete overview of the technological peculiarities of the sandwich systems developed in this work, future research will also address further aspects related to the optimization of these innovative cementitious composites, including the strengthening of the skins, or evaluating the effect of the layering time on their physical-mechanical characteristics.

Acknowledgements. The authors would like to express their sincere gratitude to Dr. Ettore Musacchi (ETRA) for the supply of the rubber aggregates used in the research. Thanks are due to the Circular and Sustainable Made in Italy Extended Partnership (MICS) funded by the European Union Next-Generation EU (Piano Nazionale di Ripresa e Resilienza (PNRR) - Missione 4, Componente 2, Investimento 1.3 - D.D. 1551.11-10-2022, PE00000004) for financial support.

References

1. Kara De Maeijer, P., Craeye, B., Blom, J., Bervoets, L.: Crumb rubber in concrete—the barriers for application in the construction industry. Infrastructures **6**, 116 (2021)
2. Thiagarajan, S., Munusamy, R.: Experimental and numerical study of composite sandwich panels for lightweight structural design. Int. J. Crashworthiness **27**(3), 747–758 (2022)
3. Valente, M., Sambucci, M., Sibai, A., Iannone, A.: Novel cement-based sandwich composites engineered with ground waste tire rubber: design, production, and preliminary results. Mater. Today Sustain. **20**, 100247 (2022)
4. Thomas, B.S., Gupta, R.C.: A comprehensive review on the applications of waste tire rubber in cement concrete. Renew. Sustain. Energy Rev. **54**, 1323–1333 (2016)
5. Richardson, A., Coventry, K., Edmondson, V., Dias, E.: Crumb rubber used in concrete to provide freeze–thaw protection (optimal particle size). J Clean. Prod. **112**, 599–606 (2016)
6. Karunarathna, S., Linforth, S., Kashani, A., Liu, X., Ngo, T.: Effect of recycled rubber aggregate size on fracture and other mechanical properties of structural concrete. J. Clean. Prod. **314**, 128230 (2021)
7. Vadivel, T. S., Thenmozhi, R., Doddurani, M.: Experimental behaviour of waste tyre rubber aggregate concrete under impact loading. Iranian J. Sci. Technol. Trans. Civ. Eng. **38**, 251 (2014)

Carbon Footprint and CO$_2$ Emissions in the Concrete-Polymer Composites Technology

Joanna Julia Sokołowska$^{(\boxtimes)}$ and Bogumiła Chmielewska

Faculty of Civil Engineering, Warsaw University of Technology, Warsaw, Poland
joanna.sokolowska@pw.edu.pl

Abstract. In the building materials industry, similarly to other industry sectors, the quantification of greenhouse gas emissions is undertaken, enabling the identification of GHG sources both for individual production processes and in total – for specific material solutions and products. While recently a lot of attention is paid to analyze carbon footprint of ordinary concrete and development of low-emission cements with significantly reduced Portland clinker content, the issue of GHG quantification in the context of concrete-like polymer composites (including concretes with polymer binders e.g. PCC or PC and concretes with significant amounts of polymer modifiers) is not recognized. This article attempts to make a preliminary assessment of the impact of the presence of polymers on the carbon footprint of such composites.

Keywords: Carbon Footprint · CO$_2$ Emissions · Polymer Carbon Footprint · Concrete-Polymer Composites · Polymer Concrete · Polymer-Cement Concrete

1 Introduction

1.1 Introduction to GWP and Carbon Footprint Concept

Ecological footprint is the method developed to measure human demands on natural capital, i.e. "the quantity of nature it takes to support people or an economy" [1]. Carbon footprint is considered as the more narrow component of the ecological footprint, and is defined in ISO 14067 as "the sum of the greenhouse gases emitted and absorbed by a product, expressed as CO$_2$ equivalents and based on a life cycle assessment". In other words, one can understand the carbon footprint as the total greenhouse gas (GHG) emissions caused directly and indirectly by an individual, society, organization, areas, products, services, etc., while carbon dioxide equivalent (CO$_2$e or CO$_2$eq or CO$_2-$e) is representing global warming potential (GWP) [2]. As different long-lived greenhouse gases (methane, nitrous oxide, halocarbons, sulphur hexafluoride, etc.) contribute to global warming to a different extent, the CO$_2$e enables to compare the emissions of gases on a common scale (it is calculated as GWP times mass of the other gas), taking into account how efficiently a given gas retains heat in the atmosphere and how long it stays in the atmosphere before it breaks down. For example, the average methane molecule, CH$_4$ stays in the atmosphere for around 12 years, much shorter than in case

L. Czarnecki et al. (Eds.): ICPIC 2023, 61, pp. 152–160, 2025.
https://doi.org/10.1007/978-3-031-72955-3_14

of CO_2 (estimated between 300 to 1,000 years or longer) yet it "captures" the heat more effectively – 1 tonne of methane released into the atmosphere would cause the same warming as ca. 27–30 tonnes of CO_2 [3]. Table 1 presents the most current values of lifetime and 100-year global warming potentials of three main greenhouse gases published in the Intergovernmental Panel on Climate Change report, IPCC AR6 (2021) [3] that are improvements upon the still commonly cited values from the older IPCC reports [4–6] due to the changing concentration of CO_2 in the atmosphere over time.

Table 1. 100-year global warming potential (GWP) relative to CO_2 for key greenhouse gases published in IPCC reports AR6 (2021) [3], AR5 (2014) [4], AR4 (2007) [5], SAR (1995) [6]

Gas name, chemical formula	Lifetime, years (AR6)	100 years GWP			
		SAR	AR4	AR5	AR6
Carbon dioxide, CO_2	300–1,000	1.0	1.0	1.0	1.0
Methane, CH_4 (fossil origin)	11.8	21.0	25.0	28.0	29.8
Methane, CH_4 (non-fossil origin)					27.2
Nitrous oxide, N_2O	109.0	310.0	298.0	265.0	273

1.2 Types of CO_2 Emissions

Regardless of the GWP values adopted for the calculations (more or less recent), there are also various methods of calculating carbon footprint. For instance an organization's carbon footprint includes the emissions caused by all its activities, including the energy consumption of buildings and means of transport, while product's carbon footprint includes emissions caused by the extraction of the raw materials from which it was made, production, use and storage or recycling after use [7]. The common classifications assumes 3 main types of emissions (depending on the level of control exercised by the particular organization) as following [8]:

- **direct emissions from activities the organization controls** (e.g. on-site combustion of fuels, emissions during production, running of a vehicle fleet),
- **indirect emissions from electricity usage** (e.g. lighting, equipment power),
- **indirect emissions from products and services that the organization does not directly control** (e.g. a company manufacturing a product is indirectly responsible for CO_2 emitted during preparation/transport of the raw materials).

Calculating all types of emissions can be a complex task. Moreover, currently there is lack of consistency in methods for calculation and reporting carbon footprint so it is difficult to compare the published values of footprints.

1.3 Exemplary CO_2 Release During the Building Materials Production

Table 2 contains exemplary data on the amount of CO_2 released during the production (direct and indirect manufacturing effects) of selected building materials and products, including cements (as cements production sector is responsible for 5–7% of the world's CO_2 emissions [9, 10]) and concrete, considered the most common building material.

Table 2. Amounts of CO_2 released when making 1 tonne of the selected building materials and products (based on [11–14] after [15, 16] and [17]).

Material	kg CO_2/t	Additional calculating notes
Steel	800–4,000	Steel reinforcement: c.a. 1,900 kg CO_2/t [13]
Glass	600–1,440	
Bricks	140–210	E.g. per brick: 0.3 [12]-0.45 [11] kg CO_2
Lime	740–780	
Cement, direct	284–912	Direct manufacturing only, including calcining
Concrete, direct and indirect	330* (800 kg/m^3 [12])	Direct and indirect effects, including: calcining, fuel, quarrying, sup-pliers, placement, etc.
Concrete, direct	65 [11]-210* [12] (200–500 kg/m^3 [12])	Direct manufacturing only, including calcining; depends on concrete strength and composition

*) Assuming an ordinary concrete density of 2,400 kg/m^3

2   Calculating Carbon Footprint of Concrete-Polymer Composites

2.1   Life Cycle Assessment of Concrete-Polymer Composites (C-PC)

As in the case of any construction material, the calculation of the carbon footprint of C-PC composites should cover the material life cycle assessment: (1) production phase followed by (2) the operation phase and (3) the post-use phase (including demolition and eventual recycling). Production phase can be additionally divided including into various sub-phases (see Table 3), depending on the qualitative composition, namely: the presence and content of cement and polymers.

In the case of polymer-modified concretes, PMC the amount of the polymer modifier is not significant (polymers are used mainly as the admixtures, so their content should not exceed 5% of cement mass), their life cycle is considered analogical to non-modified cement based concretes. Remaining C-PC (i.e. polymer impregnated concrete, PIC, polymer-cement concrete, PCC and polymer concrete, PC) life cycles are more divers and have a greater impact on the total CO_2 emissions and carbon footprint.

2.2   Polymers and Other C-PC Components CO_2 Emissions

Table 4 presents the general shares of main components of C-PC composites, i.e. polymer binder, cement and aggregates. The table does not include mixing water PMC, PIC and

Table 3. The expected life cycle assessment production sub-phases of the C-PC composites: polymer-modified concretes (PCM), polymer impregnated concretes (PIC), polymer-cement concretes (PCC) and polymer concretes (PC)

Life cycle phase	Sub-phase		Type of C-PC composite			
			PMC	PIC	PCC	PC
Production	Extraction of raw materials	Aggregates	✔□	✔□	✔□	✔□
		Cement	✔□	✔□	✔□	
	Transport of raw materials to factories/plants	Polymer admixture	✔□			
		Polymer impregnate		✔□		
	Components production	Polymer binder			✔□	✔□
		Other modifiers	✔□	✔□	✔□	✔□

PCC, nor a polymer admixture. CO$_2$e of concrete admixtures (including production and transport) is generally estimated as c.a. 220 kg CO$_2$/t [16]. In case of polymer plasticizers/superplasticizers it is estimated as 1,880 kg CO$_2$/t [11] but taking into consideration its small content in the concrete mix it can still be treated as marginal impact on total concrete emissions. As for mixing water (which is about 6–7% of the concretes total mass) CO$_2$e of tap water is estimated as c.a. 320 kg CO$_2$/t [17].

Table 4. Approximate mass contents of the main C-PC components (based on [18–20]).

Component	PMC	PIC	PCC	PC
	Content, % (by composite mass)			
Cement binder	11–16*	11–16*	9–15	0
Polymer binder	0	3–8	0.8–5	8–20
Aggregate	70–85	70–85	70–85	80–92

Cement. As mentioned earlier, cement production is considered to account for 5% of global CO$_2$ emissions. In the last decade that emissions have stabilized at around 4 billion tonnes per year, but is over 2.5 times more than it was at the beginning of the 21st century [21]. Takin into account the high emissions of cement production (up to 912 kg CO$_2$/t for CEM I [11]), cement content remain an important factor influencing total carbon footprint of C-PC composites, of which only PC is totally cement-free.

Aggregates. In PMC, PIC and PCC the aggregate constitutes about 70–85% of the mass (60–80% by volume) [18]. In the case of PC the mass content of mineral fillers is higher – c.a. 80 ÷ 90% [19], of which the traditional aggregate (sand, gravels) is about 60–80% and the remaining 20–40% is a microfiller [19, 20]. In [17] one can find general estimation of CO$_2$e of aggregates depending on their size, i.e. gravels (size > 2 mm): 4.32 kg CO$_2$/t, sand (fraction 0/2 mm): 10 kg CO$_2$/t, microsand: 110 kg

CO_2/t. Considering that the finer the aggregate, the higher the CO_2 emissions, the use of microfillers in PC noticeably increases the carbon footprint of their aggregate blends.

Polymers. Table 5 summarizes exemplary data on the amount of CO_2 released during the production of selected polymers commonly used in construction, including the polymer modifiers, impregnates and binders or co-binders of C-PC.

Table 5. Amounts of CO_2 released when making 1 tonne of the selected polymers used in building materials and C-PC composites [11, 12, 17, 22, 23]

Polymeric material/product	kg CO_2/t	C-PC polymer binder/modifier	kg CO_2/t
Acrylic paint	3,000	Epoxy (resin)	5,700–6,800
Polyethylene (LDPE, HDPE)	1,930–2,600	Epoxy (general)	4,700–8,100
Polyethylene terephthalate	1,760–2,300	Unsaturated Polyester*	3,110–3,320
Polypropylene	1,850–3,430	Vinyl-ester (BPA EP-based)	5,970
Polystyrene	3,070–3,290	Vinyl-ester (BPA based)	5,870
Polyurethane (flex./rig.)	3,610–4,990	Melamine (resin)	4,190
Polyvinyl chloride	3,100–4,400	Phenol Formaldehyde	2,980
Rubber (synthetic)	2,850–4,000	Urea Formaldehyde	2,760

*) including polyesters based on orthophthalic, isophthalic acid, maleic and DCDP

C-PC carbon footprint – examples of estimation. The above individual data were used to estimate the carbon footprint of selected C-PC composites: ordinary concrete (OC), concrete modified with polycarboxylate superplasticizer (PMC), two polymer-cement concretes with various carboxylated SBR latex co-binders in amounts of 10% (PCC-10) and 20% (PCC-20) of dry polymer in relation to the cement mass and, finally, the polymer concrete (PC) with bisphenol-A based vinyl-ester resin. With the exception of PC, concretes were to present similar strength (tested acc. to EN 12390-3) and consistence (class S3, tested acc. to EN 12350-2). Table 6 presents compositions and compressive strength, CO_2e of components and summarized "material" CO_2 emissions (without later life cycle phases impact) of the analyzed composites.

Higher compressive strength is theoretically associated with higher carbon footprint, however it is difficult to explicitly determine the actual values. According to "Circular ecology" carbon footprint calculator (based on ICE database [11]) for the analyzed OC and PMC compositions the carbon footprint is respectively 158 and 152 kg CO_2/t. Meanwhile according to the empirical model based on concrete characteristic strength ($CO_2 = \delta\sqrt{f_{ck,cyl}}$ where $\delta = 46.5$ [14] after [24]) for the analyzed concretes classified as C30/37 and C40/C50 the carbon footprint is almost twice as high – respectively 255 and 294 kg CO_2/t. Taking into account the data from the Table 6 determined for specific compositions and the most closely matched components CO_2e, in the case of OC and PMC the summarized emission was – respectively 176 and 168 kg CO_2/t, thus between the values estimated using the two abovementioned models. Moreover, these models lack the possibility to include the larger amounts of polymers used as co-binders.

Table 6 shows that in the case of PCC, where the cement content was not reduced, but polymer was added additionally, the total emission increased by 15–20% (26–33 kg CO_2/t) in comparison to reference concretes (with no SBR). Interestingly, the obtained results show how important the quantitative selection of PCC composition is – both in terms of strength and carbon footprint. A higher amount of polymer co-binder does not necessary determine the higher carbon footprint. In the case of PCC-20 where almost twice as much SBR latex was used, but less cement and much less water (the polymer binder acted as a superplasticizer) in comparison to PCC-10, practically identical carbon footprint was obtained, but a much higher strength.

Table 6. Estimated CO_2 emissions of C-PC of particular composition and compressive strength (based on the own research, partially published in [20] and [25]).

Composite type	OC	PCC-10	PMC	PCC-20	PC	CO_2 emissions
Component	Content, kg/m^3					kg CO_2/kg
Polymer binder	0	39	0	71	271	1.630/5,870
Polymer admixture	0	0	8.6	0	0	1.880
Cement	390	390	356	356	0	0.912
Water	175	175	160	99	0	0.320
Aggregate 0/2 mm	656	656	692	692	491	0.010
Aggregate 2/16 mm	1219	1219	1230	1230	997	0.004
Microfiller 0/125 μm	0	0	0	0	541	0.110
Strength (f_{cm}), MPa	45.2	46.5	55.5	54.0	111	-
Total kg CO_2/m^3	423	487	404	484	1659	-
Total kg CO_2/t	**176**	**203**	**168**	**202**	**691**	-

In case of analyzed PC the carbon footprint is much higher because of the use of large amount of vinyl-ester. However taking into consideration its very high compressive strength (110 MPa after 14 days and over 125 MPa after few years [20]), as well as tensile strength (of c.a. 20 MPa) and excellent chemical resistance (including acid resistance) [20], the use of vinyl-ester resin in such an amount (11.8% by composite mass) is justified. Especially that small-sized elements with thin cross-sections are usually made of PC and the material consumption is not as high as in the case of ordinary concrete with/without admixtures or in case of PCC.

2.3 CO_2 Sequestration in Concretes with Polymers

Calculating carbon footprint for OC can include CO_2 sequestration, i.e. concrete carbonation phenomenon taking place during the operational phase of the life cycle. As during carbonation reaction (CO_2 + $Ca(OH)_2$ → $CaCO_3$) carbon dioxide is incorporated into the near-surface layers of the hardened concrete, this can be considered a reverse emission and thus included in the total CO_2 emissions and ultimately lower the concrete carbon footprint [26]. One can expect that in PMC, such a phenomenon could also occur, though in the case of PCC, despite the significant cement content, the polymer forming a continuous phase should prevent the phenomenon of carbonation. Therefore, some studies show that even in case of PMC, using properly selected cement and admixture (e.g. polycarboxylic acid [27]) enables significant increase in the concrete anti-carbonation abilities. Therefore, what is considered an advantage in the context of the durability (tight microstructure, often completely isolated pores and therefore improved tightness), in the context of CO_2 sequestration works negatively. Off course benefits of increased durability are undeniable. Especially that the full use of the CO_2 sequestration potential is not possible even in OC due to the high risk of reinforcement corrosion and the limitation of the carbonation progress over time. The case study of CO_2 sequestration in the life cycle of concrete viaduct [26] showed that the structure was able to absorb only about 2% of total CO_2 emissions. For PIC or PCC it would be less and therefore it can be considered a marginal impact.

2.4 Using of Secondary Components: Recycled Polymers and Microfillers

There is yet no data on the C-PC made of recycled or bio-based polymers, however there are published promising results on production of other pre-cast elements made of recycled polymers. The report on the sanitary pipes production [28] showed for example that the energy consumption during production of 3m long PVC, PE and PP pipes containing 80% recycled polymers was reduced by 65–74% and CO_2 emissions – by 60–71% in comparison to the fossil-based polymers elements production. In the context of PCC and PC concretes, the authors consider the use of recycled PET polymers, which undoubtedly have a lower carbon footprint, but are hardly available on the market and so far cannot be considered as materials for use in large-scale C-PC pre-cast production.

Meanwhile the use of very fine powdered waste materials or by-products as PCC fine aggregate [29] or PC microfiller is a solution easy to implement, yet may noticeably reduce the aggregate CO_2e. The authors tested the long-term durability of PC with fly ashes – the high degree (up to 79%) of substitution the quartz powder with the fly ash enabled to reduced the composites carbon footprint even by c.a. 20 kg CO_2/t [20].

3 Summary

Despite the common belief that C-PC composites must have a very high carbon footprint, their compressive strength and durability seem to justify the use of polymers. Also, in case of PCC the total CO_2 emission depends not only on the polymer content, but the entire composition. The development of polymer processing towards the use of recycled polymers, as well as the use of aggregates from waste/secondary materials seem to be a good ways to make the C-PC production technology less burdensome for the environment. Nonetheless, there is still a need to standardize the procedures for calculating the CO_2 emissions of both components and concretes (including OC), as the currently available data only allow for rough estimates, as presented in this paper.

References

1. Wackernagel, M., Lin, D., Evans, M., Hanscom, L., Raven, P.: Defying the footprint oracle: implications of country resource trends. Sustainability **11**(7), 2164 (2019)
2. Benn, H.: Guidance on how to measure and report your greenhouse gas emissions. Department for Environment, Food and Rural Affairs, London (2009)
3. IPCC AR6: Intergovernmental Panel on Climate Change Sixth Assessment Report (2021)
4. IPCC AR5: Intergovernmental Panel on Climate Change Fifth Assessment Report (2014)
5. IPCC AR4: Intergovernmental Panel on Climate Change Fourth Assessment Report (2007)
6. IPCC SAR: Intergovernmental Panel on Climate Change Second Assessment Report (1995)
7. Wiedmann, T., Minx, J.A.: Definition of "carbon footprint". In: Ecological Economics Research Trends, Chapter 1, pp. 1–11. Nova Science Publishers, Hauppauge, NY (2008)
8. Carbon Trust Homepage, https://www.carbontrust.com, last accessed 2023/02/09
9. Środa, B.: Concrete—A Low-Emission Building Material (in Polish). SPC, Kraków (2021)
10. Anderson J., Moncaster A.: Embodied carbon of concrete in buildings, part 1: analysis of published EPD. Build. Cities **1**, 198–217 (2020)
11. Jones. C., Hammond, G.: ICE (Inventory of Carbon and Energy) database, V3.0 (2019)
12. The.CO2List.org: Amounts of CO$_2$ released when making & using products, http://www.co2 list.org/files/carbon.htm, last accessed 2023/02/13
13. Concrete CO$_2$ fact sheet. National ready Mixed Concrete Association (2012)
14. Załęgowski, K., Jackiewicz-Rek, W., Garbacz, A., Courard, L.: Carbon footprint of concrete (in Polish). Materiały Budowlane **12**(2013), 34–36 (2013)
15. Marceau M.L., Nisbet M.A., VanGeem M.G.: Life cycle inventory of Portland cement concrete. Portland Cement Association (2007)
16. Turner, L.K., Collins, F.C.: Carbon dioxide equivalent (CO$_2$-e) emissions: a comparison between geopolymer and OPC cement concrete. Constr. Build. Mater. **43**, 125–213 (2013)
17. Winnipeg Homepage, https://www.winnipeg.ca, last accessed 2023/02/15
18. Czarnecki, L.: Polymer concretes. Cem. Lime Concr. **15**(2), 63–85 (2010)
19. Mehta, P.K., Monteiro, P.J.M.: Concrete Microstructure, Properties, and Materials. McGraw Hill Professional (2013)
20. Sokołowska, J.J.: Long-term compressive strength of polymer concrete-like composites with various fillers. Materials **13**(5), 1207 (2020)
21. Asghar, R., Khan, M.A., Alyousef, R., Javed, M.F., Ali, M.: Promoting the green construction: scientometric review on the mechanical and structural performance of geopolymer concrete. Constr. Build. Mater. **368**, 130502 (2023)
22. Joshi, S.V., Drzal, L.T., Mohanty, A.K., Arora, S.: Are natural fiber composites environmentally superior to glass fiber reinforced composites? Compos. Part A. Appl. Sci. Manuf. **35**(3), 371–376 (2004)
23. Hill, C., Norton, A.: LCA database of environmental impacts to inform material selection process. JCH Industrial Ecology Ltd (2020)
24. Habert, G., Roussel, N.: Study of two concrete mix-design strategies to reach carbon mitigation objectives. Cement Concr. Compos. **31**(6), 397–402 (2009)
25. Chmielewska, B.: Adhesion strength and other mechanical properties of SBR modified concrete. Int. J. Concr. Struct. Mater. **2**(1), 3–8 (2008)
26. Woyciechowski, P.P.: Role of sequestration of CO$_2$ due to the carbonation in total CO$_2$ emission balance in concrete life. J. Constr. Mater. **2**(2021), 3–4 (2021)
27. Zhang, P., Zhang, B., Fang, Y., Chang, J.: Study on carbonation resistance of polymer-modified sulphoaluminate cement-based materials. Materials **15**(2022), 8635 (2022)
28. Recio, J., Guerrero, P., Ageitos, M., Narváez, R.: Estimate of Energy Consumption and CO$_2$ Emission Associated with the Production, Use and Final Disposal of PVC, HDPE, PP, Ductile Iron and Concrete Pipes. Univ Politécnica Catalunya, Barcelona (2005)

29. Jaworska, B., Sokołowska, J.J., Łukowski, P., Jaworski, J.: Waste mineral powders as components of polymer-cement composites. Arch. Civ. Eng. **61**(4), 199–212 (2015)

Methods for Managing the Tacit Knowledge of Employees with Long Scientific Seniority Using the Example of Research Institutions. Preliminary Assumptions

Katarzyna H. Tomiczak(✉)

Faculty of Management, Warsaw University of Technology, Warsaw, Poland
k.tomiczak@itb.pl

Abstract. Tacit knowledge management among employees with long scientific seniority within research institutions in Poland and the European Union is being discussed. The key question under analysis is: How should the tacit knowledge of employees with long scientific seniority be managed to preserve their legacy and facilitate its transfer to younger generations? As a part of a doctoral thesis, this study aims to verify and develop methods to support the management of tacit knowledge of employees with long scientific seniority in construction institutions. To achieve this, research comprising surveys and in-depth interviews will be conducted among employees of research institutions. The anticipated outcome of this study is a comprehensive method for tacit knowledge management, which includes elements such as an environment supporting knowledge transfer, cooperation techniques, and age diversity management. The insights from this research could provide a foundation for further investigation in other regions and disciplines, ultimately leading to a deeper understanding of the process of transferring tacit knowledge of senior researchers.

Keywords: Knowledge transfer · Age diversity management · Legacy preservation

1 Introduction

In the current context of technological advancement, the management of tacit knowledge has surfaced as a critical determinant of organizational success across a plethora of sectors. This process adds complexity when one considers the vast wealth of insights that senior researchers have gathered over their lengthy careers. It becomes a multifaceted process that involves safeguarding and capitalizing on seasoned expertise, thereby influencing organizational growth and dictating industry patterns and future developmental trajectories. This paper aims to introduce the concept of managing the hidden knowledge of employees with long scientific seniority in the academic construction environment–as an invitation for collaboration and discussion regarding the research concept.

Definitions of key terms used in this study have been described in detail for clarity. **The management method** "is a proven, recognized, logically structured way of solving

© The Author(s) 2025
L. Czarnecki et al. (Eds.): ICPIC 2023, 61, pp. 161–168, 2025.
https://doi.org/10.1007/978-3-031-72955-3_15

specific organizational problems" [1]. **Tacit knowledge** "is inextricably linked with a person, not easy to transfer and codify, it is a unique compilation of know-how, professional qualifications based on knowledge obtained in the process of formal education, skills (…), experiences, substantive observations, developed methods of conduct (…)" [2]. **An employee with Long Scientific Seniority**, LSS, is an employee who obtained a Ph.D. degree at least 30 years ago (approximately 1993 and earlier). **The scientific institutions** considered in this work are research institutes and polytechnics dealing with construction in Poland and in the European Union. Construction is characterized by a rather conservative and formal approach (Fig. 1), which results from the obligations involved in designing and erecting buildings—this can hinder the diffusion of knowledge. In addition, in this context, tacit knowledge can be an essential resource for further development of construction science. Formally, the study is located in the field of management sciences. Nevertheless, it is paramount to accentuate that the study brings tangible benefits to construction by promoting generational knowledge continuity and safeguarding expert wisdom intrinsic to this field (Fig. 1).

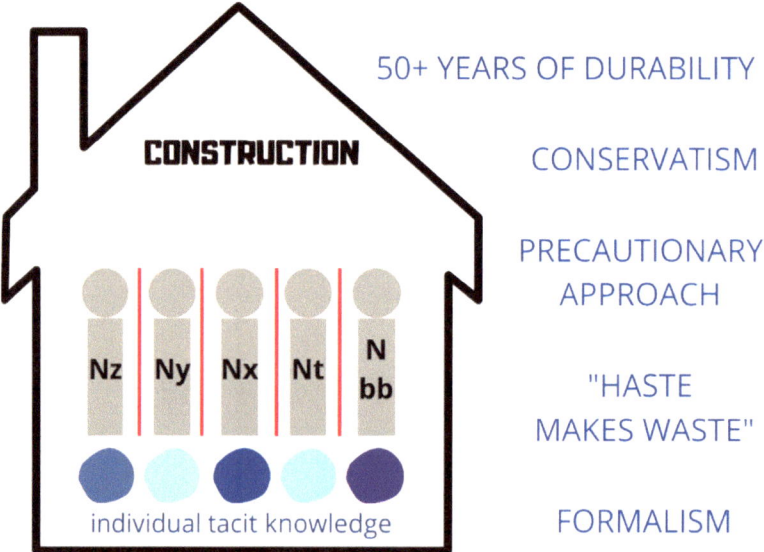

Fig. 1. Characteristics of researchers in the construction industry: Nz—a person from the Z generation (1995–2010), Ny—a person from the Y generation (1980–1994), Nx—a person from the X generation (1965–1979), Nt—a person from the T generation (1927–1945), Nbb—a person from the Baby-boomers generation (1946–1964); dates given as in [3], [own elaboration]

The subject of the doctoral dissertation is derived from the author's professional trajectory, stemming from her involvement at the Building Research Institute (ITB) since 2016. ITB is an esteemed institution with a rich almost 80-year history in pioneering construction research and maintains a diverse team of over 200 employees. The spectrum of age among the scientific staff is extensive, ranging from a 23-year-old research and technical employee to a 93-year-old professor, with the youngest assistant being 36 years old.

Two opposing mechanisms of tacit knowledge management have been recently discerned within ITB. The first pertains to the challenges of transferring knowledge from senior researchers, as their impending retirement could precipitate a significant loss of experiential knowledge. For example, a senior civil engineer might have developed an instinctive ability to detect potential design flaws that are not apparent to less experienced peers. Such intuitions cannot be easily documented or taught, but they are integral to problem-solving and innovation in construction. The significance of tacit knowledge is vast. It is a fundamental source of long-term competitive advantage, fostering innovation, enhancing operational efficiency, and driving scientific breakthroughs. Its loss, particularly from LSSs, could lead to significant setbacks, inhibiting knowledge continuity and the progress of construction research and practice. As such, effective management and transfer of tacit knowledge are crucial for the sustainability and advancement of construction institutions.

The second, in contrast, illustrates the successful acquisition of tacit knowledge via interactive observation, a process the author has experienced first-hand during her collaboration with ITB's scientific secretary, Professor Lech Czarnecki, an outstanding specialist in the field of the building materials engineering, who has accumulated unique experience and skills during his over forty-year scientific career. The author, who has been working with him on a daily basis for over seven years (certainly it is merely personal experience), has noticed that she acquires knowledge not only in the field of construction, i.e. of an expert nature, but also the one related to, for example, the general expectations of the civil engineering scientific community or the way of composing a scientific paper.

Based on these observations, the proposed research has identified its primary subjects as employees with Long Scientific Seniority (LSS), defined explicitly as those boasting a post-doctorate experience of at least 30 years in construction.

When considering the critical matter of tacit knowledge transfer from LLS employees, it is essential to take into account that systemic transformations in Poland commenced 30 years ago. These transformations spanned a broad array of economic, political, and social reforms, which also encompassed changes within the realms of education and science. This historical context could set Polish institutions apart from their European counterparts, although its influence on tacit knowledge transfer is not assumed to be deterministic.

2   Preliminary Literature Review

Knowledge transfer barriers across generations have been studied globally, including in Poland. For instance, an article by Sanei et al. [4] detailed the knowledge transfer process between generations T, BB, X, and Y in US-based construction organizations. Another study [5] conducted in Slovakia examined factors like willingness, motivation, communication, and cooperation in knowledge sharing. A doctoral thesis [6] analyzed knowledge transfer from retiring employees to successors in a mid-sized Finnish company designing and manufacturing electrical systems for the global market.

In Poland, studies by Chomątowska and Żarczyńska-Dobiesz [7] highlighted Baby Boomers' characteristics and knowledge-sharing barriers. Sidor-Rządkowska [3] addressed diversity management in modern organizations, emphasizing the benefits of implementing mentoring. Dziadek [8] provided a different perspective on intergenerational knowledge transfer challenges within modern businesses, exploring generations BB, X, Y, and Z. Her study primarily aimed to identify whether selected organizations have implemented knowledge transfer systems. The research conducted by M. Morawski [2] seems to be closest to the area of interest of the author of this paper, as it relates to the knowledge-sharing (diffusion) skills of a company's key employees–which is how employees with long scientific seniority can be defined.

Recent works, such as Rui and Ju's [9], applied relationship management theory to study intergenerational knowledge transfer among younger Chinese employees across diverse sectors. Wang et al.'s [10] paper, examining the link between workplace ostracism and knowledge-sharing behaviors within Chinese academia, is also noteworthy. Work on the borderline of the subject matter undertaken is a publication referring to knowledge loss caused by employee turnover in organizations [11].

Four main research interests related to LSS emerged from studying the literature (Fig. 2). The first is the diffusion/transfer of tacit knowledge of LSS employees, and the second is the way of managing this tacit knowledge. Another area is the management of generational diversity—the effect of cooperation of many people: from the youngest to the oldest. It is also important to identify and overcome barriers to knowledge sharing. Together, these areas shape the context of research on the tacit knowledge of LSS employees.

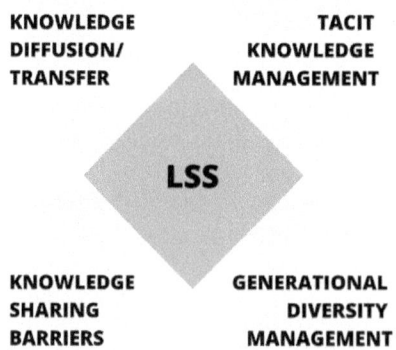

Fig. 2. Four main areas of the research context, [own elaboration]

3 Proposed Research Framework

3.1 Research Problem Definition

The primary aim of this study is to verify and develop methods to support the management of tacit knowledge of LSSs in construction institutions. Explicit knowledge, contrary to tacit knowledge, can be easily shared using scientific publications, data bases, manuals, training materials, policies and procedures or the Internet and social media. Tacit knowledge–in the form of intuition, experience, skills or reasoning–is not easy to grasp, but vital for retaining knowledge (staff rotation), strengthening cooperation between employees, improving efficiency and productivity, and strengthening innovation. The main research question of this study is: "How should the tacit knowledge of LSSs be managed to preserve and pass their legacy onto younger generations?".

3.2 Planned Research Details

Developing a comprehensive, multilevel approach that combines diverse methods and tools for tacit knowledge diffusion is paramount. This approach should be flexible, adapting to employees' varying needs and preferences across all generations. The expected outcome of this research project is a multidimensional method—potentially holistic—for supporting tacit knowledge diffusion, encapsulating elements such as:

- An environment conducive to knowledge diffusion, informed by management styles, communication methods, workspaces, and organizational structures.
- Prescribed cooperation techniques, emphasizing motivation and participatory management.
- Age diversity management, targeting the functioning of distinct employee groups.

This strategy should enhance our understanding of the needs and expectations of different generations of employees and allow for the customization of knowledge management processes accordingly.

The proposed research timeline spans one year, designated for comprehensive data collection (through surveys and interviews) both domestically and internationally. Primary research will be conducted across eight scientific institutions, split evenly between Poland and the European Union. In Poland, the investigation will involve two research institutes and two construction faculties of polytechnics. The same arrangement will apply to EU-based institutions.

For comparison, in Poland in 2022, the quality of scientific activity in the scientific discipline of civil engineering and transport was evaluated among 34 scientific units: five institutes, 16 polytechnics, and nine other higher education institutions (including AGH and ZUT). Meanwhile, 19 research institutes dealing with construction in Europe are gathered in the ENBRI—European Network of Building Research Institutes. The choice to concentrate on EU countries rather than Asian or American ones stems from their common cultural and historical contexts, which could affect tacit knowledge transfer methods. The European Union actively works to implement policy, legislation, and regulation changes, placing a notable emphasis on higher education (Europe 2020 Strategy) and fostering knowledge exchange among scientists by promoting the concept of open science (Horizon Europe).

3.3 Research Methodology

Research Methods: The doctoral project will apply the following methods:

- Reviewing and critiquing literature;
- Analyzing documents (if relevant to knowledge management or job responsibilities);
- Diagnostic surveys: using questionnaires and interviews;
- Expert panel discussions;
- Heuristic techniques like brainstorming (if applicable, based on research outcomes and changes in the situation).

The expert panel will offer valuable insights not accessible through other methods. Combining all these methods aims to give a thorough, multi-sided view of the research topic, which will help develop a more customized method to encourage the spread of tacit knowledge among LSS employees.

Sample Selection: A purposive non-random sampling strategy will ensure the most representative research results. The sample will include three study groups:

- LSS employees,
- Supervisors of LSS employees,
- Individuals working with LSS employees (both academic and administrative staff).

Key assumptions for the survey include:

- Participants will choose statements that most accurately reflect their situation;
- Most questions will be closed-ended, with open-ended questions at the end for additional information;
- LSS will be asked about their knowledge, self-assessment of their knowledge, how they use their knowledge, perceived barriers, and feelings connected to tacit knowledge;
- Leaders will be asked about their awareness of LSS employees' tacit knowledge, ways of using it, and conditions favoring the exchange and sharing of tacit knowledge.

Preliminary studies will be carried out at the Building Research Institute, while other studies will occur across all eight institutions.

Research Plan: The first year of research involves preparing research tools and conducting pilot and actual surveys. The second year focuses on the main research: running the expert panel, conducting in-depth interviews—pilot ones first, followed by the main ones, analyzing all collected data, and developing a method to promote the spread of LSS employees' tacit knowledge. The third year is set aside for implementing, testing, and validating the method and writing final conclusions and recommendations for academic institutions in the field of construction that employ LSS workers.

4 Anticipated Conclusions

The research aims to draw conclusions related to the following points:

1. Identifying key challenges and barriers in managing tacit knowledge in construction due to significant generational differences.

2. Finding the best practices and strategies for managing tacit knowledge, encouraging cooperation and knowledge sharing among different generations of researchers.
3. Developing recommendations for changes in organization structures, processes, and internal policies, to better manage tacit knowledge in construction.
4. Identifying potential benefits from consciously managing the tacit knowledge of LSS employees.
5. Exploring opportunities for using modern technologies to spread the tacit knowledge of LSS employees.

While presently focused on European Union construction institutions, the insights obtained from this study could be applied and investigated further in diverse geographical contexts, including Asian, African, or American counterparts. Similarly, the methods and tools developed through this research could extend beyond construction to other disciplines and industries. Future research could explore how tacit knowledge is managed in various fields of knowledge, such as sciences, humanities, or social sciences, and how companies across different sectors manage and promote the diffusion of tacit knowledge across generations. This broader approach could provide a comprehensive understanding of the global practices and challenges associated with tacit knowledge transfer and management.

Acknowledgements. The work is financed from the Building Research Institute's (ITB) own research fund, SN-001.

References

1. Morawski, M., Prudzienica, M.: Zarządzanie wiedzą w kreowaniu innowacji zarządczych. Wydawnictwo Uniwersytetu Ekonomicznego we Wrocławiu, Wrocław (2011). [in Polish]
2. Morawski, M.: Pracownik kluczowy w procesie dzielenia się wiedzą. Motywy, warunki, metody. Wydawnictwo Uniwersytetu Ekonomicznego we Wrocławiu, Wrocław (2017). [in Polish]
3. Sidor-Rządkowska, M.: Zarządzanie różnorodnością pokoleniową we współczesnych organizacjach. Studia i Prace WNEiZ **51**, 87–96 (2018). [in Polish]
4. Sanaei, M., Javernick-Will, A.N., Chinowsky, P.: The influence of generation on knowledge sharing connections and methods in construction and engineering organizations headquartered in the US. Constr. Manage. Econo **31**(9), 991–1004 (2013)
5. Brčić, ŽJ., Mihelič, K.K.: Knowledge sharing between different generations of employees: an example from Slovenia. Econ. Res. Ekonomska Istraživanja **28**(1), 853–867 (2015)
6. Virta, M.: Knowledge sharing between generations in an organisation - retention of the old or building the new? Lappeenranta University of Technology, Lappeenranta, Finland (2011). Available at: https://citeseerx.ist.psu.edu/viewdoc/download?doi=10.1.1.912.3975&rep=rep1&type=pdf
7. Chomątowska, B., Żarczyńska-Dobiesz, A.: Barriers of knowledge sharing by representatives of the baby boomers generation. Research on Enterprise in Modern Economy theory and practice **1**(24), 35–46 (2018)
8. Dziadek, K.: Problemy i wyzwania międzypokoleniowego transferu wiedzy we współczesnych przedsiębiorstwach. Wyniki badań. (Last accessed: 2 July 2023). [in Polish] https://zeszytyhumanitas.pl/resources/html/article/details?id=194004&language=en

9. Rui, H., Ju, H.: How does rapport impact knowledge transfer from older to younger employees? The moderating role of supportive climate. Front. Psychol. **13**, 1032143 (2022)
10. Wang, G.H., Li, J.H., Liu, H., Zaggia, C.: The association between workplace ostracism and knowledge-sharing behaviors among Chinese university teachers: The chain mediating model of job burnout and job satisfaction. Front. Psychol. **14**, 1030043 (2023)
11. Galan, N.: Knowledge loss induced by organizational member turnover: a review of empirical literature, synthesis and future research directions (Part I). Learn. Org. **30**(2), 117–136 (2023)

Alternative Binders

The Role of Polymer in Calcium Sulfoaluminate Cement-Based Materials

Ru Wang[✉]

Key Laboratory of Advanced Civil Engineering Materials of Ministry of Education, School of Materials Science and Engineering, Tongji University, Shanghai, China
ruwang@tongji.edu.cn

Abstract. In order to realize sustainable development, new types of cements were paid more attention. Calcium sulfoaluminate (CSA) cement is a kind of eco-friendly cement that has the characteristics of low carbon emission, low energy consumption, fast setting and hardening, and so on. But the main hydration product ettringite (AFt) is quite sensitive to curing conditions that makes CSA cement-based materials sensitive to temperature and ageing. Polymer plays a key role in improving the properties of CSA cement mortar. Our researches showed that styrene-butadiene copolymer (SB) could result in a big reduction of zeta potential and conductivity of the CSA cement paste, retard the very initial hydration of CSA cement but not after 3 h, and lead to the generation of more AFt and aluminium hydroxide (AH_3). With SB addition increasing, the yield stress, viscosity, thixotropy, fluidity and thus workability of CSA cement mortar were significantly improved. The mechanical strength of CSA cement mortar showed a reduction after a certain age, but when SB was added there was no reduction anymore under various curing conditions. SEM observation of the morphology accounts well for the changes in mechanical properties. The shrinkage, water capillary adsorption, and durability such as resistance to freezing and thawing cycle, carbonization and sulfate attack were also investigated. This paper reviewed the role of polymer in CSA cement-based materials taking SB as an example based on recent research work of our group.

1 Introduction

Calcium sulfoaluminate (CSA) cement, developed by China Building Materials Academy in the 1970s, is calcined with bauxite, limestone and gypsum at a relatively low temperature compared to Portland cement. The main mineral compositions of CSA cement are ye'elimite ($C_4A_3\overline{S}$), belite and anhydrite/gypsum. CSA cement is eco-friendly that has low carbon emission, low firing temperature and low energy consumption during production. At the same time, it has the characteristics of fast setting, high early strength, micro expansion or low shrinkage, and good anti-permeability. It can be applied in emergency repair, low-temperature construction, seepage control engineering, etc. [1].

Polymers are often used to modify Portland cement-based materials and good results were achieved. However, research on adding polymer to CSA cement-based materials

© The Author(s) 2025
L. Czarnecki et al. (Eds.): ICPIC 2023, 61, pp. 171–180, 2025.
https://doi.org/10.1007/978-3-031-72955-3_16

is quite limited. In order to solve the problems existed during the practical application of CSA cement, researches were carried out on the function of polymers in CSA cement-based materials. The polymers used can be divided into two types, i.e., polymer dispersions and water-soluble polymers. Cellulose ether as one of the water-soluble polymers was applied to improve the workability of CSA cement mortar, especially the water retention capacity, and some achievements were obtained [2–8]. The effect of polymer dispersions on the mechanical properties, durability, and hydration of CSA cement was analyzed in previous studies [9–13]. It is found that the addition of polymer dispersions in CSA cement mortars benefits to improve their mechanical properties. Meanwhile, it also contributes to enhancing their durability by optimizing the microstructure. However, the influence on the properties is dependent on polymer types. Generally, SB dispersion demonstrates the best for polymer-modified CSA cement mortar. Since the main hydration product AFt makes the CSA cement matrix sensitive to temperature and ageing [14–19], the effect of curing regimes on the early age and long-term performance of SB modified CSA cement mortar was also studied intensively [20–23]. This paper reviewed the effect of SB on CSA cement-based materials including the rheology behavior, hydration behavior, physical and mechanical properties based on recent researches of our group.

2 Rheology Behavior

The effect of SB dispersion on the rheology and setting behavior of CSA cement paste was investigated [13]. Amounts of 0%, 5%, 10%, 15% and 20% of SB were added to the base CSA cement paste with a constant water to cement ratio of 0.4. A series of experiments including fluidity, rheology, zeta potential, conductivity, total organic carbon (TOC), setting time and calorimetry were performed. The research showed that SB undertook negative charged carboxylic group, which could be adsorbed onto the cationic CSA cement grains as well as hydration products with positive phases, and finally resulting in a big reduction of zeta potential and conductivity of the cement paste. SB dispersion performed a good water reducing effect, which contributed to the fluidity increasing and efflux time decreasing (Fig. 1). The yield stress and thixotropy were calculated to further understand the rheology behavior of CSA cement paste with SB. The yield stress was the minimum force required for the paste to overcome before flowing. The thixotropy was represented by the hysteresis area between the up and down curves of shear stress. The results of yield stress and hysteresis area were summarized in Table 1. The decreased hysteresis area stands for a better thixotropy of SB modified cement paste, meaning that the agglomerated structure of the CSA cement paste can be gradually destroyed under the shear stress but gradually recovered once the shear stress was removed. As shown in Fig. 2, the viscosity of the control cement pastes dramatically declined at a low shear rate, which is a typical phenomenon of shear-thinning behavior. The lower viscosity was observed when SB was added into CSA cement paste, indicating that the flow resistance of cement particles decreased and finally the fluidity increased.

The zeta potential and conductivity demonstrated the electrostatic interaction between SB dispersion and CSA cement which was mainly ascribed to the adsorption of polymer particles on the cement mineral surfaces. Meanwhile, the rheological properties of cement pastes with the addition of various polymer dispersions could be highly related to the adsorption behavior of the polymers [24, 25]. The zeta potential of CSA cement particles was positive while that of SB dispersion was negative, thus SB could be easily adsorbed onto the surface of CSA cement grains via the electrostatic interaction [24–26]. The setting time of CSA cement paste was prolonged by the addition of SB. The retardation effect increased with increasing SB dosage, which agreed with fluidity, rheology, and adsorption results. The lengthened setting time could be attributed to the retardation effect of polymers on cement hydration, which will be discussed in the following.

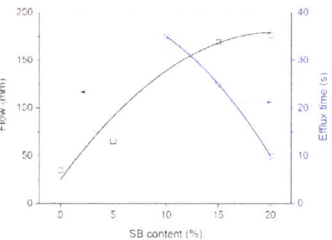

Fig. 1. Fluidity and efflux time of SB dispersion modified CSA cement paste [13]

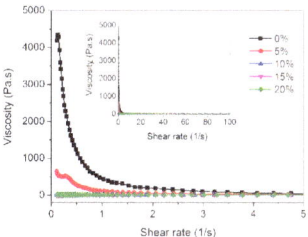

Fig. 2. Relation between viscosity and shear rate of SB modified CSA cement paste [13]

Table 1. The yield stress and hysteresis area for SB modified CSA cement paste [13]

SB content (%)	0	5	10	15	20
Yield stress (Pa)	99.81	52.55	2.58	1.17	0.78
Hysteresis area (Pa·s^{-1})	5530.19	1786.23	918.80	273.87	250.39

3 Hydration Behavior

The early hydration of CSA cement modified by SB was investigated [27]. The research showed that the addition of SB retarded the initial hydration of CSA cement before curing age of 3 h (h), at which the heat evolution was significantly decreased (Fig. 3). The results from XRD and TG analysis for those samples at 1 h showed that the formation of ettringite was strongly delayed while calcium sulfate dihydrate was formed in SB modified CSA cement paste (Fig. 4); the more amount of SB was incorporated, the less ettringite was generated. The retarded hydration of CSA cement caused by SB in the initial stage prolonged the setting time of CSA cement paste, which was further confirmed by ultrasonic measurement and calorimetry. The hydration degree of SB modified CSA cement caught up that of the control at around 3 h based on cumulative heat value from calorimetry analysis (Fig. 3 (b)). Afterwards, the hydration of SB modified CSA cement tended to be accelerated. The more SB was added, the more ettringite was generated based on the results at 6 h and thereafter. The most ettringite was formed in the paste with 20% SB. Hemicarboaluminate (Hc) was generated after 12 h in the control paste, and the addition of SB demonstrated good inhibition effect on the formation of Hc. SEM observation also confirmed above analysis that the addition of SB delayed the formation of ettringite in the initial stage while it promoted the formation and growth of ettringite crystals in the later periods.

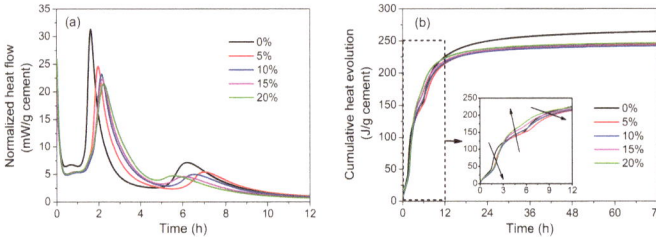

Fig. 3. Hydration heat kinetics of SB-modified CSA cement pastes: (a) heat flow and (b) cumulative heat [27].

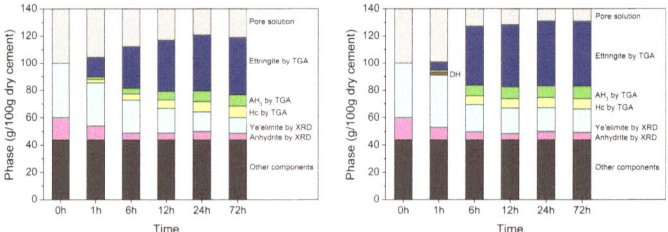

Fig. 4. Phase contents of (a) neat CSA cement paste and (b) CSA cement paste with 20% SB at various ages [27].

BSE images were widely employed to determine phase distribution and porosity in the hydrated CSA cement pastes [17, 28]. The CSA cement matrix basically consists of pores, cracks, hydration products and unhydrated cement grains in an order of black to high brightness in BSE images. It was shown in Fig. 5 (a) that the control CSA cement already had a quite dense structure after 3 d of hydration. Some unhydrated cement grains shown in the high brightness area indicated incomplete hydration. The typical hydration products exhibiting a dark grey level in BSE picture consisted mainly of ettringite and aluminum hydroxide. With addition of 10% SB (Fig. 5 (b)), less unhydrated clinker grains with low brightness background appeared after 3 d of hydration, meaning higher hydration degree in comparison with control sample. The brightness areas for unhydrated clinker grains were the least for the microstructure of CSA cement with 20% SB (Fig. 5 (c)), indicating the highest hydration degree among these three pastes. The BSE result was well agreement with XRD and TG analysis.

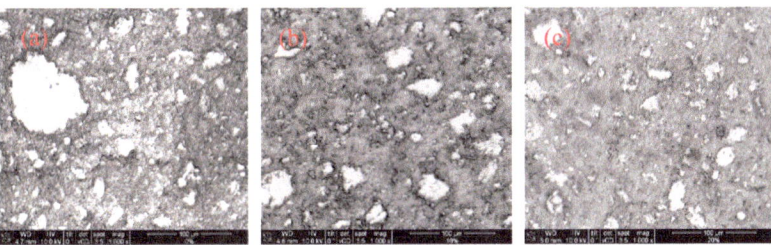

Fig. 5. BSE images of CSA cement pastes modified with (a) 0% SB, (b) 10% SB and (c) 20% SB at 3 days [27]

4   Physical and Mechanical Properties

The effect of curing regimes on the performance and microstructure of CSA cement mortar was investigated [22]. Mortars with five different ratios (0%, 5%, 10%, 15%, and 20%) of SB to CSA cement at a constant workability were prepared. Four temperatures (0 °C, 5 °C, 20 °C, and 40 °C) and three relative humidity (RH) levels varying from low (32 ± 2%) (LRH), middle (63 ± 10%) (MRH), and high (96 ± 3%) (HRH) were considered for mortar specimens curing.

The research found that CSA cement mortar cured at high temperature and HRH had the highest flexural strength at 1 day. However, the temperature showed a limited effect on the strength development of CSA cement mortar in late stages. The addition of 5% SB decreased the flexural strength. The highest flexural strength was achieved when the dosage of SB was 20%. Curing under high temperature and HRH improved the flexural strength, and this effect seemed to be much more significant for CSA cement mortar with 15% and 20% SB addition.

The tensile bond strength increased with higher SB addition; it reached the highest value when the SB content was 20%. SB content and curing temperature were the main factors influencing the tensile bond strength, while the influence from RH was relatively weak. Figure 6 showed the effect of curing conditions on the flexural and tensile bond strength of SB modified CSA cement mortar within 28 days.

SB reduced the compressive strength, however, the compressive strength tended to increased slightly with increasing SB amount. Curing temperatures of 20 °C and 40 °C brought quicker compressive strength development of the control mortar, while curing temperatures of 0 °C and 5 °C showed delayed strength. This phenomenon was more evident with HRH curing of less than 7 d. Generally, the compressive strength was much more influenced by curing temperature compared to RH. Greater strength was obtained when the SB modified mortar was cured under high temperature and HRH.

SB in CSA cement mortar led to a decrease in water capillary adsorption, but it was not true for 5% SB addition. HRH decreased the water capillary adsorption of SB modified CSA cement mortar, while high temperature tended to increase it, especially when the SB dosage was between 10% and 20%.

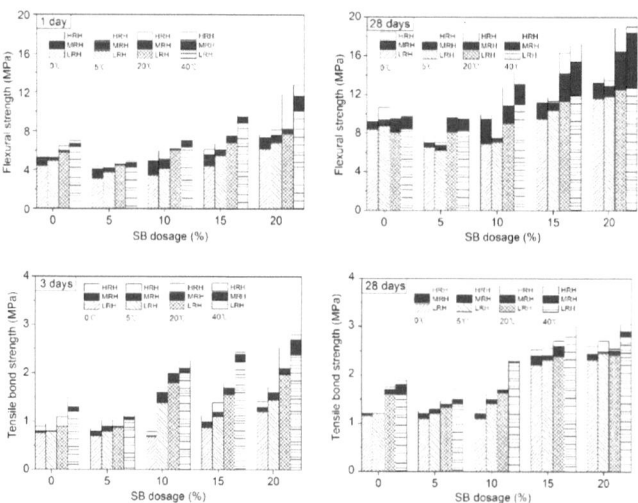

Fig. 6. Effect of curing conditions on the flexural and tensile bond strength of SB modified CSA cement mortar within 28 days [22]

The addition of SB generated continuous film between cement hydrates (Fig. 7), which was conducive to reducing the porosity of CSA cement mortar. Curing with high temperature and high RH was helpful for the formation of ettringite. The bigger size ettringite and polymer film were intertwined to make the mortar stronger. Thus, the properties of CSA cement mortar including flexural strength, tensile bond strength, and water capillary adsorption were well improved.

The long-term change in the physical and mechanical properties of SB dispersion modified CSA cement mortar as it aged from 28 to 360 days, and cured at different temperatures and relative humidities was also studied [23]. The results showed that the mechanical properties of control CSA cement mortar, including its flexural, compressive, and tensile bond strength, showed a reduction after a certain age, but its water capillary absorption was hardly affected by age. When SB was added, there was no reduction in mechanical strength anymore. The amount of SB added did matter. Addition of 5% SB had a negative effect on most properties, except for tensile bond strength. However, the properties of SB modified mortar were enhanced significantly as the amount of SB was increased from 5% to 20%. Temperature change had different effects on the properties of control mortar and SB modified mortar. High temperature was beneficial to early flexural and compressive strength development of control mortar, but caused serious strength reduction at later ages. High temperature enhanced the development of tensile

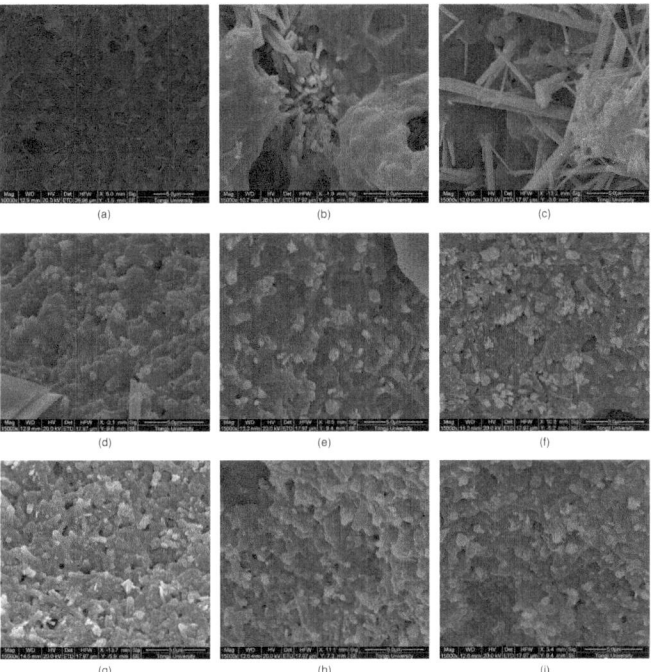

Fig. 7. Morphology of 20% SB modified CSA cement mortar cured at various temperatures and relative humidity levels for 28 days (the slices were treated with 5% HCl solution for 5 min): (a) 0 °C and LRH; (b) 0 °C and MRH; (c) 0 °C and HRH; (d) 20 °C and LRH; (e) 20 °C and MRH; (f) 20 °C and HRH; (g) 40 °C and LRH; (h) 40 °C and MRH; and (i) 40 °C and HRH [22].

bond strength of control mortar. Whereas, increasing temperature enhanced properties of SB modified mortar, including flexural, compressive, and tensile bond strength. Higher relative humidity improved all measured properties of all mortars. Figure 8 showed the effect of SB content (m_p/m_c) on the flexural and tensile bond strength development of CSA cement mortars cured at MRH and different temperatures. The shrinkage rate of CSA cement mortar modified with SB within 360 days under different curing conditions was investigated and it was found that the shrinkage rate of CSA cement mortar decreased significantly when the SB content was more than 10% in all curing conditions [21]. CSA cement mortar demonstrates well resistance to freeze-thaw cycle, carbonization and sulfate attack, and SB helped to further enhance these resistances [10].

5 Summary

The function of SB in CSA cement-based material was summarized in the paper. SB took effect on multi-performance of CSA cement. It affected rheology property, including increasing the fluidity, decreasing the viscosity, improving the thixotropy and thus improving the workability significantly. It affected the hydration behavior especially the early hydration, including retarding the very initial hydration of CSA cement, e.g., at

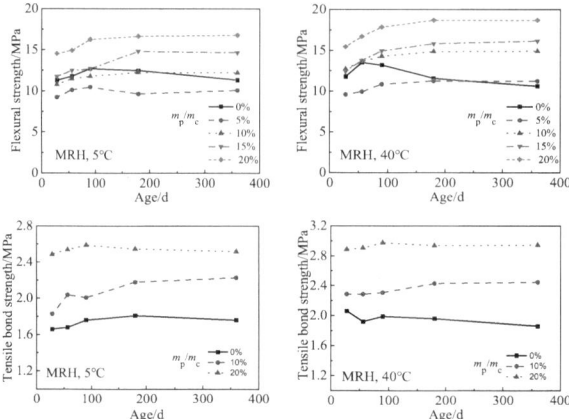

Fig. 8. Effect of SB content (m_p/m_c) on the flexural and tensile bond strength development of CSA cement mortars cured at MRH and different temperatures [23].

1 h, the formation of ettringite was strongly delayed in SB modified CSA cement paste. The retarded hydration of CSA cement caused by SB in the initial stage prolonged the setting time. The hydration degree of SB modified CSA cement caught up that of the control at around 3 h. Afterwards, the hydration of SB modified CSA cement tended to be accelerated. The more SB was added, the more ettringite was generated at 6 h and thereafter. It affected the physical and mechanical properties and durability, including increasing the flexural and tensile bond strength but not compressive strength, improving the resistance to water, carbonization, sulfate attack and freeze-thaw cycle, and especially it could inhibit long-term strength reduction. SB was helpful to improve microstructure by forming continuous films in CSA cement mortar, which finally contributed to CSA cement mortar superior properties. The function and mechanism of polymer in CSA cement-based materials is an ongoing topic.

Acknowledgement. The authors acknowledge the financial support by the National Natural Science Foundation of China (Grant No. 51872203 and 51572196) and the Top Discipline Plan of Shanghai Universities-Class I (2022-3-YB-17).

References

1. Wang, Y., Su, M., Zhang, L.: Sulphoaluminate cement. Beijing Industry University Press, Beijing (1999)
2. Ou, Z., Mao, T., Shen, Y., Liu, G.: Influence of cellulose ethers on hydration heat of different cements and single mines. Bull. Chin. Ceram. Soc. **35**(5), 1606 (2016)
3. Zhang, G., He, R., Lu, X.: Early hydration of calcium sulfoaluminate cement in the presence of hydroxyethyl methyl cellulose. J. Therm. Anal. Calorim.Calorim. **134**(3), 1429–1438 (2018)
4. Shi, C., Zou, X., Wang, P.: Influences of ethylene-vinyl acetate and methylcellulose on the properties of calcium sulfoaluminate cement. Constr. Build. Mater. **193**, 474–480 (2018)
5. Li, J., Wang, R., Li, L.: Influence of cellulose ethers structure on mechanical strength of calcium sulphoaluminate cement mortar. Constr. Build. Mater. **303**, 124514 (2021)

6. Wan, Q., Wang, Z., Huang, T., Wang, R.: Water retention mechanism of cellulose ethers in calcium sulfoaluminate cement-based materials. Constr. Build. Mater. **301**, 124118 (2021)

7. Zhang, S., Wang, R., Xu, L.: Properties of calcium sulfoaluminate cement mortar modified by hydroxyethyl methyl celluloses with different degrees of substitution. Molecules **26**(8), 2136 (2021)

8. Li, J., Wang, R., Xu, Y.: Influence of cellulose ethers chemistry and substitution degree on the setting and early-stage hydration of calcium sulphoaluminate cement. Constr. Build. Mater. **344**, 128266 (2022)

9. Brien, J.: Development of cementitious materials for adhesion type applications comprising calcium sulfoaluminate (CSA) cement and latex polymer. Theses and dissertations (2014)

10. Li, L., Wang, R., Lu, Q.: Influence of polymer latex on the setting time, mechanical properties and durability of calcium sulfoaluminate cement mortar. Constr. Build. Mater. **169**(30), 911–922 (2018)

11. Li, L., Peng, Y., Wang, R., Zhang, S.: The effect of polymer dispersions on the early hydration of calcium sulfoaluminate cement. J. Therm. Anal. Calorim.Calorim. **139**(1), 319–331 (2020)

12. Liu, X., Wang, R.: Hydration of styrene-acrylic copolymer modified calcium sulphoaluminate clinker with anhydrite. J. Chin. Ceram. Soc. **50**(2), 354–363 (2022)

13. Wang, R., Li, L.: Experimental study on the rheology and setting behavior of calcium sulfoaluminate cement paste modified with styrene-butadiene copolymer dispersion. J. Mater. Civ. Eng. **34**(4), 04022015 (2022)

14. Zhang, L., Glasser, F.: Hydration of calcium sulfoaluminate cement at less than 24 h. Adv. Cem. Res.Cem. Res. **14**(4), 141–155 (2002)

15. Liao, Y., Wei, X., Li, G.: Early hydration of calcium sulfoaluminate cement through electrical resistivity measurement and microstructure investigations. Constr. Build. Mater. **25**(4), 572–1579 (2011)

16. Berger, S., Coumes, C., Bescop, P.: Influence of a thermal cycle at early age on the hydration of calcium sulphoaluminate cements with variable gypsum contents. Cem. Concr. Res. **41**(2), 149–160 (2011)

17. Wang, P., Li, N., Xu, L.: Hydration evolution and compressive strength of calcium sulphoaluminate cement constantly cured over the temperature range of 0 to 80 °C. Cem. Concr. Res. **100**, 203–213 (2017)

18. Li, N., Xu, L., Wang, R., Li, L., Wang, P.: Experimental study of calcium sulfoaluminate cement-based self-leveling compound exposed to various temperatures and moisture conditions: Hydration mechanism and mortar properties. Cem. Concr. Res. **108**, 103–115 (2018)

19. Li, L., Wang, R., Zhang, S.: Effect of curing temperature and relative humidity on the hydrates and porosity of calcium sulfoaluminate cement. Constr. Build. Mater. **213**, 627–636 (2019)

20. Wang, R., Xu, Y.: Influence of curing temperature on physical and mechanical properties of styrene-butadiene rubber latex/sulphoaluminate cement mortar. J. Chin. Ceram. Soc. **45**(2), 227–234 (2017)

21. Wang, R., Zhang, T.: Dry shrinkage rate of styrene-butadiene copolymer dispersion/calcium sulphoaluminate cement mortar under different curing temperature and humidity. J. Build. Mater. **21**(5), 768–774 (2018)

22. Wang, R., Li, L., Xu, Y.: Influence of curing regimes on the mechanical properties, water capillary adsorption, and microstructure of CSA cement mortar modified with styrene-butadiene copolymer dispersion. J. Mater. Civ. Eng. **31**(1), 04018344 (2019)

23. Wang, R., Fan, Y., Wang, Z., Huang, T., Zhang, T.: Performance development of styrene-butadiene copolymer-modified calcium sulfoaluminate cement mortar under different curing conditions. J. Zhejiang Univ. SCIENCE A (Appl. Phys. Eng.) **22**(12), 1005–1026 (2021)

24. Lu, Z., Kong, X., Zhang, C., Xing, F., Zhang, Y.: Effect of colloidal polymers with different surface properties on the rheological property of fresh cement pastes. Colloids Surf. A **520**, 154–165 (2017)
25. Lu, Z., et al.: Influences of styrene-acrylate latexes on cement hydration in oil well cement system at different temperatures. Colloids Surf. A **507**, 46–57 (2016)
26. Guo, Y., et al.: Effect of polyacrylic acid emulsion on fluidity of cement paste. Colloids Surf. A **535**, 139–148 (2017)
27. Li, L., Wang, R.: Early hydration of CSA cement modified with styrene–butadiene copolymer dispersion. Adv. Cem. Res.Cem. Res. **33**(1), 14–27 (2021)
28. Chen, I., Hargis, C., Juenger, M.: Understanding expansion in calcium sulfoaluminate-belite cements. Cem. Concr. Res. **42**(1), 51–60 (2012)

Study and Characterization of Gypsum Mortars Made with Phenolic Melamine Polymer Wastes from the Decorative Paper Industry

Isabel Santamaría-Vicario, Belén Zurro-García, Ana María Paredes-Núñez, Carlos Junco Petrement, and Ángel Rodríguez Saiz[✉]

Higher Polytechnic School, University of Burgos, C/Villadiego s/n, 09001 Burgos, Spain
arsaizmc@ubu.es

Abstract. The proposed research studies the properties of gypsum mortars made with polymeric waste from the manufacturing process of high pressure laminated (HPL) thermosetting decorative panels, composed of cellulose paper layers impregnated with phenolic resins and melamine resins. The waste generated in the cutting, profiling and milling of the decorative panels is discarded and sent to landfill without a defined use. This research aims to contribute to the Circular Economy of Waste by recovering it as a raw material. Gypsum mortars are designed by adding different amounts of melamine waste. Subsequently, the properties of the mortars are studied following the technical prescriptions established in the European regulations. Firstly, the properties of the mortars in their fresh state are studied, such as the water/gypsum ratio, consistency, apparent density of the fresh mortar and setting time. Then, the properties of the hardened mortars are determined, such as the apparent density of the hardened mortar, mechanical resistance to bending and compression, adhesion, Shore C surface hardness and capillary absorption. Based on the results obtained in the tests, the viability of this type of waste is assessed for its use as a mineral aggregate to replace traditional aggregates, in order to obtain commercial gypsum mortars for use in masonry work, cladding, walls, or as a raw material for the manufacture of prefabricated materials. The results obtained show that the limit of gypsum substitution by melamine waste could be a maximum of 25%. New mortar formulations with lower substitutions would provide significant advantages in this type of ecological materials, in accordance with the technical requirements established by the applicable European regulations.

Keywords: Gypsum mortar · Melamine waste · Masonry mortar · Circular economy

1 Introduction

In recent decades, the world's population has been increasing [1], a circumstance that will force an increase in industrial production to satisfy people's needs [2]. This situation can have an environmental impact due to the need to use larger quantities of raw materials in industry, and the inevitable production of the waste generated [3].

© The Author(s) 2025
L. Czarnecki et al. (Eds.): ICPIC 2023, 61, pp. 181–189, 2025.
https://doi.org/10.1007/978-3-031-72955-3_17

Faced with this situation, it is necessary to seek a reasonable balance between meeting the needs of the population and protecting the environment, in accordance with the Sustainable Development Goals (SDGs) [4]. Specifically, SDG 12 provides measures for companies to consume less energy reduce polluting emissions and make efficient use of natural resources [5]. To this end, they should focus on renewable energies and the circular economy of waste by incorporating more efficient technologies into their production systems, and the reuse and recovery of waste generated in manufacturing processes [5–8].

Construction is characterized as one of the economic sectors that uses the largest amount of raw materials of natural origin, many of which are non-renewable. It is also one of the industries that consumes the most primary energy for its transformation, and is responsible for the production of significant quantities of waste that are deposited in landfills without a defined use [9, 10]. For all these reasons, it is necessary to look for sustainable and environmentally friendly alternatives, such as, for example, recovering the waste generated in the industry for the design of new construction materials [11–15]. In accordance with this objective, this document develops an investigation to assess the use of waste generated in the manufacturing process of decorative panels of cellulose paper impregnated with phenolic resins and melamine resins. Through experimental tests, we want to assess their suitability to be used as aggregates in the manufacture of gypsum mortars.

2   General Specification

The proposed research aims to valorize an industrial waste generated in the manufacturing process of decorative melamine panels, to use it as a raw material in the design of gypsum mortars for use in masonry.

2.1   Raw Materials

The following materials were used for the research:

Gypsum Type A1 composed of Hemi-hydrated Calcium Sulphate and Anhydrite, according to the specifications in EN 13279-1:2009 [16]. This type of gypsum has been provided by the company PLACO (Saint-Gobain Group). Its main characteristics are shown in Table 1.

Melamine waste from the manufacturing process of decorative laminates for the coating of particleboard panels, composed of layers of cellulose paper impregnated with synthetic polymers of phenolic resins and melamine resins. The waste has been provided by the company Tacon Decor S.L., a manufacturer of laminates and decorative coatings, located in Burgos, Spain. The offcuts were crushed in a Retsch Model SM100 mill to obtain particles smaller than 2.00 mm, in order to facilitate their incorporation into the plaster matrix, with a bulk density of 0.724 kg/m^3. (Figs. 1 and 2).

Water from the urban water supply of the city of Burgos (Spain), managed by Sociedad Municipal Aguas de Burgos, S.A.U.

Table 1. Characteristics of the Type A1 plaster.

Characteristics	Reference
Particle size	0–0.2 mm
Bulk Density	0.810 kg/m^3
CaSO$_4$ content	>92.0%
Flexural Strength	>3.5 N/mm^2
Compressive Strength	>3.0 N/mm^2
Water/Gypsum Ratio	0.75 L/kg

Fig. 1. Melamine waste. General appearance by sizes (Left); Optical microscope visualization (Right)

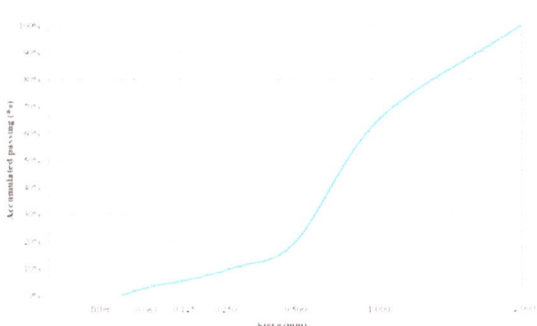

Fig. 2. Particle size distribution of ground melamine waste

2.2 Mortars Desing

The gypsum mortars made with melamine waste were designed by replacing, in volume, part of the plaster with crushed melamine waste, in order to check its effect on the properties of the dosed mixtures. Based on the reference mortar or standard mortar, made with A1 plaster, two mixtures were designed. The first is composed of 75.0% by volume of plaster and 25.0% by volume of melamine waste (75A1:25M) and the second of 50.0% plaster together with 50.0% melamine (50A1:50M). The dosage of the mixtures is shown in Table 2.

Table 2. Dosage of the mixtures

Sample	A1 (gr)	Melamine (gr)	Water (gr)	w/g
A1	1000	-	500	0.500
75A1:25M	750	223.4	404	0.415
50A1:25M	500	446.9	360	0.380

At each dosage, the necessary water was added to achieve a run-out diameter of (160 ± 5) mm on the Flow Table Method, as specified in EN 13279-2:2014 [17]. For this purpose, the gypsum plaster and the melamine waste were first mixed dry and then the necessary amount of water was added for hydration of the gypsum plaster.

3 Experimental, Results and Discussion

The gypsum mortars made with melamine waste were analyzed both in the Fresh and Hardened State, according to the requirements of the EN standard (European Committee for Standardization - CEN). The following characterization tests were carried out:

3.1 Properties of the Mortars in the Fresh State

The density in the fresh state is calculated by difference of weights of a container of known volume, according to the standard UNE 102042:2014 [18]. For each of the designed mixes, the setting time is determined by the Vicat Cone Method [17], using a standardized probe of Ø 10 mm and 100 g weight, making several successive penetrations until a depth of about (22 ± 2) mm is reached (Fig. 3). The results are shown in Table 3.

Table 3. Properties of mortars in Fresh State.

Sample	Water/binder	Density in the fresh state (kg/m^3)	Initial setting time (h:min:s)
A1	0.500	1714	0:07:23
75A1:25M	0.415	1620	0:07:30
50A1:50M	0.380	1450	0:07:30

Once melamine is added to the A1 gypsum, the mixtures obtained should be considered as a Type B1 gypsum, according to EN 13279-1:2009 [16]. For this reason, the water/gypsum ratio of mortars containing melamine is expressed as the ratio between the amount of water added and the weight of gypsum A1 plus melamine. Although, Table 3 shows that the w/g ratio decreases in mortars with melamine, Table 2 shows the amount of water dosed in each mix to obtain the design plastic consistency; part of this water is used to hydrate the gypsum and the rest to facilitate the sliding of the melamine aggregate particles on the gypsum matrix. On the other hand, due to the nature of the melamine waste, the mortars have a lower density. In terms of setting times, it should be noted that mortars with melamine waste behave similarly to standard A1 mortar.

Fig. 3. Flow table method (Left). Vicat cone method (Right)

3.2 Properties of the Mortars in the Hardened State

The flexural and compressive strength of the mortars was determined by the procedure of the standard EN 13279-2:2014 [17]. Table 4 shows the results obtained.

Table 4. Properties of mortars in Hardened State.

Sample	Flexural strength (N/mm^2)	Compressive strength (N/mm^2)	Adherence (N/mm^2)	Shore C hardness
A1	8.94	24.55	0.59	92
75A1:25M	4.65	12.35	0.43	90
50A1:50M	2.35	4.76	0.21	87

The results show that the mechanical flexural strength (Fig. 4) is reduced in the mixes incorporating the melamine waste with respect to the A1 reference mortar. As can be seen, each 25.0% of melamine added produces a reduction of the flexural strength by half (50.0%), with respect to the A1 standard mortar.

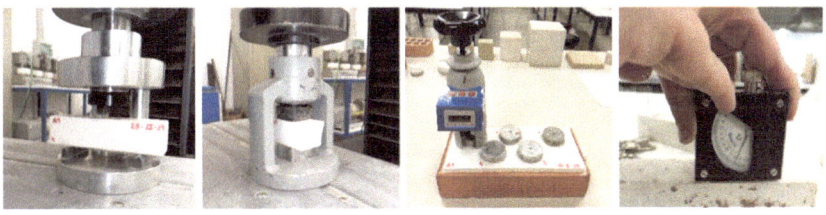

Fig. 4. Flexural strength (1); Compressive strength (2); Adhesion (3); Shore C (4)

The mechanical compressive strength follows the same behavior (Table 4). When 25.0% melamine waste is added, the strength is also halved. However, when 50.0% is added, the reduction in mechanical strength is even greater, reaching 80.0% of that of the standard A1 gypsum mortar. If we consider the mechanical compressive strength as an indication of the quality of the material, the mixture (75A1:25M) allows a strength of 12.35 N/mm^2 to be achieved, which is sufficient for many of the masonry works in which gypsum mortars are used. Finally, by applying the test procedure of EN 13279-2:2014 [17], the Shore C surface hardness was determined (Fig. 4). The results of the melamine mortars, although somewhat lower, are very similar to those of the standard A1 mortar (Table 4).

The adhesion of the mortars on a ceramic substrate also decreases as the content of melamine waste in the mixtures increases (Table 4). Replacing 25.0% of the gypsum with melamine reduces the adhesion by 27.0%, while in the case of mortar (50A1:50M) it reaches 64.5%. The type of fracture in the mortars is shown in Fig. 5. As can be seen, the five fractures in the standard A1 mortar are of the Adhesion Fracture (Type A) as specified in the standards. In the mortars made with melamine, four of the breaks are of the Adhesion Type and one of them is of the Cohesion Fracture (Type B), since part of the material remains adhered to the ceramic substrate.

Fig. 5. Adhesion test: A1 (Left); 75A1:25M (Centre); 50A1:50M (Right)

The Coefficient of Water Absorption by Capillarity (Fig. 6), determined by standard EN 13279-2:2014 [17], is very similar in the three samples analyzed, being slightly higher in mortar A1 (Table 5).

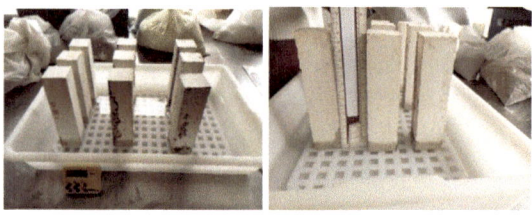

Fig. 6. Water absorption by capillarity (Left); Height of water (Right)

However, the height that the water reaches after 10 min of testing is greater in the mortar (50A1:50M), probably because the melamine favors a more extensive capillary network in the mortar. Similarly, the Total Water Absorption is similar for A1 and (75A1:25M) mortars and higher for (50A1:50M) mortars.

Table 5. Water absorption by capillarity

Sample	Absorption Coefficient $(kg/cm^2 min^{0.5})$	Height of water (cm)	Total Absorption (%)	
			(24 h)	(96 h)
A1	3.66	3.43	22.5	23.5
75A1:25M	3.17	3.31	22.9	23.6
50A1:50M	3.24	4.23	28.3	29.0

4   Conclusions

The research carried out shows a preliminary study to assess the possible use of melamine waste generated in the decorative paper manufacturing industry for particleboard panels. Mixtures of gypsum with melamine waste were designed and standardized test specimens were manufactured for testing, applying the procedures established by European regulations.

The tests carried out on mortars made with a 50.0% substitution of gypsum by melamine show a behavior that does not meet the regulatory requirements. However, mixtures with a 25.0% substitution show positive results, so that this substitution of gypsum for melamine could be considered as a maximum limit for future research with this type of waste. With this criterion, research could be reoriented by carrying out characterization studies with substitutions of less than 25.0% in order to establish the reference area in which it is possible to recover this waste with appropriate results to be used as a raw material for the manufacture of gypsum mortar for masonry.

Acknowledgements. We would like to thank the company Tacon Decor S.A. Burgos (Spain) for their collaboration in carrying out this research and the University of Burgos (Spain) for contributing to its financing.

References

1. Crist, E., Mora, C., Engelman, R.: The interaction of human population, food production, and biodiversity protection. Science **356**(6335), 260–264 (2017). https://doi.org/10.1126/science.aal2011
2. World Population Prospects.: Sumery of Results. Department of Economic and Social Affairs. United Nations, New York (2022). https://www.un.org/development/desa/pd/content/World-Population-Prospects-2022
3. Henderson, K., Loreau, M.A.: Model of sustainable development goals: challenges and opportunities in promoting human well-being and environmental sustainability. Ecol. Model. **475**, 110164 (2023). https://doi.org/10.1016/j.ecolmodel.2022.110164
4. Dietz, T., Rosa, E. A., York, R.: Environmentally efficient well-being: rethinking sustainability as the relationship between human well-being and environmental impacts. Hum. Ecol. Rev. 114–123 (2009). https://www.jstor.org/stable/24707742
5. Suchek, N., Fernandes, C.I., Kraus, S., Filser, M., Sjögrén, H.: Innovation and the circular economy: a systematic literature review. Bus. Strateg. Environ. **30**(8), 3686–3702 (2021). https://doi.org/10.1002/bse.2834

6. Sharma, P., et al.: Trends in mitigation of industrial waste: global health hazards, environmental implications and waste derived economy for environmental sustainability. Sci. Total Environ. **811**, 152357 (2022). https://doi.org/10.1016/j.scitotenv.2021.152357
7. Yang, M., et al.: Circular economy strategies for combating climate change and other environmental issues. Environ. Chem. Lett. **21**(1), 55–80 (2023). https://doi.org/10.1007/s10311-022-01499-6
8. Chioatto, E., Sospiro, P.: Transition from waste management to circular economy: the European Union roadmap. Environ. Dev. Sustain. **25**(1), 249–276 (2023). https://doi.org/10.1007/s10668-021-02050-3
9. Hu, M., Milner, D.: Visualizing the research of embodied energy and environmental impact research in the building and construction field: a bibliometric analysis. Dev. Built Environ. **3**, 100010 (2020). https://doi.org/10.1016/j.dibe.2020.100010
10. Bilal, M., Khan, K.I.A., Thaheem, M.J., Nasir, A.R.: Current state and barriers to the circular economy in the building sector: towards a mitigation framework. J. Clean. Prod. **276**, 123250 (2020). https://doi.org/10.1016/j.jclepro.2020.123250
11. Spassova, A.: Construction industry and sustainability. In: Interdisciplinary Approaches to Climate Change for Sustainable Growth (pp. 261–287). Springer International Publishing, Cham (2022). https://doi.org/10.1007/978-3-030-87564-0_15
12. Muñoz Ruiperez, C., Rodríguez Saiz, Á., Junco Petrement, C., Fiol Oliván, F., Calderón Carpintero, V.: Durability of lightweight concrete made concurrently with waste aggregates and expanded clay. Struct. Concr. **19**(5), 1309–1317 (2018). https://doi.org/10.1002/suco.201700209
13. Santamaría Vicario, I., Alonso Díez, Á., Horgnies, M., Rodríguez Saiz, Á.: Properties of gypsum mortars dosed with LFS for use in the design of prefabricated blocks. In: New Technologies in Building and Construction. Lecture Notes in Civil Engineering, vol. 258. Springer, Singapore (2022). https://doi.org/10.1007/978-981-19-1894-0_15
14. Revilla Cuesta, V., Evangelista, L., de Brito, J., Skaf Revenga, M., Manso Villalaín, J.M.: Shrinkage prediction of recycled aggregate structural concrete with alternative binders through partial correction coefficients. Cement Concr. Compos. **129**, 104506 (2022). https://doi.org/10.1016/j.cemconcomp.2022.104506
15. Alameda Cuenca-Romero, L., Arroyo Sanz, R., Alonso Díez, Á., Gutiérrez González, S., Calderón Carpintero, V.: Characterization properties and fire behaviour of cement blocks with recycled polyurethane roof wastes. J. Build. Eng. **50**, 104075 (2022). https://doi.org/10.1016/j.jobe.2022.104075
16. EN 13279-1:2009 Gypsum binders and gypsum plasters—Part 1: Definitions and requirements. European Committee for Standardization Brussels, Belgium
17. EN 13279-2:2014 Gypsum binders and gypsum plasters—Part 2: Test methods. European Committee for Standardization Brussels, Belgium
18. UNE 102042:2014 Gypsum plasters. Other test methods. Asociación Española de Normalización y Certificación, Madrid, España

Valorisation of Polyurethane Waste in Gypsum Mortar to Improve Its Circular Economy

Alba Rodrigo-Bravo[(✉)], Sara Gutiérrez-González, Verónica Calderón Carpintero, and Lourdes Alameda Cuenca-Romero

Department of Construction, University of Burgos, Burgos, Spain
arbravo@ubu.es

Abstract. The study of the behaviour of polymeric waste in building materials is of great interest. Both sectors are important and have a significant impact on the environment, so more sustainable alternatives that drive the circular economy are needed. A multi-criteria assessment on gypsum mortar with polyurethane waste from eight different industries has been carried out to analyse in depth the influence of this polymer on building materials. The methodology used studies the physico-mechanical properties of the mixtures. A "cradle to gate" Life Cycle Assessment at laboratory level is also included to evaluate and compare their environmental performance. The dosage evaluated is the one that recovers the greatest amount of waste possible while maintaining its performance above the values established in the regulation. The results of the study show that the incorporation of polyurethane waste in gypsum mortars decreases their bulk density by 2–22% in the fresh state and 7–24% in the hardened state, while flexural and compressive strengths are reduced by about one third. The environmental impact assessment of the innovative materials shows that some samples are 15–22% more environmentally friendly than the conventional one. It is concluded that the incorporation of polyurethane waste in gypsum mortar products is a viable alternative to landfill disposal or incineration, given its good technical and environmental performance.

Keywords: Polyurethane Waste · Gypsum Mortar · Life Cycle Assessment

1 Introduction

Nowadays, plastic is a material that is highly present in our economy and daily life, so that the use of traditional materials such as wood, metal, glass, stone and leather, among others, has been displaced [1]. However, numerous studies have demonstrated the serious negative effects of polymeric waste on the environment and on the health of living beings [2]. Therefore, in 2018, the European Union (EU) approved a strategic action plan with the aim of eliminating plastic pollution and accelerating the circularity of its economy, while improving the efficiency of these resources [3].

Global plastic production has reached 391 million tonnes in 2021 [4]. Polyurethane represents 5,5% of this total production, it is the 6th most demanded polymer and, within this type of plastic, polyurethane foams represent 67% of the total consumed [4, 5]. The

© The Author(s) 2025
L. Czarnecki et al. (Eds.): ICPIC 2023, 61, pp. 190–197, 2025.
https://doi.org/10.1007/978-3-031-72955-3_18

sectors with the highest consumption of this thermoset polymer are construction and automotive [6].

The volume of plastic waste generated each year stands at 250 million tonnes of which only 20% is recovered, the remaining 80% is sent to incineration, landfill, leakage or improper disposal [7]. Existing data on the management of polyurethane foam waste show a similar trend, with only 23.8% being recycled after reaching its end-of-life [8].

Recent studies have demonstrated the feasibility of building materials including polymeric wastes such as gypsum mortars that include recycled polypropylene [9], polycarbonate waste [10], recycled polystyrene [11] and polyethylene waste [12], among others.

This research line develops gypsum mortars that include recycled polyurethane waste in their composition. This helps, at the same, to reduce the amount of natural resources used in the manufacture of this construction material and to extend the life of this plastic waste. In order to improve its circularity, the waste is subjected to a mechanical shredding process. Bulk density, mechanical properties of flexural and compressive strength and environmental impact of the different samples, with an in-depth study of reusing the PUW, are analysed in this paper.

2   Materials and Methods

The raw materials, gypsum mortar mixtures and methodology used in this research are described in this section.

2.1   Raw Materials and Mixtures

The polymer-gypsum mortars developed are made of:

- Gypsum binder (type A), as per UNE-EN 13279–1 [13], with a density of 879 kg/m^3.
- Crushed polyurethane from industrial waste with a density between 40 and 142 kg/m^3, depending on the type. The different PUW are type I, type B, type P, type A, type AT, type SG, type BU and type ES.
- Water from the municipal network.
- Glass fibre with a linear density of 2400 tex and fibre diameter of 24 µm.
- Fluidifying additive.

Polymer-gypsum mortars have been designed with eight different samples of PUW (Table 1). The incorporation of polyurethane waste implies a reduction in the amount of resources (gypsum and water) used in the mix, however, it also requires a pre-crushing process.

The ideal dosage has been set up by previous research, taking into consideration the one with the highest amount of polymer valorized while maintaining the technical performance above the established standards [14]. The ratio consists of 1.5 parts of polyurethane waste (PUW) to 1 part of gypsum (by volume). The nomenclature used, G+1.5I2 as an example, refers to one part of gypsum (G) plus one and a half parts of polyurethane residue type I (by volume) with particle size less than 2 mm (1.5I2). In

Table 1. Gypsum mortar mixtures by weight (g).

Sample	Gypsum (g)	Polyurethane waste (g)	Water (g)	Glass Fibre (g)	Fluidifying additive (g)
G	1000.00	–	950.00	10.00	5.00
G+1.5I2	1000.00	97.50	1042.70	10.00	5.00
G+1.5B2	1000.00	83.00	1029.00	10.00	5.00
G+1.5P2	1000.00	271.00	1207.00	10.00	5.00
G+1.5A2	1000.00	236.00	1174.00	10.00	5.00
G+1.5AT4	1000.00	197.00	1017.00	–	5.00
G+1.5SG2	1000.00	167.50	1109.00	10.00	5.00
G+1.5BU2	1000.00	83.10	1029.00	10.00	5.00
G+1.5ES2	1000.00	1033.00	1931.40	10.00	5.00

addition, a reference gypsum mortar has been included in order to assess the performance of the new products under study.

The water-conglomerate ratio (w/c ratio) is set at 0.95, which guarantees a good workability and a suitable setting time, with the exception of the G+1.5AT4 mix, which has a ratio of 0.85 and no glass fibre, due to the specific characteristics of this type of PUW. The amount of glass fibre and additive is 1% and 0.5%, respectively, of the gypsum weight.

2.2 Characterisation

Bulk density in fresh state is determined according to standard UNE 102042 [15]. Bulk density in hardened state is obtained by dividing the mass of the samples by the known volume.

Mechanical properties are evaluated following the indications in the UNE-EN 13279–2 [16]. First, flexural strength is determined by applying a specific load in the center of the samples until failure. Compression test is then carried out on each of the specimen halves.

The environmental performance of the samples is examined using the methodology of Life Cycle Assessment (LCA). Principles, framework, requirements and guidelines are set up in the standards ISO 14040 and ISO 14044 [17, 18].

The system boundaries are "cradle to gate", which includes the raw materials acquisition phase, the transport phase and the construction phase (Fig. 1). All other stages have been excluded as the products are at a research and development stage.

The functional unit is 1 m^2 of 15 mm thick gypsum mortar coating. The inventory data are obtained from the real laboratory practice with the help of the Ecoinvent database. The LCA analysis is carried out with the software SimaPro. EN 15804 + A2 Method is the calculation methodology used to get the results. The impact categories assessed are climate change (CC) (kg CO_2 eq), photochemical ozone formation (POF) (kg NMVOC eq), particulate matter (PM) (disease inc.), acidification (A) (mol H+ eq), freshwater

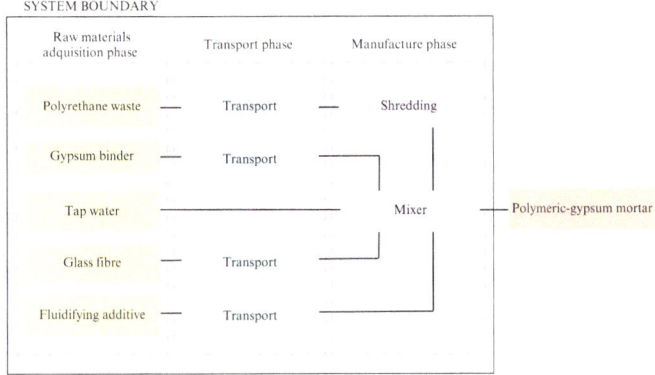

Fig. 1. System boundary for the LCA of polymeric-gypsum mortars.

ecotoxicity (EF) (CTUe), fossils resource use (RUF) (MJ) and minerals and metals resource use (RUMM) (kg Sb eq).

3   Results and Discussion

3.1   Bulk Density in Fresh and Hardened State

The determination of the bulk density in fresh and hardened state of gypsum mortars is included. The results given by tests are displayed in Fig. 2.

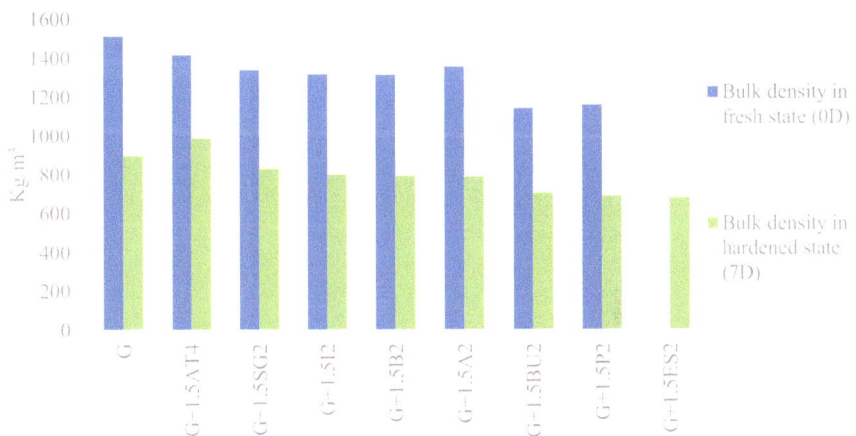

Fig. 2. Bulk density in fresh and hardened state of polymeric-gypsum mortars.

The incorporation of PUW in gypsum mortars leads to a generalised decrease in their density. In the fresh state, the density drops by 14.49% ± 9.85%; whereas in the hardened state, it falls by 15.77% ± 8.42%, with the exception of the AT4 sample whose density is 10.03% higher than the reference mortar.

Looking at the results for both states, all mortars report a loss in density from fresh to hardened state of around 39.75% ± 1.99%; with the exception of AT4 specimen which experiences a 30.39%, even if it is the only sample with a lower w/c ratio.

3.2   Flexural Strength and Compressive Strength

The data provided by the mechanical tests of the studied materials are represented in Fig. 3.

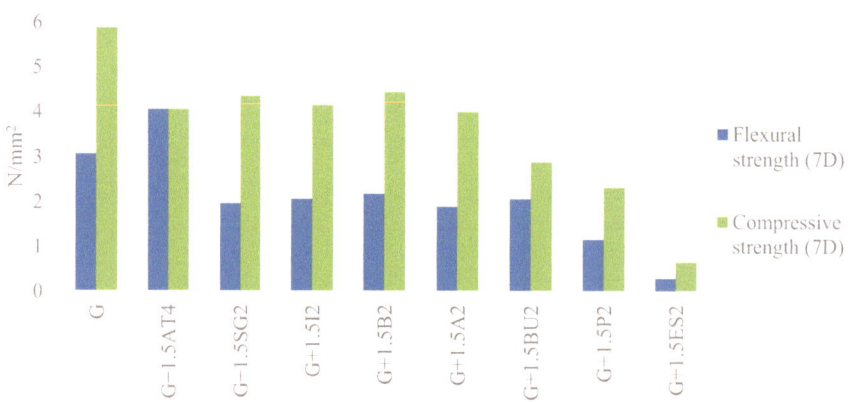

Fig. 3. Flexural strength and compressive strength of polymeric-gypsum mortars.

The polymeric-gypsum mortars experience a severe reduction in their mechanical properties. Flexural strength is reduced by 33.62% ± 4.63%, without taking into account the specimens P2, AT4 and ES2, which show a high dispersion with regard to the rest of the data. The AT4 sample improve its performance by 32.25%, while samples P2 and ES2 do not meet or are close to not meeting the minimum requirements of the standard (1 N/mm^2). Compressive strength of the developed mortars compared to the traditional one also goes down by 28.50% ± 4.18%, without considering the samples P2, BU2 and ES2, which are below or close to the limit of the regulatory requirements (2 N/mm^2).

3.3   Life Cycle Assessment

The results of the different impact categories studied for each specimen are shown in Fig. 4. The data are presented as a percentage in order to identify the specimens with the best performance in each environmental aspect. The smaller the percentage, the lower the impact. Samples G+1.5ES2 and G+1.5P2 have not been taken into account in this evaluation due to the low mechanical properties shown in the previous tests.

The G+1.5SG2 sample has the highest score in 6 of the 7 impact categories analysed, this is due to the high amount of energy required for this polyurethane waste to be shredded and the distance for transport. Specimens G+1.5A2, G+1.5B2 and G+1.5I2 have a similar performance and a slightly higher environmental impact than the reference product, because of the combination of the impact of average processing times and the relative transport distances. The mixes G+1.5AT4 and G+1.5BU2 give the best outcomes, in comparison with the reference mortar. Although the processing time of AT4 waste is high, the non-inclusion of glass fibre in its composition offsets the environmental impacts. Waste BU2 meets the ideal conditions with a low processing time and short transport distance.

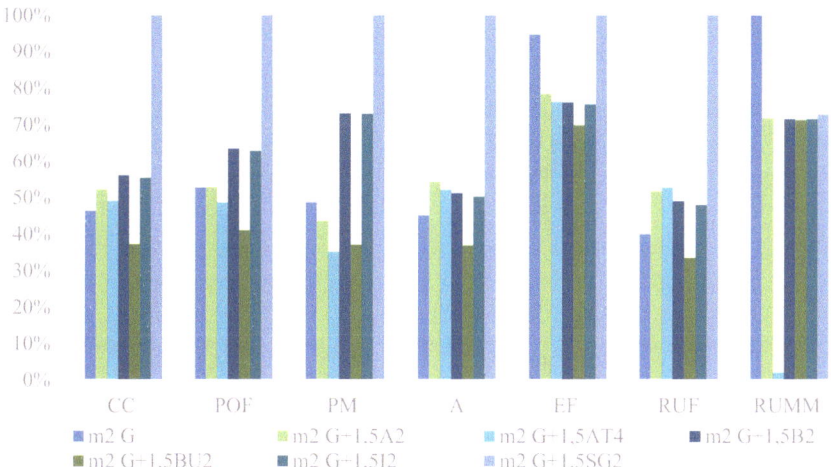

Fig. 4. Contribution of each sample to the impact categories analysed.

In order to obtain an overall assessment of the environmental performance of each specimen, all impact categories are unified. This process is known as weighting and the resulting impact category is single score (μPt). The outcomes are displayed in Fig. 5.

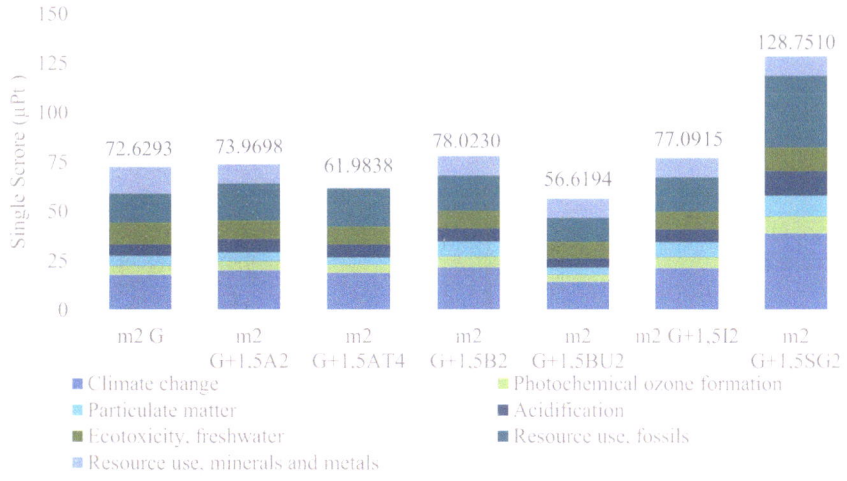

Fig. 5. Single Score of polymeric-gypsum mortars.

The single score supports the data previously discussed. In a global computation, mixtures G+1.5BU2 and G+1.5AT4 show a decrease in the overall environmental impact of 22% and 15%, respectively, compared to the reference one, while samples G+1.5A2, G+1.5B2 and G+1.5I2 are slightly above (2%, 7% and 6%, accordingly). However, the inclusion of polyurethane waste type SG2 in gypsum mortars implies an increase in environmental damage of 77%.

It has been noticed that the impact categories with the greatest weight in the single score are climate change and fossils resource use.

4 Conclusions

The current study, focused on the valorisation of polyurethane waste in gypsum mortars, concludes that:

- The manufacture of gypsum mortars with polymeric waste is feasible.
- In general, the addition of PUW in the mixtures leads to a decrease in the bulk density of the materials of between 7% and 24%.
- The mechanical properties are reduced by one third compared to the reference material, but the values are still above the normative standards.
- The environmental improvement depends on the time of processing and the distance of acquisition of the polyurethane waste. Nevertheless, samples with PUW type BU2 and AT4 show a damage reduction of 22% and 15%, respectively.
- Climate change and fossils resource use are the two environmental impact categories in which this kind of specimens cause the most damage.

Acknowledgements. The authors would like to thank the European Regional Development Fund (FEDER) (UE) (project BU070P20), the Consejería de Educación de la Junta de Castilla y León (Spain) and the European Social Fund (EU), and the Consejo General de la Arquitectura Técnica de España (CGATE) for funding the study.

This work has also been supported by the Regional Government of Castilla y León (Junta de Castilla y León) and by the Ministry of Science and Innovation MICIN and the European Union NextGeneration EU / PRTR.

References

1. Statista Distribution of Global Plastic Materials Production in 2020, by Region. Available online. https://www.statista.com/statistics/281126/global-plastics-production-share-of-various-countries-and-regions/. Last accessed 26 Sept 2022
2. Johansen, M.R., Christensen, T.B., Ramos, T.M., Syberg, K. A.: Review of the plastic value chain from a circular economy perspective. J. Environ. Manage. **302**(2022). https://doi.org/10.1016/j.jenvman.2021.113975
3. European Commission: Plastics. Available online: https://environment.ec.europa.eu/topics/plastics_en. Last accessed 26 Sept 2022
4. Plastics Europe: Plastics-the Facts 2022 (2022)
5. Gama, N.V., Ferreira, A., Barros-Timmons, A.: Polyurethane foams: Past, present, and future. Materials **11**(10) (2018). https://doi.org/10.3390/ma11101841

6. Plastics Europe: Plastics-the Facts 2021. An Analysis of European Plastics Production, Demand and Waste Data (2021)
7. The Conference Board Plastic Solid Waste Management (2021)
8. Datta, J., Kopczyńska, P., Simón, D., Rodríguez, J.F.: Thermo-chemical decomposition study of polyurethane elastomer through glycerolysis route with using crude and refined glycerine as a transesterification agent. J. Polym. Environ. **26**(1), 166–174 (2018). https://doi.org/10.1007/s10924-016-0932-y
9. Romero-Gómez, M.I., Pedreño-Rojas M.A., Pérez-Gálvez, F., Rubio-de-Hita P.: Characterization of gypsum composites with polypropylene fibers from non-degradable wet wipes. J. Build. Eng. **34**, 101874 (2021). https://doi.org/10.1016/j.jobe.2020.101874
10. Pedreño-Rojas, M.A., Morales-Conde, M.J., Pérez-Gálvez, F., Rubio-de-Hita, P.: Influence of polycarbonate waste on gypsum composites: Mechanical and environmental study. J. Clean. Prod. **218**, 21–37 (2019). https://doi.org/10.1016/j.jclepro.2019.01.200
11. del Rio Merino, M., Villoria Sáezm P., Longobardi, I., Santa Cruz Astorqui, J., Porras-Amores, C.: Redesigning lightweight gypsum with mixes of polystyrene waste from construction and demolition waste. J. Clean. Prod. **220**, 144–151 (2019). https://doi.org/10.1016/J.JCLEPRO.2019.02.132
12. Bertelsen, I.M.G., Ottosen, L.M.: Recycling of waste polyethylene fishing nets as fibre reinforcement in gypsum-based materials. Fiber. Polymer. (2021). https://doi.org/10.1007/s12221-021-9760-3
13. Asociación Española de Normalización (AENOR): Yesos de Construcción y Conglomerantes a Base de Yeso Para La Construcción. Parte 1: Definiciones y Especificaciones; UNE-EN 13279-1 (2008)
14. Gómez-Rojo, R., Alameda, L., Rodríguez, Á., Calderón, V., Gutiérrez-González, S.: Characterization of polyurethane foam waste for reuse in eco-efficient building materials. Polymers **11**(2), 359 (2019). https://doi.org/10.3390/polym11020359
15. Asociación Española de Normalización (AENOR): Yesos y Escayolas de Construcción. Otros Métodos de Ensayo; UNE 102042 (2014)
16. Asociación Española de Normalización (AENOR): Yesos de Construcción y Conglomerantes a Base de Yeso Para La Construcción. Parte 2: Métodos de Ensayo; UNE-EN 13279-2 (2014)
17. International Organisation for Standardisation (ISO): Environmental Management. Life Cycle Assessment. Principles and Framework; ISO 14040, Geneva, Switzerland (2006)
18. International Organisation for Standardisation (ISO): Environmental Management. Life Cycle Assessment. Requirements and Guidelines; ISO 14044 Geneva, Switzerland, (2006)

Recycled Brick Fines for New Alkali Activated Binder

Adèle Grellier[1,2,3], David Bulteel[1,2], and Luc Courard[3(✉)]

[1] Univ. Lille, Institut Mines-Télécom, Univ. Artois, Junia, ULR 4515 - LGCgE – Laboratoire de Génie Civil et géoEnvironnement, Lille, France
[2] IMT Nord Europe, Centre for Materials and Processes, Institut Mines-Télécom, Paris, France
[3] Université de Liège, Urban and Environnemental Engineering, GeMMe Matériaux de Construction, Liège, Belgium
luc.courard@uliege.be

Abstract. The construction industry today produces huge quantities of wastes, especially during the deconstruction and demolition of buildings. Ceramics and bricks represent a significant part of this inert waste in Belgium and Northern France. The recycling of bricks is already carried out in the form of aggregates used in road embankments. But this constitutes what is called a "downcycling" operation. The investigated way is here a valorization with higher added value in alkali-activated materials through substitution of blast furnace slag (GGBFS) by brick fines with a grain size $D50 = 20\,\mu m$. It is shown that brick fines can be a precursor equivalent to GGBFS and thus lead to mechanical performances equivalent to control even up to 50% substitution rate in brick fines. Under certain conditions of alkali-activated solution concentration, the addition of 30% brick fines can greatly improve workability time. But this leads to a decrease in mechanical performances, which is still in accordance with specific construction needs.

Keywords: bricks · alkali activated materials · GGBFS · strength · microstructure

1 Introduction

Bricks and ceramics represent around 2% of the Construction and Demolition Wastes produced in Belgium and North of France. Reusing bricks is undoubtedly better than recycling but C&DW treatment however produces large quantities of fine particles, including brick fines. Bricks are a predominantly aluminosilicate materials which, in the form of fines, can be described as a potentially pozzolanic material [1]. They can be characterized by their quantity of oxides, $SiO_2 + Al_2O_3 + Fe_2O_3\ (+ CaO) \geq 70\%$, which includes them in pozzolans and partly in glasses depending on the amount of amorphous phase [2]. Pozzolanicity and fineness are therefore considered as the parameters to characterize the activity of brick fines when used in hydraulic mixtures [3, 4].

Alkali-activated material is the result of the reaction between a solid aluminosilicate (precursor) and an alkali-activating solution. This reaction produces a hardened binder, consisting of hydrated alkali aluminosilicates. The precursors are usually aluminosilicate

L. Czarnecki et al. (Eds.): ICPIC 2023, 61, pp. 198–205, 2025.
https://doi.org/10.1007/978-3-031-72955-3_19

or calcium-rich materials such as metakaolin and ground blast furnace slag (GGBFS) respectively. The alkali-activating solution is most often sodium or potassium hydroxide, which allows the pH of the solution to be increased, with or without a silicate solution.

Brick fines, due to their high aluminosilicate content and the high percentage of amorphous phase with active silica and alumina, can be identified as a precursor for alkali-activated materials.

Previous research [5, 6] demonstrate the activating potential of brick fines to create a fully alkali-activated material. If the proportion of amorphous phase is sufficient, then the brick fines form a hardened and resistant alkali-activated material.

The use of brick fines can also overcome two disadvantages found in alkali-activated mixtures, which are (i) a very short setting time and (ii) poor workability. According to some studies, a shortening of the setting time is observed [7]. This phenomenon is linked to the larger specific surface area of the fines, which increases the reaction speed. However, the final setting time would be delayed [7]. The fineness would also result in better workability [8]. An improvement in mechanical performance is also noted, with an increase in compressive strength [9–11]. This higher strength with the addition of brick fines may be due to the presence of amorphous phase in the fines, which contributes to a better reactivity of the mix. Depending on the size of the fines, the alkaline activation can be improved. By increasing the specific surface area of the particles, the alkali-activation process is facilitated.

This paper presents the results obtained with commercial bricks and compare the effects of taking into account the possible activity of the brick fines as precursors.

2 Materials and Methods

2.1 Blast Furnace Slag

GGBFS is consisting mainly of amorphous phase. The chemical (Table 1) and physical (Table 3) characteristics of GGBFS show it is a reactive and fine material ($d_{50} = 8.5\ \mu m$) with a specific surface area of 1 m2/g. The fineness of the material allows for a sufficiently alkali-reactive powder, in addition to its mineralogical nature.

2.2 Alcali-Activating Solution

The alkali-activating solution is essential for the formulation of alkali-activated materials; it can be assimilated to the mixing water used for the hydration of cement. This solution will be composed of sodium silicate (Na_2SiO_3), sodium hydroxide (NaOH) and water. The choice of activators is mainly related to the performance that can be obtained. By coupling the efficiency of a sodium silicate and sodium hydroxide solution, it is possible to obtain very good performances with GGBFS. Two types have been used:

- Sodium silicate (GEOSIL) is formulated for alkaline activations. Sodium silicate has a molar ratio Ms = 1.7 (SiO_2/Na_2O), a density of 1.57 and a dry mass of 44%. This activator allows for slag to dissolve more slowly and for better diffusion of the reaction products between the grains.

- Sodium hydroxide is packaged in pellet form. It is the activator with the fastest kinetics: it plays an important role at a young age by rapidly dissolving the materials and bringing the ions into solution.

2.3 Brick Fines

Brick fines (BF) are prepared in order to obtain a granulometry equivalent cement (reference CEM I 52.5 N). In a first step, a jaw crusher is used to reduce the sample into aggregate form. A sieve (1 mm sieve) is then used. The by-product is then crushed in a ball mill (steel balls). Chemical and mineralogical characterizations have been performed by means of XRF and XRD (Tables 1 and 2). Bricks are a material rich in silicon, aluminium, and iron oxide, with a SiO_2/Al_2O_3 ratio equal to 6.

Table 1. Oxides in GGBFS and brick fines.

Oxides (%)	CaO	SiO_2	Al_2O_3	Fe_2O_3	K_2O	Na_2O	MgO	TiO_2	Total
Brick fine	1.7	62.8	10.4	16.3	2.1	0.6	2.2	2.4	99.3
GGBFS	42.9	38	10.8	0.5	0.3	-	6.5	0.7	99.5

The minerals found in the bricks are quartz (SiO_2), hematite (Fe_2O_3), feldspars with albite ($NaAlSi_3O_8$) and microcline ($KAlSi_3O_8$). Corundum is used as a standard to enable phase quantification. The percentage of the different minerals and phases is given in Table 2.

Table 2. Mineral phases in brick fines.

Mineral (%)	Brick fine
Quartz SiO_2	58.6
Hematite Fe_2O_3	12.8
Albite $NaAlSi_3O_8$	3.9
Microline $KAlSi_3O_8$	6.0
Cristobalite SiO_2	2.8
Amorphicity	15.9

The proportion of amorphous phase in the studied bricks is around 16%. The dissolution of amorphous phases and softer minerals such as feldspars (albite, microcline), can lead to an increase in free oxides in the mixtures which can precipitate to form C-S-H or even C-A-S-H type gels in the case of alkali-activated materials.

Pozzolanic activity can be determined by the modified Chapelle test which gives the consumption of lime by the material tested (Table 3). This factor indicates how much lime can be fixed by the fines: lime or Portlandite is the main crystallized element formed during the hydration reaction. The lime consumed by the brick fines studied is around 390–400 Ca(OH)$_2$ mg/g brick fines.

Table 3. Physical characteristics of brick fines.

Brick fine	Brick fines	GGBFS
Specific surface, BET (m^2/kg)	833	1
Water absorption (%)	1.1	-
Granulometry (μm)		
d10	1.95	1
d50	19.1	8.5
d90	56.6	30
Ca(OH)$_2$ quantity fixed (mg/g brick fines)	394	-

2.4 Preparation of Alcali-Activating Solution

Tests are firstly carried out to determine which types of activators to use and in what proportions. These tests are carried out on mixes with 0, 50 and 100% mass substitution of brick fines. Mixtures containing only brick fines do not harden, regardless of the activator concentration used. Brick fines cannot be activated on their own and a precursor must be added to the mixture. Two types of activators are finally used in the mixture: sodium hydroxide and sodium silicate.

Two factors allow the mixture to be optimized: a Na$_2$O content of between 2 and 8% and a SiO$_2$/Na$_2$O ratio of 1 to 1.5. These two parameters are calculated from the amount of GGBFS in the mixture. On the basis of the literature review and these preliminary tests, an alkali-activator solution was chosen which combines these two activators: 5% Na$_2$O fixed and a SiO$_2$/Na$_2$O ratio fixed at 1.45. The first formulations (BL) were carried out with a concentration calculated according to the amount of GGBFS in the mixture (Table 4). With these formulations, activation is associated with the amount of GGBFS and the brick fines are considered as an addition but not activatable.

Subsequently, a second series of formulations (BLM) was carried out (Table 4). The concentration of the solution was calculated according to the total amount of material (GGBFS and brick fines), so that the percentage of Na$_2$O increases with the increase in brick fines, the SiO$_2$/Na$_2$O ratio remaining fixed. As the level of brick fines increases, the amount of silicate in solution remains constant in relation to the total mass of material but increases sharply in relation to the mass of GGBFS. While it is known that optimal activation of GGBFS occurs between 2 and 8% Na$_2$O, it is expected that beyond this there will be little effect on mechanical strengths and may even lead to efflorescence and carbonation.

The solution was prepared 24 h in advance to avoid the exothermic reaction associated with the dissolution of sodium hydroxide. The sodium silicate, sodium hydroxide and water are weighed and mixed for several minutes. The solution is prepared and stored in a ventilated hood. Working with higher temperatures would result in increased reaction kinetics and therefore faster hydration: as one of the effects expected with the addition of brick fines is precisely the slowing down of the hydration, tests were carried out at room temperature.

Table 4. Design of mixes.

BL	GGBFS (g)	BF (g)	Na$_2$SiO$_3$ (g)	NaOH (g)	Water (g)	W/B	SiO$_2$/Na$_2$O	%Na$_2$O
0%	1500	0	377	14.2	550.4	0.45	1.45	5
10%	1350	150	340	13	562.7	0.45	1.45	5
20%	1200	300	302	11.5	575.2	0.45	1.45	5
30%	1050	450	265	10	587.7	0.45	1.45	5
50%	750	750	188	7.2	612.9	0.45	1.45	5
BLM	GGBFS (g)	BF (g)	Na$_2$SiO$_3$ (g)	NaOH (g)	Water (g)	W/B	SiO$_2$/Na$_2$O	%Na$_2$O
0%	1500	0	377	14.2	550.4	0.45	1.45	5
10%	1350	150	377	14.2	550.4	0.45	1.45	5.6
20%	1200	300	377	14.2	550.4	0.45	1.45	6.25
30%	1050	450	377	14.2	550.4	0.45	1.45	7.14
50%	750	750	377	14.2	550.,4	0.45	1.45	10

The high Al/Si ratio is essential for the incorporation of fines into alkali-activating mixtures with GGBFS. Alkaline activation is essentially a reaction of the aluminosilicates with the sodium silicate in the alkali-activating solution causing polymerization of these elements and giving the material its strength.

3   Results and Analysis

The behaviour of the two types of mixes BL and BLM is different. For BL formulations, the workability is impacted from 30% substitution. For BLM mixes, the setting time is affected from 50% substitution onwards and it is the end of setting that is mainly affected. The addition of brick fines to alkali-activated materials at levels above 30% extends the workability time.

Spreading tests (EN 1015-3) indicate that the workability changes little with a low addition of brick fines (below 30%). In contrast, spreading tests with continuous mixing show a change in behaviour with BL mixes and workability increases with the substitution of 30% brick fines. The "freezing" phenomenon can be delayed with a continuous mixing protocol as well as by lowering the concentration of the activating solution and increasing the brick fines content. The workability time (EN 196-3), which is a handicap of alkali-activated materials, can therefore be extended in this case.

The alkali-activation kinetics of the mixtures changes according to the mixtures. For the BL 10% formulation, the precipitation of secondary C-A-S-H is delayed by more than one day. With the BLM mixtures, at 10% brick fines, the hydration kinetics is similar or even faster than that of GGBFS. With the addition of brick fines at 50%, the induction period is slightly longer. The hydration kinetics are delayed by the addition of brick fines, but this can be compensated for by increasing the concentration of the activating solution.

Alkali-activation kinetics and hydrate generation are increased when the activator solution is more concentrated for the same brick fines substitution rate. Conversely, alkali-activation kinetics and hydrate generation decrease with increasing brick fines substitution rate for a constant activator solution concentration.

Thermogravimetric analyses show that the hydrates formed in alkali-activated GGBFS materials are identical to those formed during brick fines substitution: C-A-S-H, hydrotalcite and carbonates. The total mass loss at 90 days of the samples with 10% fines is identical to the GGBFS for the higher concentration activator solution but decreases when the activator solution is at a lower concentration. There is less hydrate formed with the samples containing 50% brick fines, especially for the BL 50% sample which is the least concentrated in activator solution.

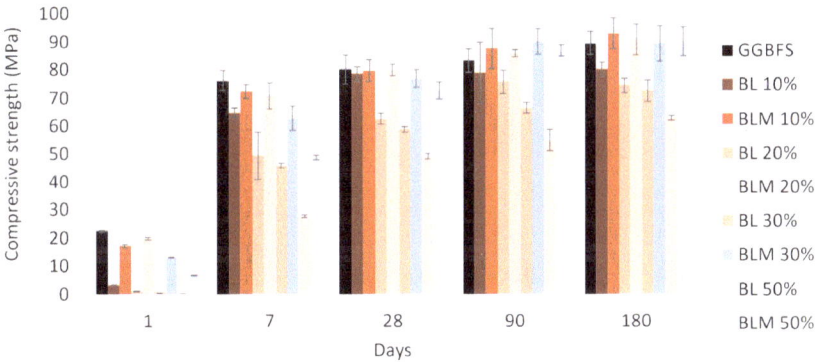

Fig. 1. Evolution of compressive strength of BL and BLM mixes with time (EN 196-1)

In terms of mechanical performance (Fig. 1), a different behaviour between the two mixes is noted. For the BL formulations, the compressive strength is very low in the early stages but improves in the long term while remaining below that for GGBFS alone. Compressive strengths exceed 60 MPa at 180 days, regardless of the level of brick fines, and above all the loss of strength is not proportional to the level of substitution, thus showing a beneficial role for brick fines. For BLM mixes, the strength increase is faster and the performance is equivalent to GGBFS, even with 50% brick fines, from 28 days. This demonstrates that brick fines can be considered as an activatable mineral addition.

Poral distribution has been evaluated by means of Mercury Intrusion Porosimetry. During phase precipitation, porosity decreases with time and the pore size distribution becomes concentrated in pores smaller than 0.1 μm in diameter after 90 days (Fig. 2). The pore size distribution is equivalent between the control and the samples with 10%

fines as well as the BLM 50% sample. The proportion of finer pores is lower for the BL 50% sample. The microstructure refinement is little impacted by the addition of brick fines when the mixtures are formulated with high concentrations of activating solution.

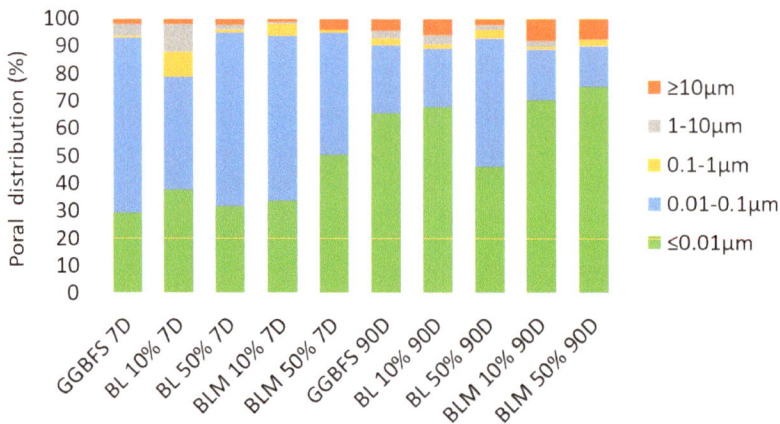

Fig. 2. Poral distribution of BL and BLM mixes at 7 and 90 days.

4 Conclusions

Results show that, for formulations with a lower concentration, the mechanical performance at a young age is very low but improves afterwards and can be partially compensated for after 90 days. This loss of strength is in favour of an increase in the workability of the material. The setting time of these formulations is longer, provided that a continuous mixing protocol is put into practice. In these formulations, the use of brick fines can also be envisaged while allowing a reduction in the activating solution used. The two types of mixes (BL and BLM) therefore have different strengths. Formulations based on the activation of brick fines with a concentration of activators calculated on the total mass of material show the best performances.

This type of mixture allows for optimal valorization and activation of the brick fines. In all cases, the development of an alkali-activated binder based on GGBFS and using a large proportion of brick fines (up to 50%) is viable. The use of brick fines allows a significant economic and ecological gain since the value of the brick fines is lower even including the cost of crushing the brick aggregates and allows the recovery of a waste product rather than its disposal.

Acknowledgements. The authors would like to thank the INTERREG FWVL for financial support through the VALDEM project "Integrated solutions for the recovery of material flows resulting from the demolition of buildings: a cross-border approach towards a circular economy".

References

1. Bediako, M., Atiemo, E.: Influence of higher volumes of Clay Pozzolana replacement levels on some technical properties of cement pastes and mortars. J. Sci. Res. Rep. **3**(23), 3018–3030 (Jan. 2014). https://doi.org/10.9734/JSRR/2014/9046
2. Baronio, G., Binda, L.: Study of the pozzolanicity of some bricks and clays. Constr. Build. Mater. **11**(1), 41–46 (Feb. 1997). https://doi.org/10.1016/S0950-0618(96)00032-3.)
3. Bediako, M.: Pozzolanic potentials and hydration behavior of ground waste clay brick obtained from clamp-firing technology. Case Stud. Constr. Mater. **8**, 1–7 (June 2018). https://doi.org/10.1016/j.cscm.2017.11.003
4. Komnitsas, K., Zaharaki, D., Vlachou, A., Bartzas, G., Galetakis, M.: Effect of synthesis parameters on the quality of construction and demolition wastes (CDW) geopolymers. Adv. Powder Technol. **26**(2), 368–376 (Mar. 2015). https://doi.org/10.1016/j.apt.2014.11.012
5. Li, L., Lu, J.-X., Zhang, B., Poon, C.-S.: Rheology behavior of one-part alkali activated slag/glass powder (AASG) pastes. Constr. Build. Mater. **258**, 120381 (Oct. 2020). https://doi.org/10.1016/j.conbuildmat.2020.120381
6. Rakhimova, N.R., Rakhimov, R.Z.: Alkali-activated cements and mortars based on blast furnace slag and red clay brick waste. Mater. Des. **85**, 324–331 (Nov. 2015). https://doi.org/10.1016/j.matdes.2015.06.182
7. Allahverdi, A., Kani, E.N.: Construction wastes as raw materials for geopolymer binders. Int. J. Civ. Eng. **7**(3), 154–160 (2009)
8. Reig, L., Tashima, M.M., Borrachero, M.V., Monzó, J., Cheeseman, C.R., Payá, J.: Properties and microstructure of alkali-activated red clay brick waste. Constr. Build. Mater. **43**, 98–106 (June 2013). https://doi.org/10.1016/j.conbuildmat.2013.01.031
9. Robayo-Salazar, R.A., Mejía-Arcila, J.M., Mejía de Gutiérrez, R.: Eco-efficient alkali-activated cement based on red clay brick wastes suitable for the manufacturing of building materials. J. Clean. Prod. **166**, 242–252 (Nov. 2017). https://doi.org/10.1016/j.jclepro.2017.07.243
10. Komnitsas, K.A.: Potential of geopolymer technology towards green buildings and sustainable cities. Procedia Eng. **21**, 1023–1032 (2011). https://doi.org/10.1016/j.proeng.2011.11.2108
11. Peyne, J., Joussein, E., Gautron, J., Doudeau, J., Rossignol, S.: Feasibility of producing geopolymer binder based on a brick clay mixture. Ceram. Int. **43**(13), 9860–9871 (Sep. 2017). https://doi.org/10.1016/j.ceramint.2017.04.169

Recycled Cement Concrete as an Eco-Friendly Aggregate in Polymer Composite – Application Feasibility

Maja Kępniak[✉]

Warsaw University of Technology, 00-637 Warsaw, Poland
maja.kepniak@pw.edu.pl

Abstract. Over the years, the development of sustainable and ecofriendly concrete has been found in the reuse of construction and demolition materials. One such waste is recycled aggregate from cement concrete structure demolition process. This paper analyzes the effect of substitution of natural stone aggregate with recycled aggregate in polymer composites. An experimental plan for the mixtures was prepared. Technological characteristics (setting course, consistency) and strength characteristics (flexural strength and compressive strength) were analyzed. The obtained results were statistically analyzed. A generalized utility function has been established. Based on it, the maximum dosage of recycled aggregate was determined without significant deterioration of technological and strength characteristics. The average compressive strength results obtained were in the range of 88.5 to 96.5 MPa. The highest compressive strength value (96.5 MPa) was obtained for the samples with the composition with the highest proportion of recycled aggregate.

Keywords: Recycled Aggregate · Eco-friendly Concrete · Polymer Concrete

1 Introduction

Increasing climate challenges and a limited supply of new natural resources for construction projects have shifted research toward sustainability. The concept of circular economy is one of the effective ways to achieve a long-term sustainable construction sector [1–4]. At the same time huge amount of construction and demolition wastes are produced every year. In Poland, these wastes are usually sent to landfills [5]. The disposal of these wastes is a severe social and environmental problem. The recycling of these wastes as aggregate to produce building composites can reduce the problem of waste and help the preservation of natural aggregate resources. Many researchers found that recycled aggregate offers a good alternative aggregate for making concrete in terms of both environment friendly and economically, also taking into account the LCA analysis [6–8].

It should also be taken into account that polymer concrete is widely use in building industry composite with its low carbon footprint compared to the cement concrete. Polymer concrete wide application results from its performing properties like high strength,

L. Czarnecki et al. (Eds.): ICPIC 2023, 61, pp. 206–211, 2025.
https://doi.org/10.1007/978-3-031-72955-3_20

excellent corrosion resistance, frost resistance, good abrasion behavior, rapid hardening and easy preparation [9–12]. There is usually no reactivity between the surrounding polymer matrix and aggregate particles [13]. Therefore, the replacement of natural aggregates in the production of polymer concrete is a very effective method of waste material waste disposal which preserves natural resources. There are many studies on the effective use of waste dusty materials [14, 15] and plastic waste [16, 17], as an ingredient in polymer concrete.

Due to the carbonation that occurs [18, 19], the use of recycled cementitious aggregate for the production of cement concrete is limited, as it could contribute to the corrosion of reinforcing steel. With polymer concrete, this problem does not occur. The use of concrete CDW aggregate to produce polymer concrete could be economically as well as environmentally beneficial. In the study presented in this paper, an attempt was made to replace part of the coarse aggregate with recycled aggregate, mainly containing crushed cement concrete.

2   Materials and Methods

The purpose of the conducted research was to determine the impact of partial substitution of the coarse aggregate by the recycled aggregate. Polymer concrete samples were prepared according to the experimental design. The prepared samples were subjected to the following tests: consistency, compressive and flexural strength and density.

2.1   Materials

The polymer used to prepare all composites presented in the study was synthetic vinyl-ester resin of low viscosity (350 ± 50 mPa·s at 25 °C) and high flexural strength and tensile strength (declared by the producer as respectively 110 MPa and 75 MPa). Therefore, concretes made from this resin should retain the high mechanical strength in long-term exploitation, even when exposed to aggressive environment.

As virgin coarse aggregates, two size fractions were used, one with aggregate size 2–4 mm (2–4N) and the other with 4–8 mm (4–8N) with a fineness modulus of 5.56 and 5.63 respectively. Both of which are from natural gravel. Furthermore, recycled coarse aggregates obtained from the demolition of concrete structures were used, so they mainly consisted of aggregates with adhered mortar. The size fraction was 4–8 mm (4–8R) with a fineness modulus of 6.95. Lastly, the fine aggregate was just a natural sand with a maximum aggregate size of 2 mm (0–2N) and a fineness modulus of 4.70.

As the microfiller the limestone powder (LP) was used. Table 1 summarizes the basic properties of the aggregates used. Figure 1 exhibits the composition of the recycled coarse aggregates according to EN 933-11. The recycled aggregate consisted mainly of crushed cement concrete (Rc = 88.5%).

On the basis of these results, they can be classified as recycled coarse aggregates from concrete demolition waste and named as RCA (recycled concrete aggregate).

Five different concrete compositions were adopted for the study, differing in the level of replacement of natural aggregate (4-8N) with recycled aggregate (4-8R) from 0% to 100%. A constant resin/microfiller ratio of 1.0 was assumed. A constant resin content of

300 kg was determined. The recycled aggregate was washed on a 0.125 mm sieve and then dried in a drying oven at 100 °C for 48 h. The mass compositions of the composites per cubic meter are summarized in Table 2.

Table 1. Basic properties of the aggregate. LP- limestone powder, 0-2N natural sand, 2-4N, 4-8N natural gravel, 4-8R recycled aggregate.

Property	LP	0-2N	2-4N	4-8N	4-8R
Density (EN 1097–6), g/cm^3	2.71	2.54	2.64	2.67	2.53
Density in owen-dry conditions (EN 1097–6), g/cm^3	-	2.13	2.46	2.60	2.25
Water absorption (EN 1097–6), %	-	7.5	2.8	1.2	4.8
Los Angeles Abrasion (EN 1097–2), %	-	-	27.0	25.0	75.0
Fines percentage (EN 933–1), %	100	1.0	0.1	0.1	0.9

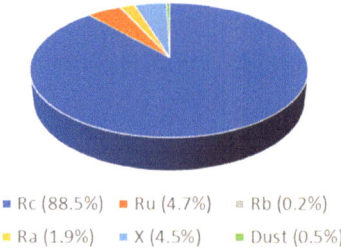

■ Rc (88.5%) ■ Ru (4.7%) ■ Rb (0.2%)
■ Ra (1.9%) ■ X (4.5%) ■ Dust (0.5%)

Fig. 1. Composition of recycled coarse aggregate according to EN 933–1 (percentage by weight)

Table 2. Composites composition

In weight	R00	R10	R20	R30	R40
Synthetic vinyl-ester resin, kg	300	300	300	300	300
Limestone powder, kg	300	300	300	300	300
Natural sand (0-2N), kg	600	600	600	600	600
Gravel (2-4N), kg	300	300	300	300	300
Gravel (4-8N), kg	600	450	300	150	0
Recycled aggregate (4-8R), kg	0	150	300	450	600
RA in total aggregate, %	**0**	**10**	**20**	**30**	**40**
Gravel replacement, %	0	25	50	75	100

2.2 Methods

The consistency of the mixture was assessed according to the procedure for measuring the plasticity of construction mortars (according to PN-EN 1015-3 standard), just after mixing the ingredients. A truncated cone (with dimensions: bottom diameter 100 mm, top diameter 70 mm, height 60 mm) was formed on the flow table. The fresh composite thus formed was subjected to 15 generative shakes by lifting and dropping the measuring table to a height of 10 mm at a rate of 1 per second). The diameter of resulting flow as then measured. To test the flexural strength of the composite, 3 rectangular specimens measuring 40 mm by 40 mm by 160 mm were made for each composition. The specimens were tested according to EN 196-1 standard. The three-point loading method was used. To test the compressive strength of the composite 6 specimens were made for each composition. The specimens were tested according to EN 196-1 standard. The compressive area was 1600 mm^2.

3   Results

Consistency testing showed that for every composition an equal composite spread of (180 ± 5) mm was obtained each time. Therefore, the use of recycled aggregate did not affect the basic technological characteristic of the mixtures. The average flexural strength results obtained were in the range of 20.5 to 21.5 MPa. The differences between the average values for different compositions are within the limits of the standard deviations of the results (Fig. 2).

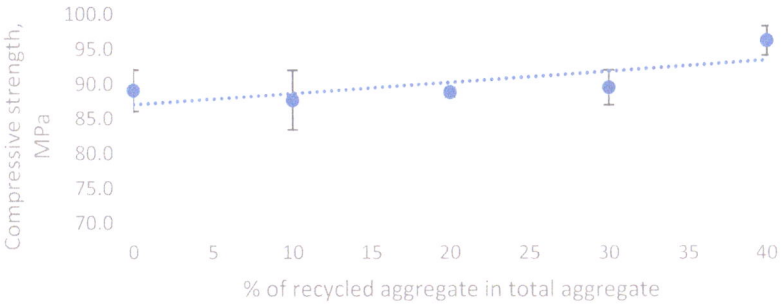

Fig. 2. Composition of recycled coarse aggregate according to EN 933-1 (percentage by weight)

The average compressive strength results obtained were in the range of 88.5 to 96.5 MPa. The differences between the average values for different compositions are within the limits of the standard deviations of the results. The highest compressive strength value (96.5 MPa) was obtained for the samples with the composition with the highest proportion of recycled aggregate. Such results were obtained despite the fact that the resistance of the recycled aggregate itself was lower than that of the natural aggregate. (Fig. 3). This was probably due to the fact that the resin binder penetrated the hardened cement slurry, increasing its strength and strengthening the resin-aggregate transition zone. This assumption was also confirmed by the nature of the failure of the specimens in the flexural and compressive test.

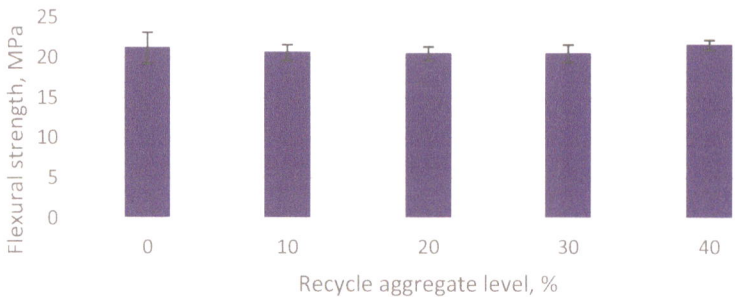

Fig. 3. Composition of recycled coarse aggregate according to EN 933-1 (percentage by weight)

4 Conclusions

Studies conducted indicate that recycled aggregate from demolition of concrete structures has great potential as an aggregate in polymer concrete. The presence of hardened cement slurry does not adversely affect the basic technological and strength characteristics of the composites. Further research is required on the microstructure. It is also necessary to conduct chemical resistance analyses. The use of recycled aggregate in polymer concretes can contribute to reducing the carbon footprint of composites and developing a closed-cycle economy.

References

1. Blomsma, F., Brennan, G.: The emergence of circular economy: a new framing around prolonging resource productivity. J. Ind. Ecol. **21**, 603–614 (2017)
2. Anwar, M., Shah, S., Alhazmi, H.: Recycling and utilization of polymers for road construction projects: an application of the circular economy concept. Polymers **13**, 1330 (2021)
3. Colangelo, F., Navarro, T.G., Farina, I., Petrillo, A.: Comparative LCA of concrete with recycled aggregates: A circular econ-omy mindset in Europe. Int. J. Life Cycle Assess. **25**, 1790–1804 (2020)
4. Kępniak, M., Załęgowski, K., Woyciechowski, P., Pawłowski, J., Nurczyński, J.: Feasibility of using biochar as an eco-friendly microfiller in polymer concretes. Polymers **14**, 4701 (2022)
5. Statista - The Statistics Portal for Market Data, Market Research and Market Studies, http://www.statista.com/. Accessed 8 Mar. 2023
6. Nuaklong, P., Sata, V., Chindaprasirt, P.: Influence of recycled aggregate on fly ash geopolymer concrete properties. J. Clean. Prod. **112**(4), 2300–2307 (2016)
7. Selva, G.M., Jagadeesh, P.: Assessment of usage of manufactured sand and recycled aggregate as sustainable concrete: a review. Mater. Today: Proc. **64**(2), 1029–1034 (2022)
8. Jiang, Y., Li, B., Liu, S., He, J., Garcia, H.A.: Role of recycled concrete powder as sand replacement in the properties of cement mortar. J. Clean. Prod. **371**, 133424 (2022)
9. Garbacz, A., Sokołowska, J.J.: Concrete-like polymer composites with fly ashes – comparative study. Constr. Build. Mater. **38**, 689–699 (2013)
10. Barbuta, M., Bucur, R.D., Cimpeanu, S.M., Paraschiv, G., Bucur-Agroecology, D., Wastes in Building Materials Industry, Chapter 3, pp. 81–99. INTECH, Croatia (2015). ISBN 978-953-51-2130-5

11. Agavriloaia, L., Oprea, St., Barbuta, M., Luca, C.: Characterization of polymer concrete with epoxy polyurethane acryl matrix. Constr. Build. Mater. **12**,190–196 (2012)
12. Kaya, A., Kar, F.: Properties of concrete containing waste expanded polystyrene and natural resin. Constr. Build. Mater. **102**, 572–578 (2016)
13. Ribeiro, M.C.S., Tavares, C.M.L., Ferreira, A.J.M.: Chemical resistance of epoxy and polyester polymer concrete to acids and salts. J. Polym. Eng. **22**(1), 27–43 (2002)
14. Kępniak, M., Woyciechowski, P., Franus, W.: Chemical and physical properties of limestone powder as a potential microfiller of polymer composites. Arch. Civ. Eng. **63**, 67–78 (2017)
15. Łukowski, P., Sokołowska, J.J., Kępniak, M.: Wstępna ocena możliwości zastosowania odpadowego pyłu perlitowego w budowlanych kompozytach polimerowych Budownictwo i Architektura **13**, 119–126 (2014)
16. Alperen, B.H., Şahin, R.: A study on mechanical properties of polymer concrete containing electronic plastic waste. Compos. Struct. **178**, 50–62 (2017)
17. Guerra, T.K., Proszek, G.J.: Polymer concrete with recycled PET: the influence of the addition of industrial waste on flammability. Constr. Build. Mater. **40**, 378–389 (2013)
18. Woyciechowski, P., Kępniak, M., Pawłowski, J.: Methodology for measuring for measuring the carbonation depth of concrete – standard and non-standard aspects. Struct. Environ. **14**(3), 89–95 (2022)
19. Tang, B., Fan, M., Yang, Z., Sun, Y., Yuan, L.: A comparison study of aggregate carbonation and concrete carbonation for the enhancement of recycled aggregate pervious concrete. Constr. Build. Mater. **371** (2023)

Geopolymer Composites with Recycled Binders

Katarzyna Kalinowska-Wichrowska[1]([⊠]), Edyta Pawluczuk[1],
Marta Kosior-Kazberuk[1], Filip Chyliński[2], Alejandra Vidales Barriguete[3],
and Carolina Pina Ramirez[4]

[1] Faculty of Civil Engineering and Environmental Sciences, Bialystok University of
Technology, Bialystok, Poland
k.kalinowska@pb.edu.pl
[2] Instytut Techniki Budowlanej, Warsaw, Poland
[3] Departamento de Tecnología de La Edificación, Escuela Técnica Superior de Edificación,
Universidad Politécnica de Madrid, Madrid, Spain
[4] Departamento de Construcciones Arquitectónicas y Su Control, Escuela Técnica Superior de
Edificación, Universidad Politécnica de Madrid, Madrid, Spain

Abstract. The application of geopolymers as an alternative to cement concretes is becoming increasingly important. The significant advantage of this composites is that, the basic ingredient is not a cement, but pozzolans such as waste materials—fly ash, fly ash slag mix, red ceramic fines, recycling cement mortar—which makes building materials more environmentally friendly. Currently the availability of blast furnace slag and high-quality fly ash is limited in Europe. At the same time, the ways for management of the concrete rubble and the construction waste are being sought, because the volume of waste materials is constantly increasing.

Therefore, the application of secondary binders extracted from the recycling of various construction waste (recycled cement mortar, red ceramic fines, fly ash-slag mix) in geopolymers was proposed. The recycled binders were introduced into geopolymer composites as a replacement of 25% by mass of primary binder (fly ash) and the 65, 75 and 85°C was the curing temperature. The process of manufacturing the recycled binders has been described and basic parameters of new binders. The tests of physical and mechanical properties of the composites such as compressive strength, flexural strength, volume density in dry state and saturated one and water absorption were performed. The microstructure of geopolymers was examined using scanning electron microscopy (SEM). The results obtained show that recycled binders obtained from the treatment of construction waste could be a valuable component of geopolymers.

Keywords: geopolymers · ceramic waste · fly ash · recycled cement mortar · fly ash-slag mix

1 Introduction

At the end of 80's the idea of sustainable development has been published by the World Commission on Environment and Development in the report "Our Common Future" [1]. According to this report *humanity has the ability to make development sustainable*

© The Author(s) 2025
L. Czarnecki et al. (Eds.): ICPIC 2023, 61, pp. 212–219, 2025.
https://doi.org/10.1007/978-3-031-72955-3_21

to ensure that it meets the needs of the present without compromising the ability of future generations to meet their own needs. That leads to wise using natural resources and recycling of materials as much as it is possible. But being sustainable should also include relatively low emission of greenhouse gases. According to European Green Deal strategy signed in 2019 the European Union has set ambitious targets to reduce greenhouse gas emissions by 55% by 2030 and achieve carbon neutrality by 2050 [2]. In order to meet these targets, the construction industry needs to find alternative materials that are more sustainable. Concrete is being used in the largest amounts in whole world, except water, what leads to increasing demands for natural resources and creates large emission of greenhouse gases [3, 4]. But production of the most important constituent of concrete—clinker, leads to very high emission of carbon dioxide. For each ton of clinker about 830 kg of CO_2 is being emitted to the atmosphere. It is about 5–8% of total industrial emission of CO_2 [5, 6]. The cement industry is responsible for a significant amount of CO_2 emissions, which is a major contributor to climate change. The process of production of clinker is still being optimized to reduce the carbon foot print but it can't reach much lower values, due to the emission of CO_2 related to the decarbonation of lime process which leads according to the stoichiometry of reaction to the emission of 600 tons of CO_2 for each ton of calcinated lime [7]. Development in the area of cement and concrete production aims to produce concretes with successively larger amounts of recycled materials and by-products. Other way of reducing the emission of CO_2 related to the production of clinker, is by decreasing its demand in cement and concrete by producing low clinker cements and concretes or by promoting other types of binders [6]. One of the alternatives of using clinker are geopolymers. Geopolymers are made by combining an aluminosilicate material with an alkaline activator solution, which then forms a solid material. Geopolymers have several advantages over traditional cement, including lower carbon emissions, higher durability, and better resistance to fire and chemicals [8]. Geopolymers might be also used in waste management. They can be used to immobilize hazardous waste, reducing the risk of contamination and pollution. Additionally, they can be used to create new materials from various types of inorganic industrial wastes and by-products. Geopolymers have the potential to revolutionize several industries and contribute to achieving the goals of the European Green Deal. They are sustainable, durable, and versatile materials that can be used in construction industry [9–11]. One of the main by-product successfully used in production of geopolymers is siliceous fly ash [12]. Nowadays high dement for good quality fly ash especially from cement and concrete industry and also decreasing production of this by-product, cause the lack of this constituent on the market. That is the reason why a new constituents for geopolymers which might replace a part of fly ash, should be applied. Using waste as a binder for geopolymers is a new direction, but one that is necessary, particularly to find ways to reuse it.

This paper presents results of tests of geopolymers in which a part of siliceous fly ash (FA) was replaced by a various types of building wastes such as ceramic waste (CW), recycled cement mortar (RCM), and also fly ash–slag mix (FAS). Aim of this article is to show that this type of waste materials might be successfully used in some conditions to replace a part of fly ash in geopolymer.

2   Materials and Methods

The fly ash (FA) that was used met the requirements of the standard EN 450–1:2012. Ceramic fines (CW) were made from grinding ceramic hollow bricks damaged during transport or production. Recycled cement mortar (RCM) was obtained after sieving the fraction < 4 mm resulting from the crushing of concrete rubble. The fly ash–slag mixture (FAS) is post-production waste that was generated as an unavoidable by-product of energy production in conventional coal-fired power plants. All additives, before use, were milled to the dust fraction < 0.063 mm.

The alkaline activator used to prepare the samples was an aqueous solution of sodium silicate and sodium hydroxide of an 8 M concentration. The mass ratio of sodium silicate (Na_2SiO_3) to sodium hydroxide (NaOH) was 2.5. In addition, standard sand 0/2 mm was used. Twelve research series were planned in the experiment, in which 25% of FA was replaced with individual additives in the form of an ash-slag mixture, ceramic dust and recycled mortar. After forming a sample of geopolymer mortar with dimensions of 40 mm x 40 mm x 160 mm, it was heated together with the mould for 24 h at temperatures of 65°C, 75°C and 85°C, respectively, and then, after demolding, it was left in laboratory conditions for 14 days. The experiment plan and the composition of additive-modified geopolymer composite mixtures is shown in Table 1.

Table 1. The experiment plan and the mix composition for each series

Series	Curing temperature, °C	Activator, g	Standard sand, g	FA, g	FAS, g	CW, g	RCM, g
1	65	225	1350	450.0			
2	75						
3	85						
4	65	225	1350	337.5	112.5		
5	75						
6	85						
7	65	225	1350	337.5		112.5	
8	75						
9	85						
10	65	225	1350	337.5			112.5
11	75						
12	85						

The flexural and compressive strength tests were performed according to EN 196–1: 2016. The water absorption test was executed by determining the percentage increase in the weight of the specimens when they saturated with water in relation to the weight of the specimen in the dry state. The volume density in a dry state and in a saturated state

were determined based on EN 1015–10:1999. The microstructural analysis of geopolymers were performed using scanning electron microscopy (SEM) model Sigma 500 VP
produced by Zeiss. Analysis in microareas were performed using Energy Dispersive
X-Ray Analysis detector (EDX) model Ultim Max 40 produced by Oxford. Samples for
SEM examinations were prepared by cutting a slice about 3 mm thick, from the inner
section of geopolimer sample. The samples were dried in oven in temperature of 40 °C
for 24 h. Than they were filled with epoxy resin in vacuum chamber. The next day the
samples were polished to receive a smooth surface. Samples were prepared in same way
as described in previous publications [13].

3   Results and Discussion

Table 2 contains physical and mechanical properties of geopolymer composites.

Table 2. The physical-mechanical properties of geopolymers

Serie	Flexural strength, MPa	Compressive strength, MPa	SAI*	Volume density		Water absorption, %
				dry state, g/cm^3	saturated state, g/cm^3	
1	2	3	4	5	6	7
1_100%FA_65	4.85	15.96	–	1.98	2.09	6.4
2_100%FA_75	4.90	21.25	–	1.97	2.02	7.4
3_100%FA_85	4.99	23.36	–	1.84	1.98	7.4
4_25%FAS_65	6.21	20.19	1.27	2.11	2.21	5.1
5_25%FAS_75	6.54	24.31	1.14	2.08	2.19	5.4
6_25%FAS_85	6.62	26.85	1.15	2.03	2.14	5.6
7_25%CW_65	4.12	11.75	0.74	1.98	2.11	6.6
8_25%CW_75	4.49	13.32	0.63	1.91	2.04	6.6
9_25%CW_85	4.69	19.33	0.83	1.85	1.98	7.2
10_25%RCM_65	2.38	13.94	0.87	2.00	2.08	3.8
11_25%RCM_75	4.19	21.85	1.03	1.93	2.02	4.7
12_25%RCM_85	4.33	22.75	0.97	1.92	2.05	6.6

*SAI—Strength Activity Index

Based on the obtained results, it was found that the highest flexural strength was
obtained by the samples of the 6_25%FAS_85 series, annealed at 85 °C (6.62 MPa) and
it was about 32% higher compared to the 3_100%FA_85 series containing only FA. The
lowest flexural strengths were shown by the series containing recycled cement mortar.
In the presence of each of the additives, an increase in flexural strength was observed
with an increase in the curing temperature from 65 °C to 85 °C, but in the presence of

FA, FAS and CW it was insignificant, while in composites with recycled mortar, it was even 80%.

As was observed before, the highest compressive strength results were obtained for the 6_25%FAS_85 series with ash and slag mix, annealed at 85 °C (26.85 MPa) and they were 15% higher compared to the series containing 100% FA (3_100%FA_85). The lowest compressive strength in the experiment was obtained for the series containing ceramic fines (CW), but it was in the presence of this additive that the highest and equal to 64% increase in compressive strength was observed with increasing the curing temperature from 65°C to 85°C. A similar increase in compressive strength (by 63%) resulting from the increase in temperature was recorded for the series with the RCM recycling mortar, in this case, the strength results were only up to 13% lower compared to those obtained for the composite with 100% FA_65. The highest Strength Activity Indices (SAI) above 1.1 were obtained for the series with FAS. It should be noted that the chemical composition of FAS is the closest to the chemical composition of FA, but it contains more aluminium and calcium oxides, as well as large amounts of silicon oxide, which are the basis for the construction of geopolymer composites forming a C-A-S-H gel [14].

The obtained results show that an increase in the curing temperature from 65 °C to 85 °C generally causes a decrease in bulk density for each composite with additives, both in the dry and saturated state. Taking into account the geopolymer composites heated at 85 °C, the highest bulk density in the dry and saturated state was obtained by composites in the 25%FAS series, respectively 2.03 and 2.14 g/cm^3, and the lowest in the 100% FA series, 1.84 and 1.98 g/cm^3. Analyzing the obtained results of tests on the weight absorption of composites, it can be concluded that the increase in the temperature of maturation of geopolymers was accompanied by an increase in the absorption of composites. Curing temperature has a significant effect on the properties of geopolymers because it affects specimen setting and hardening. Synthesized products are very sensitive to experimental conditions. The percentage of water absorption increased after curing for a certain period of time at higher temperature. Prolonged curing at higher temperatures can break down the granular structure of geopolymer mixture. This results in dehydration and excessive shrinkage due to contraction of the gel, which does not transform into a more semi-crystalline form [15, 16]. The highest water absorption of 7.4% of the mass was obtained for the 100%FA_85 series, and the lowest 3.8% of the mass for the 25%CW_65 series. This may be due to production of more compacted specimens. Fine particles are capable to fill the vacancies and produce more densified specimens.

The aim of microstructural analysis was to examine the differences between the microstructure of composites with the same composition but cured in different temperature. SEM examinations might also help to discover the causes of observed increase of mechanical properties with simultaneously decrease of volume density. Figures 1 and 2 present microstructure of geopolymers containing 25% of CW and cured at 65 °C and 85 °C, respectively.

Microstructure of analysed series 7_25%CW_65 and 9_25%CW_85 was porous at about the same level. The main difference observed between those two samples was in the shape of the C-A-S-H gel which seems to be more dense in the serie 9_25%CW_85 than

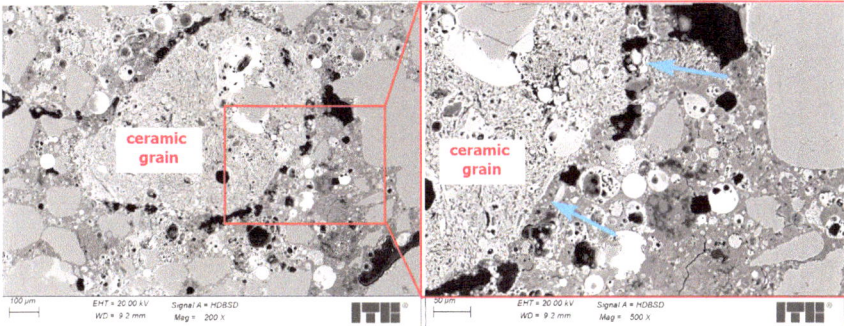

Fig. 1. Geopolymer (CW) – series 7_25%CW_65; arrows marks the transition zone

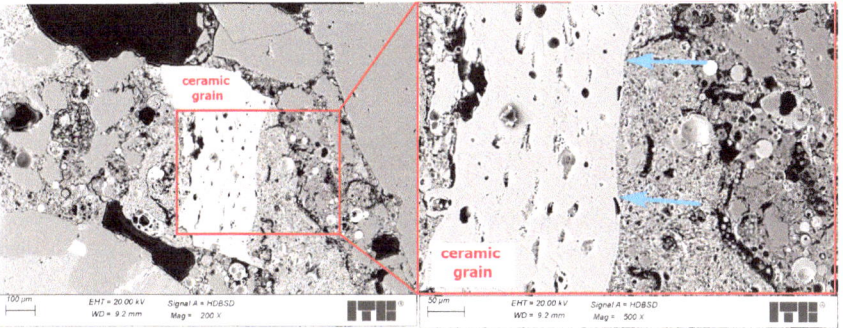

Fig. 2. Geopolymer (CW) – 9_25%CW_85; arrows marks the transition zone

in 7_25%CW_65. Also the transition zone between ceramic grains and the geopolymer gel was different. In the sample cured in lower temperature the transition zone was porous and discontinuous, but in the sample cured in higher temperature the transition was more sealed and had a better contact with the ceramic grain. The better formation and more sealed transition zone in higher temperatures and different microstructure of C-A-S-H gel might be the cause of better mechanical properties obtained by those geopolymers.

4   Conclusions

Based on the results of the conducted research, the following conclusions can be observed:

1. The curing temperature has a clear influence on the analyzed physical and mechanical properties of the composites with additives. The increase in temperature from 65 °C to 85 °C usually resulted in an increase in bending and compressive strength, a decrease in dry and saturated bulk density, an increase in water absorption by weight for all series of composites, regardless of the type of additive used.
2. The highest bending strength and compressive strength results were for composites, where 25% of the mass of the basic additive (fly ash), fly ash-slag mixture was used (the 6_25% FAS_85 series).

3. The results of the microstructural analysis showed changes in the mechanical properties of hardened geopolymers at higher temperatures. Higher temperatures resulted in a dense and sealed interphase transition zone and various microstructures of C-A-S-H gel in composites, which could have a good impact on the mechanical properties of geopolymers.
4. The use of the analysed waste additives in the amount of 25% fly ash and providing the composites with curing at a temperature of 85°C allows to obtain strength values similar to those obtained for composites where only fly ash was used.
5. The application of poor-quality waste in geopolymer composites is a good solution for their management and it is consistent with the circular economy policy.

Acknowledgements. The study was performed under the research project number WZ/WB-IIL/5/2023 funded by the Polish Ministry of Education and Science.

References

1. Keeble, B. R.: Report of the World Commission on Environment and Development: Our Common Future (1988)
2. Comisión Europea: The European Green Deal. Eur. Comm. **53**(9), 24 (2019)
3. Mudeme, L.: Cement production and greenhouse gas emission: implications for mitigating climate change (2009)
4. Flower, D., Sanjayan, J.: Green House Gas Emissions due to Concrete Manufacture. The Int. J. Life Cycle Assess. **12**(282–288)
5. Ali, M.B., Saidur, R., Hossain, M.S.: A review on emission analysis in cement industries. Renew. Sustain. Energy Rev. **15**(5), 2252–2261 (Jun 2011)
6. Antunes, M., Santos, R.L., Pereira, J., Rocha, P., Horta, R.B., Colaço, R.: Alternative clinker technologies for reducing carbon emissions in cement industry: a critical review. Materials (Basel) **15**(1), 209 (Dec 2021)
7. Sousa, V., Bogas, J.A.: Comparison of energy consumption and carbon emissions from clinker and recycled cement production. J. Clean. Prod. **306**, 127277 (Jul 2021)
8. Singh, N.B., Middendorf, B.: Geopolymers as an alternative to Portland cement: an overview. Constr. Build. Mater. **237**, 117455 (Mar 2020)
9. Nazneen, C.D. Sundeep, Vemuri, L.: Geopolymer – a Potential Alternative Binder for the Sustainable Development of Concrete Without Ordinary Portland Cement **33**, 1500–1504 (2017)
10. Toniolo, N., et al.: Fly-ash-based geopolymers: How the addition of recycled glass or red mud waste influences the structural and mechanical properties. J. Ceram. Sci. Technol. **8**(3), 411–419 (2017)
11. Horan, C., Genedy, M., Juenger, M., van Oort, E.: Fly Ash-based Geopolymers as lower carbon footprint alternatives to Portland cement for well cementing applications. Energies **15**(23), 8819 (Nov 2022)
12. Mishra, J., Nanda, B., Patro, S.K., Krishna, R.S.: Sustainable Fly Ash based geopolymer binders: a review on compressive strength and microstructure properties. Sustainability **14**(22), 15062 (Nov 2022)
13. Chyliński, F., Kuczyński, K.: Ilmenite mud waste as an additive for frost resistance in sustainable concrete. Materials (Basel) **13**(13), (2020)

14. Xie, J., Wang, J., Rao, R., Wang, C., Fang, C.: Effects of combined usage of GGBS and fly ash on workability and mechanical properties of alkali activated geopolymer concrete with recycled aggregate. Compos. Part B: Eng. **164**, 179–190 (2019)
15. van Jaarsveld, J.G.S., van Deventer, J.S.J., Lukey, G.C.: The effect of composition and temperature on the properties of fly ash- and kaolinite-based geopolymers. Chem. Eng. J. **89**, 63–73 (2002)
16. Pawluczuk, E., Kalinowska-Wichrowska, K., Jimenez, J.R., Fernandez, J,M., Suescum Morales, D.: Geopolymer concrete with treated recycled aggregates: Macro and microstructural behavior. J. Build. Eng. **44** (2021)

Chloride Diffusion and Mechanical Performances of Geopolymer Concrete with Blended Precursor

Patrycja Duży[1,2,2(✉)], Izabela Hager[1], Marta Choińska-Colombel[2], and Ouali Amiri[2]

[1] Cracow University of Technology, Kraków, Poland
patrycja.duzy@doktorant.pk.edu.pl

[2] IUT Saint-Nazaire, Research Institute in Civil and Mechanical Engineering GeM—UMR CNRS 6183, Nantes University, Nantes, France

Abstract. Geopolymer concrete is an environment-friendly material and is presently accepted as an alternative to conventional concrete. It utilizes industrial by-products like fly ash and slag to reduce CO_2 emissions associated with cement production. Despite being investigated over the decades, the application of geopolymers in construction is still very limited. Most of the research data refer to geopolymer pastes and mortars and their properties, performances, and durability. Although geopolymer concretes are well-accepted in the research community owing to their comparable or even better performances as a cement substitution.

In this paper, the precursors for geopolymer concrete preparations are blends of fly ash (FA) and ground granulated blast-furnace slag (GGBFS) in three slag proportions: 5%, 20%, and 35% expressed as a percent of FA mass. The concretes were denominated AAC5, AAC20, and AAC35, respectively. Their basic physical and mechanical characteristics were investigated, as were their transport properties of chloride ions. The ASTM C1556 test was applied to determine the chloride ions' penetration of the geopolymers. The measurements revealed a strong dependence between chloride penetration through the concrete and the precursor composition.

Keywords: Geopolymer concrete · chloride ions · Fly Ash Concrete · Alkali Activated concrete · ASTM C1556

1 Introduction

The work on material that can replace Portland cement in recent decades, has become the target of numerous studies. Cement production alone contributes between 5% to 7% of the anthropogenic CO_2 emissions worldwide [1]. Due to successive regulations limiting carbon dioxide emissions, the intensification of research in this area results in a growing base of theoretical [2] and practical [3–6] knowledge about geopolymer binders. Geopolymer concrete is a type of environmentally friendly concrete made from industrial waste materials such as fly ash, slag, and silica fume. The use of blended precursors, such as a mixture of fly ash and slag, has been shown to improve the mechanical and durability properties of geopolymer concrete.

© The Author(s) 2025
L. Czarnecki et al. (Eds.): ICPIC 2023, 61, pp. 220–229, 2025.
https://doi.org/10.1007/978-3-031-72955-3_22

Several studies have investigated the mechanical properties of geopolymer concretes with blended precursors. According to Kumar et al. [7], the compressive strength of geopolymer concrete is influenced by the type and ratio of the blended precursors used. The study showed that the use of a mixture of fly ash and slag improved the compressive strength compared to using fly ash alone. Another study by Bouziani et al. [8] investigated the effect of different proportions of fly ash and slag on the mechanical properties of geopolymer concrete. The results showed that an optimal ratio of fly ash and slag can lead to an increase in compressive strength, flexural strength, and tensile strength compared to using fly ash or slag alone.

The durability of geopolymer concrete is also affected by its resistance to chloride ion penetration. According to Raza et al. [9], the use of blended precursors can improve the resistance of geopolymer concrete to chloride ion penetration. The results showed that combining fly ash and slag in the right proportion can significantly slow down the penetration of chloride ions compared to using either one of them individually. Another study by Chen et al. [10] investigated the effect of different curing conditions on the resistance of geopolymer concrete to chloride ion penetration. The results showed that proper curing can improve the resistance of geopolymer concrete to chloride ion penetration, and the use of blended precursors can further enhance this resistance.

The study discussed in this paper examines the relationship between the precursors' compositions and the durability and mechanical strength of geopolymer concretes, as also the penetration of chloride ions. The analysis of the results was enhanced with the examination of mercury porosity data, leading to credible conclusions.

2   Materials and Methods

The research was carried out on geopolymer concretes, utilizing blends of FA (from the Połaniec power plant in Poland) and GGBFS (from Ekocem in Poland) as the precursors. The oxide compositions of both precursors are provided below (Table 1).

The activator for preparing the geopolymer binder was an aqueous solution of sodium silicate Geosil® 34417, with an additional amount of water added. Its chemical composition is presented in Table 2.

Three precursor blends were made by substituting FA with GGBFS in the ratios of 5%, 20%, and 35%, expressed as a percentage of the FA mass. These blends were used to prepare three concretes, referred to as AAC5, AAC20, and AAC35, respectively.

The design specifications for the mixtures were based on practical experiences [4, 5] and literature references [11]. Trial batches of concretes with varying slag content were prepared and subjected to preliminary testing. Considering the consistency, physical and mechanical properties of the mixtures, the final compositions of Alkali activated concretes were established. Each of the recipes, outlined in Table 3, can be defined by the following attributes:

- Water/binder (w/b) ratio = 0,37
- Alkaline Solution/binder ratio = 0,53
- Amount of paste in concrete = 300 dm3/m3

Where:

Table 1. Chemical compositions of FA and GGBFS

wt.%	SiO$_2$	Al2O3	Fe$_2$O$_3$	CaO	MgO	SO$_3$	K$_2$O	Na$_2$O	P$_2$O$_5$	TiO$_2$	Mn$_3$O$_4$	Cl$^-$
FA	52.30	28.05	6.32	3.05	1.71	0.28	2.51	0.76	0.69	1.35	0.07	–
GGBFS	39.31	7.61	1.49	43.90	4.15	0.51	0.36	0.47	–	–	–	0.04

Table 2. Chemical composition of Geosil® 34417

Characteristic	Unit	Woellner Geosil® 34417
Na_2O content	wt.%	16.74
SiO_2 content	wt.%	27.5
Density	g/cm^3	1.552
Viscosity	mPa*s	470
Weight ratio (WR = wt.% SiO_2/wt. Na_2O)	–	1.64
Molar ratio (MR = mol SiO_2/mol Na_2O)	–	1.70

Water - a mass of water added and contained in Geosil® 34417.
Binder - a sum of FA and GGBFS (by mass).
Alkaline Solution - a mass of diluted Geosil® 34417.

Table 3. Composition of Alkali Activated concretes.

[kg/m^3]	AAC5B	AAC20B	AAC35B
FA	336.9	292.3	244.1
GGBFS	17.7	73.1	131.4
Alkaline Solution + water	189.4	195.1	200.5
Sand 0/2	662.4	662.4	662.4
Basalt 2/8	708.9	708.9	708.9
Basalt 8/16	648.4	648.4	648.4

The samples were made following the guidelines of EN 206 + A2 [12]. They were left in molds for 24 h and then taken out and stored in a laboratory environment (18 ± 2 °C) under plastic wrap to prevent water evaporation. Compressive and splitting tensile strength were performed according to PN-EN 12390-3 [13] and PN-EN 12390–6 [14] standards. The experiments were conducted on cubic specimens with sides measuring 100 mm, at 28 and 180 days.

The study of the chloride diffusion coefficient utilized the ASTM C1556 test method. 360-day-old cylindrical specimens, with dimensions of Ø11 x 5 cm, were cut from larger Ø11 x 22 cm cylinders. The samples were prepared by coating them with epoxy resin and soaking them in distilled water before being placed in a test cell filled with a 16.5% sodium chloride solution for 35 days.

After 35 days of immersion in the 16.5% aqueous NaCl solution, the test specimens were taken out of the solution and allowed to air-dry for 24 h in laboratory conditions. Then, the specimens were ground with the 'Germann Instruments' Profile Grinder to produce obtain samples. The grinding process was performed in parallel layers to the exposed surface, with a maximum depth of 5 mm for each layer, ensuring that the weight of the powder sample was at least 10 g for each layer.

Investigations of the acid-soluble chloride contents of the powder samples were performed according to ASTM C1152 [15] method with the use of 5 g instead of 10 g recommended. The process of measuring the chloride content involved taking two 50 ml samples from a solution that was prepared from each powder sample. Each of the solutions had a volume of 250 ml. The average value of chloride content was calculated using a potentiometric titration machine, with a silver nitrate solution as recommended in the ASTM C114 standard [16]. The apparent chloride diffusion coefficient (D_a) and surface concentration of chloride ions (C_s) were calculated by applying Eq. 1 to the chloride profile data obtained through non-linear regression analysis using the method of least squares, as per the ASTM C1556 standard.

$$C(x, t) = C_s - (C_s - C_i) \times erf\left(\frac{x}{\sqrt{4 \times D_a \times t}}\right) \tag{1}$$

where:

C(x, t) chloride concentration, measured at depth x and exposure time t, mass %,

C_s projected chloride concentration at the interface between the exposure liquid and test specimen that is determined by the regression analysis, mass %,

C_i initial chloride-ion concentration of the cementitious mixture prior to submersion in the exposure solution, mass %,

x depth below the exposed surface (to the middle of a layer), m,

D_a apparent chloride diffusion coefficient, m²/s,

T the exposure time, s,

erf the error function.

To fit Eq. (1) to experimentally obtained points MATLAB® code was used.

Tested materials were also subjected to a porosity analysis using Mercury Intrusion Porosimetry (MIP). It is a method used to measure the pore size distribution of materials. The measurement is performed by injecting mercury under high pressure into the sample and measuring the pressure required to intrude the mercury into the pores. The pressure at which the mercury begins to penetrate the sample is recorded, and this value is used to determine the pore size of the sample. Specimens' sizes were about 1 cm³. The results were used to interpret the chloride ion penetration results and as the supplementary characteristics of investigated materials.

3 Results and Discussion

3.1 Mechanical Characteristics

The findings of the compressive and tensile strength evaluations indicate that replacing FA with GGBFS had a positive impact on the mechanical properties of AA concretes.

The test outcomes demonstrate a nearly linear correlation between the content of GGBFS and the compressive and splitting tensile strength after 28 days.

However, results from tests performed after 180 days show a similar trend, although with slightly reduced precision. The results for both compressive strength and splitting tensile strength tests reveal an increase in values over time.

The results showed a remarkable increase in values for the AA concretes that had a low GGBFS content (in particular, the AAC5 average compressive strength rose from 22.5 to 41.0 MPa and the average splitting tensile strength increased from 3.10 to 4.85 MPa). The results are presented in Figs. 1 and 2.

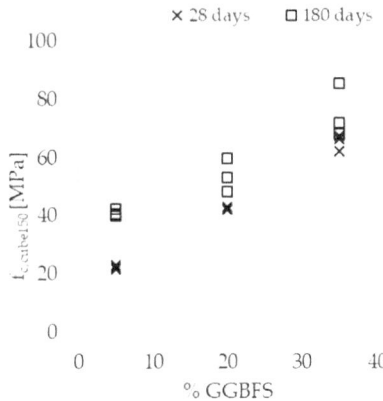

Fig. 1. AACs' compressive strength after 28 and 180 days

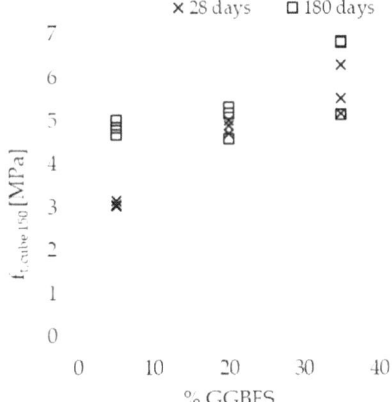

Fig. 2. AACs' splitting tensile strength after 28 and 180 days

3.2   Chloride Diffusion Coefficient (Da)

The tests performed provided the profiles of chloride content in the materials. Figures 3–5 display the chloride profiles of the AACs and the curves fitted, as per Eq. (1), to the experimental data points.

Figures 3, 4, and 5 demonstrate the positive impact of adding GGBFS on the depth of chloride penetration. The specimen containing 5% slag has the highest concentration of

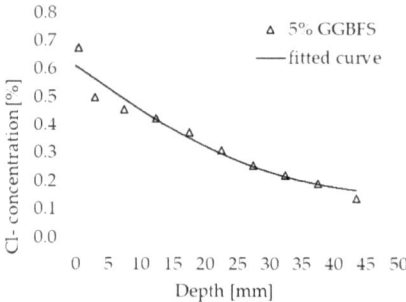

Fig. 3. Chloride profile of AAC5 and curve fitted to the experimental results

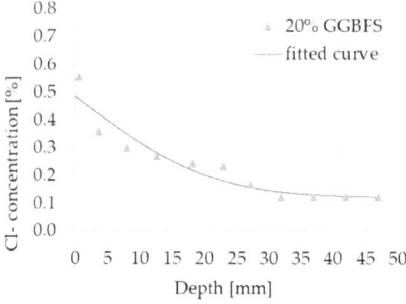

Fig. 4. Chloride profile of AAC20 and curve fitted to the experimental results

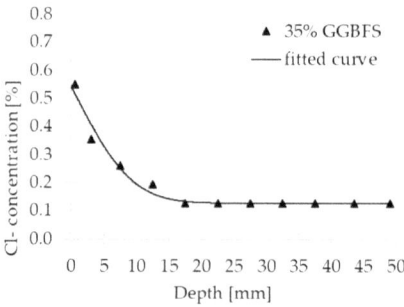

Fig. 5. Chloride profile of AAC35 and curves fitted to the experimental results

chloride ions. The curve fitting determined that the surface chloride concentration was 0.61% (the percentage of chloride ion mass in the mass of the powder sample), whereas, for AAC20 and AAC35, the values were 0.49% and 0.54% respectively. The concretes with the greatest amount of GGBFS showed a rapid decrease in chloride concentration, reaching 0.13% in the fifth layer at a depth of 17.5 mm. In contrast, the decrease in chloride ions was slower in concretes with low slag content. For AAC20 and AAC5, the chloride concentration reached 0.13% only at depths of 32 mm and 43.5 mm respectively.

The characteristics of porous materials, such as cement concrete and AAC, are largely determined by the volume and size distribution of their pores. The values of total porosity, compressive and splitting strength, as also chloride diffusion coefficients are presented in Table 4.

Table 4. Values of f_c, f_t, D_a, and total porosity for AACs

Parameter	AAC5	AAC20	AAC35
$f_{c(180\ days)}$ [MPa]	43.16	56.62	79.18
$f_{t(180\ days)}$ [MPa]	5.11	5.29	6.60
D_a [m^2/s]	90.67	44.38	8.55
Porosity [%]	14.28	9.18	7.30

The total porosity of hardened AA concretes is strongly associated with precursors' compositions. When it comes to the overall pore content of the materials, elevating the level of GGBFS substitution for FA reduces porosity by nearly 50%. Total porosity of AAC5, AAC20 and AAC35 are 14.28%, 9.18% and 7.30%, respectively.

Resistance to chloride ions' aggression is a complex phenomenon strongly associated with total porosity and pore size distribution but also tortuosity and chloride binding capacity. The chemical properties of the material, which are the subject of further studies, may also affect the penetration of diseased ions.

Noushini A. [11] defined the superior value of chloride diffusion coefficient for geopolymer concretes applied in chloride environments as 14 x 10–12 m2/s. That recommendation was based on experimental results according to standard ASTM C1556 [17]. Values of diffusion coefficient for AAC35 seem to be promising and encourage further research on the material.

Certainly, the variable content of GGBFS in precursors' compositions triggered changes in the structure's development process. Except for the slag addition, AACs' components were matching as well as hardening and storage conditions. Therefore the results presented in this paper are closely related to the precursor used. In the case of such assumptions, it can be concluded that the addition of slag affects the pore structure which contributes to the features dependent on porosity.

4   Conclusions

The study results suggest that the inclusion of GGBFS has a substantial impact on the mechanical characteristics and chloride resistance of geopolymer concrete. The tests conducted demonstrate that it is feasible to achieve desirable properties for geopolymer concrete without the need for heating the material.

The conducted research allows for drawing general conclusions:

- The results of the study demonstrate that the use of GGBFS in Alkali Activated concrete reduces the porosity significantly.

- The use of higher levels of GGBFS in the mix with FA of geopolymer concrete's binder significantly enhances its mechanical strengths, such as compressive and splitting tensile strength
- An increase in the amount of blast furnace slag enhances the ability of geopolymer concrete to withstand attacks from chloride ions.
- It has been observed that there is a nearly linear correlation between the amount of GGBFS in the composition of the material and its compressive strength and splitting tensile strength.
- The trend of improvement in the mechanical properties of the materials, particularly for concretes with a low amount of GGBFS, becomes more apparent over time.

The outcomes of the study suggest the need for continued exploration to comprehend the chloride penetration process in geopolymer concrete and the effect of incorporating blast furnace slag on the material's physical and mechanical characteristics.

References

1. Meyer, C.: The greening of the concrete industry. Cement Concr. Compos.Concr. Compos. **31**, 601–605 (2009)
2. Provis, J.L., Van Deventer, J.S.J.: Alkali activated materials, RILEM state art reports, RILEM TC 224-AAM vol. 13; Springer: Berlin, Germany (2014)
3. Law, D.W., et al.: Long term durability properties of class F fly ash geopolymer concrete. Mater. Struct.Struct. **48**, 721–731 (2014)
4. Hager, I., Sitarz, M., Mróz, K.: Fly-ash based geopolymer mortar for high-temperature application - effect of slag addition. J. Clean. Prod. **316**, 128168 (2021)
5. Sitarz, M., Hager, I., Choińska, M.: Evolution of mechanical properties with time of fly-ash-based geopolymer mortars under the effect of granulated ground blast furnace slag addition. Energies **13**, 1135 (2020)
6. Duży, P., Sitarz, M., Adamczyk, M., Choińska, M., Hager, I.: Chloride ions' penetration of fly ash and ground granulated blast furnace slags-based alkali-activated mortars. Materials **14**, 6583 (2021)
7. Kumar, S., Pazhani, K.C., Ravisankar, K.: Studies on fly ash and slag blended geopolymer concrete. J. Struct. Eng.Struct. Eng. **43**, 303–310 (2016)
8. Bouziani, N., Sabir, B. B., Mahfoud, A. Influence of fly ash and slag proportions on the mechanical properties of geopolymer concrete. Constr. Build. Mater. **235**, 117854. https://doi.org/10.1016/j.conbuildmat.2019.117854. (2020)
9. Raza, M., Suthar, J.S., Mehta, Y.R.: Chloride ion resistance of geopolymer concrete made with blended precursors. J. Clean. Prod. **258**, 120537 (2020). https://doi.org/10.1016/j.jclepro.2020.120537
10. Chen, J., Yu, Y., Fan, Y.: Study on curing effect on chloride ion resistance of geopolymer concrete with blended precursors. Materials **12** (18), 2980. https://doi.org/10.3390/ma1218 2980. (2019)
11. Noushini, A., Castel, A.: Performance-based criteria to assess the suitability of geopolymer concrete in marine environments using modified ASTM C1202 and ASTM C1556 methods. Mater. Struct.Struct. **51**, 146 (2018)
12. PN-EN 206+A2:2021-08 Concrete -- Specification, performance, production and conformity, PKN/KT 274, Warsaw, Poland (2021)
13. PN-EN 12390-3:2019 Badania betonu -- Część 3: Wytrzymałość na ściskanie próbek do badań

14. PN-EN 12390-6:2011 Badania betonu -- Część 6: Wytrzymałość na rozciąganie przy rozłupywaniu próbek do badań
15. ASTM C1152-20 Standard Test Method for Acid-Soluble Chloride in Mortar and Concrete
16. ASTM C114–22 Standard Test Methods for Chemical Analysis of Hydraulic Cement
17. ASTM C1556-11a. Standard test method for determining the apparent chloride diffusion coefficient of cementitious mixtures by bulk diffusion; ASTM international: West Conshohocken, PA, USA (2016)

The Effects of Calcium and Phosphate Compounds on the Mechanical and Microstructural Properties of Fly Ash Geopolymer Mortars

Piotr Prochoń[1,1(✉)], Tomasz Piotrowski[1], Luc Courard[2], and Zengfeng Zhao[3]

[1] Warsaw University of Technology, Warsaw, Poland
piotr.prochon@pw.edu.pl
[2] Urban and Environmental Engineering, University of Liège, Liège, Belgium
[3] College of Civil Engineering, Tongji University, Shanghai, People's Republic of China

Abstract. Phosphorus and calcium compounds are present in the chemical composition of byproducts from coal and biomass combustion. They may have an influence on the microstructure and the mechanical properties through the specific bonds in polymeric aluminosilicates - geopolymers. Results proved that the 5% of CaO added to high-silica fly ash geopolymer increases material density and mechanical properties. Phosphate compounds available in the biomass fly ash have a negative effect on geopolymer mortars by increasing porosity and decreasing their compressive strength.

Keywords: Phosphate · Calcium additives · Geopolymers · mechanical properties

1 Introduction

Geopolymers are a kind of inorganic polymer material with a three-dimensional network structure ranging from amorphous to semicrystalline [1]. Initially, geopolymers were strictly defined as alkali-aluminosilicate (AAS) geopolymers produced by the reaction of aluminosilicate precursors (such as clay minerals, metakaolin, fly ash, volcanic ash, slag, and so on) with an alkali activator. Geopolymer, however, can be obtained by two different routes, such as in an alkaline medium (sodium silicate or sodium hydroxide solution) described above and in an acidic medium with phosphoric acid or humic acids [2]. When sodium or potassium silicate is used as a hardener, the geopolymer network is based on poly (sialate), whereas phosphoric acid geopolymers are based on poly (phospho-siloxane). Both systems have been examined separately, with mechanical and microstructural properties as well as reaction processes documented. Geopolymers derived from phosphoric acid solutions were shown to have higher compressive strengths than those derived from sodium silicate solutions [3].

Different calcium sources (slag, limestone, wollastonite, etc.) have been used to investigate their effects on geopolymer composition and mechanical properties [4–6].

© The Author(s) 2025
L. Czarnecki et al. (Eds.): ICPIC 2023, 61, pp. 230–238, 2025.
https://doi.org/10.1007/978-3-031-72955-3_23

Geopolymer cement can be a mixture of polysialate or sodium, calcium aluminosilicate hydrate (N, C-A-S-H), and calcium silicate hydrate (C-S-H). This system involves the substitution of sodium by calcium during the condensation reaction and the formation of a distinct calcium phase, which is beneficial for strength development [7, 8]. Some researchers, such as Kang et al. [10] analyzed calcium phosphate compounds for producing an inorganic polymer that can be used as bioactive materials for bone tissue scaffolds. Such a system is interesting as calcium could substitute sodium or potassium in the network, and phosphates could replace some silicates during the formation of the geopolymer network, leading to a hybrid system based on Ca, Na-poly(sialate), and poly(phospho-siloxane) networks [11].

However, no literature discusses the influences of phosphate components available in biomass fly ashes on the mechanical and microstructural properties of fly ash-based geopolymers. This work aims to investigate fly ash geopolymer mortars with modified calcium levels and the effects of calcium phosphate compounds on their mechanical and microstructural properties.

2 Materials and Methods

The study was performed on geopolymer mortars designed with coal fly ash (RFA) as geopolymerization precursor, standard siliceous natural sand in accordance with EN 196–1 and alkaline activators - sodium hydroxide (NaOH) and calcium additives - quick lime (CaO) or biomass fly ash (BFA). According to Table 1, RFA demonstrates the chemical characteristics needed for Class F fly ash in EN 450–1 and ASTM C618 (the total amount of silica, aluminum, and iron oxides is roughly 83.07%). Due primarily to their contribution to the strength development following alkali-activation, Si and Al elements at high concentrations are essential to the geopolymerization process [12]. Biomass fly ash (BFA) was chosen for its low primary oxide content (less than 1%), but high phosphorus levels (more than 27%).

All geopolymer mortar mixes were prepared and formed according to a procedure modified from EN 196–1 as follows:

- Adding geopolymerization precursor to the mixer and operating the mixer at low RPM [1] (rotational movement - 140 ± 5 RPM-1);
- Pouring the sand in at an even rate for the first 30 s of mixing;
- Adding alkaline activator solution at an even rate for the next 30 s of mixing (the "zero time" for setting time measurement);
- Switching the mixer to high RPM (rotational movement 285 ± 10 RPM-1) and continuing mixing for 30s more.

A 5M NaOH solution was used to activate the fly ash. Calcium additions (CaO, biomass fly ash - BFA) replaced precursors by 2%, 5%, and 7% of their mass, respectively (Table 2). After 24 h, samples were demoulded and stored in laboratory conditions (20 ± 2 °C and 60% relative humidity) until testing. During the conditioning of the samples, no temperature curing was applied. The specimen references are based on general notation RFA-Z#, where "RFA" is the coal fly ash and "Z#" is a symbol of calcium additive – CaO (C), BFA (B) - with the rate of substitution "#".

Table 1. Chemical compound of fly ashes

Fly ash type/ wt%	SiO_2	TiO_2	Al_2O_3	Fe_2O_3	MnO	MgO	CaO	Na_2O	K_2O	P_2O_5	Na_2O_{eq}	LOI	Total
RFA	50.8	1.1	26.1	6.2	0.1	2.6	2.7	1.0	3.5	0.2	3.2	5.5	100.0
BFA	0.4	0.0	0.1	0.2	0.1	1.2	31.0	1.2	16.2	27.7	11.9	18.0	96.00

Table 2. Mortar mix design

Mix type	Fly Ash (g)	Sand (g)	NaOH pellets l(g)	Calcium additive (g)	Water (g)
RFA-N5-R	450.0	1350.0	60.0	0.0	256.0
RFA-N5-C2	441.0	1350.0	60.0	9.0	256.0
RFA-N5-C5	427.5	1350.0	60.0	22.5	256.0
RFA-N5-C7	418.5	1350.0	60.0	31.5	256.0
RFA-N5-B2	441.0	1350.0	60.0	9.0	256.0
RFA-N5-B5	427.5	1350.0	60.0	22.5	256.0
RFA-N5-B7	418.5	1350.0	60.0	31.5	256.0

After 7, 28, and 56 days, the mortars were tested for flexural and compressive strengths. The test was carried out on a Controls and Instron hydraulic press with a loading rate of 0.33 MPa/min and a sensitivity of 100kN for compressive strength and a loading rate of 0.017 MPa/min and a sensitivity of 5 kN for flexural strength. The result was the average of three flexural strength tests and six compression strength tests.

The Thermal Field Emission Scanning Electron Microscope (FE-SEM Zeiss EVO-40) and Carl-Zeiss EDS analyzer were used to analyze microstructure of mortar samples under high vacuum and 5 kV acceleration.

Samples porosity for chosen samples was measured by micro computed tomography (μCT). μCT test was performed on an XRADIA XCT-400 tomograph at a lamp voltage of 150 kV and a current of 60 A. 900 x-ray images were obtained with exposure time 8 s and resolution of 25 m. Cross-sections of the sample were reconstructed using the Feldkamp algorithm, resulting in over 900 virtual X-Y cross-sections covering a two-dimensional sample area.

3 Results

At 7, 28, and 56 days, the flexural strength of all CaO mortars was higher than the reference samples (Fig. 1). The increase in strength was further noticed with increasing CaO concentration, allowing for the best flexural results for RFA-N5-C7 (3.8 MPa). RFA mortars containing BFA are at the other end of the spectrum, with no flexural results at 7 days and values below 0.5 MPa at 28 days due to inadequate gepolymerization. The RFA replacement with BFA also had a significant impact on flexural strength growth after 56 days, decreasing from 53% to 73% compared to the RFA-N5-C5.

The Fig. 2 depicts the evolution of compressive strength with increasing calcium content. Compressive strength was less than 2 MPa for the majority of samples after 7 days. The RFA reference sample RFA-N5-R (0.3 MPa) and those with BFA content RFA-N5-B (2,5,7) produce the worst results (0.21 MPa, 0.18 MPa and 0.19, respectively). Only RFA mortars containing 5% and 7% CaO demonstrated greater early compressive strength values (3.0 MPa and 3.7 MPa, respectively). It is consistent with the findings of Buchwald et al., who found that adding $Ca(OH)_2$ to a geopolymer improved the

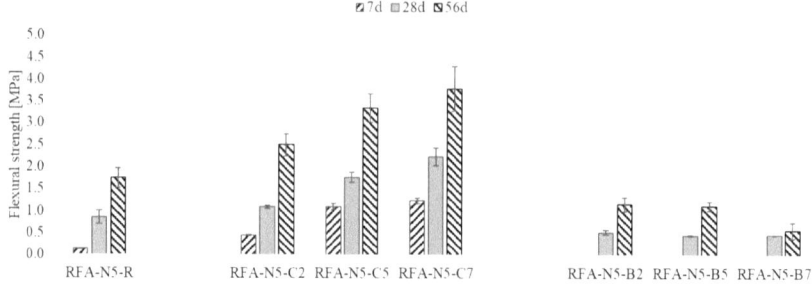

Fig. 1. Flexural strengths of RFA-based mortars

material's early mechanical performance [13]. The lower RFA reference mortar values were expected because alkali-activated fly ash has a relatively modest rise in mechanical characteristics with time when cured in ambient conditions [14]. RFA mortars with a greater CaO content outperformed reference mortars in terms of strength (CFA-N5-R and RFA-N5-R). The compressive strength growth tendency in CaO mortars is similar to the flexural strength development trend. This conclusion is consistent with recent studies on the increased strength of AAMs with increased calcium concentration [13]. It was also shown that after 56 days, mortars made with 5% and 7% CaO had higher compressive strengths, reaching a peak value close to 10 MPa among all tested fly ash mortars.

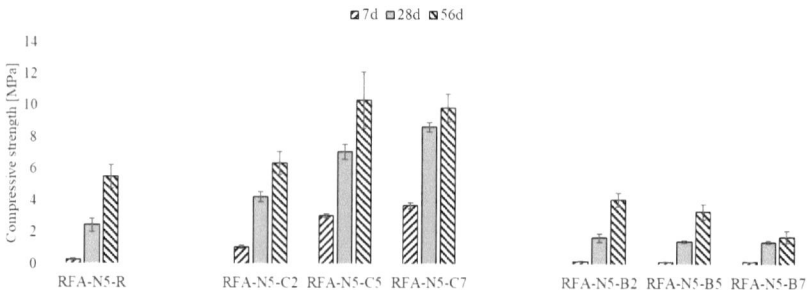

Fig. 2. Compressive strengths of RFA-based mortars

In comparison to samples activated with CaO, the mortars activated with high amounts of BFA gave reduced strengths at 28 days, as seen in Fig. 2. At 28 and 56 days, none of the RFA binders with BFA reached the reference value (2.4 MPa and 5.5 MPa, respectively). When compared to RFA-N5-R, compressive strength values at 56 days decreased by 68% when BFA content increased to 7% in RFA mortars.

The μCT test clearly presented differences in pore structure between RFA-N5-R, RFA-N5-C5 and RFA-N5-B5. The RFA-N5-C5 mortar was almost two times less porous than other samples. It had the lowest total porosity (2.74%) with mean pore size 440 μm. s (Fig. 3 b). In RFA-N5-B5 mortar, vast amount of closed spherical pores was observed.

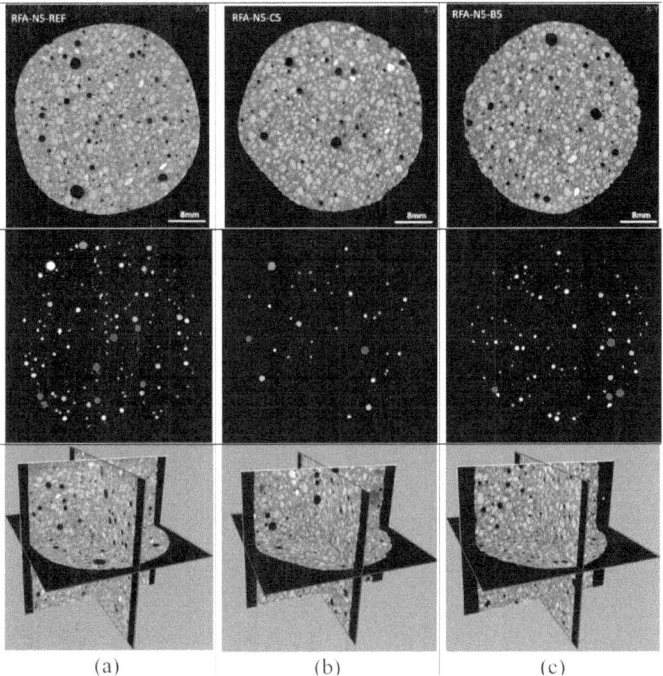

Fig. 3. The cross-sections and pores structures from μCT test: (a) RFA-N5-R; (b) RFA-N5-C5; (c) RFA-N5-B5

This structure can be connected with a production of amorphous calcium phosphate due to high levels of calcium and phosphate in BFA [15]. RFA mortar with BFA content had the highest total porosity (4.29%).

Figure 4 shows SEM images of alkali-activated RFA mortars. The surface morphology of RFA-N5-C5 (Fig. 4c, d) revealed substantially more dense microscale structures than other samples. It can be attributed to faster pozzolanic reaction rates and increased solidification rates of alkali-activated fly ash with calcium addition. According to Lee and Van Deventer [15], the presence of calcium in fly ashes can provide extra nucleation sites for precipitation of dissolved ions and cause fast hardness.

The RFA mortar microstructures revealed reacted amorphous microspheres and some partially reacted fly ash spheres throughout the matrix. During the dissolving of fly ash particles in specific parts of the RFA-N5-C5 sample, some micro shaped pores appeared in the matrix. At higher magnification (Fig. 4 b, f), it is clear that the fly ash spheres in RFA-N5-R and RFA-N5-B5 have not been completely dissolved. That can be justified for the reference sample due to insufficient curing temperature [1]. SEM scans of the sample RFA-N5-B5 revealed that fly ash grains were encased in a thin coating of another chemical substance. EDS analysis revealed that the material comprises phosphorus and is related to amorphous calcium phosphate [16]. Because BFA contains calcium and phosphate compounds, it appears that the retarding effects in binding are linked to the phosphate salts, as stated by Kalina [17].

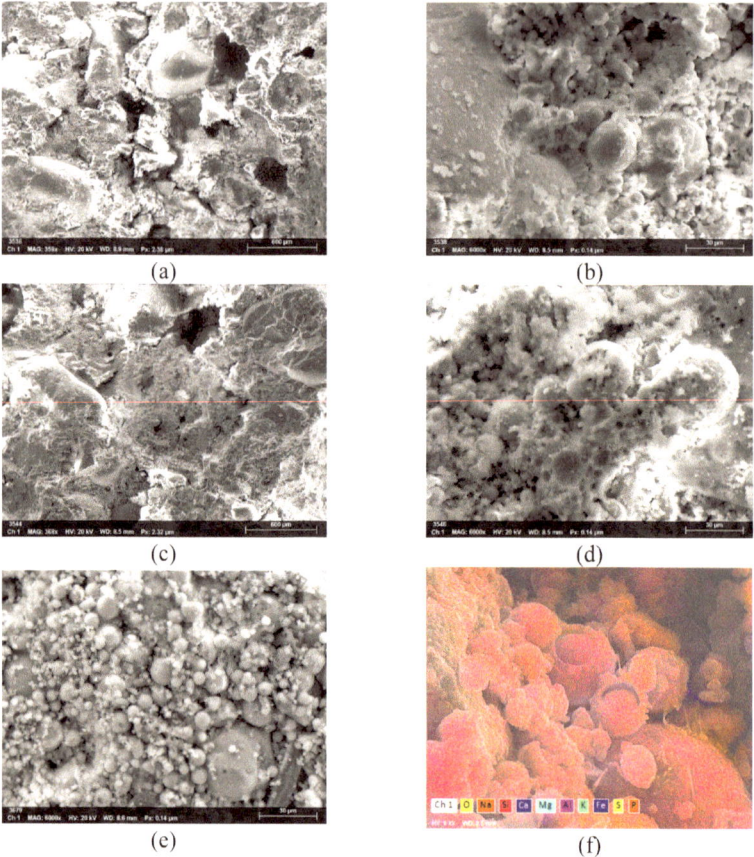

Fig. 4. SEM pictures of RFA mortars alkali-activated: (a,b) N5-R; (c,d) N5-C5 and (e,f) N5-B5

4   Conclusions

The following conclusions from the research described above can be drawn:

- Adding at least 5% of CaO to the geopolymer mortar can enhance material density, and mechanical properties. Mortars activated with sodium hydroxide and quicklime offered the highest mechanical properties suiting to M5 mortar requirements.
- Biomass fly ashes can be used as a high-calcium activator, when added to RFA based mortars. Mortars activated with sodium hydroxide and BFA could be used as general purpose mortar with lower mechanical properties.
- The retarding effect of biomass fly ash (BFA) is caused by its phosphorus compound that envelopes in a thin layer fly ash particle, postponing the activation reaction of the fly ashes in alkali-solution.

Acknowledgements. The authors would like to acknowledge the Regional Government of Wallonia (Belgium) and the European Regional Development Fund for their financial support through

ECOLISER (Eco-binders for Soil treatment, Waterproofing membranes and Roads) research project (2016–2020). They are also grateful to Wallonia Brussels Internation and the government of Poland for their support in scientific cooperation "*CarBoFLY Carbonation and biomass fly ashes for new concretes*". The authors would like to also express their appreciation for the support of the National Science Center Project Preludium 16 No 2018/31/N/ST8/02276 "*The influence of phosphorus oxide on development of microstructure, binding and mechanical properties of polymeric aluminosilicate composites from byproducts of coal and biomass combustion*".

References

1. Duxson, P., Fernández-Jiménez, A., et al.: Geopolymer technology: the current state of the art. J. Mater. Sci. **42**, 2917–2933 (2006)
2. Davidovits, J.: Geopolymers. Chemistry & Application., 3rd ed. Saint-Quentin, France, p. 6133. (2011)
3. Tchakouté, H.K., Rüscher, C.H.: Mechanical and microstructural properties of metakaolin-based geopolymer cements from sodium waterglass and phosphoric acid solution as hardeners: A comparative study. Appl. Clay Sci. **140**, 81–87 (2017)
4. Prochon, P., Zhao, Z., Courard, L., et al.: Influence of Activators on Mechanical Properties of Modified Fly Ash Based Geopolymer Mortars. Materials **13**, 1033 (2020)
5. Blaise, N., Ndigui, B., Emmanuel, Y., et al.: Effect of limestone dosages on some properties of geopolymer from thermally activated halloysite. Constr. Build. Mater. **217**, 28–35 (2020)
6. Chunjie, Y., Ping, D., Zuhua, Z., et al.: Durability performances of wollastonite tremolite and basalt fiberreinforced metakaolin geopolymer composites under sulfate and chloride attack. Construction and Building MaterialsVol. **134**, 56–66 (2017)
7. Susan, A.B., Nan, Y., John, L.P., Jiakuan, Yang.: "One-part geopolymers based on thermally treated red mud/NaOH blends. J. Am. Ceram. Soc. **98** no (2015)
8. John, L.: Susan ABernal "Binder chemistry-blended systems and intermediate C content." IA activated materials: S report, RILEM T 224-A, 125144 (2013)
9. Wang, X., Schröder, H.C., Müller, W.E.G.: Biocalcite, a multifunctional inorganic polymer: Building block for calcareous sponge spicules and bioseed for the synthesis of calcium phosphate-based bone. Beilstein J. Nanotechnol. **5**, 610–621 (2014)
10. Kang, Z., Zhang, X., Chen, Y., et al.: Preparation of polymer/calcium phosphate porous composite as bone tissue scaffolds. Mater. Sci. Eng., C **70**, 1125–2113 (2014)
11. Tchakouté, H.K., Fotio, D., Rüscher, C.H., et al.: The effects of synthesized calcium phosphate compounds on the mechanical and microstructural properties of metakaolin-based geopolymer cements. Constr. Build. Mater. **163**, 776–792 (2018)
12. Soutsos, M., Boyle, A.P., Vinai, R., et al.: Factors influencing the compressive strength of fly ash based geopolymers. Constr. Build. Mater. **110**, 355–368 (2016)
13. Dombrowski, K., Buchwald, A.: Weil: M The influence of calcium content on the structure and thermal performance of fly ash based geopolymers. J. Mater. Sci. **42**, 3033–3043 (2006)
14. Hou, Y., Wang, D., Zhou, W., Lu, H., Wang, L.: Effect of activator and curing mode on fly ash-based geopolymers. Journal of Wuhan University of Technology-Mater Sci Ed **24**, 711–715 (2009)
15. Lee, W.K.W., van Deventer, J.S.J.: Effects of anions on the formation of aluminosilicate gel in geopolymers. Ind. Eng. Chem. Res. **41**, 4550–4558 (2002)
16. Tas, A.C.: Calcium metal to synthesize amorphous or cryptocrystalline calcium phosphates. Mater. Sci. Eng., C **32**, 1097–1106 (2012)
17. Kalina, L., Bílek, V., Novotný, R., et al.: Effect of Na3PO4 on the hydration process of alkali-activated blast furnace slag. Materials **9**, 395 (2016)

Effect of Polymer Mortar Modification Using Eco-friendly Biochar on Microstructure

Kamil Załęgowski$^{(\boxtimes)}$ and Maja Kępniak

Faculty of Civil Engineering, Warsaw University of Technology, Warsaw, Poland
kamil.zalegowski@pw.edu.pl

Abstract. The construction sector should have much to offer in terms of helping to achieve circular economy goals, among others the use of waste materials. The example of such materials is biochar, a black porous and carbon-rich matter that could be converted from various waste biomass. A biochar could be utilized as microfiller in polymer concretes. This application of biochar is promising due to good interfacial bonding with polymer, no reactivity between surrounding polymer matrix and filler particles and fact that even fillers with irregular particles and large specific surface area could be utilized in polymer matrix. These create real opportunity to effectively dispose waste materials as a replacement of natural aggregates in polymer concrete technology. The presented paper is a second part of the research concerning the utilization of ecofriendly biochar in polymer composites conducted by authors. To better understand the impact of modification by biochar, already performed tests were supplemented by measurements of ultrasonic pulse velocity and quantitative analysis of microstructure.

Keywords: biochar · UPV · microstructure · image analysis

1 Introduction

As climate change is threatening the world and society grows exponentially, more and more waste and greenhouse gases being generated, environmental sustainability is being questioned. It is fundamental and urgent to develop a sustainable economy, actively promoting the efficient use of resources, highlighting product, component and material reuse. The construction sector should have much to offer in terms of helping to achieve circular economy goals, among others the use of waste materials – byproducts originating from industrial processes [1–3]. The example of such materials is biochar, a black porous and carbon-rich matter could be converted from various waste biomass, such as wood waste, agricultural waste, food waste, manure waste, and municipal/industrial sludge [Biochar as construction materials for achieving carbon neutrality]. According to the guidelines of the European Biochar Certificate [4] biochar is defined as charcoal resulting from biomass pyrolysis, a process in which organic substances decomposed at temperature ranging from 350 °C to 1000 °C in a low oxygen environment. Biochar has attracted the attention of researchers due to its unique properties, such as high surface area, high porosity, functional groups, high cation exchange capacity and stability, making it suitable for various applications [5, 6].

© The Author(s) 2025
L. Czarnecki et al. (Eds.): ICPIC 2023, 61, pp. 239–247, 2025.
https://doi.org/10.1007/978-3-031-72955-3_24

Numerous studies have concerned incorporation of biochar into cement composites in terms of cement hydration, mechanical properties, durability, microstructure, rheological properties, and carbon sequestration etc. [7–12], and the results of these studies have been very promising.

A biochar could be also utilized as microfiller in polymer concretes. Powder waste materials are often used as an essential microfiller in polymer concretes [13–15]. This application of biochar is promising due to good interfacial bonding with polymer, like other carbon-based fillers [16]. It is also well known that usually there is no reactivity between surrounding polymer matrix and filler particles [17]. In contrast to cement composites, even fillers with irregular particles and large specific surface area could be utilized in polymer concretes [18, 19]. These create real opportunity to effectively dispose waste materials as a replacement of natural aggregates in polymer concrete technology.

Important aspect of polymer concrete modification with biochar or any other material is its influence on the essential properties of fresh and hardened composite. Modification cannot negatively influence basic properties of hardened composite, what could threaten is quality and possibility of its application as a construction material. The influence of biochar incorporation as a partial replacement of quartz powder and/or resin has been already investigated by the authors [20]; tests conducted included course of setting, consistency, flexural and compressive strength. The results have shown that the curing time of two composites containing resin, quartz powder and biochar in vol. proportions 75:20:5 and 75:15:10 was the longest - the curing time about 5h in comparison to about 2h for pure resin, and about 3h for resin containing 25% by vol. of quartz powder. The highest temperature was noticed in case of composition with 10% by vol. of biochar, and it was just few degrees of Celsius higher than temperature measured for being next in line composition with 5% by vol. of biochar. To assess the consistency of mixes the flow table method acc. to EN 1015-3 was used. It was observed that increase in level of quartz powder substitution with biochar caused decrease in flow diameter. Despite a drop in flowability designed mixes could be considered as sufficiently workable, as it was possible to tightly fill molds. In terms of flexural strength, the composites have not shown statistically important differences. The obtained results were within standard deviations of the measurements made for individual compositions. Compressive strength have appeared more sensitive for presence of biochar. It has been concluded that the introduction of biochar into the polymer matrix yields a slight reduction in compressive strength. This effect is less pronounced when biochar replaces part of the polymer (a decrease of 5–7%) than when it replaces part of the quartz filler (a decrease of 10–15%). The dosage level of biochar up to 5% does not result in the significant reduction of compressive strength.

The presented paper is a second part of the research concerning the utilization of ecofriendly biochar in polymer composites conducted by authors. To better understand the impact of modification by biochar, authors decided to supplement already performed tests by measurements of ultrasonic pulse velocity and quantitative analysis of microstructure by image analysis method.

2   Materials and Methods

Composites in the study were polymer mortars prepared using synthetic vinyl-ester resin of low viscosity (350 ± 50 mPa·s at 25 °C), high flexural strength and high tensile strength (110 MPa and 75 MPa respectively, as declared by manufacturer). The chemical formula of vinyl-ester were presented in Fig. 1. Quartz powder and biochar were used as a microfillers. The analysis with a laser particle size analyzer (Horiba, Irvine, CA, USA) showed that about 85% of the particles of biochar were below 120 μm in diameter [20], and could be considered a microfiller particles. The biochar grain refinement differed significantly from that of the quartz powder, a traditional filler of polymer composites. Its grains were mostly not spherical in shape but had smooth surface without much branching that could increase specific surface area and resin demand (Fig. 2). Final mortars ingredient was fine aggregate in the form of CEN standard sand EN 196-1.

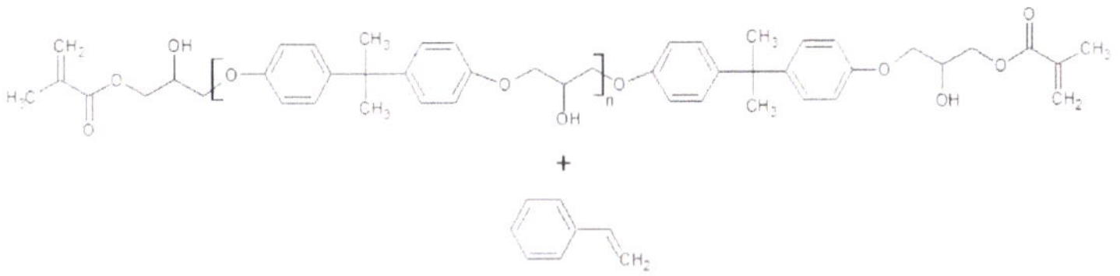

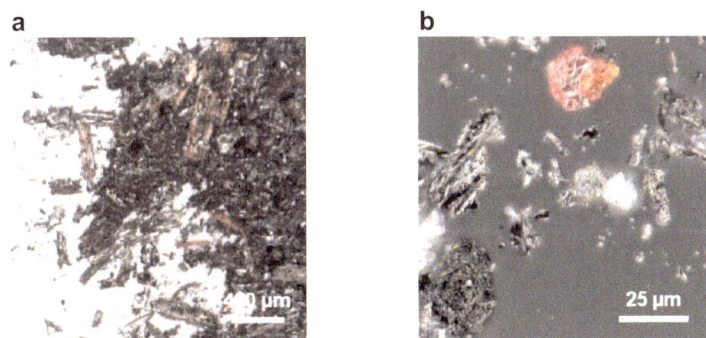

Fig. 1.   Vinyl-ester (VE) resin before cross-linking [22]

a   b

Fig. 2.   Optical microscope view of biochar: (a) overall picture; (b) shape of the single grain [20]

To determine the effect of biochar on polymer composites an experimental plan basing on the standards for ternary mixtures was prepared. The experimental plan was intended for the component proportions in the micro slurry. Therefore, the amount of resin (R), biochar (B), and quartz powder (Q) were applied as input variables. The mass of sand was constant for all mixes. While, the maximal and minimal masses of the individual components were determined in liminary studies, so to ensure the mixes will be workable enough to form samples. Consequently, the experimental plan had limitations (Fig. 3). In the experimental plan, the volumetric variation in the dosage of components was assumed. The detailed composition of composites used is given in Table 1.

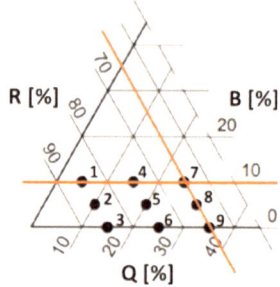

Fig. 3. Experimental plan with limitations (orange lines): R - resin, B - biochar, Q - quartz powder

Table 1. Compositions of analyzed mortars according to the experimental plan: R - resin, Q - quartz powder, B - biochar and S - sand

Composition number	Volume proportions [%]			Mass proportions [kg/m³]			
	R	Q	B	R	Q	B	S
1	85	5	10	451	60	79	1458
2		10	5		120	390	
3		15	0		180	0	
4	75	15	10	398	180	79	
5		20	5		239	39	
6		25	0		299	0	
7	65	25	10	345	299	79	
8		30	5		359	39	
9		35	0		419	0	

The process of sample preparation was divided into four steps: (1) manual mixing of the microfillers - quartz powder and biochar; (2) adding the microfillers mix to the resin and manual mixing; (3) mechanical mixing of the fillers and resin mix with quartz sand; and (4) forming samples by placing the ready composite mix in two layers in molds and compacting on the vibrating table by 5s.

To fully assess the influence of biochar on the polymer mortars the measurements already made in the first part of the research [20], were supplemented by measurements of ultrasonic pulse velocity and quantitative analysis of microstructure by computer image analysis.

The procedure of samples preparation for image analysis involved (1) cutting out slices of dimensions 10 mm x 40 mm x 40 mm from samples of each of 9 composites, (2) cold mounting in colored epoxy resin under lowered pressure, (3) three steps grinding and (4) final polishing. After this procedure, 2D images of microstructure were acquired on a scanner with a resolution of 800 DPI (1 pixel equals about 0,02 mm). The images were subjected to computer processing using photo editing software with a purpose to obtain the most precise binary image of black pores, which are of interest, on white background of aggregate and cement paste. In this study preparation of images for the analysis was the combination of contrast, brightness, gamma modulation and color saturation, followed by selection of pores base on the color and ending with binarization. The computer aided quantitative description of the microstructure of concretes was achieved by specialized software, developed at the Faculty of Materials Science and Engineering of Warsaw University of Technology. The image analysis was applied to calculate the volume fraction of air voids in material – V_V, and the surface area of pores in a volume unit of a material – S_V. The detailed information about the image analysis method and stereological parameters is presented elsewhere [23].

The ultrasonic testing was done by direct method (transmission method) using a digital ultrasonic flow detector and piezoelectric transducers of 100 kHz central frequency. To ensure adequate acoustic coupling between the concrete surface and the special head, commercial coupling gel was applied. The ultrasonic measurements included the determination of the propagation time of ultrasonic impulse between emitter and receiver, and then calculation of the wave velocity by dividing the distance between transducers by the time of travel. The signals were registered using specialized program and to reduce random error each signal was averaged 10 times. The UPV was determined 2 times per sample and including fact that 2 samples per each composition were tested, finally 4 UPV results per composition were computed. The full description of the ultrasonic methods and UPV calculations are presented elsewhere [24].

3 Results and Discussion

3.1 Ultrasonic Measurements

Direct ultrasonic method was used to determine the ultrasonic pulse velocity (UPV) (Table 2). It was concluded that the increase in biochar content in fillers mix led to decrease in ultrasonic pulse velocity, the most significant in case of composites containing 75% of resin by vol. (Fig. 4). The empirical results for all three contents of resin fit the regression lines with at least good - determination coefficient ($r^2 > 0.87$). Although should be noticed that some of the results were characterized by quite high standard deviation and were within standard deviations of the measurements made for another composites.

3.2 Quantitative Analysis of Microstructure

Quantitative analysis of microstructure was performed by image analysis method, the calculated stereological parameters are given in Table 2. Quite very strong relation between relative pore volume V_V and biochar content in fillers mix was observed only for composites containing 85% of resin by vol. ($r^2 = 0.99$). In case of composites with 75% of resin the share of biochar has no influence on the V_V – the regression line is almost flat. Whereas for the composites containing 65% of resin, in which the share of fillers were the greatest, there was no relation of V_V with share of biochar in total mass of microfillers.

Table 2. The results of ultrasonic pulse velocity and quantitative image analysis

Composition number	Volume proportions [%]			UPV [m/s]	Stereological parameters	
	R	Q	B		V_V [%]	S_V [mm^{-1}]
1	85	5	10	2835 ± 43	4,79 ± 0,69	0,34 ± 0,04
2		10	5	2920 ± 103	3,06 ± 0,60	0,23 ± 0,02
3		15	0	3053 ± 98	2,11 ± 0,35	0,18 ± 0,02
4	75	15	10	2928 ± 137	3,68 ± 0,38	0,32 ± 0,03
5		20	5	3200 ± 184	3,45 ± 0,34	0,23 ± 0,01
6		25	0	3240 ± 152	3,48 ± 0,32	0,25 ± 0,02
7	65	25	10	3038 ± 19	4,01 ± 0,23	0,32 ± 0,01
8		30	5	3070 ± 84	2,63 ± 0,38	0,24 ± 0,02
9		35	0	3086 ± 92	3,55 ± 0,29	0,31 ± 0,03

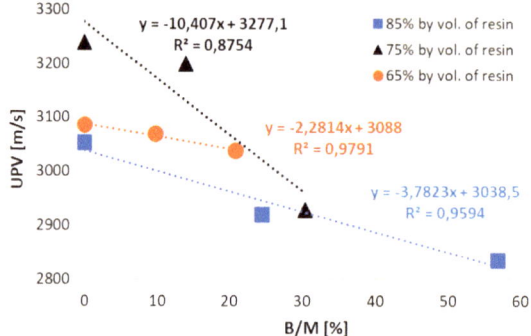

Fig. 4. The ultrasonic pule velocity (UPV) as a function of share of biochar (B) in total mass of fillers (M) and resin content

The changes in the value of S_V in the function of share of biochar in total mass of microfillers is similar to changes in V_V. Clear, very strong relation could be noticed in case of compositions with 85% of resin by vol. ($r^2 = 0,99$). While for the lower contents of the resin (75% and 85%) and the higher content of the microfillers there seem to be no relation between S_V and share of biochar (Figs. 5 and 6).

4   Conclusions

The aim of this paper was to analyze how the increase in content of biochar in polymer mortar may influence the ultrasonic pulse velocity as well as stereological parameters of microstructure calculated by image analysis method. It was concluded that the increase in biochar content in fillers mix led to decrease in ultrasonic pulse velocity, the most significant in case of composites containing 75% of resin by vol. The increase in content of

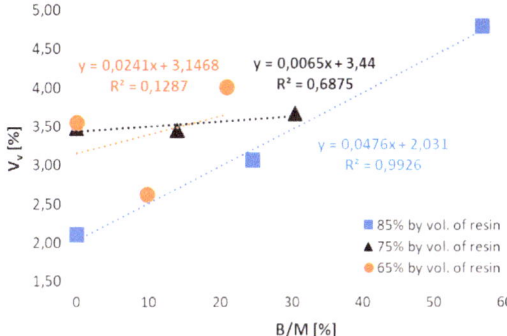

Fig. 5. The volume fraction of air voids in material (V_V) as a function of share of biochar (B) in total mass of microfillers (M) and resin content

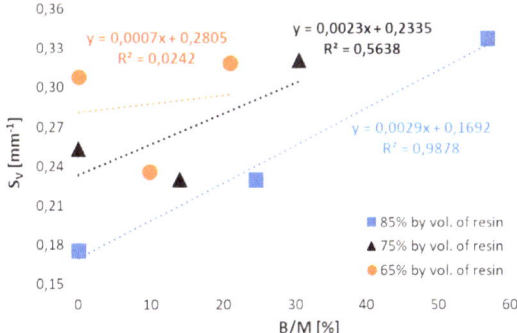

Fig. 6. The surface area of pores in a volume unit of a material (S_V) as a function of share of biochar (B) in total mass of microfillers (M) and resin content

biochar in composites containing 85% of resin by vol. Led to increase in volume fraction of pores, and thus to increase in relative surface area of pores. When the total content of microfillers increases above 15%, the stereological parameters do not exhibit any relation with biochar content. Consequently, these could suggesting that microstructure became more heterogeneous.

Summarizing the results obtained in both parts of the research, the addition of biochar does not cause a significant deterioration of mechanical and physical properties, so its utilization in polymer mortars should not be an obstacle in application of such the composites in construction sector.

References

1. The role of cement and concrete in the circular economy. CEMBUREAU, Belgium (2016)
2. Al-Hamrani, A., Kucukvar, M., Alnahhal, W., Mahdi, E., Onat, N.C.: Green concrete for a circular economy: a review on sustainability, durability, and structural properties. Materials **14**(2), 351–387 (2021)

3. Marsh, A., Velenturf, A., Bernal, S.A.: Circular economy strategies for concrete: implementation and integration. J. Clean. Prod. **362**(4), 132486 (2022)
4. https://www.european-biochar.org. Last accessed 20 Feb 2023
5. Aman, A.M.N., Selvarajoo, A., Lau, T.L., Chen, W.H.: Biochar as cement replacement to enhance concrete composite properties: a review. Energies **15**, 7662 (2022)
6. Xie, Y., et al.: A critical review on production, modification and utilization of biochar. J. Anal. Appl. Pyrolysis **161**, 105405 (2022)
7. Praneeth, S., et. al.: Accelerated carbonation of biochar reinforced cement-fly ash composites: Enhancing and sequestering CO_2 in building materials. Construct. Build. Mater. **244**(30):118363 (2020)
8. Restuccia, L., Ferro, G.A., Suarez-Riera, D., Sirico, A., Bernardi, P., Belletti, B., Malcevschi, A.: Mechanical characterization of different biochar-based cement composites. Proc. Struct. Integ. **25**, 226–233 (2020)
9. Danish, A., Mosaberpanah, M.A., Salim, M.U.: Reusing biochar as a filler or cement replacement material in cementitious composites: a review. Constr. Build. Mater. **300**, 124295 (2021)
10. Zhu, X., et al.: Bonding mechanisms and micro-mechanical properties of the interfacial transition zone (ITZ) between biochar and paste in carbon-sink cement-based composites. Cement Concr. Compos. **139**, 105004 (2023)
11. Javed, M.H., Sikandar, M.A., Ahmad, W., Bashir, M.T., Alrowais, R., Wadud, M.B.: Effect of various biochars on physical, mechanical, and microstructural characteristics of cement pastes and mortars. J. Build. Eng. **57**, 104850 (2022)
12. Mensah, R.A., et al.: Biochar-added cementitious materials - a review on mechanical, thermal, and environmental properties. Sustain **13**, 9336 (2021)
13. Gupta, S., Kashani, A., Mahmood, A.H., Han, T.: Carbon sequestration in cementitious composites using biochar and fly ash—effect on mechanical and durability properties. Constr. Build. Mater. **291**, 123363 (2021)
14. Wang, J., Dai, Q., Guo, S., Si, R.: Mechanical and durability performance evaluation of crumb rubber-modified epoxy polymer concrete overlays. Constr. Build. Mater. **203**, 469–480 (2019)
15. Bulut, H.A., Sahin, R.A.: Study on mechanical properties of polymer concrete containing electronic plastic waste. Compos. Struct. **178**, 50–62 (2017)
16. Zhang, Y., et al.: Biochar as construction materials for achieving carbon neutrality. Biochar **4**(59):8005 (2022)
17. Seco, A., Echeverría, A.M., Marcelino, S., García, B., Espuelas, S.: Durability of polyester polymer concretes based on metallurgical wastes for the manufacture of construction and building products. Constr. Build. Mater. **240**, 117907 (2020)
18. Sosoi, G., Barbuta, M., Serbanoiu, A.A., Babor, D., Burlacu, A.: Wastes as aggregate substitution in polymer concrete. Procedia Manuf. **22**, 347–351 (2018)
19. Kępniak, M., Woyciechowski, P., Franus, W.: Chemical and physical properties of limestone powder as a potential microfiller of polymer composites. Arch. Civ. Eng. **63**, 67–78 (2017)
20. Kępniak, M., Załęgowski, K., Woyciechowski, P., Pawłowski, J., Nurczyński, J.: Feasibility of using biochar as an eco-friendly microfiller in polymer concretes. Polymers **14**:4701 (2022)
21. Zalegowski, K., Piotrowski, T., Garbacz, A., Adamczewski, G.: Relation between microstructure, technical properties and neutron radiation shielding efficiency of concrete. Constr. Build. Mater. **235**, 117389 (2020)
22. Sokołowska, J.J.: Technological properties of polymer concrete containing vinyl-ester resin waste mineral powder. J. Build. Chem. **1**, 84–91 (2016)
23. Zalegowski, K., Piotrowski, T., Garbacz, A.: Influence of polymer modification on the microstructure of shielding concrete. Materials **13**, 498 (2020)
24. Zalegowski, K.: Assessment of polymer concrete sample geometry effect on ultrasonic wave propagation. Materials **14**(23), 7200 (2021)

Admixtures and Additives

Effect of Polymer Content on Properties of Polymer Cement Mortar

Katsunori Demura$^{(\boxtimes)}$ and Toshikatsu Saito

Nihon University, 1 NakagawaraTamura-Machi, TokusadaKoriyama, Fukushima, Japan
demura.katsunori@nihon-u.ac.jp

Abstract. In this paper, the volume fraction of polymer as a solid in the polymer cement mortar (PCM) is defined as the polymer content. The effect of the polymer content on the properties such as the flexural and compressive strengths, water permeability, carbonation and chloride ion penetration of PCM is discussed. As a result, the equation of the effective factor (F) for the properties of the PCM by using the water-cement ratio (W/C), polymer content (Vp), and volume fractions of air (Va) and sand (Vs) is established as "$F = (1-W/C)(1 + AVp)(1-Va)(1 + 5Vs)$". The empirical constant (A) is the 4, -6, -8, -6 and -4 for the flexural strength, compressive strength, water permeability, carbonation depth and chloride ion penetration depth of PCM, respectively. The equation for estimating the properties of PCM by using the effective factor and the properties of cement mortar (Plain) is established as "$Pp = B(FP_0) + C$". Where, P_P, F, and P_0 are the properties of PCM, the effective factor and the properties of Plain with the same W/C, sand-cement ratio and curing condition as the PCM. The empirical constants B and C in this equation are depending on the type of polymer used and curing condition.

Keywords: Polymer Content · Effective Factor · Property · Estimating Equation

1 Introduction

In general, the properties of polymer cement mortar are discussed by the effectiveness of polymer-cement ratio (P/C). However, P/C is the mass ratio of polymer solid to the cement in PCM. Therefore, the volume fraction of the polymer solid is affected by the cement content of PCM. However, PCM is the composite material with cement, sand, water and polymer. The properties of such composite material are generally explained by the volume fraction of materials used.

In this paper, the volume fraction of the polymer in PCM as a solid is defined as the polymer content. The effect of the polymer content on the properties of PCM is discussed. The equation of the effective factor for the properties of PCM by using the water-cement ratio (W/C), the polymer content, the volume fractions of air and sand of PCM, and the estimating equation of its flexural strength, compressive strength, water permeability, carbonation depth and chloride ion penetration depth are established by using the effective factor and those properties of Plain with the same W/C, sand-cement ratio (S/C) and curing condition as PCM.

© The Author(s) 2025
L. Czarnecki et al. (Eds.): ICPIC 2023, 61, pp. 251–259, 2025.
https://doi.org/10.1007/978-3-031-72955-3_25

2   Materials

2.1   Cement and Fine Aggregate

Ordinary portland cement and standard sand were used for the mix proportions of PCMs.

2.2   Redispersible Polymer Powders and Anti-Forming Agent

Four types of redispersible polymer powder (PAE-N, PAE-C, SA, VA/VeoA) with silicone-type anti-forming agent were used. Their properties are listed in Table 1.

Table 1. Properties of redispersible polymer powder for cement modifier.

Type of polymer	Apparent density (g/cm^3)	Glass-transition temp., Tg (°C)	Non-volatile matter (%)	Type of charge
PAE-N	0.5	0	99 ± 1	Non-ion
PAE-C	0.5	0	99 ± 1	Cation
SA	0.5	0	99 ± 1	—
VA/VeoA	0.53	0	99 ± 1	—

3   Testing Procedures

3.1   Preparation of Specimens

Table 2 shows the mix proportions of Plain with the polymer content of 0%. According to JIS A 1171 (Test Methods for Polymer Cement Mortar), PCM were mixed by addition of the redispersible polymer powder to Plain. The combinations of the mix proportioning factors of PCMs are shown in Table 3. The polymer content was calculated by using the density of polymer solid of 1.06 g/cm^3 as an approximate value.

Mortar specimens having the size of 40x40x160 mm were molded and then given following curing conditions;

a) Standard Cure (W5D21): 2d-moist (20°C, 90%(RH)), 5d-water (20°C) and 21d-dry (20°C, 60%(RH)) cure
b) Dry Cure (D26): 2d-moist (20°C, 90%(RH)) and 26d-dry (20°C, 60%(RH)) cure

3.2   Tests for Air Content, Strength and Properties of Durability

The tests for air content, flexural and compressive strengths, water permeability, and accelerated carbonation and chloride ion penetration for 28d and 56d were conducted by JIS A 1171 and JIS A 6205 (Corrosion Inhibitor for Reinforcing Steel in Concrete).

Table 2. Mix Proportions of Plain (unmodified cement mortar).

S/C	W/C (%)	Mix proportion by volume (%)			
		Cement	Sand	Water	Air
2	45	19.8	48.5	28.1	3.6
3	45	15.6	57.1	22.1	5.3
	50	15.3	56.2	24.2	4.4
	55	15.0	55.1	26.1	4.0
	60	14.6	53.9	27.8	3.8

Table 3. Combinations of mix proportioning factors of PCMs using redispersible polymer powders.

Type of polymer	S/C*	W/C (%)*	Polymer content (%)**
PAE-N, PAE-C, SA,	2	45	0, 2, 4, 6
VA/VeoA	3	45, 50, 55, 60	

Notes * S/C: Sand-cement ratio by mass, W/C: Water-cement ratio by mass
** Polymer content is calculated by volume

4 Test Results and Discussion

4.1 Strength Properties

Figures 1 and 2 show the effect of the polymer content on the flexural and compressive strengths of W5D21-Cured PCMs.

The same tendency of the effect of the polymer content on those strengths of D26-cured PCMs is recognized.

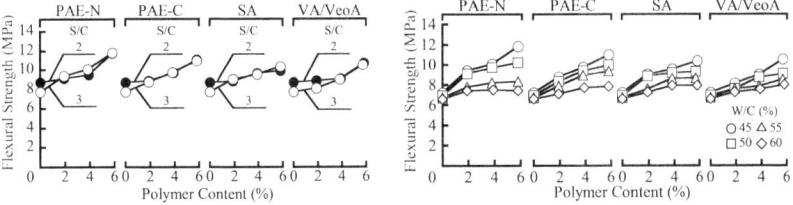

Fig. 1. Polymer content vs. flexural strength of W5D21-cured PCMs with variations of S/C at W/C of 45% and W/C at S/C of 3.

From those results, the effectiveness of mix proportioning factors on the flexural and compressive strengths of PCM is explained as follows regardless of the curing conditions;

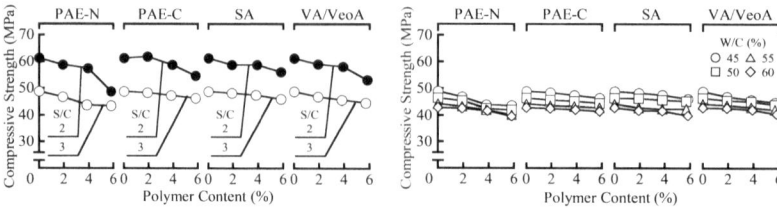

Fig. 2. Polymer content vs. compressive strength of W5D21-cured PCMs with variations of S/C at W/C of 45% and W/C at S/C of 3

  i) PCM with large S/C shows higher flexural strength and lower compressive strength.
 ii) PCM with lower W/C shows higher flexural and compressive strengths.
iii) In increasing the polymer content of PCM, the flexural strength is increased, but the compressive strength is decreased.

PCM is prepared by the modification with polymer addition to Plain. Therefore, the properties of PCM may be explained by inconsideration of the effectiveness of the mix proportioning factors mentioned above and air content, and the properties of Plain as a base material of PCM.

In this paper, the water-cement ratio (W/C), the volume fractions of polymer (Vp), air (Va) and sand (Vs) are intervened as mix proportioning factors for effecting the properties of PCM.

Following equations are proposed as the effective factors for the flexural and compressive strengths of PCM.

a) Flexural strength-effective factor (S_f)

$$S_f = (1 - W/C)(1 + 4Vp)(1 - Va)(1 + 5Vs) \tag{1}$$

b) Compressive strength-effective factor (S_c)

$$S_c = (1 - W/C)(1 - 6Vp)(1 - Va)(1 + 5Vs) \tag{2}$$

The flexural and compressive strengths of PCM may be explained by following formulas:

$$\sigma_{f_p} = f_{(\sigma_{f_0} \cdot S_f)} \tag{3}$$

$$\sigma_{c_p} = f_{(\sigma_{c_0} \cdot S_c)} \tag{4}$$

where, σ_{f_p}, σ_{c_p}: Flexural and compressive strengths of PCM.

σ_{f_0}, σ_{c_0}: Flexural and compressive strengths of Plain with same S/C, W/C and curing condition as PCM.

Figures 3 and 4 show the multiplied values of the flexural and compressive strengths of Plain and those strength-effective factors vs. the flexural and compressive strengths of PCM. There is good correlation between the multiplication values and the strength-effective factors and the strengths of PCMs.

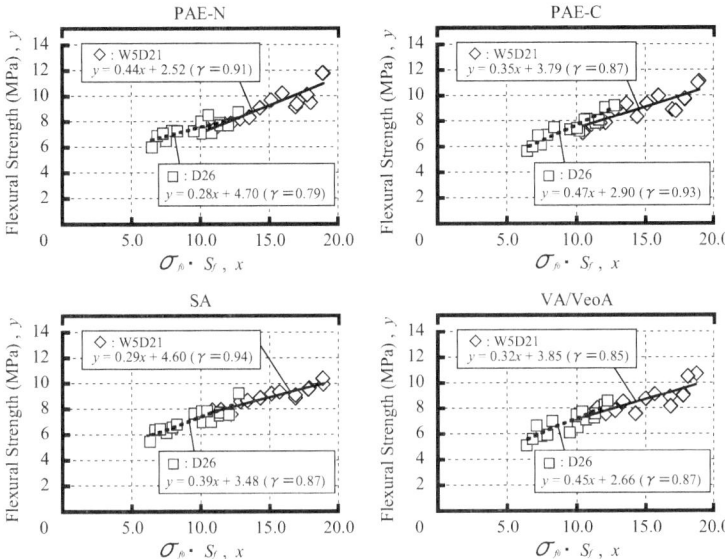

Fig. 3. Multiplied value of flexural strength of plain and flexural strength-effective factor vs. flexural strength of PCMs.

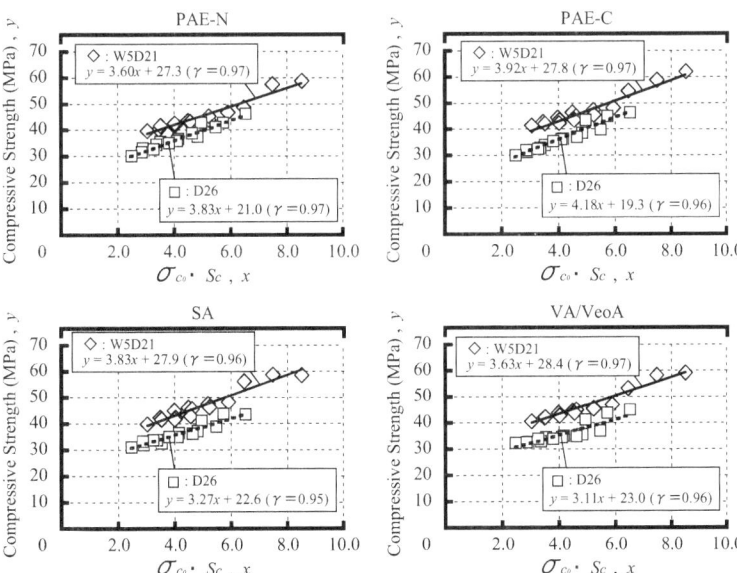

Fig. 4. Multiplied value of compressive strength of Plain and compressive strength-effective factor vs. compressive strength of PCMs.

Therefore, the flexural and compressive strengths of PCM may be explained by the following general equations;

$$\sigma_{f_p} = A(\sigma_{f_0} \cdot S_f) + B \tag{5}$$

$$\sigma_{c_p} = C(\sigma_{c_0} \cdot S_c) + D \tag{6}$$

where, A, B, C and D are empilical constants.

4.2 Properties of Durability

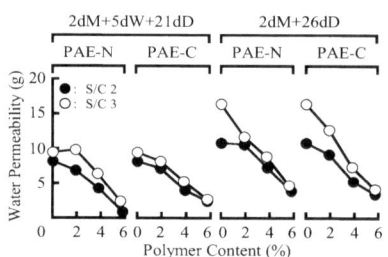

Fig. 5. Polymer content vs. water permeability of PCMs.

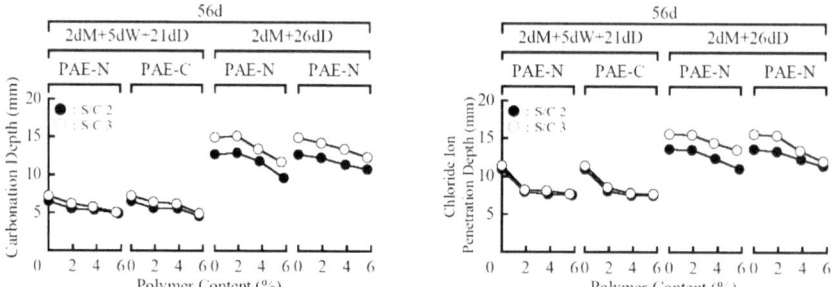

Fig. 6. Polymer content vs. carbonation and chloride ion penetration depths of PCMs at test period of 56d.

In the bases of the discussion results for the strength properties, the effective factors for the water permeability, carbonation depth and chloride ion penetration depth of PCMs using the redispersible polymer powders of PAE-N and PAE-C are discussed.

Figures 5 and 6 show the polymer content vs. the water permeability, carbonation depth and chloride ion penetration depth at the test period of 56d of PCMs. The water permeability, carbonation depth and chloride ion penetration depth of PCMs are decreased with an increase in the polymer content.

In the consideration of the effect of the polymer content on the water permeability, carbonation depth and chloride ion penetration depth, following equations are introduced for the effective factors for those properties of PCMs.

$$F_w = (1 - W/C)(1 - 8Vp)(1 - Va)(1 + 5Vs) \tag{7}$$

$$F_c = (1 - W/C)(1 - 6Vp)(1 - Va)(1 + 5Vs) \qquad (8)$$

$$F_{cl} = (1 - W/C)(1 - 4Vp)(1 - Va)(1 + 5Vs) \qquad (9)$$

Where, F_w : water permeability-effective factor, F_c : carbonation depth-effective factor, F_{cl} : chloride ion penetration depth-effective factor.

Figures 7–9 show the multiplied values of the values of water permeability (W_0), carbonation depth (C_0), chloride ion penetration depth (Cl_0) of Plain and the effective factors calculated by the equations of (7) to (9) vs. those properties of PCMs. Here, Plain has same W/c, S/C and curing condition of PCM. There is good correlation between the multiplication values and the properties of PCMs.

Therefore, the water permeability (W_p), carbonation depth (C_p) and chloride ion penetration depth (Cl_p) of PCM may be explained by the following general equations;

$$Wp = E(W_0 \cdot Fw) + G \qquad (10)$$

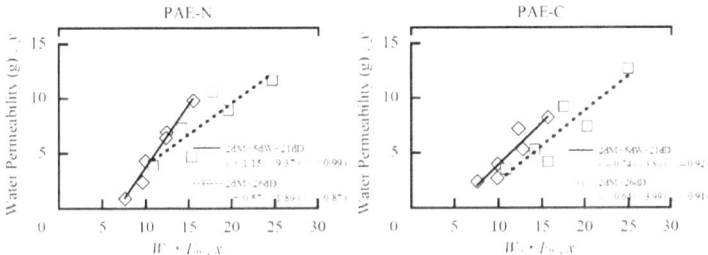

Fig. 7. Multiplied value of water permeability-effective factor and water permeability of Plain vs. water permeability of PCMs.

$$C_p = H(C_0 F_c) + I \qquad (11)$$

$$Cl_p = J(Cl_0 F_{cl}) + K \qquad (12)$$

where, E, G, H, I, J and K are empilical constants.

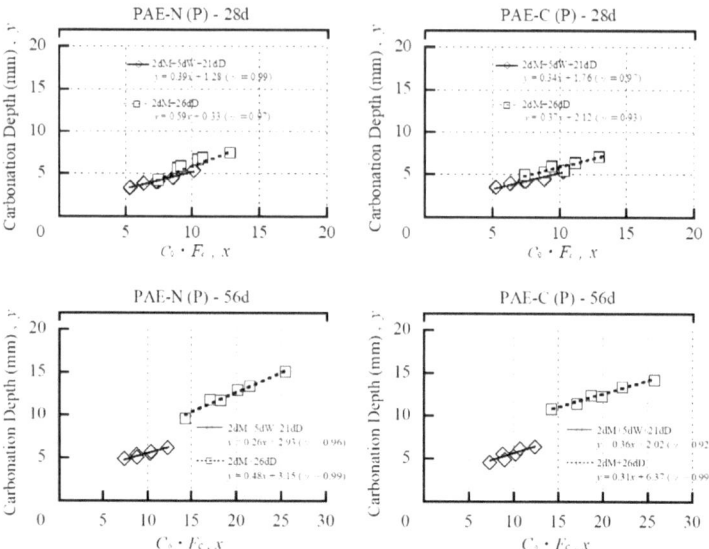

Fig. 8. Multiplied value of carbonation depth-effective factor and its depth of Plain vs. carbonation depth of PCMs at test periods of 28d and 56d.

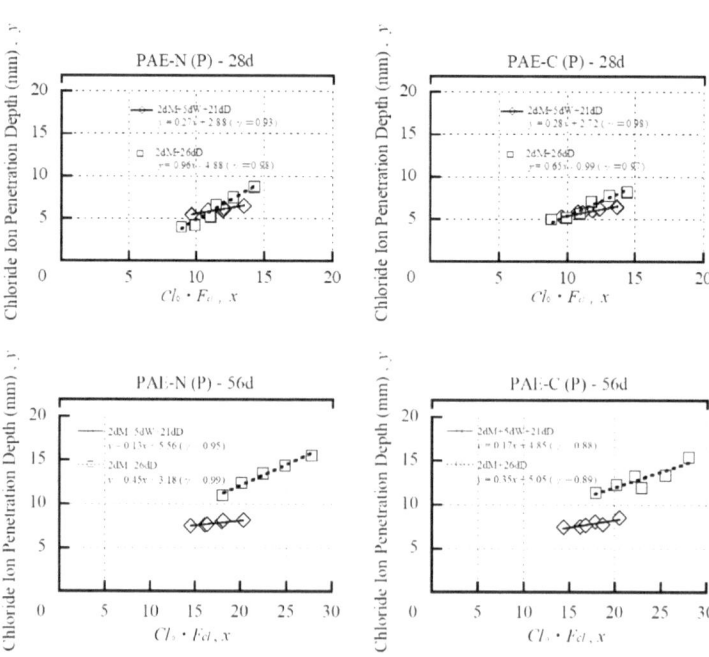

Fig. 9. Multiplied value of chloride ion penetration depth-effective factor and its depth of Plain vs. chloride ion penetration depth of PCMs at test periods of 28d and 56d.

5 Conclusions

The equations for the effective factors for the strength, water permeability, carbonation depth and chloride ion penetration depth of PCM are established as the equations of (1), (2), and (7) to (9). Such propertiess of PCM may be estimated by using the equations of (5), (6), and (10) to (12). The empilical constants of those equations are depending on the type of polymere used and curing condision.

[This paper is the summary of two papers prepared by the authors in Japanese; Nishida, A., Saito, T., Demura, K., Gakiya, M.,: Effect of polymer content on flexural and compressive strengths of polymer-modified mortars. Cement science and concrete technology 72, 99–105(2019), and Demura, K., Saito, T., Takeda, M.,: Eeffect of polymer content on resistance to water permeability and durability of polymer-modified mortars, Cement science and concrete technology 73, 95–102(2020).]

Effect of Polymer Paste Content on the Porosity and Strength of Pervious Polymer Concrete

Jung Heum Yeon[1($\boxtimes$)], Yeoung-Geun Choi[2], Cheol-Jae Yang[2], and Kyu-Seok Yeon[2]

[1] Ingram School of Engineering, Texas State University, San Marcos, TX, USA
jung.yeon@txstate.edu
[2] iCONTEC ENC Co., Ltd, Anyang, South Korea

Abstract. This study investigates the effect of polymer paste content on the porosity and strength of pervious polymer concrete made of unsaturated polyester resin, fly ash filler, and crushed coarse aggregate. The porosity (total porosity and connected porosity) and strength (compressive and flexural strengths) for different polymer paste contents were investigated. The polymer paste content was chosen as an experimental variable because it determines the cost-effectiveness and has a significant impact on various material properties. The results showed that the total and connected porosity fell between 37.5–8.8% and 34.2–7.2%, respectively, when the polymer paste content increased from 7 to 19.5 wt.%. The porosity tended to decrease as the polymer paste content increased. The compressive and flexural strengths ranged from 14.5 to 41.5 MPa and 4.3 to 16.1 MPa, and the strengths increased as the paste content increased. In particular, the strengths were much higher than those of many existing studies on conventional portland cement concrete due to the enhanced adhesion of the polymer binder upon the addition of the cross-linking agent.

Keywords: Previous Polymer Concrete · Unsaturated Polyester Resin · Polymer Paste Content · Porosity · Strength

1 Introduction

Pervious concrete is a sort of lightweight concrete, which is also called porous concrete, permeable concrete, or no-fines concrete [1]. Most of the voids in pervious concrete are large and interconnected because the coarse aggregates are in point-to-point contact via mortar or paste, resulting in a lower unit weight than conventional cement concrete. Coarse aggregate is the main component of the skeleton structure of pervious concrete, which bears a substantial amount of load [2]. In particular, one of the primary advantages is that the hydrostatic pressure applied to pervious concrete is relatively low, typically 1/3 of that applied to conventional cement concrete [3]. In addition, previous concrete undergoes less drying shrinkage due to its larger void size, and the moisture movement by capillary action is minimal. More importantly, the unit weight of the pervious concrete is quite low, which makes it lightweight and has excellent insulation performance. A former study reported that the life-cycle cost (LCC) of a surface course made with

© The Author(s) 2025
L. Czarnecki et al. (Eds.): ICPIC 2023, 61, pp. 260–267, 2025.
https://doi.org/10.1007/978-3-031-72955-3_26

pervious concrete is lower than that made with conventional concrete since the cost required for underline pipes can be significantly reduced when using pervious concrete [4]. Given the benefits, pervious concrete has been widely used for water-permeable blocks and pavement, sound-absorbing concrete, and eco-concrete. Despite such advantages, pervious concrete has been limitedly used for non-structural applications because it inherently has a low strength, and the strength even becomes lower as the porosity increases.

The main goal of this study is to develop pervious concrete using polymeric resin as a binder in place of conventional portland cement to improve the strength characteristics while ensuring the highly porous nature of typical pervious concrete. Particularly, this study focuses on assessing the effect of polymer paste content on the strength and porosity of porous polymer concrete made of unsaturated polyester (UP) polymeric resin. The outcomes of this study will contribute to the commercialization of porous polymer concrete for various structural and non-structural applications.

2 Materials and Methods

2.1 Materials

Unsaturated polyester resin.
An ortho-type UP resin (Aekyung Chemical Co., Ltd., Korea) added with a cobalt-based curing accelerator was used. The UP resin had a specific gravity of 1.13 at 25 °C, viscosity of 3.0 ± 0.4 at 25 °C, acid value of 20.0, gel time of 7–11 min at 25 °C, and styrene content of 40%.

Initiator.
An initiator composed of 55% MEKPO and 45% DMP (KeumJungCo., Ltd., Korea) was used. Because the UP used in this study was premixed with an accelerator, the hardening reaction began upon the addition of the initiator.

Cross-linking agent.
To enhance the bond between the polymer matrix and inorganic aggregate, silane (Hansol Chemical, Korea) was used as a cross-linking agent, of which specific gravity and viscosity were 1.03 and 2.41 mm^2/sec, respectively.

Filler.
Fly ash was added as a filler to formulate polymer paste. The fly ash had a specific gravity of 2.20, SiO_2 content of 51.9%, loss on ignition of 3.3%, and specific surface of 3,648 cm^2/g.

Crushed coarse aggregate.
The crushed coarse aggregate (GyeonginMaterials Inc., Korea) used in this study had a grain size of 5–20 mm, which was dried to keep the moisture content below 0.5 wt.%. The coarse aggregate had a specific gravity of 2.63, absorption capacity of 2.67%, abrasion resistance of 24.3%, and soundness of 3.95%. The result of a sieve analysis is presented in Table 1.

Table 1. Sieve analysis of crushed coarse aggregate

Sieve size (mm)	Passing (%)
2.5	0
5	0.3
10	29.5
20	99.0
25	100.0

2.2 Mixing

The polymeric resin content often falls within the range of 10–20 wt.% of polymer concrete [5], and the optimum resin content depends on the properties of the aggregate. As the aggregate size becomes smaller, more polymer content is required since the specific surface area of the aggregate increases [6, 7]. In this study, only 3–8% polymeric binder content was used to achieve the target void ratio of 10–35%, while the mixing ratio between the polymeric binder and filler was fixed at 1:1.5 by weight. Table 2 summarizes the mixture proportions of the pervious polymer concrete tested in this study. Wet mixing was done at 60 rpm for 3 min after dry mixing coarse aggregate and filler for 3 min.

Table 2. Mixture proportions of pervious polymer concrete

Polymer paste content (wt.%)	Target porosity (%)	Polymeric binder (kg/m^3)	Fly ash (kg/m^3)	Coarse aggregate (kg/m^3)	Unit weight (kg/m^3)
7.0	35	48	72	1,600	1,720
9.5	30	67	101	1,600	1,768
12.0	25	88	132	1,600	1,820
14.5	20	108	162	1,600	1,870
16.0	15	132	198	1,600	1,930
19.5	10	156	234	1,600	1,990

2.3 Specimen Preparation

Polymer concrete was placed in three layers for cylindrical specimens (Φ100 mm × 200 mm) while placed in two layers for cube specimens (200 mm × 200 mm × 60 mm). All the specimens were cured for 7 days at 25 ± 2°C and 50–60% RH. The top surface of the cylindrical specimens was flattened with polymer paste to avoid possible eccentricity when loaded as shown in Fig. 1. Three replicates were prepared for each test.

Fig. 1. Appereance of flattened loading area

2.4 Methods

Total porosity.

The total porosity was computed as follows [8, 9]:

$$V_p = \left(1 - \frac{W_2 - W_1}{V}\right) \times 100 \tag{1}$$

where V_p is the total porosity (%), W_1 is the weight of the saturated specimen underwater, W_2 is the weight of the specimen oven dried at $60 \pm 3°C$ for 24 h, and V is the volume of the specimen (cm^3).

Connected porosity.

The connected porosity is key to ensuring the required permeability and strength. The connected porosity can be computed as follows [8, 10]:

$$V_{cp} = \left(1 - \frac{W_3 - W_1}{V}\right) \times 100 \tag{2}$$

where V_{cp} is the connected porosity (%), W_1 is the weight of the saturated specimen underwater, W_3 is the weight of the saturated surface dry-conditioned specimen, and V is the volume of the specimen (cm^3).

Compressive strength.

$\Phi100$ mm $\times$ 200 mm cylindrical specimens were tested at 7 days as per the test procedures specified in KS F 2405:2010.

Flexural strength.

Flexural strength was measured at 7 days using 200 mm $\times$ 200 mm $\times$ 60 mm cube specimens in accordance with KS F 4419:2016.

3 Results and Discussion

3.1 Porosity

Figure 2 shows that the total porosity gradually decreased from 37.5 to 8.8% as the polymer paste content increased from 7 to 19.5%. This is because as the polymer paste content increases, more voids are filled with polymer paste. Mounika et al. [11] reported that the total porosity was about 13–28% of the total volume of concrete (no-fines concrete) when the cement-aggregate ratio was between 1:3 and 1:9. Another study by Muthaiyan et al. [12] revealed that the total porosity fell between 18.68–28.70% and 18.19–26.47% for pervious cement concrete and pervious fly ash-cement concrete, respectively. Liu et al. [13] found that the total porosity of pervious concrete made with 5–20 mm natural and recycled aggregates fell between 15 and 33%. Xia et al. [14] used 15–30 mm artificial gravel aggregate to fabricate grass-planting concrete, and the effective porosity was found to be 20–31.8%. Sriravindrarajah et al. [15] used two single-sized natural coarse aggregates ranging from either 5 to 13 mm or 13 to 20 mm in pervious concrete and showed that the total porosity was 26.8–28.1% for 13–20 mm aggregate and 33.3–36.1% for 5–13 mm aggregate. ACI 522R [16] reported that the total porosity of pervious concrete usually ranges from 15 to 35%.

Figure 2 also indicates that the connected porosity decreased from 34.2 to 7.2% as the polymer paste content increased from 7 to 19.5%, which showed a very similar trend to the total porosity. The connected porosity was approximately 91.2–81.8% of the total porosity. A former study by Muthaiyan [12] revealed that the connected porosity was 17.85–27.23% for pervious cement concrete and 15.01–24.79% for pervious fly ash-cement concrete. Yao et al. [2] reported a 14–32% connected porosity for pervious concrete made with 9.5–26.5 mm coarse aggregate. Some discrepancies in measured porosity were noted between this study and the previous studies. The possible reason appeared to stem from the type and content of paste/mortar and the maximum size and gradation of coarse aggregate used.

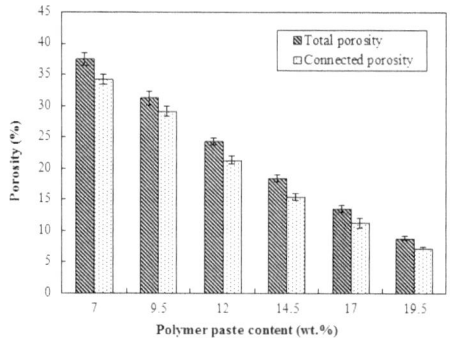

Fig. 2. Effect of polymer paste content on the porosity

3.2 Strength

Figure 3 shows the effect of polymer paste content on the compressive and flexural strengths. The compressive strength increased from 14.5 to 41.5 MPa as the polymer paste content increased from 7 to 19.5%. Geethanjali et al. [15] reported that the 28-day compressive strength of pervious concrete was 16.85 MPa. Liu et al. [13] found that the compressive strength of porous concrete with 5–20 mm natural aggregate and recycled aggregate fell between 7 and 25 MPa. Yao et al. [2] reported a 28-day compressive strength of 3–12 MPa for porous concrete made with 9.5–26.5 mm coarse aggregate. Lori et al. [3] reported a compressive strength of 23.45 MPa for pervious concrete containing 60% copper slag. Lang et al. [18] demonstrated that the 28-day compressive strength of magnesium phosphate cement steel slag pervious concrete was between 29.5 and 36.0 MPa.

As can also be seen in Fig. 3, the flexural strength tended to increase from 4.3 to 16.1 MPa as the polymer paste content increased from 7 to 19.5%. The flexural strength was about 29.6–38.7% of the compressive strength. Given the flexural strength of pervious concrete from many former studies typically fell within the range between 1.44 and 1.88 MPa [12], 0.3 and 1.7 MPa [2], 6.02 and 7.75 MPa [17], 3.15 and 3.80 MPa [9], 1 and 3.8 MPa [2], and 5.0 and 8.0 MPa [18], The flexural strength of the pervious polymer concrete developed in study was significantly higher (14.5 MPa to 41.5 MPa) than that of the previous research studies.

Consequently, the compressive and flexural strengths observed in the present study were much superior to those reported in the previous studies since the polymeric binder provided a stronger bond strength, and the cross-linking agent further improved the bond.

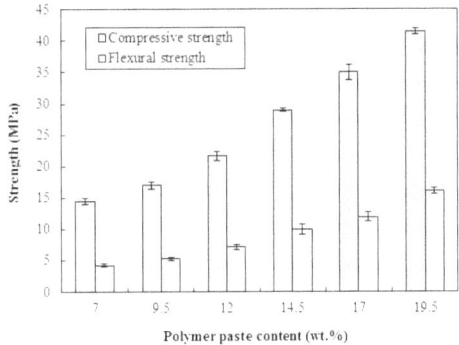

Fig. 3. Effect of polymer paste content on the strengths

3.3 Porosity vs. Strength

The strength of pervious concrete is closely related to its total porosity [19]. Figure 4 presents the relationship between the total porosity and strength of the pervious polymer concrete. As noted in the results, there was an inverse exponential relationship between the strength and total porosity with a high coefficient of determination (R^2) of 0.9927; as the total porosity increased, the compressive strength was dramatically reduced. This trend was similar to the results of a previous study by Jia Hao et al. [20].

4 Conclusions

This study investigated the effect of polymer paste content on the strength and porosity of pervious polymer concrete made with an unsaturated polyester (UP) resin. The following conclusions can be drawn from the findings of this study.

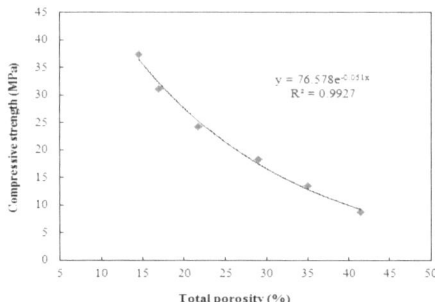

Fig. 4. Relation between total porosity and compressive strength

- As the polymer paste content increased, the total and connected porosity substantially decreased. The connected porosity was approximately 81.8–91.2% of the total porosity.
- The porosity appears to be closely related to the type and content of paste/mortar and the maximum size and gradation of coarse aggregate used.
- The strength of the previous polymer concrete was found to be much higher than that of conventional pervious cement concrete. The strength tended to increase as the polymer paste content increased. The flexural strength was about 29.6–38.7% of the compressive strength.
- The strengths of porous polymer concrete were much superior to those of conventional pervious cement concrete because the polymeric binder and cross-linking agent enhanced the bond between the polymer matrix and aggregate.
- A strong exponential relationship existed between the porosity and strength with a high coefficient of determination.

References

1. Yu, F., Sun, D., Hu, M., Wang, J.: Study on the pore characteristics and permeability simulation of pervious concrete based on 2D/3D CT images. Constr. Build. Mater. **200**, 687–702 (2019)
2. Yao, A., Ding, H., Zhang, X., Hu, Z., Hao, R., Yang, T.: Optimum design and performance of porous concrete for heavy-load traffic pavement in cold and heavy rainfall region of NE China. Adv. Mater. Sci. Eng. **2018**, Article ID 7082897 (2018)
3. Concrete construction staff. Features: No-Fines Concrete. https://www.concreteconstruction.net/business/no-fines-concrete_o. Last accessed 7 Mar 2023
4. Chen, X.D., Wang, H., Najm, H.: Environmental assessment and economic analysis of porous pavement at sidewalk. LCA Symposium, University of Illinois Urbana-Champaign , Illinois, USA (2017)
5. Vipulanandan, C., Dharmarajan, N.: Flexural behavior of polyester polymer concrete. Cem. Concr. Res. **17**(2), 219–230 (1987)
6. Ferreira, A.J.M.: Flexural properties of polyester resin concretes. J. Polym. Eng. **20**(6), 459–468 (2000)
7. Ribeiro, M.C.S., Tavares, C.M.I., Figueiredo, M., Ferreira, A.J.M., Fernandes, A.A.: Bending characteristics of resin concretes. Mater. Res. **6**(2), 247–254 (2003)
8. Japan Concrete Institute. Technical Committee Report on Eco-concrete, [in Japanese] (1995)

9. Lori, A.R., Hassani, A., Sedghi, R.: Investigating the mechanical and hydraulic characteristics of pervious concrete containing copper slag as coarse aggregate. Constr. Build. Mater. **197**, 130–142 (2019)
10. Montes, F., Vlavala, S., Haselbach, L.M.: A new test method for porosity measurements of portland cement pervious concrete. J. ASTM Int. **2**(1), 1–13 (2005)
11. Mounika, P., Srinivas, K.: Mechanical properties of no-fines concrete for pathways. Int. J. Eng. Tech. **4**(2), 68–81 (2018)
12. Muthaiyan, U.M., Thirumalai, S.: Studies on the properties of pervious fly ash–cement concrete as a pavement material. Cogent Eng. **4**(1), 1–17 (2017)
13. Liu, J., Ren, F., Quan, H.: Prediction model for compressive strength of porous concrete with low-grade recycled aggregate. Materials **2021**(14), 3871 (2021)
14. Xia, Q., Jiang, H., Wang, J., Liu, Y., Cui, L.: Physical and mechanical properties of grass-planting concrete for river revetment projects. In: The 6th International Conference on Environmental Science and Civil Engineering, IOP Conf. Series: Earth and Environmental Science 455 (2020)
15. Sriravindrarajah, R., Wang, N.D.H., Ervin, L.J.W.: Mix design for pervious recycled aggregate concrete. Int. J. Concr. Struct. Mater. **6**(4), 239–246 (2012)
16. A.C.I. (ACI) C. 522, 522R-10 Report on Pervious Concrete. American Concrete Institute, Farmington Hills, MI (2010)
17. Geethanjali, M.S., Manonmani, M.E.B., Sowmya, P., Suvetha, T., Balakumar, V.: Experimental study of pervious (no-fine) concrete. Int. J. Sci. Eng. Res. **11**(3), 83–86 (2020)
18. Lang, L., Duan, H., Chen, B.: Properties of pervious concrete made from steel slag and magnesium phosphate cement. Constr. Build. Mater. **197**, 130–142 (2019)
19. Lian, C., Zhug, Y., Beecham, S.: The relationship between porosity and strength for porous concrete. Constr. Build. Mater. **25**(11), 4294–4298 (2011)
20. Jia Hao, L., Chin Lian, F., Hejazi, F., Azline, N.: Study of properties and strength of no-fines concrete. IOP Conference Series: Earth and Environmental Science 357, Sustainable Civil and Construction Engineering Conference, Univ. Putra Malaysia, Kuala Lumpur, Malaysia (2018)

Study on the Use of Glass By-Products for Sustainable Polymer-Modified Mortars

Nikol Žižková[✉], Jakub Hodul, and Rostislav Drochytka

Faculty of Civil Engineering, Brno University of Technology, Brno, Czech Republic
`zizkova.n@fce.vutbr.cz`

Abstract. This investigation is focused on the observation of changes in the properties of polymer-modified cement mortars caused by the addition of recycled glass. The current requirements for reducing CO_2 emissions in the production of cement composites, are also forcing the producers of polymer-modified mortars (PMMs) to use alternative materials, such as silica-rich supplementary materials. Selected types of recycled glass with pozzolanic behavior were specifically ground (particle size below 63 μm) and used as a partial cement substitute (10 wt.%, 20 wt.% and 30 wt.% substitution of Portland cement). In order to explain the obtained results and garner new knowledge of the microstructure of the mixtures being studied, the following tests were performed: scanning electron microscopy (SEM) observation, differential thermal analysis (DTA) and high-pressure mercury intrusion porosimetry. The findings show that the finely ground recycled glass has high potential to be used as an effective cement replacement for PMM materials, that are currently used in large amounts, mainly in the rehabilitation of concrete structures.

Keywords: Polymer-Modified Mortars · Cement Substitution · Glass By-products

1 Introduction

The disposal of waste materials presents a complicated issue around the world [1]. An increasing interest in the use of recycled materials and more environmentally friendly alternative sources of building materials is the aim of promoting sustainable and green construction [2]. The growing environmental requirements, increasing scarcity of landfill sites and depleting sources of raw materials is leading to an increased use of recycled materials [3]. The goal of the current ambitious environmental challenge, known as the European Green Deal, is to achieve net zero emissions of greenhouse gases by 2050 [4]. The measures needed to achieve this goal will need to apply to all industries, including construction. The use of recycled materials as supplementary cementing materials (SCM) with lower carbon dioxide emissions can reduce environmental impact and promote sustainable construction [5].

Millions of tons of waste glass are being produced worldwide every year. Although glass can be easily recycled, and has been recycled for many years, it is still disposed of

© The Author(s) 2025
L. Czarnecki et al. (Eds.): ICPIC 2023, 61, pp. 268–279, 2025.
https://doi.org/10.1007/978-3-031-72955-3_27

in significant quantities as waste glass [6]. Once the glass becomes designated as waste it is disposed of in landfill sites, which is not only environmentally destructive but very expensive as well. The use of milled waste glass as partial replacement of cement could be an important contribution to sustainable development [7]. Supplementary cementing materials like fly ash, silica fume or fine glass powder are useful in improving the mechanical and durability properties of cement-based composites [8]. Being amorphous and containing large quality of silicate, finely ground glass can be used as a pozzolanic material [9, 10]. A particle size of 75 μm or less is reported to be favorable for pozzolanic reaction [8]. Several studies confirm that fine glass powder can even be used as a suppressor, as well as SCM, to mitigate alkali-silica reactions (ASR). To avoid cracking due to deleterious ASR it is recommended the grain size of the glass is reduced to 300 μm or less [6, 9]. Patel et al. [11] reported that very fine glass powder (particle size below 45 μm) replaced cement at 20 wt.% without compromising the quality parameters. According to Parghi and Alam [12], recycled glass powder can be used as an effective supplementary material in cement mortar with 25% optimal replacement of cement. Calcium silicate hydrate gel (C-S-H) formed during the pozzolanic reactions leads to the refinement of pores and the reduction of porosity [13]. Zheng [9] found that the pozzolanic reaction of soda-lime glass not only consumes portlandite to form C-S-H but also leads to a decrease in monosulfate. Dehghan et al., [14] investigated the use of glass fibers from waste glass fiber reinforced polymers (GFRPs). Among GFRPs, E-glass is the most common reinforcement and represents the vast majority of the commercial market. The chemical composition for E-glass used in general applications such as GFRPs are outlined in ASTM D578, see Table 1.

Table 1. Certified chemical composition for glass fiber products used in general applications [14].

Compound	% by weight
B_2O_3	0–10
CaO	16–25
Al_2O_3	12–16
SiO_2	52–62
MgO	0–5
$Na_2O + K_2O$	0–2
TiO_2	0–1.5
Fe_2O_3	0.05–0.8
Fluoride	0–1

Some of the recyclable glass shards that do not meet the requirements of the glassworks and would end up in landfill sites can be used for different applications, e.g., for the production of foam glass for thermal and acoustic insulation [15]. This material is also produced in the Czech Republic and used in the building industry as foam glass boards and granulates.

In order to get closer to the circular economy, the government of the Czech Republic approved a secondary raw materials policy, that is in line with the European raw materials strategy. The main vision of the secondary raw materials policy of the Czech Republic is "turning waste into resources" and is focused on ten important commodities and sources of secondary raw materials, including waste glass [16]. Unfortunately, despite support for the conversion of waste into resources, part of the untreated waste glass still remains in the waste category and is landfilled. Container glass accounts for approximately 25% of the glass marketed in the Czech Republic.

2 Materials

The disposal of waste materials presents a complicated issue around the world [1]. An increasing interest in the use of recycled materials and more environmentally friendly alternative sources of building materials is the aim of promoting sustainable and green construction [2].

The aim of this experiment was to investigate changes in the properties of polymer-modified cement mortars caused by an addition of two types of recycled glass. This study uses waste E-glass (fibers) and the rest is from foam glass production. Both types of waste glass have to be ground to particle sizes below 63 μm. The pozzolanic activity was measured based on the amount of calcium oxide reacting with milled E-glass and foam glass, see Table 2. Oxide contents are specified in Table 3.

Table 2. Pozzolanic activity of used waste glass.

Type of waste glass	mg $Ca(OH)_2$/ 1 g pozzolan-active waste glass
E-glass	652
Foam glass	830

Table 3. Content of main oxides of used waste glass.

Compound	E-glass [wt. %]	Foam glass [wt. %]
SiO_2	53.1	71.1
CaO	21.1	9.56
Al_2O_3	14.0	1.83
MgO	0.58	1.88
Na_2O	0.12	12.80
K_2O	0.50	0.776
TiO_2	0.39	0.077
Fe_2O_3	0.32	0.404

Specimens were prepared with ordinary Portland cement CEM I as binder (OPC), CEN standard sand (according EN 196-1) and redispersible polymer powder on the basis of ethylene/vinyl acetate (EVA). OPC was substituted with waste glass by 10 wt.%, 20 wt.% and 30 wt.%. Table 4 shows the composition of the mixtures.

Table 4. Composition of verified mortar's mixtures.

Mixture ID	OPC [g]	CEN sand [g]	Foam glass [g]	E-glass [g]	EVA [g]	Water [ml]
REF	450	1350	-	-	-	225
FG10	405	1350	45	-	-	225
FG20	360	1350	90	-	-	225
FG30	315	1350	135	-	-	225
E-G10	405	1350	-	45	-	225
E-G20	360	1350	-	90	-	225
E-G30	315	1350	-	135	-	225
E/REF	450	1350	-	-	36	225
E/FG10	405	1350	45	-	36	225
E/FG20	360	1350	90	-	36	225
E/FG30	315	1350	135	-	36	225
E/E-G10	405	1350	-	45	36	225
E/E-G20	360	1350	-	90	36	225
E/E-G30	315	1350	-	135	36	225

3 Methods

Specimens of PMMs were made with the dimensions of $40 \times 40 \times 160$ mm and were tested for physical-mechanical properties and examined to study their microstructure. The determination of flexural and compressive strength of PMMs was performed according to the standard EN 1015–11:2019 – methods of test for mortar for masonry – Part 11: determination of flexural and compressive strength of hardened mortar. Specimens were tested after 28, 90, and 90 days + 25 freezing cycles (FCs). Prior to cyclic freezing and thawing, the test specimens were saturated with water for 24 h by immersion in a water bath at $+20$ °C $\pm$ 3 °C so that the water was at least 3 cm above the surface of the samples. They were frozen immediately after saturation with water. Freezing and thawing of test specimens was performed in FCs based on the standard ČSN 72 2452 – mortar frost resistance test. One FC consisted of four hours of freezing at -20 °C $\pm$ 3 °C and two hours of defrosting. At the end of 25 FCs, the test specimens were measured, weighed and bulk density determined. The set of beams was then tested for flexural and compressive strength.

Another variable that was monitored immediately after the preparation of the samples was the course of temperatures during the hydration of mortars, which indicates the course of hydration reactions during the change of structure and the formation of hydration products. To monitor the temperatures, comparability and possible repeatability, a constant ambient temperature is necessary. This was ensured by the air conditioning cabinet and was 20 °C. The temperature monitoring itself was performed using a Testo 177-T4 measuring control panel using thermocouple probes type K - nickel-chromium-nickel [17].

The microstructure of selected specimens was examined, using differential thermal analysis (DTA), high-pressure mercury intrusion porosimetry and a scanning electron microscope (SEM), in order to gain a better understanding of the obtained strength test results.

4   Results and Discussion

The results of compressive and flexural strength performed at an age of 28 days, 90 days and 90 days + 25 FCs are shown in Figs. 1, 2, 3 and 4. To determine the flexural strength, 3 test beams with the dimensions of $40 \times 40 \times 160$ mm were produced and subsequently the compressive strengths of the 6 halve-beams were determined.

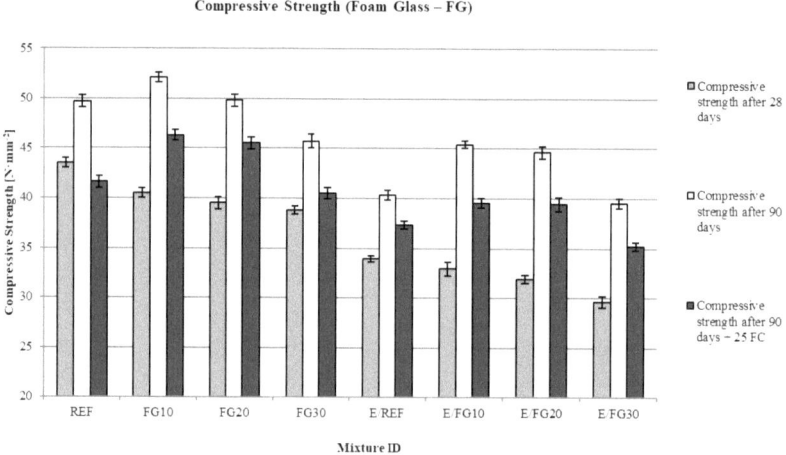

Fig. 1. Compressive strength of mixtures containing foam glass.

Changes in compressive and flexural strength of PMMs is dependent on the age and FCs showing an improvement up to 20 wt.% substitution of Portland cement by grounded waste E-glass. According to authors [11], very fine glass powder (particle size below 45 μm) can replace cement at 20 wt.% without compromising the quality parameters. This study confirmed that glass particle size below 63 μm is sufficient in the case of 20 wt.% substitution of Portland cement and leads to increased compressive and flexural strength, especially after 90 days and 90 days + 25 FCs. The maximum improvement in the compressive strength of approximately 12% was achieved when 10% replacement was used for PMM (marked as E/FG10) compared to the reference mixture (marked as E/REF) at 90 days. The use of waste glass powder as a partial replacement of Portland cement contributes to the long-term strength development of mortars and PMMs due to the slower rate of pozzolanic reaction compared to the hydration of Portland cement.

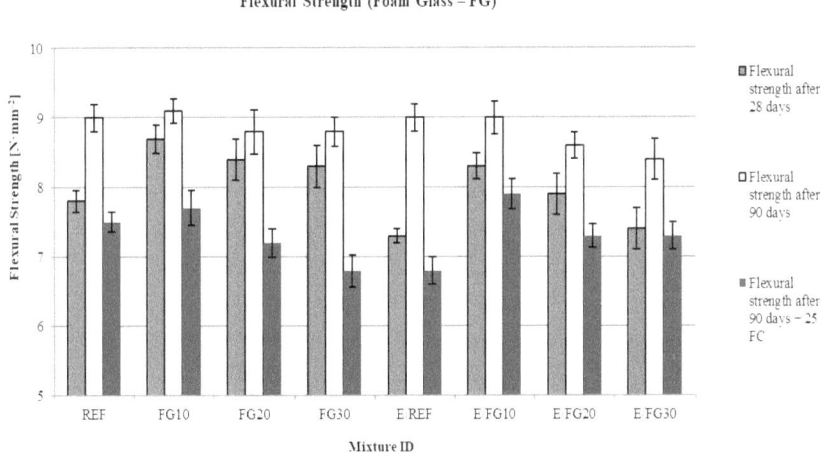

Fig. 2. Flexural strength of mixtures containing foam glass.

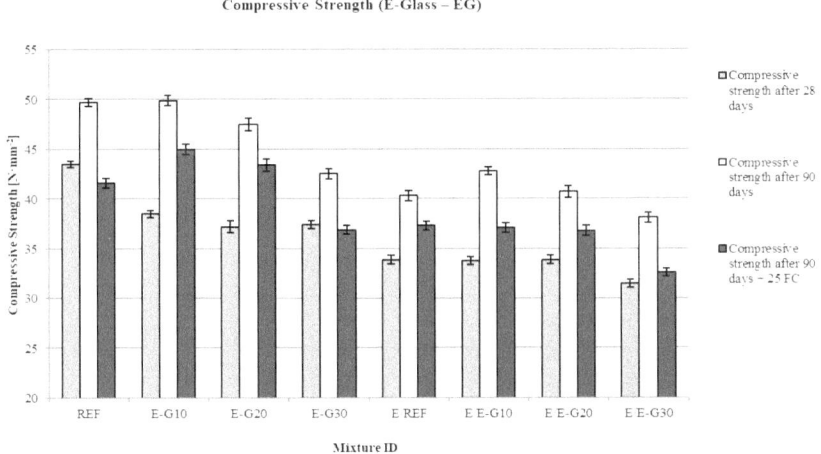

Fig. 3. Compressive strength of mixtures containing E-glass.

The results of hydration temperature evolution during first 24 h are shown in Fig. 5. It is seen that the hydration peak was reduced and retarded by the partial cement substitution. Glass, similar to other SCMs, generates lower temperatures during the early stages of curing. This effect can be beneficial in the design of the mortar to avoid shrinkage.

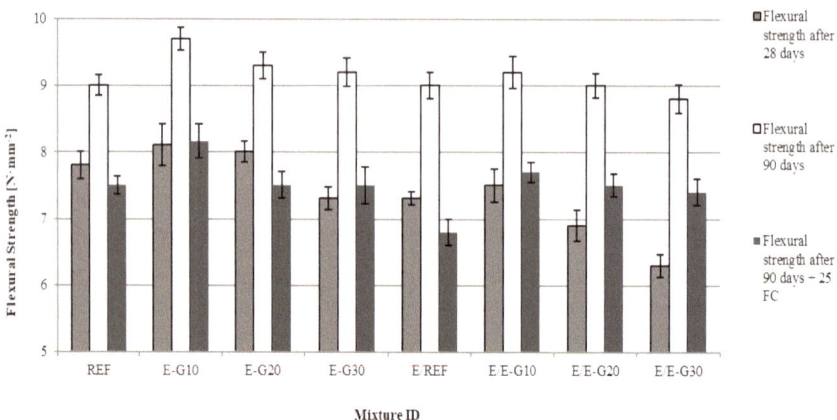

Fig. 4. Flexural strength of mixtures containing E-glass.

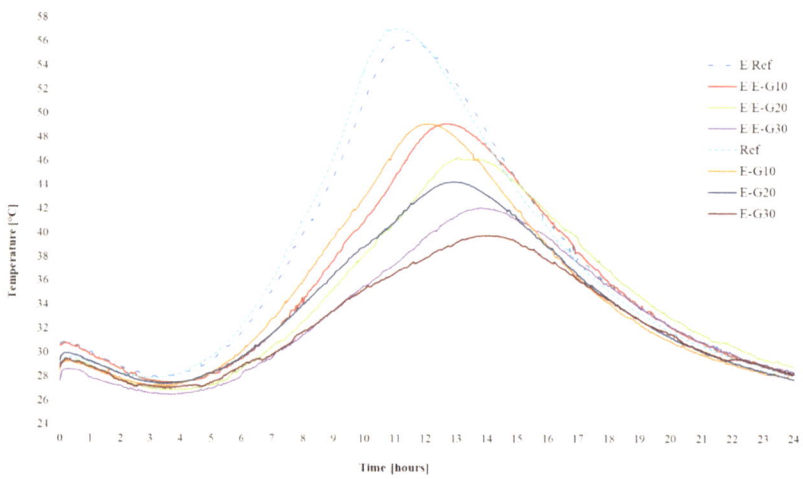

Fig. 5. Hydration temperature evolution of mortars and PMM containing grounded waste E-glass.

According to the results of DTA analysis shown in Table 5, PMMs with the use of grounded waste E-glass contained the highest amount of calcium silicate hydrate gel. The pozzolanic reaction of grounded glass consumes portlandite to form C-S-H phases [8], thus enhancing resistance to aggressive chemicals.

The selected materials were examined using SEM imaging. Figure 6 shows an example of mortar REF, Fig. 7 shows an example of polymer-modified mortar E/REF and Fig. 8a–b the sample of PMM containing grounded waste E-glass (marked as E/E-G20 with 20 wt.% substitution of Portland cement). The glass powder converts calcium hydroxide into a calcium silicate hydrate gel, and therefore the newly formed C-S-H

Table 5. Results of DTA – mass loss content of selected mixtures after 90 days + 25 FC.

Mixture ID	Mass loss corresponding to decomposition [%]	
	C-S-H	Ca(OH)$_2$
REF	2.959	1.613
E-G20	3.299	1.682
E/REF	3.234	0.445
E/E-G20	4.523	0.397

phases were observed as a benefit of the pozzolanic behavior of the grounded E-glass used.

Fig. 6. SEM photomicrographs of mortar REF after 90 days + 25 FCs with visible portlandite and ettringite crystals.

It can be seen from Fig. 9 that the addition of ground waste E-glass shows the reduced porosity compared with reference mixtures (without cement replacement). A reduction in porosity and pore refinement is expected to enhanced durability such as improving freeze–thaw resistance and increasing resistance to aggressive environments.

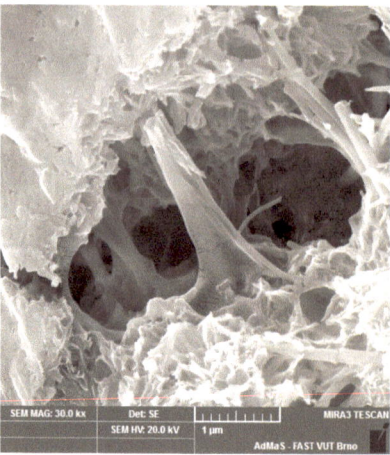

Fig. 7. SEM photomicrographs of mortar E/REF after 90 days + 25 FCs with visible polymeric film.

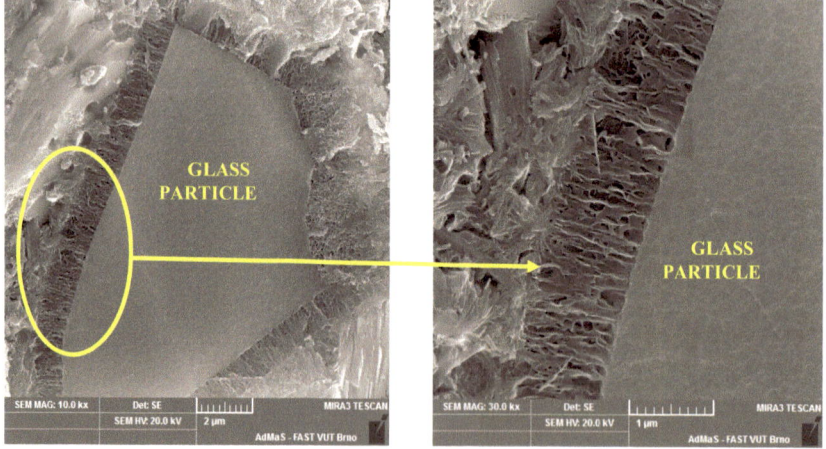

Fig. 8. (a) SEM photomicrographs of PMM sample marked as E/E-G20 after 90 days + 25 FCs with visible E-glass grain. (b) Detail of newly formed C-S-H phase on the surface of E-glass particle.

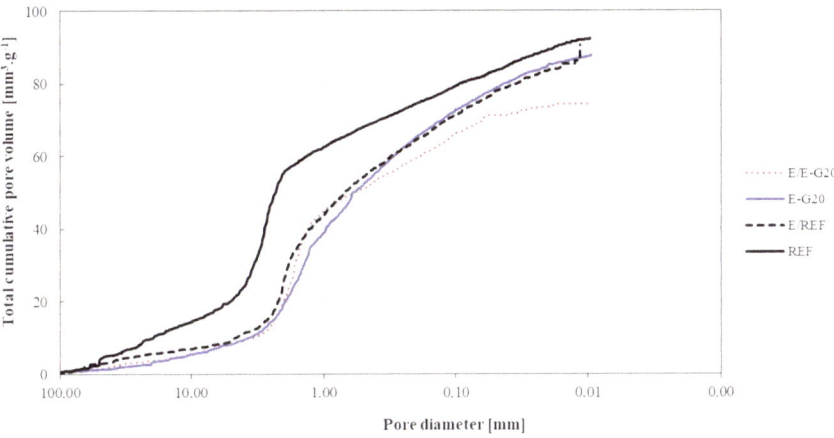

Fig. 9. Porosity of the mortars.

5 Conclusion

In this study, changes in the properties of cement mortars and polymer-modified cement mortars (PMMs) using two types of glass powder were investigated. The following findings were recorded:

- Up to 20% of cement can be replaced by both types of waste glass powder used. For PMMs, this replacement appears to be more effective compared to mortars without EVA-modification.
- In the case of tested PMMs, at 10 wt.% and 20 wt.% substitution of Portland cement, increasing compressive and flexural strength, especially after 90 days and 90 days + 25 FCs compared with reference mixture (without cement substitution).
- The increase in the content of the C-S-H phase in PMMs was confirmed by DTA and C-S-H phases indicated on the surface of the glass particle by scanning electron microscopic analysis. The increase in the content of C-S-H phase also contributed to the reduction in porosity and refinement of pores in mixtures containing glass powder.
- The use of waste glass powder as a partial replacement of Portland cement in polymer-modifies cement mortars can have a significant impact on curing kinetics. However, the additional production of C-S-H phase as a result of the pozzolanic reaction typically occurred at a slower rate compared to the Portland cement hydration enhanced the long-term strength development and freeze–thaw resistance of PMMs.

The findings show that finely ground recycled glass, that no longer meets the requirements of glassworks for the reproduction of glass, has high potential to be used as an effective cement replacement for PMM materials, that are currently used in large amounts, mainly in the rehabilitation of concrete structures. Using waste glass in this way not only supports the circular economy but improves important properties such as durability of cementitious mortars and especially polymer-modified cement mortars as well.

Acknowledgement. This research was supported by the project of Technology Agency of the Czech Republic Nr. CK03000240 "Development of cement composites and process parameters for 3D printing of elements complying the requirements of traffic constructions".

References

1. Schwarz, N., Cam, H., Neithalath, N.: Influence of a fine glass powder on the durability characteristic of concrete and its comparison to fly ash. Cement Concr. Compos. **30**, 486–496 (2008)
2. Kamali, M., Ghahermaninezhad, A.: An investigation into the hydration and microstructure of cement pastes modified with glass powders. Constr. Build. Mater. **112**, 915–924 (2016)
3. Nassar, R., Soroushian, P.: Strength and durability of recycled aggregate concrete containing milled glass as partial replacement for cement. Constr. Build. Mater. **29**, 368–377 (2012)
4. The European Green Deal, https://ec.europa.eu/info/strategy/priorities-2019-2024/european-green-deal_en, last accessed 2022/02/01
5. Salim, M.U., Mosaberpanah, M.A.: Mechanical and durability properties of high-performance mortar containing binary mixes of cenosphere and waste glass powder under different curing regimes. J. Mater. Res. Technol. **13**, 602–617 (2021)
6. Letelier, V., Bustamante, M., Olave, B., Martínez, C., Ortega, J.M.: Properties of mortars containing crumb rubber and glass powder. Dev. Built Environ. **14**, 100131 (2023)
7. Islam, S.G.M., Rahman, M.H., Kazi, N.: Waste glass powder as partial replacement of cement for sustainable concrete practice. Int. J. Sustain. Built Environ. **6**, 37–44 (2017)
8. Schwarz, N., Neithalath, N.: Influence of a fine glass powder on cement hydration: comparison to fly ash and modeling the degree of hydration. Cem. Concr. Res. **38**, 429–436 (2008)
9. Zheng, K.: Pozzolanic reaction of glass powder and its role in controlling alkali-silica reaction. Cement Concr. Compos. **67**, 30–37 (2016)
10. Dvořák, K., Dolák, D., Dobrovolný, P.: The improvement of the Pozzolanic properties of recycled glass during the production of blended Portland cements. Procedia Eng. **180**, 1229–1236 (2017)
11. Patel, D., Shrivastava, R., Tiwari, R.P., Yadav, R.K.: Properties of cement mortar in substitution with waste fine glass powder and environmental impact study. J. Build. Eng. **27**, 100940 (2020)
12. Parghi, A., Alam, M.S.: Physical and mechanical properties of cementitious composites containing recycled glass powder (RGP) and styrene butadiene rubber (SBR). Constr. Build. Mater. **104**, 34–43 (2016)
13. Nahi, S., Leklou, N., Khelidj, A., Oudjit, M.O., Zenati, A.: Properties of cement pastes and mortars containing recycled green glass powder. Constr. Build. Mater. **262**, 120875 (2020)
14. Dehghan, A., Peterson, K., Shvarzman, A.: Recycled glass fiber reinforced polymer additions to Portland cement concrete. Constr. Build. Mater. **146**, 238–250 (2017)
15. Couto da Silva, R., Puglieri, F.N., Chiroli, D.M.G., Bartmeyer, G.A., Kubaski, E.T., Tebcheran, S.M.: Recycling of glass waste into foam glass boards: a comparison of cradle-to-gate life cycles of boards with different foaming agents. Sci. Total. Environ. **771**, 145276 (2021)
16. The secondary raw materials policy of the Czech Republic, www.mpo.cz, last accessed 2022/02/01
17. Zach, J., Sedlmajer, M., Hroudová, J., Nevařil, A.: Technology of concrete with low generation of hydration heat. Procedia Eng. **65**, 296–301 (2013)

Effects of Polymers and Other Material Components on Electrical Resistivity of Cement Mortar

Mikio Wakasugi[1]($\boxtimes$), Takuya Fukui[1], Toshiyuki Kanda[1], and Katsunori Demura[2]

[1] Chemical Construction, 5-5, Uozakihamamachi, Higashinada-ku, Kobe 658-0024, Japan
m.wakasugi@chemical-koji.co.jp

[2] Nihon University, 1 Nakagawara, Tokusada, Tamura-machi, Koriyama 963-8642, Fukushima, Japan

Abstract. In the repair of the reinforced concrete structures deteriorated due to the chloride corrosion of rebars, it is desirable that the electrical resistivity of the patching repair mortar is equal to or lower than that of the existing concrete, considering the case of adopting the cathodic protection in the future. In general, cementitious patching repair mortars contain polymer components, which are thought to increase the resistivity, but there are also components other than polymer components that affect the resistivity. Such components without polymer include admixture components such as ground granulated blast furnace slag and fly ash, cement components such as ultra rapid hardening cement, and corrosion inhibitor components such as lithium nitrite. In this study, the effects of those components on the resistivity of cementitious patching repair mortars up to the age of 26 weeks are clarified. As a result, the electrical resistivity of cementitious patching repair mortar depends greatly on the types and amounts of their components.

Keywords: Patching repair mortar · Electrical resistivity · Polymer cement mortar · Admixture · Ultra-rapid hardening cement · Lithium nitrite

1 Introduction

In the repair of the reinforced concrete structures deteriorated due to the chloride corrosion of rebar, patching repair mortars are selected in consideration of the future adoption of the cathodic protection. It is desirable that the electrical resistivity is less than the same level as the existing concrete [1]. Therefore, the resistivity of the cementitious patching repair mortar is desirable to be $50k\Omega \cdot$ cm or less [2]. On the other hand, there are various types of commercially available cementitious patching repair mortar depending on the purpose of application. In past studies, since the resistivity of each commercial product differs, it is necessary to measure the resistivity when applying it to the cathodic protection [3]. In particular, the influence of the polymer component on the resistivity is a concern. In addition, although it has been reported that the admixture of fly ash increases the resistivity over a long period of time [4], there are few reports that have examined the effects of the type and amount of each component constituting

© The Author(s) 2025
L. Czarnecki et al. (Eds.): ICPIC 2023, 61, pp. 280–288, 2025.
https://doi.org/10.1007/978-3-031-72955-3_28

the patching repair mortar in detail. In this study, three types of polymer dispersion, an ordinary Portland and three types of ultra-rapid hardening cement as cement, ground granulated blast furnace slag (GGBS) and fly ash (FA) as an admixture, and lithium nitrite aqueous solution (LN) as a corrosion inhibitor were used. Mortar specimens were prepared and the effects of these components on the resistivity of cementitious patching repair mortars were investigated.

2 Experimental Method

2.1 Materials Used

Tap water was used as mixing water, and standard sand specified by JIS R 5201 "Physical testing methods for cement" was used as fine aggregate. Three types of polymer dispersion, which are typical in Japan [5], were used, ethylene vinyl acetate (EVA), styrene butadiene rubber (SBR), and polyacrylic ester (PAE). Polymer cement mortar (PCM) was prepared with three levels of polymer cement ratios of 5, 10, and 20%. Table 1 shows the properties of the polymer dispersions used for this study. Table 2 shows the property of ordinary Portland and 3 types of ultra-rapid hardening cements (UHC) used in the study. UHC-I is a calcium sulfoaluminate cement, UHC-II is an amorphous calcium aluminate cement, and UHC-III is a calcium aluminate cement. The main components that impart ultra rapid hardening properties are different [6].

Table 1. Properties of Polymer Dispersions

Type	Appearance	Solid content (%)	Density (g/m^3)	pH	Viscosity (mPa · s)
EVA	milky liquid	45.3	1.04	6.2	960
SBR	milky liquid	45.1	1.03	9.5	300
PAE	milky liquid	45.0	1.02	8.0	1010

Table 2. Properties of ordinary Portland and ultra-rapid hardening cements

Type	Density (g/cm^3)	Blaine* (cm^2/g)	Chemical composition (%)					
			SiO$_2$	Al$_2$O$_3$	Fe$_2$O$_3$	CaO	MgO	SO$_3$
OPC	3.14	3070	20.3	5.2	2.9	64.3	0.9	2.1
UHC-I	3.01	4690	14.4	13.7	2.0	54.5	1.4	11.7
UHC-II	3.00	5400	15.1	9.7	2.1	57.5	0.7	10.1
UHC-III	2.98	6050	11.2	17.2	1.9	52.4	0.8	10.3

*Blaine: Blaine specific surface area

Table 3 shows the properties of GGBS and FA used in the investigation of the admixture composition. The properties of GGBS were measured according to JIS A

6206 "Ground granulated blast-furnace slag for concrete", and the properties of FA were measured according to JIS A 6201 "Fly ash for use in concrete". The amounts of GGBS added were 16%, 46% and 66%, and the amounts of FA added were 8%, 16% and 26%. These mixing ratios are values corresponding to Class A, Class B and Class C in JIS A 5211 "Portland blast-furnace slag cement" and JIS R 5213 "Portland fly-ash cement", respectively. In order to study the components of the corrosion inhibitor, an aqueous solution of lithium nitrate (LN) with a solid content of 40 wt% (blue transparent liquid, pH 9.2, density 1.24) was used, and the unit amount was adjusted to 55 kg/m^3, and OPC, GGBS-46 and FA-16 were used as cements.

Table 3. Properties of admixture material

Type	Density (g/cm^3)	Blaine* (cm^2/g)	Chemical composition (%)					
			MgO	SiO$_2$	SO$_3$	Moisture	Ig. Loss	Cl
GGBS	2.91	4160	5.81	—	0.01	—	0.03	0.004
FA	2.29	4070	—	63.6	—	0.2	2.5	—

*Blaine: Blaine specific surface area

2.2 Preparation and Curing of Specimen

Weighing, mixing and molding conformed to JIS R 5201. Sand cement ratio of 3 and a water cement ratio of 0.5 (both mass ratios) were used as the basis for the mix proportions. Mixing amount of each specimen shows Table 4, 5, 6, and 7. The specimen size was 4 × 4 × 16 cm, and the curing was performed in accordance with JIS A 1171 "Test methods for polymer-modified mortar" for 2 days after molding. After demolding, it was cured in water at a temperature of 20 °C for 5 days and then in air at a temperature of 20 °C for 21 days. Tables 4, 5, 6, and 7 show the mixing amount of the mortar specimens.

Table 4. Mixing amount of OPC and PCM specimen (Unit g)

Material	OPC	EVA			SBR			PAE		
	0%	5%	10%	20%	5%	10%	20%	5%	10%	20%
Polymer	0	50	100	200	50	100	200	50	100	200
Water	225	198	170	115	198	170	115	198	170	115

Ordinary Portland Cement 450g and Standard Sand 1350g were common and constant

2.3 Measurement Method of Resistivity

After 28-day curing, the specimens were subjected to outdoor exposer at Kobe in Japan. For conditioning, the specimens were stored at 20 °C-room for 3 days before the measurement of the resistivity. Resistivity were measured at 4,8, 13 and 26 weeks after curing,

Table 5. Mixing amount of OPC and UHC specimen (Unit g)

Material	OPC	UHC-I	UHC-II	UHC-III
Cement	450	450	450	450
Sand	1350	1350	1350	1350
Water	225	225	225	225

Table 6. Mixing amount of mixed cement specimen (Unit g)

Material		OPC	GGBS			FA		
		0%	16%	46%	66%	8%	16%	26%
Cement		450	378	243	153	414	378	333
Admixture	GGBS	0	72	207	297	–	–	–
	FA	0	–	–	–	36	72	117

Standard Sand 1350g, Water 225g were common and constant

Table 7. Mixing amount of LN-containing mixed cement specimen (Unit g)

Material		OPC	OPC-LN	GGBS46	GGBS46-LN	FA16	FA16-LN
Cement		450	450	243	243	378	378
Admixture	GGBS	0	0	207	207	–	–
	FA	0	–	–	–	36	72
Lithium Nitrite		–	124	–	124	–	124
Water		225	151	225	151	225	151

Standard Sand 1350g was common and constant

using an AC earth resistance tester (MCMILLER 400A, manufactured by MCMILLER, USA) as shown in Fig. 1. Conductive gel was applied to both ends of the $4 \times 4 \times 16$cm specimen, and the specimen was kept for 10 min [7]. In addition, the data consistency was confirmed by checking the values measured by the JSCE-G 581 "Test method for electrical resistivity of concrete by four electrode method".

3   Result and Discussion

3.1   Effect of Polymer Components on Resistivity

Figures 2, 3, 4 show the relationship between the resistivity and type of polymer and P/C of PCM at 4, 8, and 13 weeks. The resistivity of PCM tended to increase with an increase in the aging and P/C. However, at 4weeks, which is the condition of the standard test, all PCMs except PAE P/C = 20% were 50 k$\Omega \cdot$ cm or less. Therefore, it was confirmed that

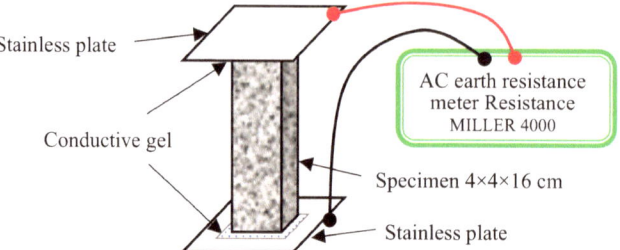

Fig.1. Outline of Measurement Method of Electrical Resistivity

the resistivity of PCM is not as high as generally assumed. However, the resistivity of PCM continued to increase after 4 weeks, and at 13 weeks all PCM of P/C 10% and P/C 20% became greater than 50 kΩ · cm. It seems that the measuring age of the standard test method should be a little longer than 4 weeks.

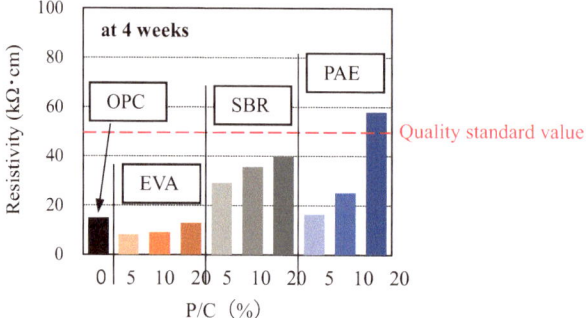

Fig. 2. Relationship between resistivity and type of polymer and P/C of PCM at 4 weeks

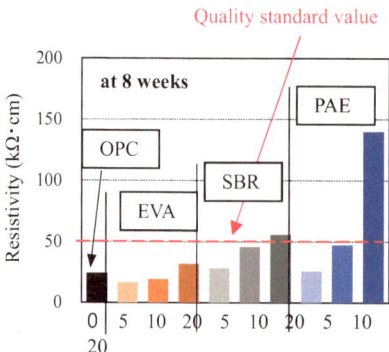

Fig. 3. Relationship between resistivity and type of polymer and P/C of PCM at 8weeks

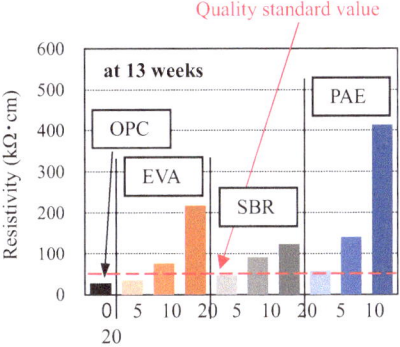

Fig. 4. Relationship between resistivity and type of polymer and P/C of PCM at 13weeks

Resistivity of PCM is also affected by the hydrophilic and hydrophobic functional groups contained in the molecular structure shown in Fig. 5 and by the hydrophilic-lipophilic balance (HLB) value of the surfactant contained as a subcomponent [8]. It is presumed that the resistivity of the EVA was lowered due to the water retention of the surfactant contained [8]. However, the EVA also showed high resistivity at P/C = 20% at 13 weeks, which is presumed to be due to progress in drying of the polymer phase.

On the other hand, in SBR and PAE, effect of the hydrophobic functional group is considered to have increased the resistivity. However, SBR uses a nonionic surfactant with a high HLB value, and it is thought that this action improves the water retention, resulting in a lower resistivity than the PAE [8]. It is presumed that PAE showed the highest resistivity because it does not contain hydrophilic functional groups or strongly hydrophilic surfactants.

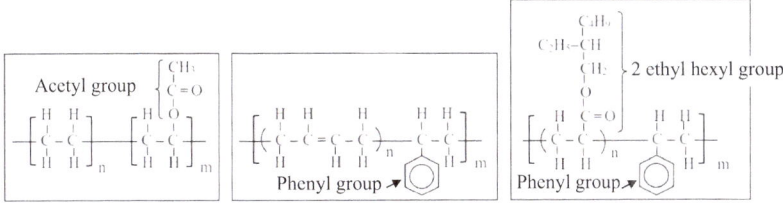

Fig. 5. Molecular structure of polymer

3.2 Effect of Cement Components on Resistivity

Figure 6 shows the temporal change in resistivity of mortars using OPC, UHC-I, UHC-II and UHC-III from 4 weeks to 26 weeks. OPC is almost constant at 30–70 kΩ · cm, and UHC-I is almost the same as OPC up to 13 weeks, and then increased to 220 kΩ · cm. UHC-II is similarly same as OPC until 13 weeks, and then increased to 420 kΩ · cm. UHC-III shows the resistivity of 150 kΩ · cm at 4 weeks, about three times that of OPC, and then it increased with age to a very large resistivity of about 1000 kΩ · cm. It has been reported that the three types of ultra-rapid hardening cement differ in the type and amount of calcium aluminate that imparts ultra-rapid hardening properties, resulting in different types and amounts of calcium aluminate hydrates that are produced, as well as different microstructural structures [9], and these factors are presumed to have influenced the increase in resistivity with age.

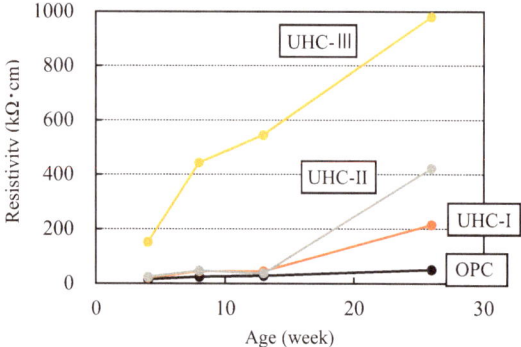

Fig. 6. Temporal change in resistivity of ordinary and ultra fast hardening cement mortars

3.3 Effect of Admixture Components on Resistivity

Figure 7 shows the change in the resistivity with GGBS and FA replacement at each age. The resistivity of mortar mixed with any admixture increased with the age. GGBS66 showed the most largest resistivity exceeding 200 kΩ · cm at 26 weeks. In any cases, the increase in resistivity is large from 13 to 26 weeks. Comparing the effects of GGBS and FA on resistivity, the effect of GGBS was slightly greater than FA. Since GGBS has latent hydraulic property and FA consumes calcium hydroxide in the hardening material by pozzolanic reaction, it is considered that the ion concentration in the pore solution decreased and the resistivity of each mortar increased [4].

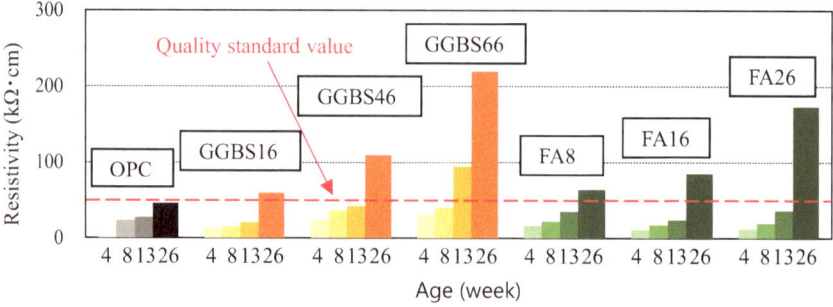

Fig. 7. Change in resistivity with GGBS and FA replacement

3.4   Effect of Lithium Nitrite on Resistivity

Figure 8 shows the change in resistivity when 55 kg/m³ of lithium nitrite is added to OPC, GGBS46 and FA16. Lithium nitrite makes the mortar hygroscopic and increases the moisture content. It is presumed that the mortar containing nitrite ions and lithium ions improves the ionic conductivity and reduces the resistivity. The effect of lithium nitrite addition on resistivity was a 20–30% decrease in OPC, a 20–30% decrease in GGBS46 and FA16 up to 13 weeks, but 40–50% decrease at 26 weeks.

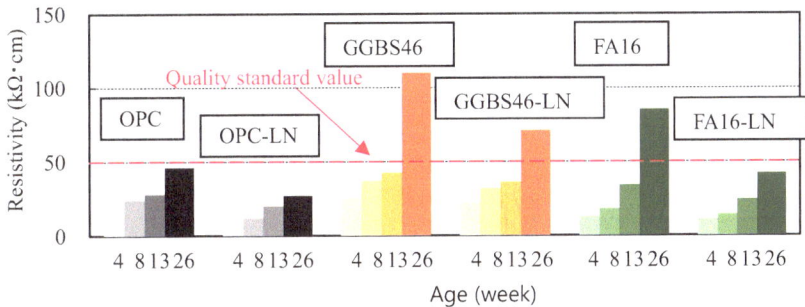

Fig. 8. Change in resistivity with LN addition

4   Conclusion

The following findings were obtained as a result of this study.

- The resistivity of PCM is increased with an increase in the age and P/C. The degree of increase with increasing P/C differs depending on the polymer component, being the smallest in the SBR and the largest in the PAE.
- The resistivity of CSA based and am-CA based ultra rapid hardening cement mortars is similar to that of OPC mortars up to 13 weeks, and then increases. The CA based ultra rapid hardening cement mortar shows a marked increase in resistivity from 4 weeks of age, showing the highest resistivity.
- The resistivity GGBS and FA mortar is similar to that of OPC mortar at the early age, but it increases as the age progresses and the mixing ratio increases. It increases greater in mortar mixed with GGBS than FA.
- The addition of LN to OPC and mixed with GGBS and FA have the effect of reducing the resistivity.

From the above, it is clear that the electrical resistivity of cementitious patching repair mortar depends greatly on the types and amounts of their components. In addition, we plan to continue this study in the future because the effect changes with the age.

References

1. Coastal Development Institute of Technology: Port Concrete Structure Repair Manual p. 38 (2018)
2. Tokyo Port Terminal Co., Ltd.: Pier Deterioration Investigation and Repair Manual, p. 64 (2012)
3. Naito, H., Moriya, S., Kawamata, K., Minagawa, H.: Investigation on electrical resistivity of various cement-based cross-section repair materials. Concrete Engineering Annual Proceedings **30**(2), 595–600 (2008)
4. Sato, M., Sakai, T., Minagawa, H., Hisada, M.: Changes in resistivity of concrete and mortar mixed with fly ash over time. Proceedings of the Japan Concrete Institute **32**(1), 695–700 (2010)
5. Demura, K.: Polymers for mixing with cement. Concrete Engineering **26**(3), 85–90 (1988)
6. Hara, H., Mori, T., Higuchi, T., Morioka, M.: Fundamental properties of mortar using ultra rapid hardening cement with different hardening components. Cement Concrete Papers **69**, 154–160 (2015)
7. Naito, H., Moriya, S., Minagawa, H., Kawamata, K.: Research on electrical resistivity of cement-based cross-section repair materials. Annual Report of Penta-Ocean Construction Technology **39**(4), 1–12 (2009)
8. Muroi, S.: Introduction to Polymer Latex, Kobunsha, pp. 30–34, 125–169 (1983)
9. Nikaido, Y., Ito, T., Sakai, E., Daimon, M.: Ettringite Formation and Microstructure in Initial Hydration of Ultra Rapid Hardening Cement. Proceedings of the Japan Concrete Institute **17**(1) (1995)

Effect of the Cellulose Ether on Water Loss of the Calcium Sulphoaluminate Cement Mortars

Chuanchuan Guo[1], Qin Wan[1], Ru Wang[1(✉)], Bo Chen[1], and Ning Chen[2]

[1] Key Laboratory of Advanced Civil Engineering Materials of Ministry of Education, School of Materials Science and Engineering, Tongji University, Shanghai, China
ruwang@tongji.edu.cn
[2] Shanghai Research Institute of Building Sciences Co., Ltd, Shanghai, China

Abstract. Cellulose ether (CE) is widely used in cement-based materials because of its good water retention capacity that can improve the workability of the fresh mortars significantly. However, in the high temperature conditions, the CE modified cement mortars sometimes are easy to lose their good workability, which may be due to the change of the water-retention capacity of CE. This work investigates the changes in the water loss rate (WLR) of the CE modified calcium sulphoaluminate cement (CSAC) mortars at 20 °C, 40 °C, 60 °C and 80 °C respectively, and the effect of the types and contents of the CE was also considered. Additionally, isothermal calorimeter and ^{1}H low-field NMR were carried out to monitor the changes of chemically bound water and CE adsorbed water content during the reaction. The results show that the WLRs of the CE modified CSAC mortars changes with temperature and the types and contents of the CE. These changes are mainly based on the fact that CE affects the state and relative contents of water molecules in mortar, and the microstructure of the CSAC mortars.

Keywords: cellulose ether · calcium sulphoaluminate cement mortar · water loss rate · degree of substitution · different temperatures

1 Introduction

Compared with the Portland cement (PC) the calcium sulphoaluminate cement (CSAC) have advantages of such as rapid setting and early strength [1–7]. As a result, the characteristics of CSAC determine that it is better suited to mechanical construction because it is difficult to complete the operation in a short time by relying on labor [8, 9]. However, due to the high reactivity of ye'elimite in CSAC, its workability decreases significantly with time, so that the efficiency of mechanized construction has been greatly affected. In order to improve the mechanized construction efficiency, cellulose ether (CE) is usually added to cement-based materials to improve the workability of fresh mortar [10–14]. This is mainly because the incorporation of CE can reasonably distribute the state and quantity of water in the mortar. But in high temperature conditions CE modified CSAC mortar sometimes is easy to lose its good workability owing to the change of the water retention capacities of the CE. Therefore, it is necessary to understand the water loss of the CE modified CSAC mortars to evaluate the water retention effect of CE.

L. Czarnecki et al. (Eds.): ICPIC 2023, 61, pp. 289–295, 2025.
https://doi.org/10.1007/978-3-031-72955-3_29

The object of this work is to investigate the water loss rate (WLR) of CE modified CSAC mortar at different temperatures. The samples were placed at four curing temperatures including 20 °C, 40 °C, 60 °C and 80 °C to measure the water loss rate within 12 h. Moreover, isothermal calorimeter and ^{1}H low-field NMR were used to determine the heat flow and the state and quantity of water in CSAC paste during hydration, respectively.

2 Experimental

2.1 Raw Materials and Mix Proportions

The cement used in this study is 42.5 grade CSAC, its detailed index can be seen from our previous paper [15]. A hydroxyethyl cellulose (HEC) and two hydroxyethyl methyl cellulose (HEMC) with high and low degree of substitution (DS) were utilized for the test, which were labeled as HEMC-H and HEMC-L. The DS of HEC, HEMC-H and HEMC-L are 2.5, 1.9 and 1.6 respectively. Tap water and the standard sand conforming to GB/T 17671–1999 were applied for the experiment. CE dosing is 0.05%, 0.1%, 0.2%, 0.3%, 0.4% and 0.5% by weight of cement respectively and the mortar without CE is a control sample. The sample was named as a combination of cellulose ether and the dosage, such as HEC-0.05%. The cement to sand ratio is 1:3, and the water to cement ratio is 0.54.

2.2 Determination of Water Loss Rate (WLR) and Air Content Measurement

The mixing method refers to Chinese standard GB/T 17671–1999. The freshly mixed mortar was immediately placed in a glass dish. After weighing the initial weight of the mortar and the glass dish, the dishes filled with mortar were placed in an environment of 20 °C, 40 °C, 60 °C or 80 °C until the weight test. The curing age is 15 min, 30 min, 45 min, 1 h, 2 h, 4 h, 6 h and 12 h. The WLR of the mortar at a certain time is equal to the water loss during the time divided by the initial water content of the mortar. Besides, the air content of the CSAC mortar is determined by instrument method according to the standard JGJ/T 70-2009.

2.3 Isothermal Calorimetry Test and 1H Low-Field NMR Experiment

The hydration heat measurement refers to literature [10]. The ^{1}H low-field NMR experiment refers to literature [15]. From the test, the transverse relaxation time (T_2) was obtained. It is reported that the relative content of CE adsorbed water can be expressed by the peak area of CE adsorbed water (A_{CE}) [15]. In the same way, the A_{CE} value of the sample is used to evaluate the relative content of CE adsorbed water in this study.

3 Results and Discussion

3.1 WLR of CE Modified CSAC Mortar

The WLR of CE modified CSAC mortar with curing time at 20 °C is plotted in Fig. 1. There is no obvious change in the development trend of WLR of all samples in the first 1 h, but then, the WLR of CE modified CSAC mortar is related to the content and type of

CE. Specifically, for the same type of CE, the WLR of the samples broadly increases first and then decreases with the CE content at the same curing time (such as 12 h). Besides, for different types of CE, the minimum amount of CE to retain water (lower than the control sample in WLR) is different. The dosage of HEC is the least, 0.2%, followed by 0.3% of HEMC-L, and HEMC-H is the most, 0.4%. The reason may be that CE affects the hydration of CSAC paste, the water absorption of CE, and the pore structure of the sample.

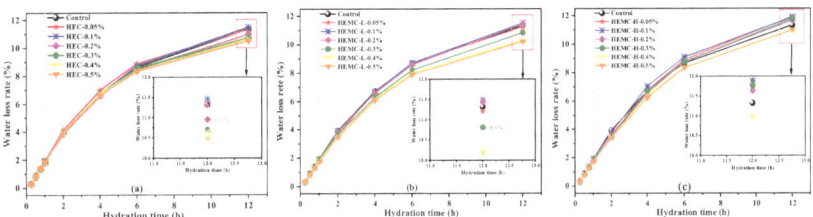

Fig. 1. The WLR of HEC modified CSAC mortar within 12 h at 20 °C

3.2 The Change of WLR of CE Modified CSAC Mortar at Different Temperatures

Figure 2 presents the WLR of 0.5% content CE modified CSAC mortar at 20 °C, 40 °C, 60 °C and 80 °C. Without considering 60 °C, the WLR increases rapidly with the rise of temperature, especially in the first 1 h. However, at 60 °C, the WLR of the CSAC mortar is lower than that at 40 °C in the early time. This is inconsistent with the expected results. As we all known, the gelation temperature of CE is between 50–90 °C. Consequently, the CE used in the CSAC mortar may also undergo gelation at 60 °C, which greatly increases the viscosity of the suspension slurry and improves the water retention capacity of the CSAC mortar in the plastic stage. Further, it can be seen from Fig. 2 that there are differences in WLR of CSAC mortar containing different types of CE at high temperature. The WLR of HEC-0.5% is still the lowest, followed by HEMC-L-0.5%, and the WLR of HEMC-H-0.5% is the highest. The result is in agreement with the above case at 20 °C, but the difference among CEs is bigger.

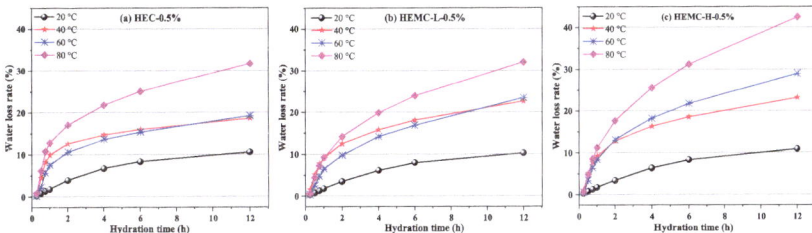

Fig. 2. The WLR of 0.5% content CE modified CSAC mortar at 20 °C, 40 °C, 60 °C and 80 °C

3.3 Hydration Heat Analysis

Figure 3 presents the heat flow of CSAC paste containing different dosage HEMC-L (a) and different kinds of CE (b) within 12 h [10]. As presented, four peaks appear on the heat flow curve of the control, corresponding to the dissolution, transformation, AFt primary formation and AFt secondary formation, respectively. The effect of CE content on the first three exothermic peaks is not so significant. But with the increase of HEMC-L content, the fourth exothermic peak is advanced and the peak becomes higher. This shows that HEMC-L promotes the CSAC hydration after 1 h. Similarly, four peaks appear on each heat flow curve in Fig. 3 (b). Compared with the control, CE advances and increases the fourth exothermic peak. The results show that CE promotes the CSAC hydration. However, the promotion degree is different, HEMC-L has the strongest promotion effect, followed by HEMC-H, and HEC. In general, the promotion of hydration can lead to more free water into bound water. In this case, the free water that can evaporate in the mortar is relatively less, which may lead to a decrease in WLR.

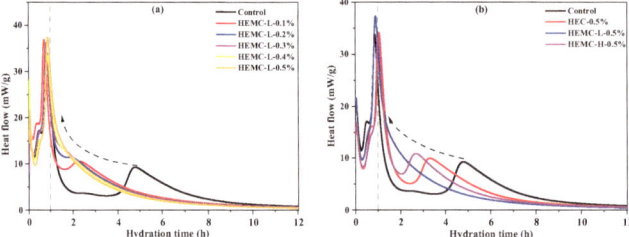

Fig. 3. Heat flow of CSAC paste with different dosage HEMC-L (a) and different kinds of CE (b)

3.4 Relative Content of CE Adsorbed Water (ACE)

The A_{CE} alteration of HEC (a), HEMC-L (b) and HEMC-H (c) modified CSAC paste with hydration time is shown in Fig. 4. From it, the A_{CE} value of the same type CSAC paste basically increases with the content of CE at the same time scale, but it gradually decreases with the hydration time, and the decline trend to different type of CE is distinctive. As shown in Fig. 4 (a), there is a significant decline in the A_{CE} (HEC) value before 2 h. After that, the A_{CE} value tends to be stable, but on the whole, it is still that the paste with high HEC content has a larger A_{CE} value. This shows that HEC always absorbed water in the first 12 h, and the higher the content, the more water absorption. From Fig. 4 (b) and (c), the A_{CE} value of the HEMC modified paste decreases sharply within 1 h and then approaches to 0, indicating HEMC loses water absorption after 1 h of cement hydration. In theory, the water retention capacity of CSAC mortar increases with the A_{CE} value, because the relative quantity of internal evaporation water of the mortar becomes less.

With the increase of CE content, the CSAC hydration was promoted, and the relative content of CE adsorbed water also increased. As a result, the WLR should have decreased with CE dosage. However, the fact is that the WLR is higher than or close to that of the control at low content of CE (HEC $\leq$ 0.1%, HEMC-L $\leq$ 0.2%, HEMC-H $\leq$ 0.3%). The reason may be that the incorporation of CE affects the pore structure of the CSAC mortar, especially the capillary pores. It is reported that the cumulative capillary pore volume of mortar increases first and then decreases with the CE content [11]. The more capillary pores, the higher WLR should be. From this perspective, the CSAC mortar with low content of CE do not play a water retention effect compared with the control, while the water retention effect becomes stronger with the increase of CE content at high dosage.

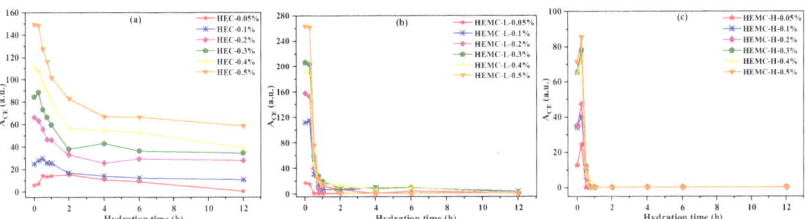

Fig. 4. The A_{CE} of HEC (a), HEMC-L (b) and HEMC-H (c) modified CSAC pastes

3.5 Air Content

The incorporation of CE can introduce air pores into the mortar during mixing [11, 14]. Through tests, the air content of the control, HEC-0.5%, HEMC-L-0.5% and HEMC-H-0.5% is 2.2%, 11.7%, 18.2% and 21.2% respectively. It shows that the incorporation of CE has a strong air-entraining effect, and the air-entraining effects of the three CEs are different. Obviously, HEMC has better air-entraining effect than HEC, and the air-entraining effect of HEMC-H is the best. Commonly, the mortar with high air content has low bulk density and high porosity after hardening. The mortar with large porosity may be more likely to lose water.

Comprehensively, although HEC is not as strong as HEMC in promoting CSAC hydration, it has obvious water absorption effect within 12 h, and the porosity of HEC modified mortar is lower than that of HEMC modified mortar. Thus, the WLR of CSAC mortar containing HEC is lower than that containing HEMC under the same conditions. For HEMC modified CSAC mortar, both HEMC-L and HEMC-H basically lose their water absorption after the mortar hardening, but HEMC-L is superior to HEMC-H in promoting the CSAC hydration, and the porosity of HEMC-L modified mortar is lower than that of HEMC-H modified mortar. Accordingly, the WLR of CSAC mortar containing HEMC-L is lower than that containing HEMC-H.

4 Conclusions

The WLR of CSAC mortar changes regularly due to the influence of the content and type of CE and the temperature. The WLR generally increases first and then decreases with the CE content. The WLR of HEMC modified CSAC mortar is higher than that of

HEC modified mortar, and the WLR of HEMC-H modified mortar is the highest. With the increase of temperature, the WLR of CE modified CSAC mortar increases rapidly. But the CE in mortar may be subjected to gelation at 60 °C, resulting in a WLR in the plastic stage lower than that in 40 °C.

CE affects the CSAC hydration, the CE absorbed water content, the porosity of mortar, and thus alters the WLR of the CSAC mortar. CE promotes the CSAC hydration, and in terms of promotion effect, HEMC-L > HEMC-H > HEC. With the increase of CE dosage, the content of CE adsorbed water increases. The water absorption effect of HEC is better than that of HEMC. Different types of CEs have various air-entraining effects, at the same dosage, HEMC-H > HEMC-L > HEC.

Acknowledgements. The authors acknowledge the financial support by the National Natural Science Foundation of China (Grant No. 51872203) and the Top Discipline Plan of Shanghai Universities-Class I (2022-3-YB-17).

References

1. Lin, L., Wang, R., Lu, Q.: Influence of polymer latex on the setting time, mechanical properties and durability of calcium sulfoaluminate cement mortar. Constr. Build. Mater. **169**, 911–922 (2018)
2. Wang, R., Fan, Y., Wang, Z., Huang, T., Zhang, T.: Performance development of styrene-butadiene copolymer-modified calcium sulfoaluminate cement mortar under different curing conditions. J. Zhejiang Univ. Sci. A. **22**, 1005–1026 (2021). https://doi.org/10.1631/jzus.A2000526
3. Liu, X., Wang, R.: Hydration of styrene-acrylic copolymer modified calcium sulphoaluminate clinker with anhydrite. J. Chin. Ceram. Soc. **50**, 354–363 (2022)
4. Wang, R., Li, L.: Experimental study on the rheology and setting behavior of calcium sulfoaluminate cement paste modified with styrene-butadiene copolymer dispersion. J. Mater. Civ. Eng. **34** (2022). https://doi.org/10.1061/(ASCE)MT.1943-5533.0004154
5. Wang, R., Li, L., Xu, Y.: Influence of curing regimes on the mechanical properties, water capillary adsorption, and microstructure of CSA cement mortar modified with styrene-butadiene copolymer dispersion. J. Mater. Civ. Eng. **31**, 04018344 (2019)
6. Li, L., Peng, Y., Wang, R., Zhang, S.: The effect of polymer dispersions on the early hydration of calcium sulfoaluminate cement. J. Therm. Anal. Calorim. **139**, 319–331 (2020)
7. Li, L., Wang, R.: Early hydration of CSA cement modified with styrene-butadiene copolymer dispersion. Adv. Cem. Res. **33**, 14–27 (2021). https://doi.org/10.1680/jadcr.19.00038
8. Guo, C., Wang, R.: Using sulphoaluminate cement and calcium sulfate to modify the physical–chemical properties of Portland cement mortar for mechanized construction. Constr. Build. Mater. **367**, 130252 (2023)
9. Guo, C., Wang, R.: Influence of calcium sulfoaluminate cement on early-age properties and microstructure of Portland cement with hydroxypropyl methyl cellulose and superplasticizer. J. Build. Eng. **45**, 103470 (2022). https://doi.org/10.1016/j.jobe.2021.103470
10. Wang, R., Liu, K., Wan, Q.: Effect of cellulose ethers with hydroxyethyl group on early hydration of CSA cement. J. Build. Mater. **25**, 836–842 (2022)
11. Li, J., Wang, Z., Huang, T., Wang, R., Wang, S.: Influence of HEMC on properties of sulphoaluminate cement mortar. J. Build. Mater. **24**, 199–206 (2021)

12. Zhang, S., Wang, R., Xu, L.: Properties of calcium sulfoaluminate cement mortar modified by hydroxyethyl methyl celluloses with different degrees of substitution. Molecules **26**, 2136 (2021)

13. Li, J., Wang, R., Xu, Y.: Influence of cellulose ethers chemistry and substitution degree on the setting and early-stage hydration of calcium sulphoaluminate cement. Constr. Build. Mater. **344**, 128266 (2022). https://doi.org/10.1016/j.conbuildmat.2022.128266

14. Li, J.: Influence of cellulose ethers structure on mechanical strength of calcium sulphoaluminate cement mortar. Constr. Build. Mater., 8 (2021)

15. Wan, Q., Wang, Z., Huang, T., Wang, R.: Water retention mechanism of cellulose ethers in calcium sulfoaluminate cement-based materials. Constr. Build. Mater. **301**, 124118 (2021)

Utilisation of Hydrophobic Agents for Water-Repellent Cement Screeds Intended for External Thermal Insulation Composite Systems

Jakub Hodul[(✉)], Lenka Mészárosová, and Nikol Žižková

Faculty of Civil Engineering, Brno University of Technology, Brno, Czech Republic
hodul.j@fce.vutbr.cz

Abstract. The paper discusses the possibilities of hydrophobization of cement screeds intended primarily for external thermal insulation composite systems. The hydrophobic character of the mixtures both with and without the addition of hydrophobic agents was investigated. The following hydrophobic agents were used: zinc and magnesium stearate, sodium oleate, two types of mixed product of stearates and oleates, micronized wax and a silane hydrophobic agent. The mixture containing 0.6% zinc stearate and 0.6% silane showed the lowest water absorption and high flexural and compressive strength. The microstructure of the selected mixtures was also monitored using a scanning electron microscope, where the pores covered with the layer of polymer admixture were observed.

Keywords: hydrophobization · cement screed · water repellent · ETICS

1 Introduction

It is highly important to provide sufficient protection for the whole composition when designing the external thermal insulation composite system (ETICS), which is exposed to adverse weather conditions. It is particularly important to protect it from water influence. The hydrophobic concrete surface is characterized by a water contact angle of more than 90° [1]. It can be achieved by hydrophobic admixture, surface treatment [2], pore structure closure by sealant agents or impregnation [3]. The second method to increase water resistance is indirect hydrophobization by adhesive coating or hydrophobic composition adjustment of the internal structure [4]. To increase the hydrophobicity of the material, it is essential to use compounds with low surface free energy [5]. The surface wettability is mainly determined by the chemical composition and microscopic geometry of the material's surface [6]. The chemical composition can be classified into two categories: inorganic and organic [2]. The long non-polar hydrocarbon chains of some compounds lead to increased hydrophobicity of such materials (for example stearates, oleates and micronized wax). Stearates are chemical compounds derived from stearic acid, a long-chain saturated fatty acid. The reaction of stearic acid with zinc oxide forms the zinc stearate. Magnesium stearate is produced by the reaction of magnesium oxide

© The Author(s) 2025
L. Czarnecki et al. (Eds.): ICPIC 2023, 61, pp. 296–306, 2025.
https://doi.org/10.1007/978-3-031-72955-3_30

(or magnesium hydroxide) with stearic acid. The clusters of stearate molecules and long non-polar hydrocarbon chains predetermine the hydrophobicity. Sodium oleate is a sodium salt of oleic acid with sodium hydroxide. The structure of the oleate molecule, which has a long non-polar hydrocarbon chain and polar carboxylate group, leads to increased water resistance. The micronized wax has, due to its small particles (from 1 to 10 μm), special hydrophobic properties. Small particle size allows for a larger surface area, which can be exposed to liquid and be more effective in water repellence. The nature of micronized wax can lead to poor dispersion or compatibility with aqueous solutions. Highly efficient hydrophobic properties were also recorded with well-known silanes [7, 8], siloxanes, and their mixtures. They are composed of silicon and hydrogen atoms, with one or more organic group (alkyl or aryl) attached to the silicon atom. Silicon-hydrogen bonds in silanes can react with surfaces containing hydroxyl groups to form a covalent bond, which can further enhance the hydrophobic barrier by enlarging the contact angle and coarsening the pore surface [9]. The silane impregnation can provide a long-term hydrophobic effect even in applications where the long-term durability is demanded (over 20 years) [10]. Superhydrophobic coating treatment based on polydimethylsiloxane can reduce the water absorption of the foundry dust/Portland cement-based composites by more than 76% [11], and capillary water absorption of specimens impregnated with waterborne silane-based hydrophobic agents can suppress 5.4% of the value for the untreated mortars [12]. The process of molecular absorption can be adjusted by the structure of organic functional groups, number of hydrolytic groups, surface energy, and cross-linking degree of silanes [13].

2   Materials and Methods

The cement CEM I 42.5R was used as a binder. Grounded limestone (GL), with a high content of calcite and siliceous sand (SS), with particle size under 2 mm (Fig. 1) and content of SiO_2 98.5%, Fe_2O_3 0.95%, Al_2O_3 0.18% and TiO_2 0.21%, was used as the filler. Ethyl vinyl acetate, cellulose ether (as a stabilizer) and de-foaming agent were used as additives in the mixtures to increase the adhesion of material to the surface and durability.

The different polymers (zinc stearate (ZS), magnesium stearate (MS), sodium oleate (SO), mixed product of stearates and oleates (MPSO), micronized wax (MW), and silane hydro-phobic agent (SH), in amounts of 0.2, 0.4 and 0.6% for all) were used as internal hydrophobization (Table 1). The reference mixture (REF) did not contain any hydro-phobic agent.

The water permeability (according to EN 12086), coefficient of water absorption (accord. to EN 1015-18), water absorption coefficient due to capillary action (EN 12808-5), bulk density (EN 1015-10), three-point flexural and compressive strength (EN 13892-2), adhesive bond strength (ČSN 73 2577), shrinkage by the dilatometer, crack size of the reinforcing layer detected by the tensile test scanning electron microscopy, differential thermal analysis and mercury intrusion porosimeter were determined.

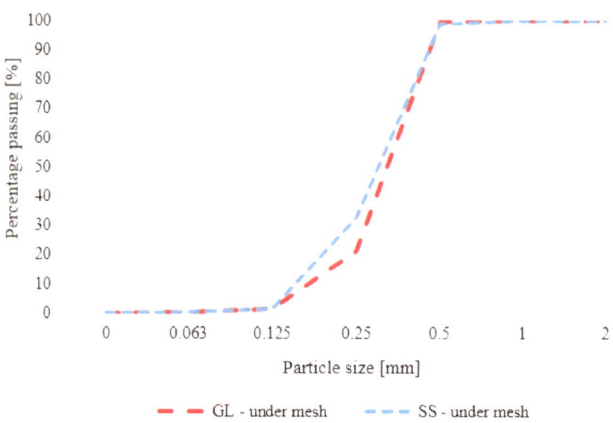

Fig. 1. Particle size of the fillers.

3   Results and Discussion

3.1   Bulk Density

The bulk density of samples was highly similar (approximately 1500 kg/m^3), the exception being one of the mixtures with stearates and oleates (MPSO II) used as the hydrophobic agent (Fig. 2).

3.2   Flexural Strength

The results show that the screeds with the sodium oleate content led to decrease of flexural strength (Fig. 3). The presence of sodium in the structure of cement matrix can positively impact the workability of the cementitious material, but it can also lead to deterioration [14]. Sodium can with its alkali properties, contribute to the formation of the alkali–silica reaction, which leads to a weaker structure [15]. The degradation of the structure is more noticeable after 90 days, when the strength decreases are more obvious.

3.3   Compressive Strength

A similar trend to the flexural strength was observed with determination of compressive strength – see Fig. 4. Mixtures containing sodium oleate additives showed a significant decrease in compressive strength. This could be caused by improper presence of the sodium in the structure and weakening of the compactness of the microstructural bonds. The screeds with higher w/c ratio (MPSO) showed lower strength because a more porous structure was created, so the screeds were less resistant to mechanical stress.

3.4   Water Vapour Permeability

The low permeability of concrete is the most important factor for protecting the material against sulphates, acids, carbonation, frost, alkali-aggregate reaction, efflorescence and

Table 1. Composition of the mortar's mixtures

	Hydrophobic agent	SS	LS	CEM	Additive	Water	w/c ratio
	[% wt.]	[% wt.]	[% wt.]	[% wt.]	[% wt.]	[% wt.]	
REF	0.00	16.60	40.40	38.00	5.00	24.00	0.63
ZS 0.2	0.20	16.63	40.48	38.08	5.01	24.00	0.63
ZS 0.4	0.40	16.67	40.56	38.15	5.02	26.70	0.70
ZS 0.6	0.60	16.70	40.64	38.23	5.03	26.70	0.70
MS 0.2	0.20	16.63	40.48	38.08	5.01	26.00	0.68
MS 0.4	0.40	16.67	40.56	38.15	5.02	26.00	0.68
MS 0.6	0.60	16.70	40.64	38.23	5.03	24.00	0.63
SO 0.2	0.20	16.63	40.48	38.08	5.01	26.70	0.70
SO 0.4	0.40	16.67	40.56	38.15	5.02	29.40	0.77
SO 0.6	0.60	16.70	40.64	38.23	5.03	31.30	0.82
MPSO I. 0.2	0.20	16.63	40.48	38.08	5.01	30.00	0.79
MPSO I. 0.4	0.40	16.67	40.56	38.15	5.02	30.00	0.79
MPSO I. 0.6	0.60	16.70	40.64	38.23	5.03	32.00	0.84
MPSO II. 0.2	0.20	16.63	40.48	38.08	5.01	26.00	0.68
MPSO II. 0.4	0.40	16.67	40.56	38.15	5.02	25.30	0.66
MPSO II. 0.6	0.60	16.70	40.64	38.23	5.03	26.00	0.68
MW 0.2	0.20	16.63	40.48	38.08	5.01	24.00	0.63
MW 0.4	0.40	16.67	40.56	38.15	5.02	24.00	0.63
MW 0.6	0.60	16.70	40.64	38.23	5.03	24.00	0.63
SH 0.2	0.20	16.63	40.48	38.08	5.01	24.00	0.63
SH 0.4	0.40	16.67	40.56	38.15	5.02	24.00	0.63
SH 0.6	0.60	16.70	40.64	38.23	5.03	23.00	0.60

other cement-based materials [16]. The lowest water vapour permeability was reached with the samples containing magnesium oleate – see Fig. 5.

3.5 Coefficient of Capillary Absorption

When the cementitious material is more hydrophobic, it takes longer to reach the saturated water absorption stage [7]. In accordance with this finding were the lowest values of capillary absorption coefficient being observed in samples with micronized wax, silane and both magnesium-stearate-based hydrophobic agents – see Fig. 6. The sodium oleates did not behave as well as the other compounds.

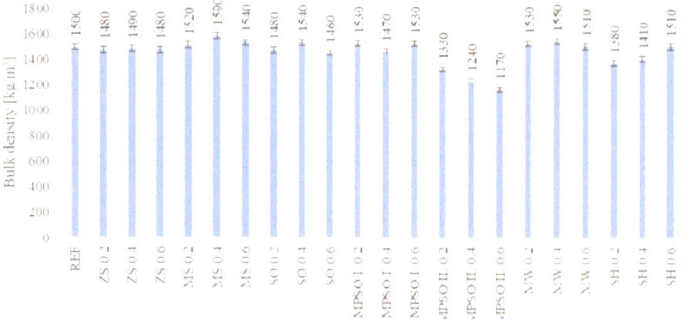

Fig. 2. Results of bulk density.

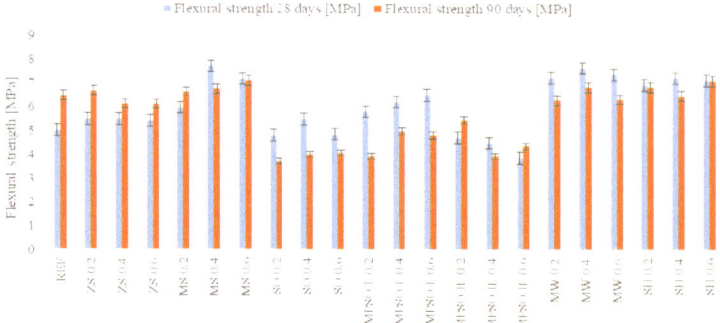

Fig. 3. Results of the three-point flexural strength test.

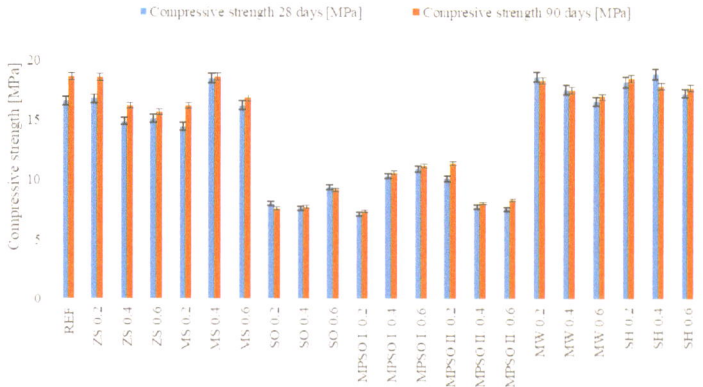

Fig. 4. Results of the compressive strength test.

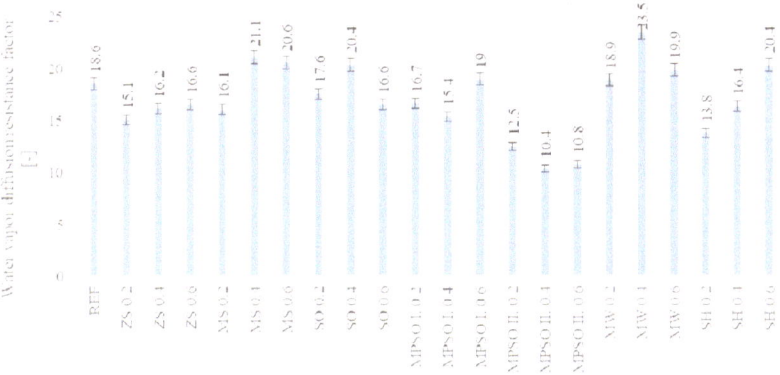

Fig. 5. Results of the water vapour permeability test.

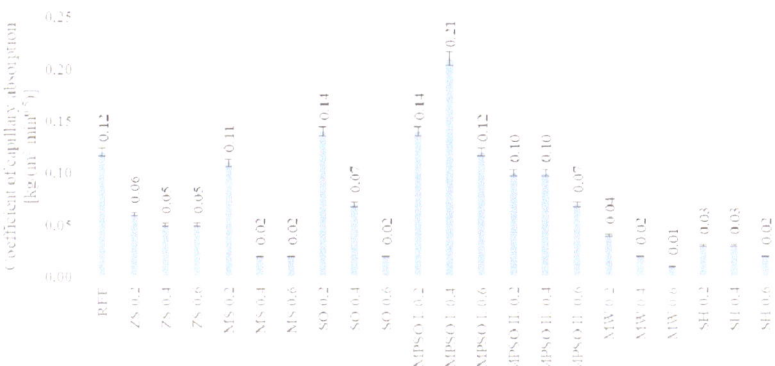

Fig. 6. Results of the coefficient of capillary absorption.

3.6 Water Absorption

It is seemed from Fig. 7 that the hydrophobic admixtures rapidly decreased the water absorption of samples with hydrophobic agents based on both zinc and magnesium stearate, sodium oleate, micronized wax and a silane hydrophobic agent.

Both types of mixed product of stearates and oleates increased the water absorption coefficient. The mixture of stearates and oleates can be used as an emulsifier in the presence of water or other polar solvents. The water absorption coefficient is closely related to the coefficient of capillary absorption: the higher the water absorption, the higher the capillary absorption.

3.7 Mineralogy and Microstructure

The XRD analysis and scanning electron microscope (SEM) identified the minerals calcite and portlandite and the C-S-H phases (Figs. 8, 9, 10 and 11). The pores of

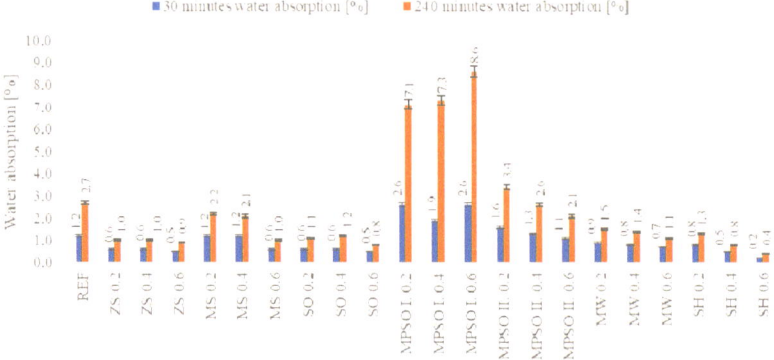

Fig. 7. Results of the water absorption.

samples with hydrophobic agent were covered with the layer of polymer. This aligns with Liang's conclusion that the hydrophobic components create the barrier for water intrusion in cement-based materials by covering the pores with a layer of low surface energy materials [4].

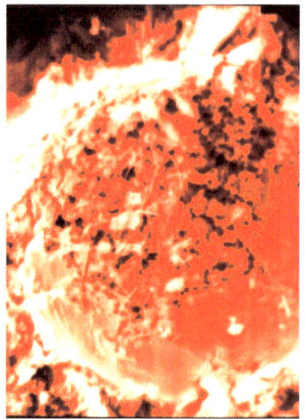

Fig. 8. The SEM image of needle-like and reticular C-S-H phases formed inside the pore structure in the sample MPSO II and covered by the layer of hydrophobization agent: magnification 1 500×
.

According to Zhang et al. [17], the hydrophobic agent does not have an obvious effect on the hydration degree, amount and structure of hydration products. The current study results are consistent with these findings.

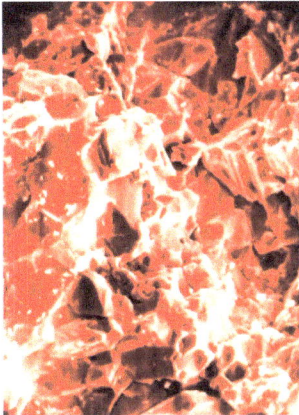

Fig. 9. The SEM image of a well crystallized trigonal structure of portlandite in the sample SH: magnification 900×.

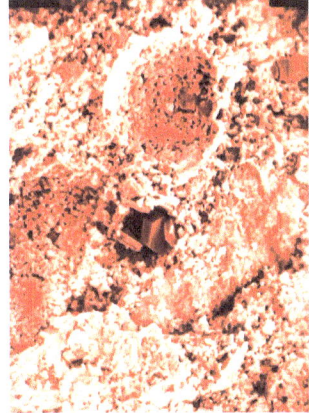

Fig. 10. The SEM image of a partially filled pore of the cementitious matrix in the sample MPSO: magnification 900×.

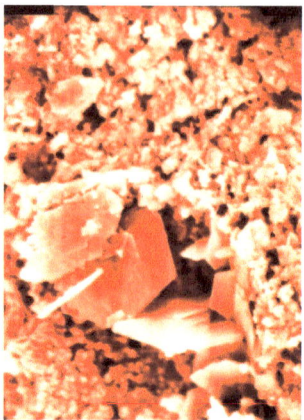

Fig. 11. The SEM image of calcite crystals in a fully filled pore in the sample MS: magnification 1 000×.

4   Conclusion

The addition of hydrophobic agents positively influenced the hydrophobic properties of the screeds. Hydrophobization based on sodium oleates had an adverse effect on the physical and mechanical properties and significantly affected the reduction of strength; however, this did improve in the long term. Highly favourable results regarding mechanical parameters and hydrophobic properties were achieved by the use of silane-based admixtures. These admixtures did not affect either the compressive or flexural strength, causing significant water absorption and capillary absorption, and in larger amounts increased water vapour permeability.

Acknowledgement. This paper was created with the financial support of the Czech Science Foundation (GACR), Standard project No. 22-08888S "Increasing the durability of cement composites using water-based hydrophobization".

References

1. Szymańska, A., Dutkiewicz, M., Maciejewski, H., Palacz, M.: Simple and effective hydrophobic impregnation of concrete with functionalized polybutadienes. Constr. Build. Mater. **315** (2021). https://doi.org/10.1016/j.conbuildmat.2021.125624
2. Pan, X., Shi, Z., Shi, C., Ling, T.C., Li, N.: A review on concrete surface treatment Part I: types and mechanisms. Constr. Build. Mater. **132**, 578–590 (2017). https://doi.org/10.1016/j.conbuildmat.2016.12.025
3. Wang, F., Lei, S., Ou, J., Li, W.: Effect of PDMS on the waterproofing performance and corrosion resistance of cement mortar. Appl. Surf. Sci. **507** (2019/2020). https://doi.org/10.1016/j.apsusc.2019.145016
4. Liang, C., et al.: Fabrication of bulk hydrophobic cement-based materials with ultra-high impermeability. J. Build. Eng. **63**, 105492 (2023). https://doi.org/10.1016/J.JOBE.2022.105492

5. Zhang, K., Xu, F., Gao, Y.: Superhydrophobic and oleophobic dual-function coating with durablity and self-healing property based on a waterborne solution. Appl. Mater. Today **22**, 100970 (2021). https://doi.org/10.1016/j.apmt.2021.100970

6. Yao, H., Xie, Z., Huang, C., Yuan, Q., Yu, Z.: Recent progress of hydrophobic cement-based materials: preparation, characterization and properties. Constr. Build. Mater. **299**, 124255 (2021). https://doi.org/10.1016/j.conbuildmat.2021.124255

7. Zhang, B., Li, Q., Niu, X., Yang, L., Hu, Y., Zhang, J.: Influence of a novel hydrophobic agent on freeze–thaw resistance and microstructure of concrete. Constr. Build. Mater. **269**, 121294 (2021). https://doi.org/10.1016/J.CONBUILDMAT.2020.121294

8. Carette, J., Delsaute, B., Milenković, N., Lecomte, J. P., Delplancke, M.P., Staquet, S.: Advanced characterisation of the early age behaviour of bulk hydrophobic mortars. Constr. Build. Mater. **267** (2021). https://doi.org/10.1016/j.conbuildmat.2020.120904

9. Chen, J., Zhang, Y., Hou, D., Yu, J., Zhao, T., Yin, B.: Experiment and molecular dynamics study on the mechanism for hydrophobic impregnation in cement-based materials: a case of octadecane carboxylic acid. Constr. Build. Mater. **229**, 116871 (2019). https://doi.org/10.1016/J.CONBUILDMAT.2019.116871

10. Christodoulou, C., Goodier, C.I., Austin, S.A., Webb, J., Glass, G.K.: Long-term performance of surface impregnation of reinforced concrete structures with silane. Constr. Build. Mater. **48**, 708–716 (2013). https://doi.org/10.1016/j.conbuildmat.2013.07.038

11. Wang, F., et al.: Preparation and properties of foundry dust/Portland cement based composites and superhydrophobic coatings. Constr. Build. Mater. **246**, 118466 (2020). https://doi.org/10.1016/j.conbuildmat.2020.118466

12. Xue, X., et al.: A systematic investigation of the waterproofing performance and chloride resistance of a self-developed waterborne silane-based hydrophobic agent for mortar and concrete. Constr. Build. Mater. **155**, 939–946 (2017). https://doi.org/10.1016/j.conbuildmat.2017.08.042

13. Zhao, J., Gao, X., Chen, S., Lin, H., Li, Z., Lin, X.: Hydrophobic or superhydrophobic modification of cement-based materials: a systematic review. Compos. B Eng. **243**, 110104 (2022). https://doi.org/10.1016/J.COMPOSITESB.2022.110104

14. Gholizadeh-Vayghan, A., Rajabipour, F.: Quantifying the swelling properties of alkali-silica reaction (ASR) gels as a function of their composition. J. Am. Ceram. Soc. **100**, 3801–3818 (2017). https://doi.org/10.1111/jace.14893

15. Sant, G., Kumar, A., Patapy, C., Le Saout, G., Scrivener, K.: The influence of sodium and potassium hydroxide on volume changes in cementitious materials. Cem. Concr. Res. **42**(11), 1447–1455 (2012). https://doi.org/10.1016/j.cemconres.2012.08.012

16. Matar, P., Barhoun, J.: Effects of waterproofing admixture on the compressive strength and permeability of recycled aggregate concrete. J. Build. Eng. **32**, 101521 (2020). https://doi.org/10.1016/j.jobe.2020.101521

17. Zhang, H., Mu, S., Cai, J., Chen, R.: The impact of carboxylic acid type hydrophobic agent on compressive strength of cementitious materials. Constr. Build. Mater. **291**, 123315 (2021). https://doi.org/10.1016/J.CONBUILDMAT.2021.123315

Influence of Aging Condition on the Hydration and Setting Performance of Cement Paste in the Presence of Triethanolamine

Zichen Lu[✉], Zhiwei Liu, Liheng Zhang, and Zhenping Sun

Key Laboratory of Advanced Civil Engineering Materials of Ministry of Education, School of Materials Science and Engineering, Tongji University, Shanghai, China
luzc@tongji.edu.cn

Abstract. Effect of aging conditions on the hydration and setting performance of cement paste with the addition of TEA were investigated through the combined techniques of calorimetry, Vicat test, XRD and TGA. It is found that, along with the increased aging time and RH, an obvious formation of AFt, AFm and CH can be found due to the pre-hydration of clinker. Besides, under the RH of 90%, the pre-hydration within the first day was significantly improved, and this process was continuously developed in the next 7 days through the strong carbonation of CH. Further analysis indicates that the pre-hydration can retard the cement hydration and increase the setting time of cement paste. Compared to the reference, a prolonged induction period and an increased setting time were observed for the aged samples with the addition of TEA. Besides, the commonly observed flash setting performance under the TEA dosage of 0.5 wt.-% was eliminated when the samples were pre-hydrated under 90% RH, which indicates the pre-hydration can alleviate the strongly accelerated aluminate phase hydration caused by TEA.

Keywords: Aging · Cement hydration · Setting time · Triethanolamine

1 Introduction

Triethanolamine (TEA) is a grinding agent and setting modifier during the production and application of ordinary Portland cement (OPC) with the advantage of low dosage and high efficiency [1]. However, one problem that restricts its wide application is the high sensitivity of setting and hardening performance of cement paste caused by TEA [2]. Many researchers have reported that, along with the increased dosage of TEA, the setting time of cement paste was first decreased, then increased and then followed by a flash setting [3]. Beside the dosage, many other parameters could also affect the performance of TEA, such as the sulfate type and content in cement [4] and the preparation method of cement paste [5] etc. Even though many parameters affecting the setting performance of cement paste with the addition of TEA have been investigated, a huge variation in the setting time of cement paste can still be observed in our former study. Considering the only difference between these experiments was the aging of cement, it is necessary to uncover the mechanism of how the aging of the cement could affect the interaction of

© The Author(s) 2025
L. Czarnecki et al. (Eds.): ICPIC 2023, 61, pp. 307–314, 2025.
https://doi.org/10.1007/978-3-031-72955-3_31

TEA and cement particles, which in the end substantially changes the hydration process and setting performance of cement paste.

The aging of cement can be regarded as pre-hydration process, which is a continuously physicochemical reaction of mineral phases in cement under a certain relative humidity (RH). Some studies investigate the threshold values of atmospheric humidity at which each cement mineral phase turns to sorb water vapor. Free lime reacts with water vapor at a very low 14% RH, and C_3A starts to take up water vapor at 55% RH. While, the silicate phases, mainly including C_2S and C_3S start to sorb small amounts of water vapor at the RH of 63% and 64% [6–8]. Vektaris [9] found that white cement is more resistant to pre-hydration compared to OPC due to the fewer amounts of C_3A and free lime.

The pre-hydration not only changes the properties of cement particles, but also affects its interaction with the chemical admixtures. M. R. Meier [10] compared the dispersing performance of three different superplasticizers (BNS, PCE and casein) on the fresh and aged cement. All three superplasticizers showed decreased dispersing performance in cement after 1 d aging and BNS exhibited the biggest reduction. Sun [11] found an increased initial fluidity in cement after being exposed at 20 ± 2 °C and 85%-90% RH for 4 d compared to that of fresh cement. Contrary to the many publications on the interaction of superplasticizer with aged cement, limited research focus on the performance of TEA in aged cement.

Hence, the impact of different aging conditions on the hydration process and setting time of cement paste with the addition of TEA was studied. These results, on the one hand, deepen the understanding of the impact of different aging conditions on the properties of cement with the addition of TEA. On the other hand, it can provide practical guidance on the application of TEA in cement stored under different conditions.

2 Materials and Methods

2.1 Materials

Table 1. Chemical and mineralogical compositions of the fresh cement

Oxides	Al_2O_3	CaO	Fe_2O_3	MgO	SiO_2	SO_3	TiO_2	LOI
Value/%	5.6	57.1	4.6	2.1	24.5	3.3	0.4	1.3
Phases	C_3S	C_2S	C_4AF	C_3A	Anhydrite	Hemihydrate	Gypsum	$CaCO_3$
Value/%	59.6	17.7	7.3	9.2	1.6	1.8	2.1	0.2

PI 42.5 type OPC (Fushun Orcel Technology Co., Ltd.) was used in this study. The chemical and mineralogical compositions of cement are shown in Table 1. The aging process of the cement was carried out at 25 °C and RH of 30%, 55% and 90% in a climatic chamber (CH-150R, Dongguan Tude Environment Testing Equipment Co. Ltd.) for 1, 3, and 7 d. During the aging process, the cement particles were spread over an acrylic

plate with a thickness of ~ 1mm. The cement particles were turned over and remixed every 6 h. After the aging process, the aged cement was milled into its original fitness in agate mortar by hand for further experiments. Double-deionized (DI) water was utilized to prepare cement paste, TEA was provided by Sinopharm Chemical Reagent Co., Ltd with an analytical purity exceeding 99%.

2.2 Methods

Cement paste was prepared by mixing cement, water and TEA for 1 min with a hand mixer (MFQ4080, Bosch). The TEA dosage was 0.1% and 0.5% by weight of cement. The water-to-cement ratio was 0.41. A constant ambient temperature of 25 °C was kept.

The setting times of pastes were determined according to Chinese National Standard GB/T 1346–2011. The hydration process of pastes with and without TEA was monitored by an isothermal calorimeter (TAM air, TA instrument). The external mixing method was used. The heat flow was normalized based on the amount of cement. XRD and TGA were used to characterize the consumption of mineralogical phases and the formation of hydration products after aging. The radiation (Cu Kα) was generated with 40 kV and 40 mA, and the samples were scanned in a range from 5 to 40°. Phase quantification was determined by using α-Al_2O_3 as the internal standard. Rietveld analysis was performed with the software HighScore Plus 4.8 (PANalytical). For the TGA measurement, a heating rate of 10 °C/min from room temperature to 1000 °C was applied under a N_2 atmosphere with a flow rate of 40 mL/min (STA 449C, NETZSCH).

3 Results and Discussion

3.1 Characterization of the Aged Cement

In order to verify the effect of aging on the interaction of TEA with cement, a thorough characterization of the cement after aging under different conditions was conducted. QXRD was applied for the different samples, as shown in Table 2. Under the 30% RH, a slight decrease in the mineralogical phases and an increase in hydration products were observed with the increased curing age. Along with the increased RH to 55% and 90%, more clinker phases were consumed and a higher amount of hydration products were detected. Moreover, carbonation was found for samples aging in 55% RH for 7 d and 90% RH for 1 d and 7 d.

The pre-hydration and carbonation during the aging process were also quantified by TGA (Fig. 1). The fresh cement exhibits only a minor mass loss of 1.6%, which is mainly related to the decomposition of hydration products portlandite and ettringite in the temperature range from 105 to 640 °C, while the decomposition of $CaCO_3$ in the temperature range from 650 to 1000 °C is negligible, which matches well with the QXRD data. Similar to the fresh cement, samples aged at 30% RH for 1 d and 7 d exhibit nearly no changes in mass loss, which proves again 30% RH may not induce a strong pre-hydration of cement. However, when the RH increased to 55%, the mass loss slightly increased along with the aging time. Notably, an obvious decomposition of $CaCO_3$ was found after aging in 55% RH for 7 d, which proves the considerable

Table 2. Composition of fresh and aged cement under different aging conditions (wt.%)

Phase	Ref	30RH		55RH		90RH	
		1 d	7 d	1 d	7 d	1 d	7 d
C$_3$S	59.6	59.5	59.1	59.1	58.8	56.6	47.1
C$_2$S	17.7	17.5	17.3	17.2	17.0	13.8	9.8
C$_4$AF	7.3	7.3	7.2	7.0	6.6	5.7	4.2
C$_3$A	9.2	9.1	8.9	8.5	6.2	5.1	2.9
Anhydrite	1.6	1.5	1.3	1.3	0.9	-	-
Hemihydrate	1.8	1.6	1.2	1.3	0.6	-	-
Gypsum	2.1	2.3	2.9	2.2	1.4	0.9	-
CaCO$_3$	0.2	0.3	0.4	0.5	1.4	1.6	4.6
CH	0.1	0.4	0.6	0.5	0.9	0.6	0.7
AFt	0.1	0.1	0.1	0.2	0.7	1.1	1.4
AFm	0.2	0.2	0.3	0.4	0.8	0.8	1.0

carbonation as indicated by the XRD results in Table 2. Along with the increased RH to 90%, an obvious weight loss was found at around 100 °C after aging 1 d, which is caused by the large formation of C-S-H and AFt (Table 1). Besides, an obvious weight loss was found at around 450 °C for samples aged under 90% RH for 1 d, which indicates the fast pre-hydration. However, after aging 7 d, the decomposition of CH cannot be observed, but a sharp weight loss was found for the decomposition of CaCO$_3$.

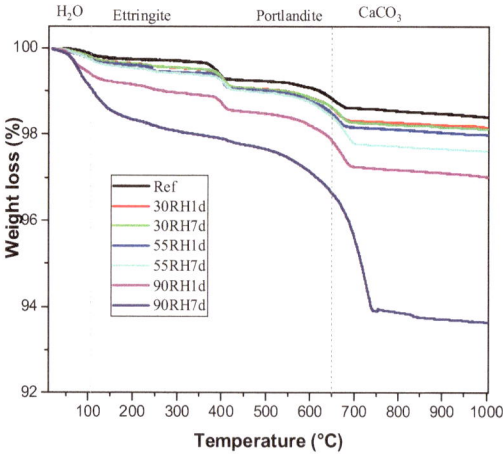

Fig. 1. TGA of the fresh and aged cement under different aging conditions

3.2 Hydration of the Aged Cement with the Addition of TEA

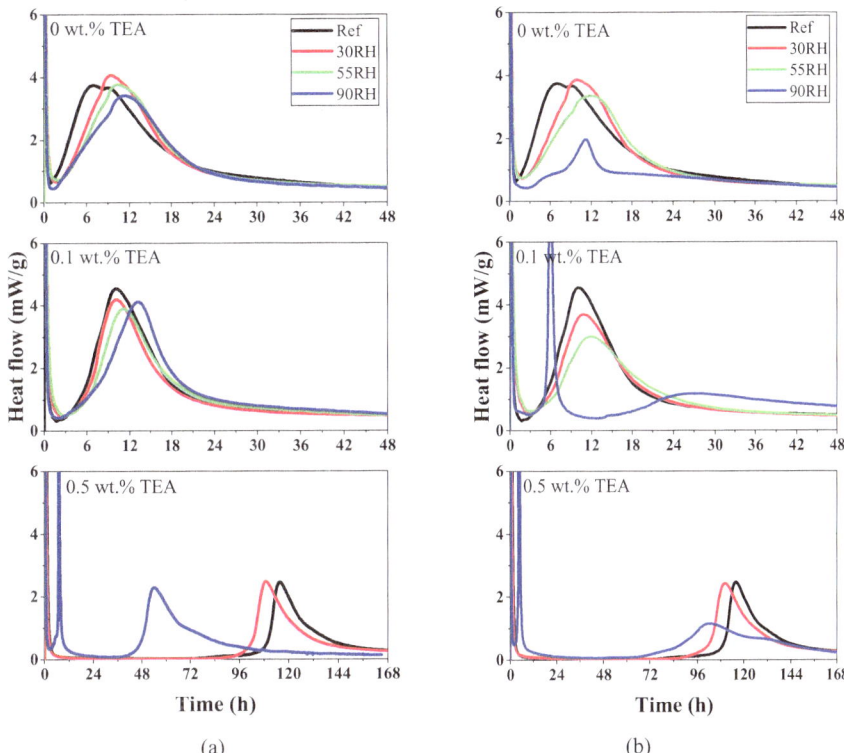

(a)   (b)

Fig. 2. Heat flow of the fresh and aged cement with and without the addition of TEA under different aging conditions; (a) Aged for 1 d; (b) Aged for 7 d.

The hydration process of the fresh and aged cement with TEA was measured and the results are shown in Fig. 2. Without the addition of TEA, a prolonged induction period was observed in all aged samples, which is proportional to the increased aging time and RH. Moreover, except for samples aged at 30% RH, a decreased main hydration peak was found, especially for samples aged at 90% RH for 7 d.

For the samples with 0.1 wt.% TEA but aged under different conditions, a prolonged induction period to different extents was observed. Compared to the samples without TEA, the presence of TEA significantly delayed the occurrence of the main hydration peak of the fresh cement. However, nearly no difference can be found for the aged samples, regardless of the aging time of 1 d or 7 d. For the samples aged 7 d under 30% and 55% RH, the presence of TEA not only prolongs the induction period but also significantly reduces the main hydration peak. However, when the RH increased to 90%, a fast aluminate phase reaction can be observed before the hydration of silicate phase. It indicates that the sulfate balance was broken through the addition of TEA and pre-hydration can increase the sensitivity of the cement hydration process to the

presence of TEA. Along with the further increased TEA dosage to 0.5 wt.%, compared
to the reference without aging, the aluminate phase reaction was retarded, which in turn
resulted in a faster silicate phase reaction.

3.3   Setting Time of the Aged Cement with the Addition of TEA

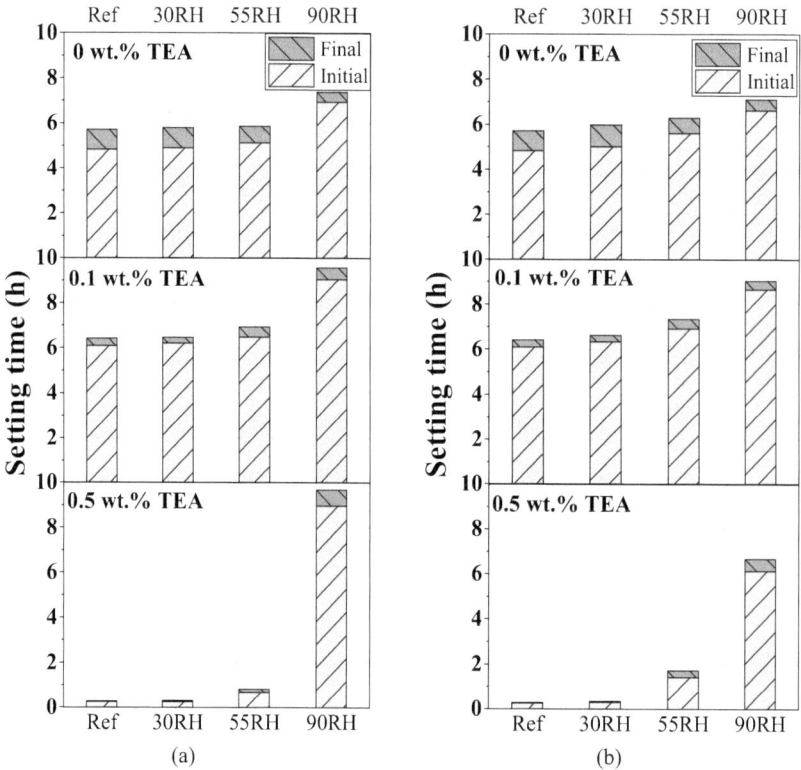

Fig. 3. Setting times of the fresh and aged cement with and without the addition of TEA under
different aging conditions; (a) Aged for 1 d; (b) Aged for 7 d.

The setting times of the fresh and aged cement with and without the addition of TEA
under different aging conditions were measured and the results are shown in Fig. 3.
For the condition without the addition of TEA, cement aged under 30% RH for 1 d
exhibits a similar setting time as fresh cement. Besides, the setting time of cement paste
was prolonged along with the further increased RH during aging. Samples exhibited
a strong delay in setting time for about 1.5 h after only 1 day of storage in 90% RH
due to the pre-hydration of cement grains. Furthermore, the effect became pronounced
with the increased aging times. Compared to the samples without the addition of TEA,
the samples in the presence of TEA show a similar trend, namely the setting time was
prolonged along with the increased RH and the aging time. However, two phenomena

need to be noted. Firstly, the commonly observed flash setting performance under the TEA dosage of 0.5 wt.% was eliminated when the samples were pre-hydrated under 90% RH, which indicates the pre-hydration can alleviate the strongly accelerated aluminate phase hydration caused by TEA. Secondly, for the samples aged for 7 d under the RH of 90%, a reduced setting time was observed when the TEA dosage was increased from 0.1 wt.% to 0.5 wt.%. Considering the calorimetric curves shown in Fig. 2, it can be inferred that both the setting of cement paste with TEA dosage of 0.1 wt.% and 0.5 wt.% was controlled by the accelerated aluminate phase reaction. Because of the strong acceleration on the aluminate phase along with the increased TEA dosage, the sample aged for 7 d under 90 RH shows a shorter setting time.

4 Conclusion

The combined effect of RH and aging time on the hydration process and setting time of aged cement with and without the addition of TEA was investigated. Based on the results above, the following conclusions can be obtained.

- Aging of cement can lead to the pre-hydration of cement, which results in the formation of a certain amount of hydration products. Besides, with the increased RH and aging time, strong carbonation can be observed.
- The pre-hydration can increase the sensitivity of the cement hydration process to the presence of TEA, which can result in a sulfate imbalance.
- Both the aluminate phase and the silicate phase reactions can be suppressed through the process of pre-hydration, which results in a prolonged setting time with increased RH and aging time.
- Due to the retarded aluminate phase reaction through the pre-hydration, the commonly observed flash setting performance was eliminated under the TEA dosage of 0.5 wt.%.

Acknowledgements. We sincerely thank the support from Shanxi Yuncheng Science and Technology Bureau (project "Development of specialized admixtures and auxiliary techniques for efficient shotcreting under complex and severe tunnel construction environment") and the National Natural Science Foundation of China (Project No. 52208282). In addition, the authors would like to deeply thank the Key Laboratory of Concrete Functional Materials of Yuncheng City and Shanxi Jiawei New Material Co., Ltd for their contribution to this research.

References

1. Lu, Z., Peng, X., Dorn, T., Hirsch, T., Stephan, D.: Early performances of cement paste in the presence of triethanolamine: Rheology, setting and microstructural development. J. Appl. Polym. Sci. **138**(31), 50753 (2021)
2. Lu, Z., Lu, J., Liu, Z., Sun, Z., Stephan, D.: Influence of water to cement ratio on the compatibility of polycarboxylate superplasticizer with Portland cement. Constr. Build. Mater. **341**, 127846 (2022)
3. Lu, Z., et al.: Towards a further understanding of cement hydration in the presence of triethanolamine. Cem. Concr. Res. **132**, 106041 (2020)

4. Hirsch, T., Lu, Z., Stephan, D.: Effect of different sulphate carriers on Portland cement hydration in the presence of triethanolamine. Constr. Build. Mater. **294**, 123528 (2021)
5. Lu, Z., Peng, X., Liu, Z., Sun, Z., Stephan, D.: Influence of mixing speed on the hydration and setting performance of cement paste in the presence of triethanolamine. Constr. Build. Mater. **385**, 131490 (2023)
6. Dubina, E., Plank, J., Black, L.: Impact of water vapour and carbon dioxide on surface composition of C3A polymorphs studied by X-ray photoelectron spectroscopy. Cem. Concr. Res. **73**, 36–41 (2015)
7. Dubina, E., Wadsö, L., Plank, J.: A sorption balance study of water vapour sorption on anhydrous cement minerals and cement constituents. Cem. Concr. Res. **41**(11), 1196–1204 (2011)
8. Dubina, E., Black, L., Sieber, R., Plank, J.: Interaction of water vapour with anhydrous cement minerals. Adv. Appl. Ceram. **109**(5), 260–268 (2010)
9. Vektaris, B., Kaziliūnas, A., Striūgienė, I.: Ageing of dry cement mixes for finishing purposes. Mater. Sci. **19**(3), 326–330 (2013)
10. Meier, M.R., Napharatsamee, T., Plank, J.: Dispersing performance of superplasticizers admixed to aged cement. Constr. Build. Mater. **139**, 232–240 (2017)
11. Sun, Z., Shui, L., Yang, H., Liu, Y., Ji, Y.:. Effect of prehydration on cement performance and cement-polycarboxylate superplasticizer interaction. J. China Ceram. **44**(5) (2016)

Repair and Protection of Concrete Structures

Applications of Concrete-Polymer Composites: Where Are We Now and Where We Are Going?

Jung Heum Yeon[1](✉), Yeoung-Geun Choi[2], and Kyu-Seok Yeon[2]

[1] Ingram School of Engineering, Texas State University, San Marcos, TX, USA
jung.yeon@txstate.edu
[2] iCONTEC ENC Co., Ltd, Anyang, South Korea

Abstract. This study aims to overview the current state of concrete-polymer composite (CPC) applications and proposes ways to develop and generalize the use of CPC in practical applications. According to the literature, polymer-modified concrete (PMC) is mainly formulated using SBR or epoxy-based latex and used for repair and overlays. Polymer concrete (PC) uses unsaturated polyester resin as a polymer binder and is mainly used for precast products. Polymer-impregnated concrete (PIC) research was actively conducted in the 1970s and 1980s, but currently, it is not easy to find except for a few applied studies. This study also presents the challenges and suggestions for developing and generaling CPC use in real-world practice.

Keywords: Concrete-Polymer Composites (CPC) · Polymer-Modified Concrete (PMC) · Polymer Concrete (PC) · Polymer-Impregnated Concrete (PIC) · Applications · Repair and Overlay · Precast Products

1 Introduction

Concrete-polymer composites (CPC) are generally classified into the following three types by the principal of their process technology [1, 2]: (1) polymer-modified concrete (PMC), (2) polymer concrete (PC), and (3) polymer-impregnated concrete (PIC). PMC is a composite material made by partially replacing and strengthening the cement hydrate binders of conventional concrete with polymeric modifiers or admixtures. PC is a composite material made by entirely replacing the cement hydrate binders of conventional concrete with polymeric binders or liquid resins. PIC is a composite material that impregnates hardened cement concrete with monomeric impregnants. Herein, the definition of "concrete" encompasses "mortar" that does not incorporate coarse aggregates.

Historically [1–6], PMC was registered as a British patent by Cresson in 1923, and a fundamental study was conducted by Geist et al. in 1953. Later, practical research and development on polymer-modified concrete began in the 1960s worldwide, including in the US, Russia, Germany, Japan, and the UK. PC has a shorter history than PMC, but research and development were conducted in Russia, the United States, Germany, and Japan in the late 1950s to the early 1960s. PIC has the shortest history among CPCs. The processing technology for PIC was developed based on the concept of wood-polymer

© The Author(s) 2025
L. Czarnecki et al. (Eds.): ICPIC 2023, 61, pp. 317–330, 2025.
https://doi.org/10.1007/978-3-031-72955-3_32

in the late 1960s. Therefore, CPC has a history of 60–70 years; however, if based on Cresson's patent registration, it could also be considered 100 years. This study aims to overview the current state of CPC applications and suggest future directions to move.

2   Polymer Materials for CPC and Their Benefits

2.1   Polymer-Modified Concrete

Polymer materials [1, 7, 8]**.**

- Latex polymers

  - Synthetic rubber latex: SBR, etc.
  - Resin latex: EVA, epoxy, etc.

- Emulsified polymers
- Water-soluble polymers: Polyvinyl alcohol
- Re-dispersible polymers: Ethylene vinyl acetate

Benefits [6]**.**

- Rapid curing
- High strength (compressive, tensile, and flexural strength)
- Good abrasion resistance
- Good adhesion
- Long-term durability
- Low permeability and water tightness

2.2   Polymer Concrete

Polymer materials [6, 9]**.**

- Unsaturated Polyester resin
- Epoxy resins
- Methyl methacrylate
- Urea formaldehyde
- Poly Urethane Resins
- Furan Resins
- Phenol formaldehyde, etc.

Benefits [6]**.**

- Rapid curing
- High strength (compressive, tensile, and flexural strength)
- Good abrasion resistance
- Good adhesion
- Long-term durability
- Low permeability and water tightness

2.3 Polymer Impregnated Concrete

Polymer materials [10].

- Styrene
- Methyl methacrylate (MMA)
- Butyl acrylate
- Acriloniyrite
- Epoxies and their copolymer combinations
- Polyester

Benefits [9].

- Partial impregnation: durability and chemical resistance
- Full impregnation: structural properties (strength, elastic module, etc.)

3 Cases Studies

3.1 Polymer-Modified Concrete

PMC applications in previously published literature and technical data include industrial floorings, bridge and pavement overlays, integral waterproofings, decorative coatings, repair materials, anticorrosive linings, deck coverings, and grouting works [1, 3, 9]. Currently, PMC is mainly used for patching and repairing, grouting and installing tile, waterproofing and flooring, road and bridge overlay, etc. PMC was used in the form of concrete only for pavement and bridge overlay applications, while the rest were used in the form of mortar. As a polymer material, SBR or epoxy-based latex is often used. Notably, PMC is used to restore existing structures rather than for new construction. Also, among the three CPCs, it is the easiest to use. However, PCM differs depending on the type of polymer material used for modification. It is generally advantageous to improve physical properties such as adhesion or waterproofness, but it makes hard to improve the mechanical properties such as compressive strength and modulus of elasticity.

Patch and repair.
See Figs. 1 and 2.

Fig. 1. Floor patch [11]

Fig. 2. Concrete repair [7]

Fig. 3. Grouting for base plate [12]

Fig. 4. Installing tiles [7]

Grouting and installing tile.
See Figs. 3 and 4.

Waterproofing and flooring.
See Figs. 5 and 6.

Fig. 5. Waterproofing [13]

Road and bridge overlay.
See Figs. 7 and 8.

Fig. 6. Floor covering [14]

Fig. 7. Pavement overlay [15]

Fig. 8. Bridge deck overlay [15]

3.2 Polymer Concrete

Applications of PC presented in previously published literature and technical data include floorings (including decorative finishings), pavements, anticorrosive linings, pipes for sewage and irrigation systems, artificial marbles, repair to corrosion-damaged concrete, prestressed concrete, nuclear power plants, electrical or industrial construction, marine works, prefabricated structural components (acid tanks, manholes, drains, etc.), highway barriers, waterproofing of structures, sewage works and desalination plants, kerbstones, precast slabs for bridge decks, marine works, and high voltage insulator [1, 6, 9]. PC is currently mainly used for pipes, manholes, trenches and channels, curb drains and rainwater sumps, artificial marble claddings and bathroom sinks, troughs and heating plates, trash and waste containers, overlay and repairing, etc. As a polymer material, unsaturated polyester resin is widely used. Polymer concrete is mainly used to manufacture precast products rather than cast-in-place applications. Although some use PC for overlaying or repairing existing concrete pavement, PC has a significant difference in material properties, such as modulus of elasticity and coefficient of thermal expansion from conventional cement concrete, which must be considered.

Pipes.
See Figs. 9 and 10.

Fig. 9. Polymer concrete jacking pipes [16]

Fig. 10. FRP-reinforced polymer composite pipes [17]

Manholes.
See Figs. 11 and 12.

Fig. 11. Polymer concrete manhole for sewage pipeline [16]

Trenchs and channels.
See Figs. 13 and 14.

Fig. 12. Polymer concrete manhole for communication line [18]

Fig. 13. Polymer concrete trench [19]

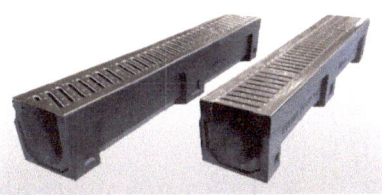

Fig. 14. Road rainwater polymer concrete drainage channel [18]

Curb drains and rainwater sumps.
See Figs. 15 and 16.

Fig. 15. Polymer concrete curb drain [20]

Artificial marble claddings and bathroom sinks.
See Figs. 17 and 18.

Fig. 16. Polymer concrete rainwater sumps [21]

Fig. 17. Polymer concrete artificial marble claddings [22]

Fig. 18. Polymer concrete bathroom sinks [23]

Troughs and heating plate.
See Figs. 19 and 20.

Fig. 19. Polymer concrete trough for pig house [24]

Trash and waste containers.
See Figs. 21 and 22.

Fig. 20. Polymer concrete heating plate for pig house [25]

Fig. 21. Polymer concrete trash can [26]

Fig. 22. Decorative waste containers made with polymer concrete [27]

Overlay and repair.
See Figs. 23 and 24.

Fig. 23. Polyester polymer concrete overlay [28]

3.3 Polymer-Impregnated Concrete

PIC applications presented in previously published literature and technical data include bridge decks, irrigation structures, structural members, marine functions, nuclear power

Fig. 24. Repairing work with polymer concrete [6]

plants, sewage disposal works, Ferro cement products, waterproofing, flooring (dairy farm product buildings), tanneries and chemical factories, etc. [10, 29]. However, follow-up studies were hardly conducted since some used PIC in highways and dams in the 1970s and 1980s [30–33]. Only a few case studies in permanent forms [34] and concrete spacers [35] were reported recently. The reason for the less popularity of PIC is that the processing technology of PIC is complicated, and high thermal energy is required for its applications, which is costly. Also, it is hard to measure the polymer impregnation depth (especially in the field), which makes contractors hard to control the quality.

Permanent form.
See Figs. 25 and 26.

Fig. 25. Installment of polymer-impregnated concrete permanent form (PICPF) [34]

Fig. 26. PICPF after casting concrete [34]

Spacers.
See Fig. 27.

Fig. 27. Polymer-impregnated concrete spacers [35]

4 Proposal for Development and Generalization of CPC

4.1 Challenges

Slow propagation compared to its long history.
As mentioned earlier, CPC was researched and developed in the 1950s and 1960s and has a history of 60–70 years. Despite such a long history, the propagation of these technologies has been very slow. The possible reason stems from the conservative nature of the construction industry, but it might also be true that CPC technology has not been appealing to the relevant industries. No significant progress has been made in terms of the quantity or quality of publications on the R&D of CPC. These are undoubtedly important challenges that we CPC researchers should face.

Cost ineffectiveness.
It should be understood that one of the primary limitations CPC is cost. PC is still much more expensive than conventional cement concrete. The cost of polymers can range from 10 to 100 times that of portland cement, even considering the specific gravity unit volume cost of the polymer [5]. As such, CPC is still disadvantageous compared to other construction materials in terms of price [4]. However, since the comparison between the cost of polymers and portland cement by unit is unjust, cost-effectiveness should be compared, and research and development should be conducted in a direction that can improve it.

A bias against the merits of CPC.
CPC is different from conventional concrete in manufacturing methods and performance. Compared to conventional concrete, CPC definitely has excellent performance, such as high strength and good durability. Thus, CPC can be effective for repairs, structural applications, and architectural components. However, it should always be considered that polymer concrete also has various disadvantages.

Understanding polymer properties and manufacturing challenges.
The three types of CPC significantly differ in properties because they use different processing technologies. In addition, because the type, addition amount (injection amount), mixing ratio (mixing ratio of the main material, hardener, etc.), and curing method (curing temperature) lead to significantly different material properties, it is important to choose a suitable type, addition amount, mixture proportions, and curing regime. This complexity makes the field applications of CPC difficult. Eventually, CPC applications required high skills and precise work.

4.2 Suggestions

Advancing practical value of research outcomes.

Most studies on CPC are premised on field applications. Applied research is essential for the rapid dissemination and generalization of research results. Impractical research results sometimes confuse users and end up not getting attention. Even fundamental research should focus on advancing practical contributions and values to avoid these undesirable outcomes. In other words, practical research should be conducted while clarifying the purpose of the research. In particular, field application research should be able to provide practical information to subsequent researchers or users through follow-up research.

Advancing cost-effectiveness.

CPC has not been widely adopted mainly due to its high cost. However, recent progress has greatly reduced the cost, and its use has gradually spread. However, CPC cannot be cheaper than conventional concrete, no matter how much the cost is reduced. Therefore, comparing the unit price is contradictory. However, it is necessary to seek to reduce the cost of polymer materials through the development (modification) of low-cost polymers and the optimization (minimization) of polymer usage. Also, it is necessary to find a way to increase cost-effectiveness by minimizing production cost and maximizing performance. A study on the cost-effectiveness analysis method of CPC is also one of the necessary tasks.

Raising awareness of CPC.

Although CPC is expensive and difficult to manufacture, limiting its widespread application, it has advantages over conventional concrete. Despite these advantages and a long history of development, it is still not recognized by field engineers. Therefore, it is very important to disseminate technologies that raise awareness and appeal to users through conferences, workshops, fairs, etc. In addition, to advance the technology of CPC, it is necessary to expand the research base by increasing the pool of researchers. In addition, it is necessary to create an environment where researchers in different fields can actively collaborate.

5 Conclusions

This study reviewed the current state of CPC applications, which has a 60–70 years of research and development history, and sought the development direction to move forward. The results are summarized as follows:

1) PMC is used for patching and repairing, grouting and installing tiles, waterproof flooring, and road and bridge overlays. As a polymer material, SBR or epoxy-based latex is widely used. PMC is commonly adopted for reconstruction rather than new construction.
2) PC is currently adopted for pipes, manholes, gutters, flumes, curb drains, rain gutters, artificial marble cladding and bathroom sinks, troughs and soleplates, trash and waste containers, overlays and repairs. As a polymer material, an unsaturated polyester resin is widely used. PC is primarily used to manufacture precast products rather than cast-in-place.

3) Only a few PIC case studies have been reported since the 1970s and 1980s. The reason for the less popularity of PIC appears to stem from the complicated processing technology and high thermal energy requirement.

4) However, CPC is taking its place as an essential material in the construction industry. The current challenges of CPC applications include slow propagation compared to a long history, expensive materials, bias towards the advantages of CPC, and difficulties in manufacturing and quality control. More applied, practical, cost-effective research, awareness raising, and research base expansion were proposed as countermeasures.

References

1. Chandra, S., Ohama, Y.: Polymers in Concrete. CRC Press (1994)
2. Czarnecki, L.: Polymer concretes. Cement Wapno Beton **15**(2), 63–85 (2010)
3. Czarnecki, L., Łukowski, P.: Polymer-cement concretes. Cement Wapno Beton **15**(5), 243–258 (2010)
4. Fowler, D.W.: State of the art in concrete polymer materials in the U.S., Proceedings of the 12th International Congress on Polymers in Concrete, Kangwon National. Univ., Chuncheon, Republic of Korea, 29–36 (2007)
5. Fowler, D.W.: Polymers in concrete: a vision for the 21st century. Cement Concr. Compos. **21**(5–6), 449–452 (1999)
6. https://gharpedia.com/blog/polymer-concrete-pros-cons-uses-and-properties/
7. https://theconstructor.org/concrete/polymer-modified-mortar/28917/
8. https://www.paramvisions.com/2023/03/what-is-polymer-modified-mortar-their.html
9. https://www.civilengineeringweb.com/2020/05/what-is-polymer-sulphur-infiltrated-concrete.html
10. https://www.engineeringenotes.com/concrete-technology/polymer-impregnated-concrete/polymer-impregnated-concrete-uses-and-properties-concrete-technology/31908
11. https://www.ctscement.com/datasheet/FLOOR_PATCH_Datasheet_DS_222_EN?c=FLOORING&t=Homeowners
12. https://www.civillead.com/what-is-grouting/
13. https://www.de.weber/en/waterproofing-mortars
14. https://usa.sika.com/en/construction/floor-covering/flooring-levelers-patches.htm
15. https://www.wagman.com/specialized-services/rapid-set-latex-modified-concrete/
16. http://tokai.e-const.jp/tansangas.html
17. https://www.hfiber.com/home/biz/pipe_item.asp
18. https://polycon.co.kr
19. https://www.alibaba.com/product-detail/Manufacturer-250-320-mm-CO-Polymer_62544473167.html?spm=a27aq.27059075.6360844600.157.66651d6b3SKX1o
20. https://www.environmental-expert.com/products/yete-model-yt150-500kb-polymer-curb-drains-726792
21. https://www.environmental-expert.com/products/polycon-polymer-concrete-rainwater-sump-647455
22. https://www.stonecontact.com/products-a667976/pure-white-nano-white-artificial-marble-slabs-nano-crystallized-stone
23. https://www.rdmarble.com/cast-polymer-products/
24. https://www.acofunki.com/products/troughs/polymer-concrete-troughs
25. https://www.acofunki.com/fileadmin/standard/aco-acofunki-com/images/download/brochures/components/B-8021_GB_Heating_plate.pdf

26. https://www.parktables.com/50-gal-fiberglass-trash-can-reinforced-polymer-concrete
27. https://fortecomposites.com/applications.html
28. https://www.kwikbondpolymers.com/products/PPC-1121/
29. https://theconstructor.org/concrete/polymer-impregnated-concrete-properties-applications/17114/
30. Fowler, D.W., Houston, J.T., Paul, D.R.: Polymer-impregnated concrete for highway applications, pp. 114–121. Center for highway research, The University of Texas at Austin, Research Report (1973)
31. Meyer, S.: Concrete-polymer composite materials and its potential for construction, urban, waste utilization and nuclear waste storage. Copolymers, Polyblends, and Composites Chapter **37**, 431–441 (1975)
32. Smoak, G.W.: Polymer impregnation and polymer concrete repairs at Grand Coulee Dam. Polym. Concr. Uses Mater. Prop. ACI Spec. Publ., SP-89 (1985)
33. Cady, P.D., Weyers, R.E., Manson J.M.: Field performance of deep polymer impregnation. J. Transport. Eng. **113**(1) (1987)
34. Bhutta, M.A.R., Maruya, T., Tsuruta, K.: Use of polymer-impregnated concrete permanent form in marine environment: 10-year outdoor exposure in Saudi Arabia. Constr. Build. Mater. **43**, 50–57 (2013)
35. Saeed, H.H.: Properties of polymer impregnated concrete spacers. Case Stud. Construct. Mater. **15** (2021)

Recent Application of Concrete Polymer Materials for Highway Bridges in Japan

Makoto Kawakami[1(✉)], Katsunori Demura[2], Mikio Wakasugi[3], Fujio Omata[4], and Shinya Satoh[5]

[1] Akita University, 4-7-14, Yabasehoncho, Japan
kawakami@gipc.akita-u.ac.jp
[2] Nihon University, 1 Nakagawara, Tokusada, Tamura-Machi, Koriyama, Japan
[3] Chemical Construction Co., Ltd., 5-5, Uozakihamamachi, Higashinada-Ku, Japan
[4] Kensetsu Toso Co., Ltd., 2-6-1, Horiuchi Bld.7F, Kaji-Cho, Chiyoda-Ku, Japan
[5] Suncoh Consultans. Co., Ltd., 1-8-9, Kameido, Koto-Ku, Japan

Abstract. Recently, the deterioration and renovation of the aged bridges are urgent issue. Approximately 30% of more than 700,000 highway and road bridges in Japan have been in service for more than 50 years. The performance and function of them are required equal to be better than those at the beginning of construction and extend the service life. In addition to above, Japan is also a highly seismic country and there have been frequent damages of the bridges due to earthquakes. Therefore, securing the resilience of the bridges corresponding to robustness, redundancy, resourcefulness and rapidity is strongly demanded. In this study, the current status and practical countermeasures for the bridges to ensure the required performance and function, and to enhance the resilience using concrete polymer materials were investigated and discussed. Concrete polymer materials, which have high strength, high early strength development and high durability, are effective for repair and strengthening to sustain the current performance of structures. The main construction contents are as follows;

- It is a construction to replace the damaged reinforced concrete floor slab with a more durable floor slab. Treatment of joints of precast prestressed concrete slab is included.
- It is a construction to install high-performance floor slab waterproofing on the waterproof layer.
- In order to improve the durability of the bridge, it is a construction to attach reinforcing members to the girder.
- Seismic retrofit of concrete piers such as steel jacketing is one of main reinforcement technologies.

Keywords: resin concrete · salt damage · concrete slab · repair · seismic reinforcement

© The Author(s) 2025
L. Czarnecki et al. (Eds.): ICPIC 2023, 61, pp. 331–339, 2025.
https://doi.org/10.1007/978-3-031-72955-3_33

1 Introduction

More than 700,000 highway and road bridges have been constructed and serviced in Japan. About 9000 km of expressways has been built since 1962. The most advanced technologies of the time for the highway infrastructures were applied. However, deterioration of the structures over time is unavoidable and materials degradation such as neutralization, salt damage and alkali-silica reaction has been exposed during service. Furthermore, those structures were severely damaged by the unexpected external forces such as earthquake and tsunami. Maintenance, which is essential for appropriate service of structures, is changing from corrective maintenance initially to preventive maintenance recently.

In this paper, causes and characteristics of degradative damage of bridges were firstly investigated. Then, practical application of polymer materials, which are indispensable for maintenance of structures, was studied. Those were a repair of deterioration of concrete slab, measures against degradation of water exposure and seismic reinforcement. Finally, efforts to extend service life of bridges were discussed.

2 Salt Damage and Countermeasures

The main factors of deterioration damage are shown in Table 1 [1]. Salt damage due to de-icing agent, use of sea sand and airborne salt and transient loads are significant. There are many combined deterioration such as damages due to deicing agent and heavy loads, and deicing agent and ASR. Studded tires posed health hazards and its use was prohibited in 1990. After that, the amount of de-icing agent used, increased rapidly and spurred the salt damage. Measures against salt damage are epoxy coating on the concrete structures to suppress salt penetration, use of epoxy-coated rebar and polyethylene-coated strand. Recently the use of stainless steel has increased by assuming long-term service over 125 years. Furthermore, polymer cement mortar mixed with acrylic resin mortar is also used as cross section repair material to improve the adhesion to the concrete members and to control crack occurrence due to shrinkage.

Table 1. The factors of deterioration damage of highway bridge

Degradation damage factor	Percentage of deteriorated bridges	Note
De-icing agent	51.6%	In some cases, there are overlaps, such as 6.6% due to de-icing agent and vehicle traffic volume, and 3.4% due to de-icing agent and alkali silica reaction
Vehicle traffic volume	31.6%	
Sea sand	8.8%	
Alkali silica reaction	4.5%	
Airborne salt	3.5%	

3 Structural Corrosive Environment Subjected to Water

There are unavoidable structural gaps such as both ends of girder and arch bridges, arch bridge from the view point of thermal movement, and the limitations of structural calculation ability as shown in Fig. 1.

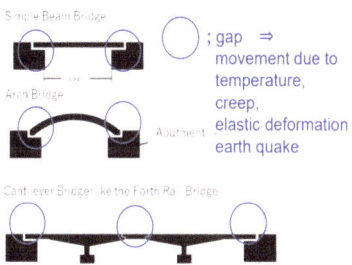

Fig. 1. Unavoidable structural gaps

Steel and rubber expansion joints which are normally water proofing are installed at those gaps. The damage of the end of girder, bearing and the top of abutment due to leakage of water from expansion joints [2] are often observed as shown in Fig. 2. Especially in snow and cold regions, leaking water contains chloride ion from pavement, which causes severe corrosion. The degradation due to water of exposure around drainage basin is also severe. Measures against degradation of water exposure, are installation of rubber water sealing at the opening of expansion joint, and bearing sealing used by transparent silicone resin of high flexibility and deformability as shown in Fig. 3.

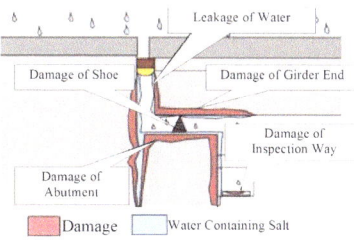

Fig. 2. Leakage from expansion joint

Fig. 3. Bearing sealing used by transparent silicone resin

4 Deterioration of Concrete Slab and Its Countermeasures

Deterioration of concrete slab is diverse. Fatigue damages due to vehicle load and excessive loading are remarkable in addition to salt damage. Those typical pattern are axial and longitudinal cracks and sinking of concrete floor slab. Furthermore, waterproofing of concrete slab was normally not installed before 1987 and most old slabs up to that time caused the segmentation of slabs as shown in Fig. 4. As slab fatigue was accelerated by rain water, blocks of coarse aggregates due to outflowing of cement from the upper surface of slab by polishing action between coarse aggregates and cement mortar are observed.

As some repair methods for slabs, epoxy resin crack injection, three axial vinylon fiber mesh lining with epoxy resin for prevention of concrete pieces falling [3], and sectional restoration with polymer cement mortar have been applied over 40 years [4].

Typical reinforcement measures of slabs are thickening upper or lower surface of slab by polymer cement mortar, and steel plate and carbon fiber sheet are bonded by epoxy resin as shown in Fig. 5.

Fig. 4. Segmentation of slabs due to fatigue damage

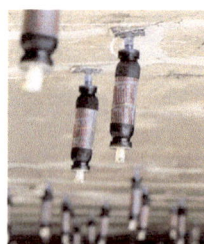 

Fig. 5. Crack injection, and fiber mesh lining with epoxy resin for prevention of concrete pieces falling

Recently reinforced concrete slabs which are severely deteriorated or in service over 30 years, are sequentially updated. Prestressed concrete slab units are prefabricated and jointed. Many joint methods for precast PC slabs have been proposed. One of the excellent construction methods is solved by use of polymer cement mortar or epoxy resin mortar having strong adhesion. Injection of these polymer materials into narrow width of 25 mm to achieve continuity is enable to eliminate the weakness of joints of units as shown in Fig. 6 [5].

In addition to conventional rubber and asphalt-based waterproofing materials, the fast curing and sprayable urethane resin [6] is proposed and achieved as shown in Fig. 7.

Fig. 6. Injection of polymer materials into joint parts between PC slab units

Fig. 7. The fast curing and sprayable urethane resin

5   Earthquake Damage and Seismic Reinforcement

Japan is an earthquake-prone country and damages caused by repeated large earthquakes such as Kobe in 1995, Great East Japan Earthquake in 2011 and Kumamoto in 2016. New modes of bridge damages one after another have been observed; those damages are shearing failure of piers, unseating of bridges, and collapse of the piers with multiple rocking columns as shown in Fig. 8.

In order to prevent these bridge damage, seismic retrofit of existing concrete piers has been executed.

The steel jacketing method used by steel plate and epoxy resin is adopted to increase the strength and rigidity as shown in Fig. 9. The application of carbon/aramid fiber sheet is also adopted when the clearance and weight increase were limited, and the

Fig. 8. Collapse and unseating of bridges

sheets in longitudinal and transverse are effective to increase the flexural strength and shear strength, respectively. Furthermore, the hinges at the top and bottom of rocking piers were fixed by polymer cement mortar to resist the seismic forces in the transverse direction to the bridge axes. In some cases, the multiple rocking piers were improved to the wall type piers by use of reinforced concrete.

A great variety of devices are installed to prevent bridge falling due to earthquake. Those devices are connecting superstructure and substructure by use of shock absorbing chains, steel, steel bracket and concrete block as protuberance. In order to install these devices, steel bolts are anchored in existing concrete pier/abutment by use of epoxy resin. Rubber is regularly used as buffer to moderate the impact of earthquake, and thermosetting polyester elastomer is also used.

The excellent performance of the above seismic reinforcement was confirmed by the earthquakes occurred after countermeasures.

Fig. 9. Typical jacketing methods and connecting super- and sub-structure by shock absorbing chain

6   Initial Defects and Subsequent Troubles

The reinforced concrete slab was designed and constructed from the view point of rationality. The minimum thickness of the slabs was 16cm since 1964 to 1968. Damages due to extreme thinness of slabs continued. Currently the thickness of slabs is specified to more than 25cm from the reflection of the original inappropriate design. Sometimes poor construction such as insufficient reinforcing bar cover and honeycomb of concrete causes the defects in concrete structures.

In construction of post tension system prestressed concrete, occurrence of poor filling as non-filling of grouting into cable sheath becomes serious problems. That is a cause of corrosion and fracture of prestressing steel due to entering water to sheath.

The concrete ceiling panels suspended from the top of the concrete tunnels fell on the cars in 2012. The upper part of the bulkhead plate was connected by the adhesive anchor to the top of tunnel as shown in Fig. 10 (a) and two ceiling plates were installed on the lower part of bulkhead plate. The detail of bolt connection to the concrete through roof hanger plate and CT steel are shown in Fig. 10(b).

Several undesirable factors should be mentioned; creep and deformation of adhesives, use of short bolt length of 110mm, and upward construction. Currently this type structures are not adopted.

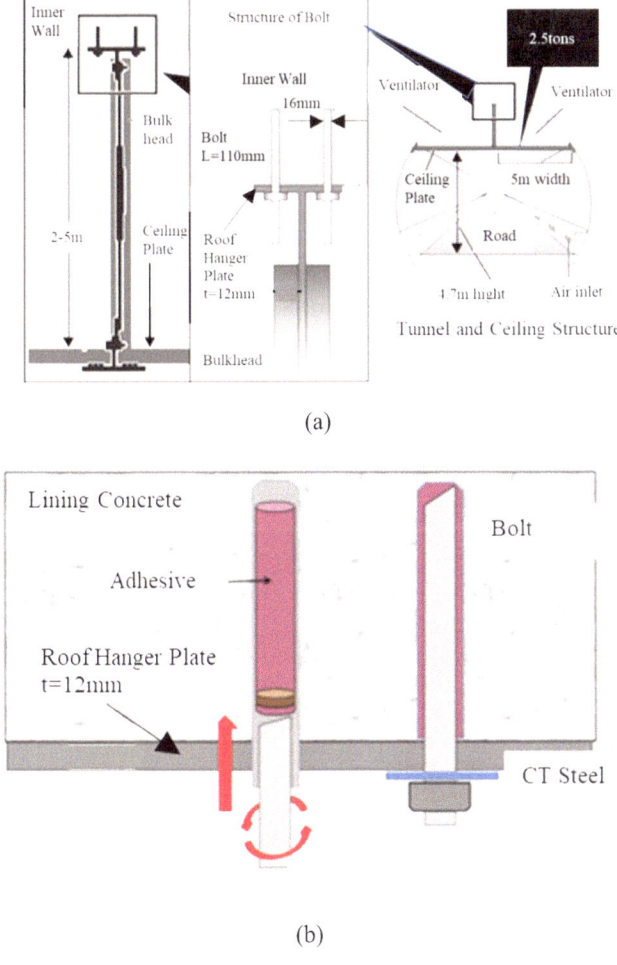

(a)

(b)

Fig.10. (a) Installation of concrete ceiling panel (b) Adhesive Anchor

After this tunnel accident, all highway bridge and tunnel undergo regular inspections and diagnosis every five years. Seismic reinforcement and preventing concrete pieces falling have promoted by the preventive maintenance.

The deterioration such as salt damage and fatigue of concrete slab in above occurs independently.

Furthermore, combined deterioration due to multiple factors of salt, ASR and loads is concerned actually and individual measures in the application of polymer materials are required.

Even if repair and reinforcement are accomplished, re-deterioration with time and unexpected external forces will act, and regular and proper maintenance are essential for longevity of structures.

7   Conclusions

Practical application of concrete-polymer composites for repair and reinforcement of highway bridges in Japan was overviewed and investigated. The main results obtained are summarized in the following:

- In Japan, the highway bridges had been constructed and in serviced under breakdown maintenance for about 50 years since 1962. Main deterioration is seen in reinforced concrete slab due to salt damage, fatigue due to cyclic loading and excessive loading.
- Degradation of water exposure such as leakage of water from the expansion joint to girder ends, bearing, and abutment, and water around drainage to slabs is remarkable, and polymer materials have been applied to properly countermeasures.
- The typical performance due to application of polymer materials was crack repair, sectional restoration and surface coating. However, slabs of the serious and combined deterioration such as segmentation of bridge deck and fatigue damage must be demolished and renewed.
- After great earthquake damage and tunnel crest collapse accident, preventive maintenance such as seismic reinforcement and preventing delamination based on the deterioration diagnosis by periodic inspection has been adopted since 2012.
- Deterioration will progress normally due to environment, material degradation and unexpected external force, even if the repair and reinforcement are executed, and so appropriate regular maintenance is essential to ensuring a long-term service life.

References

1. https://www.e-nexco.co.jp/assets/pdf/pressroom/committee/02_reference02.pdf (in Japanese)
2. Kawakami, M., Omata, F., Toyoda, A., Kato, S.: Silicone resin enclosing method applied for the maintenance of steel bearings. 10th International Congress on Polymers in Concrete, pp.743–750 (2018)
3. Kawakami, M., Omata, F., Ono, S.: The advanced countermeasures of prevention for concrete pieces falling at concrete structures in Japan. Proceedings of the 6th Asian Symposium on Polymers in Concrete, pp. 177–182 (2009)

4. Japan Concrete Institute, Concrete Crack Investigation, Repair and Reinforcement Guidelines-2013-, Reference case 1 Evaluation of bridge slabs repaired by epoxy resin injection after 40 years, pp.427–431 (in Japanese)
5. Kyokuto-Takamiya Corporation; ELSS JOINT catalog
6. Matsui, T., et.al: Outline of guideline for highway bridge deck waterproofing system 2016: concrete engineering. Jpn. Soc. Civ. Eng. **55**(7), 563–569 (2017) (in Japanese)

Ionic Conductive Polyesters—Assessing the Risk of Corrosion in Steel-Reinforced Concrete

Oliver Weichold$^{(\boxtimes)}$

Institute of Building Material Science, RWTH Aachen University, Aachen, Germany
weichold@ibac.rwth-aachen.de

Abstract. Sensors based on ion-conducting polymers are a reliable alternative to conventional metallic sensors. Formulated as 2K resin, they are quick and easy to install and cost-effective, so that larger sensor arrays with improved accuracy are affordable. The present systems are based on poly(ethylene oxide) or poly(propylene oxide) containing unsaturated polyesters doped with lithium perchlorate and are cross-linked on site with styrene. The curing reaction proceeds even at 0 °C and tolerates the presence of water. The best system in this series exhibits a resistivity of 194 Ω·m, which is several orders of magnitude lower than conventional polymers, but also several orders of magnitude higher than metals. The values are sufficient to accurately reproduce the progress of corrosion currents measured with conventional sensors and to detect changes in the humidity of concrete specimen.

Keywords: Unsaturated polyester · Sensor · Monitoring

1 Introduction

Resistivity sensors in concrete used to assess the moisture content of the cement matrix or the corrosion potential of the reinforcement usually consist of metal rods or rings. For retrofit applications, these need to be embedded in a grout mortar to establish the electrical coupling between concrete and the metallic sensor surface. Installations following this procedure, apart from being labour and, with it, cost intensive, suffer from two common problems: (i) the grout mortar has a different pore system than the existing concrete and the water introduced by the mortar changes the moisture content in the volume surrounding the sensor. Both change the resistivity and it is unclear, when reliable reading can be obtained; (ii) at the steel-mortar interface, the conductor type changes from electron transport in steel to ion transport in concrete. However, this interface is far from perfect, which increases the overall resistivity and the susceptibility to failure.

One possibility to circumvent these problems is to use conducting polymers as sensor material. However, electrical conductivity is a property rarely associated with polymers. Common polymers show resistivities in the range of 10^5 to 10^{12} Ω·m [1] and are, thus, used to make e. g. wire insulation and sockets. However, electrically conductive polymer composites have been known since the 1950s [2]. Initially, fine metal powders were

© The Author(s) 2025
L. Czarnecki et al. (Eds.): ICPIC 2023, 61, pp. 340–346, 2025.
https://doi.org/10.1007/978-3-031-72955-3_34

incorporated into thermoplastic polymers at high loadings and the electrical conductivity was the result of a percolating network of particles inside an insulating matrix [3]. Due to the high filler content, these materials had poor mechanical properties and were hard to process. Bulk electrical conductivity in polymer materials was first demonstrated by Heeger, Shirakawa, and MacDiarmid in 1977 using doped poly(acetylene)s, where the conductivity originates in the transport of electrons along the delocalised π-electron system [4]. These materials are usually obtained as brittle films and require very controlled conditions during their synthesis. While these two are electron conductors, Wright started to follow a rather different approach in 1975. By dissolving lithium salts in poly(ethylene oxide)s, he discovered polymer materials with ionic conductivity [5], in which the ions move through the polymer in a hopping process.

Using polymers as sensor material would improve the concrete-sensor interface, as polymer resins easily adapt to rough surfaces and the transition from electron to ion transport occurs at the steel-polymer interface, which is much smoother and more adherent. In order to substitute conventional retrofit sensors in concrete structures with conducting polymers, the application needs to be simple, curing should be fast and little affected by environmental conditions such as temperature and moisture, and the material must be resistant to alkaline hydrolysis. These requirements and constraints rule out metal-filled thermoplastics and intrinsically conductive polymers as well as common epoxy, polyurethane, and silicone resins. The choice, therefore, fell on ion-conducting unsaturated polyester (UP) resins containing poly(ethylene oxide) segments and dissolved lithium ions which, which can be formulated as 2K resins and harden under a wide range of environmental conditions. The following article presents some of our results on the development of ion-conducting polymer resins and their initial lab-scale test as corrosion sensors in steel-reinforced concrete.

2   Results and Discussion

The ion-conducting polymer resins to be used as sensor materials are based on unsaturated polyesters, which are prepared by first reacting poly(ethylene oxide), PEO, or poly(propylene oxide), PPO, with maleic anhydride (Fig. 1). Both PEO and PPO form the conductive blocks along the chains, while maleic anhydride serves as connection to the cross-linker. In order to avoid high-molecular weight polymers, which would complicate subsequent handling, the theoretical degree of polymerisation was adjusted to 6 by using a calculated excess of PEO or PPO. Both polyesters were individually doped with lithium perchlorate to achieve ion conductivity and then cross-linked using styrene and the redox initiating system methyl ethyl ketone peroxide (MEKP)/cobalt(II) 2-ethylhexanoate. Cross-linking is necessary to obtain mechanical stability. Details concerning the synthesis of the materials can be found in [6].

Ionic conductivity in polymers is based on the solubility of small metal cations in the matrix and the movement of these cations along and between the chains [7]. In the well-documented system poly(ethylene oxide)/Li$^+$, the solubility of the lithium ions is accomplished by their coordination to the oxygen atoms of the polyether. As a result, the atomic ratio of oxygen atoms in the chain to mobile lithium ions (O/Li$^+$) plays an important role. In previous reports for similar systems, a V-shaped course of the resistivity

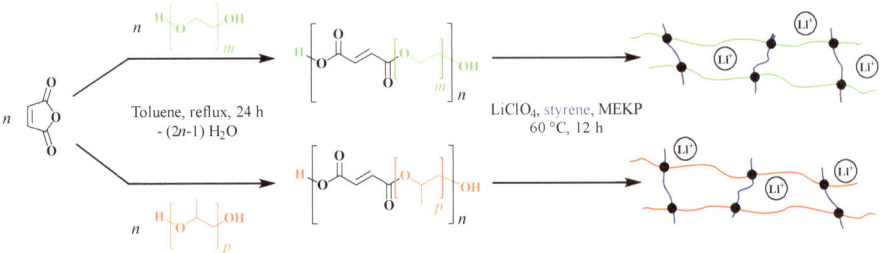

Fig. 1. Preparation of poly(ethylene oxide)-, top row, and poly(propylene oxide)-based, bottom row, unsaturated polyesters, which are then transformed into ion-conducting resins by doping with LiClO$_4$ and cross-linking with styrene.

with the O/Li$^+$-ratio was reported [8]. The reason for the initial decrease in resistivity with increasing Li$^+$ concentration is that more mobile charges facilitate the current flow. Beyond a certain concentration, the ions hinder each other, which results in an increase in the resistivity. For the present poly(ethylene oxide)-containing unsaturated polyesters, the minimum was found at O/Li$^+$ = 50 (Fig. 2, circles) [9], which is in accordance with literature [8].

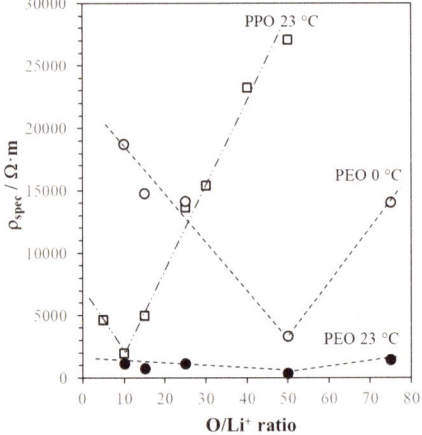

Fig. 2. Resistivity of cross-linked, lithium-doped PEO- (circles) and PPO-based (squares) unsaturated polyesters as function of the atomic ratio of oxygen atoms in the chain to lithium ions (O/Li$^+$). The dashed lines are not trendlines, but simply guides for the eye.

However, for the structurally related resins based on poly(propylene oxide), the minimum was found at O/Li$^+$ = 10 (Fig. 2, squares). The reason for this might be the additional methyl group of poly(propylene oxide), which imposes a larger helix diameter. As a result, more lithium ions are needed to overcrowd the larger helix and hamper ion movement. Two further observations can be drawn from Fig. 2: (i) as expected, the resistivity decreases with increasing temperature (cf. Figure 2, PEO 0 °C and PEO 23 °C), which is a direct consequence of the increased mobility and (ii) the minimum

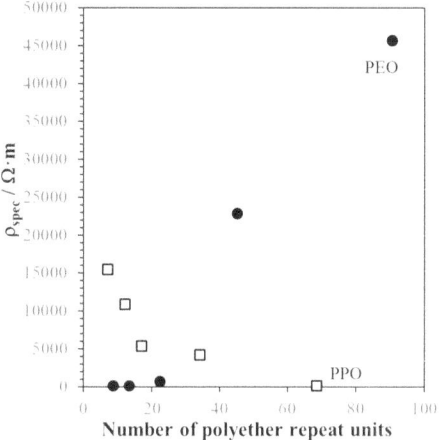

Fig. 3. The resistivity of cross-linked, lithium-doped unsaturated polyesters based on PEO (circles) and PPO (squares) as a function of the molecular weight.

in the poly(propylene oxide) system (1941 Ω m) is lower than that of the poly(ethylene oxide) system at 0 °C, but approx. 5 times higher than that at 23 °C (401·Ω m). Due to its lower polarity, the poly(propylene oxide) matrix is a poorer solvent for lithium ions than poly(ethylene oxide), resulting in higher resistivities.

Since the systems contain polymers as conductive blocks, it was interesting to investigate the influence of the block length on the resistivity of the cross-linked materials (Fig. 3). For the poly(ethylene oxide)-based systems at O/Li$^+$ = 50 (Fig. 3, circles), the resistivity increases only moderately from approx. 200 Ω m for PEO blocks with 9 repeat units to 740 m for 22 repeat units. Up to this chain length, the PEO starting materials are liquid. The subsequent drastic increase is due to crystallisation of the polymer, which hampers the movement of the ions [10]. In contrast, the resistivity of lithium-doped poly(propylene oxide)-based unsaturated polyesters was found to decrease with increasing molecular weight from 15.4 kΩ m for PPO blocks with 7 repeat units to 194.3 Ω m for blocks with 69 repeat units (Fig. 3, squares). On the molecular level, the movement of ions along the chain (intrachain hopping) is faster than the change from one chain to another (interchain hopping) [11]. An increase in the molecular weight of the conductive block should, therefore, allow ions to travel longer distances along the same chain, rather than having to switch from one chain to the next (interchain hopping) with a loss of time. However, this effect is normally counteracted by the increased viscosity of high molecular-weights polymers, which hinders ion movement. The reason why this is not observed in the present system remains unclear.

Despite certain positive aspects of the poly(propylene oxide) system, application studies were run using the poly(ethylene oxide)-based unsaturated polymers at O/Li$^+$ = 50. As the mixtures cross-link by radical polymerisation, it is not possible to formulate a 1K system with considerable shelf-life. Rather, the two components of the redox initiator (peroxide and catalyst) were individually mixed with solutions of the unsaturated polyester in styrene. Fabrication of the sensor from there is quite simple: the two solutions are simultaneously injected into a prepared drill hole be means of a commercial 2K

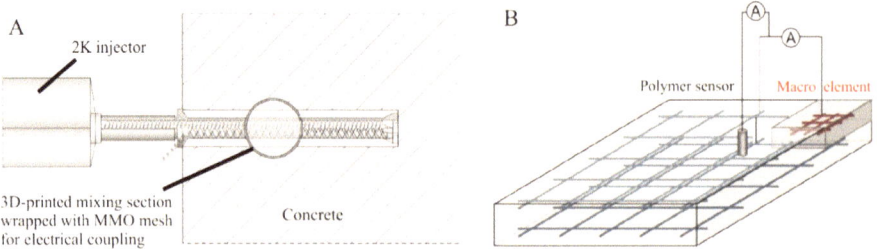

Fig. 4. A: Prototypical application set. Commercial 2K injector delivers the components through a 3D-printed mixing section into the drill hole. A mixed metal-oxide coated titanium-mesh serves as electrical coupling. B: Concrete specimen containing chloride-induced macro element and polymer sensor.

applicator through a 3D-printed mixing section (Fig. 4A). The latter remains in the hole and serves a support for a mixed metal-oxide coated titanium-mesh, which provides the electrical contact with the polymer. The sensor is ready to use as soon as a sufficient degree of cross-lining is obtained.

For the initial lab trials, a 75 cm × 75 cm × 15 cm concrete slab with two layers of reinforcement is prepared from CEM I suitable for exposure class XD3 according to DIN EN 206–1. In one of the corners, an area of 25 cm × 25 cm of the upper reinforcement layer is left out during casting. Here, a reference electrode FORCE ERE20 is mounted and the section is then filled with a concrete mixture containing 3 wt% NaCl. Two polymer sensors are installed at distances of 18 and 24 cm to the macro element (Fig. 4B).

Before assessing the performance of the polymer sensors it is important to monitor the cross-linking reaction under application-relevant conditions. It has already been discussed in several places that ion mobility has an enormous influence on the resistivity of the final product. As cross-linking changes the macroscopic appearance of the mixture from viscous liquid to rubbery, the resistivity of the mixture increases as the cross-linking reaction proceeds. As can be seen from Fig. 5, the cross-linking reaction can be accomplished over a wide temperature range and accelerates with increasing temperature. The mixture even works at 0 °C, but takes about a day to be ready to use. At ambient temperatures, reliable readings can be obtained after approx. 8 h, while at 60 °C, cross-linking is essentially completed within 90 min. The presence of water, which is usually a challenge for thermoset resins and an absolute no-go for conventional ion-conducting systems, hardly affects the cross-linking reaction. Water lowers the recorded resistivity most likely as a result of an increased ion mobility at the concrete-polymer interface.

One of the samples from Fig. 4B is stored at 21 °C and 60 r. h. for 8 months (Fig. 6A), while a second sample is first stored at 21 °C and 85% r. h. for 96 days, after which the humidity is raised to 95% r. h (Fig. 6B). Shown are the measured macro-element currents between the depassivated and passive reinforcement as well as the short-circuit currents between the total reinforcement and the respective sensor (distance 18 and 24 cm). A plot with two reference axes was chosen to allow a better qualitative comparison of the current curves. As can be seen in Fig. 6A, the actual macro-element current between active and passive reinforcement decreases continuously over time. This is presumably

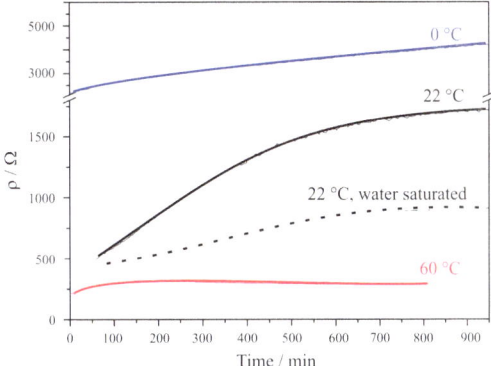

Fig. 5. Cross-linking of unsaturated polyesters with styrene in concrete as function of the environmental conditions.

due to the comparatively dry storage and corresponds to the desired desiccation according to principle 8.3 of DIN EN 1504–9.

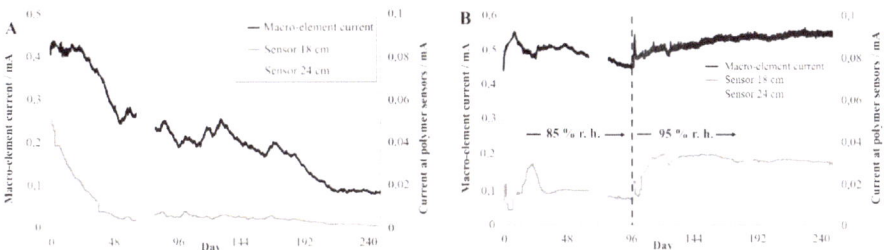

Fig. 6. Development of the recorded current in a specimen stored at 21 °C and 60% r. h. (A) and 85%/95% (B) over the course of 8 months.

    In more humid climates (Fig. 6B), the corrosion current remains high from the beginning and increases significantly after the humidity has been increased. In both cases, the polymer sensors allow a good qualitative estimation of the actual good qualitative estimation of the actual macro-element fluxes. The exact distance of the sensors to the corrosion site obviously plays a subordinate role for qualitative considerations. Signs of a loss of function of the sensors, e. g. due to loss of adhesion, etc., could not be detected in the period under the test conditions.

3 Conclusions

The following conclusions can be drawn from the above results:

- Ion-conducting polymers based can be prepared as ready-to-use 2K resin based on unsaturated polyesters.
- The resins cure under a wide range of environmental conditions in concrete, even in water-saturated drill holes.

- Resistivities as low as 194.3 Ω m can be obtained, which is in the order of weakly doped silicon, and sufficient to monitor steel corrosion in concrete.
- Polymer sensors accurately reproduce the corrosion current curve measured with conventional sensors, albeit at significantly lower current levels.

Acknowledgements. The work was funded by the Federal Ministry of Economic Affairs and Energy through the ZIM programme (Zentrales Innovationsprogram Mittelstand) under Grant No. ZF2669716KM4. The author thanks Pia Sassmann and Christian Helm for technical assistance.

References

1. Gulrez, S.K.H., et al.: A review on electrically conductive polypropylene and polyethylene. Polym. Comp. **35**(5), 900–914 (2014)
2. a) Coler, M. A.: US 2,761,849 (1956); b) Coler, M. A. US 276854 (1956)
3. Bhattacharya, S.K., Chaklader, A.C.D.: Polym. Plast. Technol. Eng.. Plast. Technol. Eng. **19**(1), 21–51 (1982)
4. Chiang, C.K., et al.: Electrical-Conductivity in Doped Polyacetylene. Phys. Rev. Lett. **39**(17), 1098–1101 (1977)
5. Wright, P.V.: Electrical conductivity in ionic complexes of poly(ethylene oxide). Br. Polym. J.Polym. J. **7**(5), 319–327 (1975)
6. Sassmann, P.B., Weichold, O.: Synergistic effects in cross-linked blends of ion-conducting PEO-/PPO-based unsaturated polyesters. Ionics **27**, 3857–3867 (2021)
7. Maitra, A., Heuer, A.: Cation transport in Polymer Electrolytes: a microscopic approach. Phys. Rev. Lett. **98**, 227802 (2007)
8. Fang, B., Hu, C.P., Ying, S.K.: Structure and ionic conductivity of graft polyester networks containing lithium perchlorate. Eur. Polym. J.Polym. J. **29**(6), 799–803 (1993)
9. Sassmann, P.B., Weichold, O.: Preparation and characterisation of ion-conductive unsaturated polyester resins for the on-site production of resistivity sensors. Ionics **25**, 3971–3978 (2019)
10. Shi, J., Vincent, C.A.: The effect of molecular weight on cation mobility in polymer electrolytes. Solid State Ion. **60**(1), 11–17 (1993)
11. Ratner, M.A., Shriver, D.F.: Ion transport in solvent-free polymers. Chem. Rev. **88**(1), 109–124 (1988)

Chemically Resistant Concrete Coating Systems with Secondary Raw Materials

Jakub Hodul[✉], Rostislav Drochytka, and Tomáš Žlebek

Faculty of Civil Engineering, Brno University of Technology, Brno, Czech Republic
hodul.j@fce.vutbr.cz

Abstract. New types of highly chemically resistant coating systems, mainly developed for concrete and metal substrates were subject to experimental testing and evaluation within the project. Secondary raw materials, including solidified hazardous waste (neutralization sludge (NS)), were used as microfillers. The three-layer polymer coating systems, applied using spray technology, were tested at two quality levels – one with a high content of solidification products, and the other with a low content. The microstructure of the epoxy coatings, including an observation of the degree of contamination of the polymer matrix, was investigated using scanning electron microscopy (SEM). It was demonstrated that the substitution of some of the primary filler with a solidification product does not result in the deterioration of the properties of the coating system, such as its adhesion to concrete or chemical resistance.

Keywords: Polymer Coating · Microfiller · By-products · Solidification · Chemical Resistance · Epoxy Resin

1 Introduction

Today, the pressure on the quality and durability of building materials and construction works continues to grow. This creates space for the development of new, progressive materials such as, for example, polymeric coatings. Coating materials, intended for use in industrial operations, are subject to particularly demanding requirements based on the extreme conditions that exist within most industries [1]. Special coatings that can withstand extreme stress plus the requirements for high mechanical strength and chemical resistance are mostly created using a polymer matrix and a suitable filler to form a composite with excellent properties [2].

As industry continues to develop, there is also the generation of significant quantities of industrial waste and secondary raw materials. Many of these wastes contain a high proportion of pollutants and are classified as hazardous waste (HW). The continuous production of HW, has led to efforts to reuse these waste products. This is a more ecologically friendly solution to the issue rather than just sending it to landfill [3, 4]. Currently, there is pressure to re-use these wastes and efforts are under way to develop building materials that incorporate, to the greatest extent, waste and secondary raw materials while maintaining the existing properties of the materials, or improving them.

© The Author(s) 2025
L. Czarnecki et al. (Eds.): ICPIC 2023, 61, pp. 347–355, 2025.
https://doi.org/10.1007/978-3-031-72955-3_35

This trend can have a positive effect from both an economic and an ecological point of view. HW and secondary raw materials, suitable for use in coatings, can be used as specially modified (solidified) fillers, which can be effectively used to replace primary raw materials [5]. There are a number of different types of hazardous waste that contain large amounts of pollutants and dangerous substances. These are, for example, cement dust, neutralization sludge (NS) generated by the surface treatment of metals, waste from the incineration of municipal solid waste, waste from the production of elements such as zirconium dioxide and many others. Through an appropriate special treatment of the HW (solidification), the resultant solidification products could then be used as a suitable filler in polymer coatings [6]. Epoxy systems are the most commonly used chemically resistant coatings, they have a long service life, adhere very well to most surfaces and are stable at temperatures up to 100 °C [7]. One disadvantage of solvent-free epoxy resins is that they must be applied to a dry surface [8].

The assumption that treated HW may be used in polymer protective systems (coatings) must first be subject to proper verification, which is the aim of this paper. In the future, it will be necessary not only to solve issue around the storage of these types of waste, but also to find further uses.

2   Materials

2.1   Polymer Binder

As the coating systems under test are designed to be used in demanding conditions, such as sewers, where there is a chemically aggressive environment, it is necessary to use the highest quality of polymer materials available. Used epoxy resin with technical abbreviation IN-ER (commercial name IN-EPOX 4090, manufactured by IN-CHEMIE Technology Ltd. (Olomouc, Czech Republic)) is a 2-component, colourless polymer. It stands out for its good mechanical and chemical resistance, high UV stability, fast polymerization, minimal odour and easy application. It does not contain thinners, benzyl alcohol or nonylphenol. The mixing ratio of the components, A (epoxy resin) to B (hardener) was 2.3: 1. The processing time at +25 °C is approximately 25 min and it is fully cured after 7 days. The humidity of the substrate during application cannot exceed 6% and the relative humidity of the ambient atmosphere cannot be higher than 75%. Component A (epoxy resin) contains: epoxy resin, (Alkoxymethyl)oxirane (alkyl C12-C14), formaldehyde and oligomeric reaction products with 1-chloro-2,3-epoxypropane and phenol products along with 1-chloro-2,3- epoxypropane and phenol. Component B (hardener) contains: Formaldehyde, polymer with N-(3-aminopropyl)-1-3-propanediamine, Carbo-monocyclic alkylated mixture of poly-aza-alkanes, hydrogenated and Polyamine adduct.

2.2   Materials Used for Microfiller Preparation

Pre-treated Hazardous Waste (Neutralization Sludge - NS). It is a by-product that arises from the surface treatment of metal elements. Neutralization sludge is produced by the neutralisation of acid waste water, in most cases a whitewash suspension (a product made from hydrated lime and water) [9]. According to the Waste Catalogue [10], this specific HW can be classified in the group, 19 02 05, sludges from physical/chemical treatment that contain dangerous substance. The NS needed to be treated, first by drying at 105 °C for 24 h, and then ground to a suitable particle size ($< 100 \mu m$) before its use as a microfiller. The NS contained high levels of lead and nickel. The specific weight of the treated sludge was 2850 kg/m3 and the specific surface area was 5400 cm^2/g.

Fly Ash (FA). This is the ash that is generated by the fluid bed combustion process within a thermal lignite power plant. It is contaminated with ammonium ions as a result of the de-nitrification of the flue gases using the selective non-catalytic reduction method. The concentration of ammonium ions (NH_3) was 30.110 ppm, the specific surface area was 627 m^2/kg and the density was 2872 kg/m^3. The chemical composition of the fly ash was: 36% SiO_2, 20% Al_2O_3, 19% CaO and 6% Fe_2O_3. This secondary raw material served as a solidification agent during the dry solidification of NS.

Quartz Flour (QF). High-quality quartz flour with a SiO_2 content of more than 99% is produced by grinding treated and dried sand in non-ferrous mills. Quartz flour has a high degree of purity, good chemical and mechanical resistance, good resistance to UV radiation and weathering, a consistent grain structure and a grain size of up to 0.1 mm (larger particle size than silica fume). Silica flour was used as the filler in the coating systems that were used as a reference within the study. It was also used in small quantities as a solidifying agent and in the top layer of the coating system.

Glass Flakes (GF). These are very thin flat plates with a smooth surface, a thickness of $7 - 100 \mu m$ and a length of $10 - 2000 \mu m$. They are used as a filler in protective coatings to increase chemical resistance, hardness and abrasion resistance. The flakes, due to their shape, form dense obstacles in the material that prevent the penetration of water or chemicals. Glass flakes were used in the top layer of the coating systems to increase the chemical resistance. The specific weight of the flakes was 2470 kg/m^3. They contain approximately 70% SiO_2, 5% Al_2O_3, 10% Na_2O, 6% CaO, and 3% MgO.

2.3   Preparation and Composition of Fillers

The first step was to solidify the HW through the dry homogenization method. This method mixes the dry components in closed containers at 80 revolutions per minute, they are then placed in a homogenizer for 24 h. Quartz powder and fly ash served as the solidifying agents. By using the dry homogenization method, the particles of treated NS were coated with the solidifying agents. The next step saw the milling of the homogenized mixture in a laboratory vibratory disc mill. Finally, the prepared filler was passed through a 63 μm sieve. In this way, fillers containing 10% and 50% of treated HW (F-NS10, F-NS50) were prepared, these were then used in the middle (second) layer of the polymer coating systems. The composition of the fillers used is shown in Table 1. The reference filler only contained quartz flour, the F-GF filler, used in the top layer, also contained glass flakes. The basic parameters of the fillers, such as the specific gravity and specific surface area, are presented in Table 2. The granulometry of the fillers is shown in Fig. 1 – it can be seen that the F-GF filler contained glass flakes with an edge length of up to 600 μm. SEM photomicrographs of the fillers can be seen in Fig. 2.

Table 1. Composition of the fillers in wt.% used in the layers of coating systems.

Mark of filler	Neutralization sludge	Fly ash	Quartz flour	Glass flakes
F-NS10	10	40	50	–
F-NS50	50	30	20	–
F-GF	–	40	50	10
QF (REF)	–	–	100	–

Table 2. Specific parameters of the fillers.

Parameter	F-NS10	F-NS50	F-GF	QF
Specific gravity [g·cm^{-3}]	2.89	2.86	2.72	2.68
Specific surface [cm^2·g^{-1}]	10950	13260	6380	4660

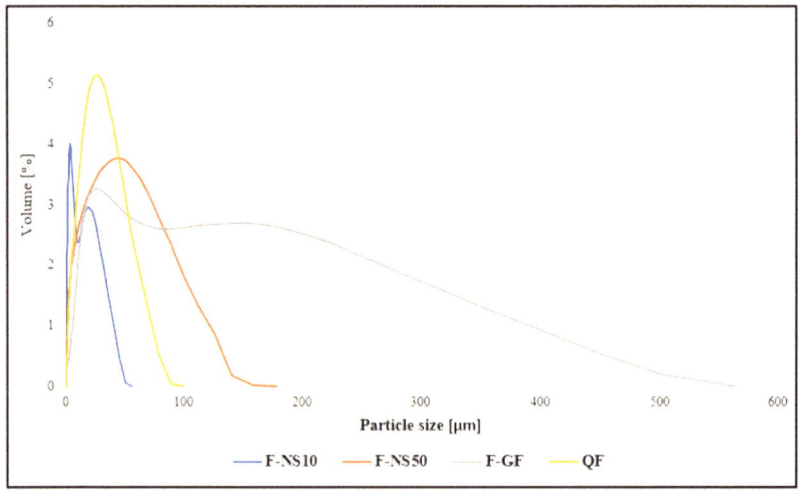

Fig. 1. Particle size distribution of the fillers.

2.4 Composition of Coating System's Layers

The pre-treated HW was used as a filler in the lower layers of the coating system, so that in the event of a breach of the top coat, there would be no possibility of a release of pollutants into the surrounding environment. The upper layers (top coats) only contained fillers based on QF, GF and FA. As the filler is perfectly incorporated into the polymer matrix, there is no release of hazardous substances into the surrounding environment. The CS-HW10 coating system used a filler with an NS content of 10% in the second layer, and the CS-HW50 used a filler with an NS content of 50%. The same IN-ER epoxy resin was used for all the coatings tested. The composition of the coating systems

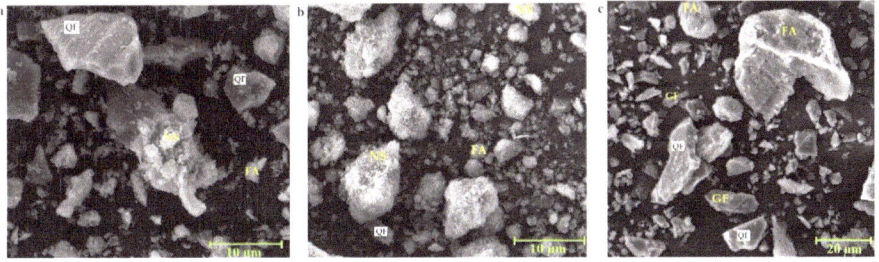

Fig. 2. SEM photomicrographs of the fillers: a) F-NS10; b) F-NS50; c) F-GF.

is shown in Table 3. Preferred dry thickness of the base layer was 200 – 250 μm, and for the top coat it was 150 – 200 μm.

Table 3. Composition of the layers of the coating system.

Type of coating system	Coating system layer	Filler	Binder
CS-HW10	1 (primer)	5% NS	95% IN-ER
	2 (base coat)	30% F-NS10	70% IN-ER
	3 (top coat)	30% F-GF	70% IN-ER
CS-HW50	1 (primer)	5% NS	95% IN-ER
	2 (base coat)	30% F-NS50	70% IN-ER
	3 (top coat)	30% F-GF	70% IN-ER
CS-REF	1 (primer)	5% QF	95% IN-ER
	2 (base coat)	30% QF	70% IN-ER
	3 (top coat)	–	100% IN-ER

3 Methods

3.1 Adhesion

Testing of the adhesion of the coating systems to a dry and clean concrete surface was performed according to EN ISO 4624 [11] using a pull off adhesion tester, Elcometer 506-20D. The coating systems were applied evenly with a suitable brush on the concrete paving block measuring 30x30 cm. The concrete surface was smooth out of formwork. The individual layers of the coating system were applied after the previous layer had hardened (24 h). After polymerization of the entire coating system, metal dolls with a diameter of 20 mm were glued using a two-component epoxy glue on the surface of top coat. Pull off tests were performed 7 days after the application of the coating materials to the substrate, and again after the chemical resistance test. The adhesion was tested from each coating system in three places.

3.2 Chemical Resistance

To test the chemical resistance of the coating systems, 4 different aggressive liquids were applied to their surfaces: 15% HCl, 15% CH_3COOH, 30% H_2SO_4 and 30% HNO_3. The liquid solutions were poured onto the surface of the coating systems using a plastic funnel, which were then sealed to prevent any leakage. The area of top coat, on which the aggressive medium acted, was 80 cm^2. The chemically aggressive environment was left on the surface of the test coating for 28 days. The samples were then visually assessed. The pull-off test was performed and any disruption of the individual layers of the coating systems was also investigated using a Keyence VHX-950F digital microscope.

3.3 Microstructure (SEM)

The degree of incorporation of the NS particles into the polymer matrix was investigated using scanning electron microscopy (SEM), with a TESCAN MIRA3 XMU microscope. The cohesion of the individual layers of the coating system was also investigated using SEM.

4   Results and Discussion

4.1 Adhesion and Chemical Resistance

Figure 3 presents a graphical representation of the adherence of the coating systems to the concrete substrate. It can be observed that after the surface of the coating had been exposed to 15% HCl and 30% H_2SO_4 for 28 days, there was no significant reduction in the adhesion to concrete. For the samples exposed to a 15% solution of CH_3COOH (acetic acid), it was not possible to determine an adhesion value, the coating system showed such significant signs of degradation that it was not even possible to attach the necessary test target. A significant decrease in adhesion was recorded for samples exposed to the 30% solution of HNO_3, which showed an approximate decrease in adhesion of 80% in comparison to those samples that had not been exposed to an aggressive environment. The 30% HNO_3 solution penetrated the structure of the concrete and the lack of cohesiveness of the coating systems was evident. In all cases, except for those samples exposed to 30% HNO_3, the failure mode was seen in the underlying concrete. The bonding strength between the coating and concrete substrate is an important indicator of the service performance of a coating system [12].

Based on an evaluation of the images from the digital microscope (Fig. 4), it can be concluded that the coatings demonstrated good resistance to inorganic acids (15% HCl and 30% H_2SO_4), where only minimal penetration of the aggressive solutions into the top layer was visible and there was no disruption of the cohesion between the individual layers or with the concrete - see Fig. 4a. The most pronounced degradation of the epoxy-based coating systems was caused by a 15% solution of CH_3COOH. Formic acid and acetic acid are known to be aggressive to epoxy coatings. The immersion of epoxy coatings in acetic acid significantly affects the chemical resistance of the coating - bubble formation associated with undercoat corrosion [13]. As can be seen in Fig. 4b, the coating system was severely damaged and the acetic acid solution penetrated into

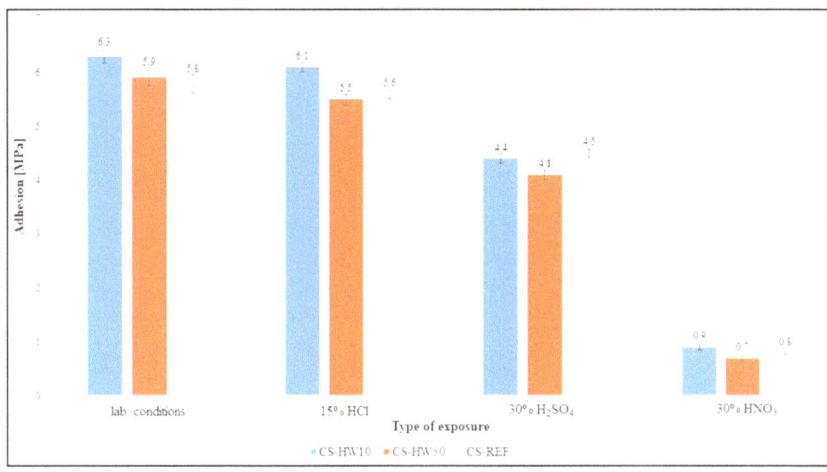

Fig. 3. Results of coatings system's adhesion on concrete substrate.

the concrete substrate. Obvious signs of degradation could also be seen after exposure to the 30% solution of HNO₃, for comparison see Fig. 4c. Nitric acid was attacking the same phase of the coating (epoxy matrix) like the acetic acid, but the solution of HNO₃ reacted only on the surface of top coat. To allow a comparison, Fig. 4d shows an undamaged coating system that has been stored within a laboratory environment and not exposed to a chemically aggressive environment.

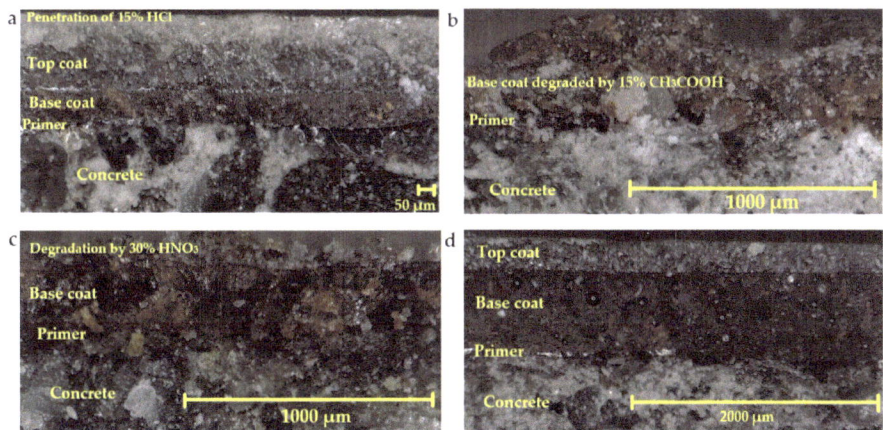

Fig. 4. The CS-HW50 coating system after 28 days of chemical exposure: a) 15% HCl; b) 15% CH₃COOH; c) 30% HNO₃; d) reference – laboratory conditions.

4.2   Microstructure (SEM)

In Fig. 5a, it is possible to observe a perfect connection between the first layer of the CS-HW50 coating system (primer) and the base layer. The secondary raw material used, in the form of a microfiller, including solidified HW, did not have a negative effect on the cohesion between the individual layers of the coating system. The successful incorporation of the NS particles in the epoxy matrix can be seen in Fig. 5b. A strong contact zone is visible between the NS particles and the epoxy binder, which should guarantee that pollutants from the filler are not released into the environment, even after the coating system has been in service for a long period of time in demanding conditions.

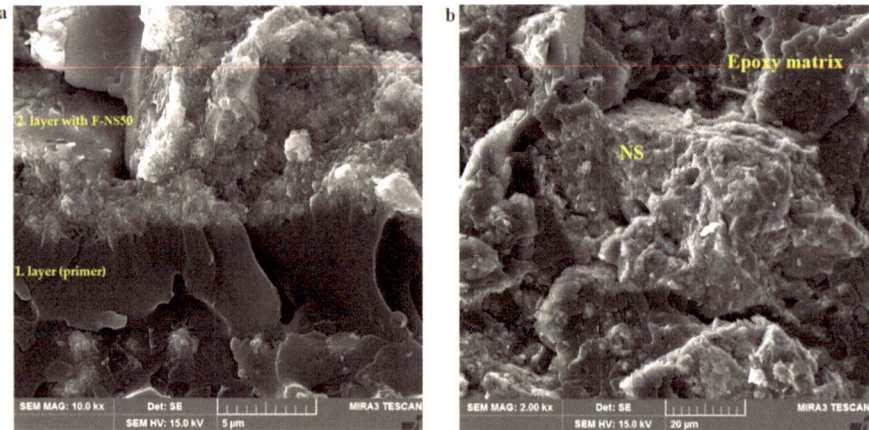

Fig. 5. SEM photomicrographs of the coating system CS-HW50: a) transition between the primer and base coat; b) incorporation of the NS particles in the epoxy matrix.

5   Conclusion

As part of the research, it was proven that the particles of the treated HW (solidified neutralization sludge) were perfectly incorporated into the structure of the epoxy matrix and in practice there should be no release of hazardous substances into the surrounding environment. The coating systems tested showed a high degree of adhesion to the concrete substrate and good cohesion between the individual layers. This was confirmed using a digital microscope and SEM. The coating systems under test were found to have poor chemical resistance to acetic and nitric acids. The chemical resistance of the coating systems that contained solidified hazardous waste was not inferior to the reference CS-REF coating system that only contained primary raw materials. The coating system that included a higher HW (CS-HW50) content demonstrated almost the same properties as the CS-HW10 coating system.

Acknowledgements. The paper was prepared with the financial support of The Technology Agency of the Czech Republic (TA CR) project No. FW03010107 "Development and research of new materials for polymer rehabilitation sprays".

References

1. De Belie, N., Lenehan, J.J., Braam, C.R., Svennerstedt, B., Richardson, M., Sonck, B.: Durability of building materials and components in the agricultural environment, Part III: Concrete Structures. J. Agric. Eng. Res. **76**(1), 3–16 (2000)
2. Ohama, Y.: Polymer-based materials for repair and improved durability: Japanese experience. Constr. Build. Mater. **10**(1), 77–82 (1996)
3. Joseph, A.M., Snellings, R., Van den Heede, P., Matthys, S., De Belie, N.: The use of municipal solid waste incineration ash in various building materials: a Belgian point of view. Materials **11**(1), 141 (2018)
4. Pappu, A., Saxena, M., Asolekar, S.R.: Solid wastes generation in India and their recycling potential in building materials. Build. Environ. **42**(6), 2311–2320 (2007)
5. Hodul, J., Mészárosová, L., Žlebek, T., Drochytka, R., Dufek, Z.: Impact of aggressive media on the properties of polymeric coatings with solidification products as fillers. Coatings **9**(12), 793 (2019)
6. Massardier, V., Moszkowicz, P., Taha, M.: Fly ash stabilization-solidification using polymer-concrete double matrices. Eur. Polymer J. **33**(7), 1081–1086 (1997)
7. Dahalan, E.N.E., Sofian, A.H., Abdullah, A., Noor, N.M.: Corrosion behavior of organic epoxy-zinc coating with fly ash as an extender pigment. Materials Today: Proceedings **5**(10), 21629–21635 (2018)
8. Baheti, V., Militky, J., Mishra, R., Behera, B.K.: Thermomechanical properties of glass fabric/epoxy composites filled with fly ash. Compos. B Eng. **85**, 268–276 (2016)
9. Dohnálková, B., Drochytka, R., Hodul, J.: New possibilities of neutralisation sludge solidification technology. J. Clean. Prod. **204**, 1097–1107 (2018)
10. Commission of the European Communities Eurostat, Guidance on classification of waste according to EWC-Stat categories, Supplement to the Manual for the Implementation of the Regulation (EC) No 2150/2002 on Waste Statistics, p. 28, Eurostat, Luxembourg (2002)
11. ISO 4624:2016; Paints and Varnishes—Pull-Off Test for Adhesion. International Organization for Standardization (ISO) Technical Committee ISO/TC35/SC9: Geneva, Switzerland (2016)
12. Szymanowski, J.: Evaluation of the adhesion between overlays and substrates in concrete floors: literature survey, recent non-destructive and semi-destructive testing methods, and research gaps. Buildings **9**, 203 (2019)
13. Marrota, A., Faggio, N., Ambrogi, V., Mija, A., Gentile, G., Cerruti, P.: Biobased furan-based epoxy/TiO$_2$ nanocomposites for the preparation of coatings with improved chemical resistance. Chem. Eng. J. **406**, 127107 (2021)

Towards the Use of Waste Limestone Powder as a Filler for Epoxy Coatings in Floors: Research on Mechanical Properties

Agnieszka Chowaniec-Michalak[(✉)], Sławomir Czarnecki, and Łukasz Sadowski

Wrocław University of Science and Technology, Wroclaw, Poland
agnieszka.chowaniec@pwr.edu.pl

Abstract. This paper presents an analysis of the mechanical properties of modified epoxy coatings used as epoxy floors. Waste mineral powder (limestone powder) was used as filler for the epoxy coating. Epoxy resin mixtures were made with waste limestone powder in amounts ranging from 0% to 29% of the mixture mass. Then, four mechanical properties were tested: hardness with the Shore D durometer, tensile and flexural strength with a standard testing machine, and pull-off strength by the pull-off method. The use of waste limestone powder as filler for epoxy coatings resulted in an improvement in hardness by 5%, does not significantly change the pull-off strength, but a deterioration of tensile strength by 6–27% and flexural strength by 18–38%. However, the modified epoxy coating still meets the standard requirements for epoxy floors. Therefore, waste limestone powder can be used in practice as filler for epoxy floor coatings. This solution allows the recycling of mineral powders, reduces the consumption of harmful epoxy resin and lowers the cost of the coating.

Keywords: Epoxy resin · Recycling · Adhesion

1 Introduction

The search for fillers for building materials is a common practice among manufacturers and scientists. This is often dictated by the desire to reduce material costs and waste disposal. Civil engineering opens up wide possibilities for recycling and finding applications for materials previously considered waste. An example is waste fly ash, which is now widely used in the production of concrete. Epoxy floor coatings are a material worth looking for fillers. Epoxy coatings are readily used in industrial construction because they create a surface with high chemical resistance and are resistant to persistent dirt. However, epoxy resin is a relatively expensive material and its components are harmful to the environment. As shown by the toxicity analysis in [1], epoxy resin contains more than 50% of ingredients that are harmful to organisms, especially aquatic ones, can

The original version of the chapter has been revised: The funding information has been updated. A correction to this chapter can be found at https://doi.org/10.1007/978-3-031-72955-3_64

L. Czarnecki et al. (Eds.): ICPIC 2023, 61, pp. 356–364, 2025.
https://doi.org/10.1007/978-3-031-72955-3_36

damage organs or may cause cancer. Epoxy resin ingredients are also irritating to the skin, eyes and respiratory tract. Adding 29% of mineral powder can reduce the content of harmful ingredients in the epoxy coating by 20 percentage points [1]. Therefore, it is worth looking for fillers for epoxy coatings, especially fillers that are currently waste.

In recent years, many studies have been carried out on the modification of epoxy resins with bio-waste. Among the tested bioadditives to epoxy resins, the following can be mentioned: coconut fibers [2] biosilica and biocarbon from rice husk [3], waste hemp-derived carbon fibers [4], waste eggshell [5], orange peel [6] and waste hemp fibres [7]. Some waste additives affect the properties of epoxy resin important for the coating, e.g. waste lignin and salicylate alumoxane nanoparticles improves tensile strength and hardness [8], walnut shells improve hardness and reduce tensile strength [9], and on the other hand in [10] it is shown that hazelnut shell improve tensile strength. Other waste epoxy resin additives tested include: polypropylene plastic wastes, which improved hardness [11], waste polycarbonate, which improved hardness, impact strength and wear resistance [12] and fly ash, which improved hardness, tensile strength, flexural strength and wear resistance [13]. The addition of fly ash can improve the bond strength of the epoxy resin [14].

According to the latest waste statistics data published by Eurostat [15], more than 23% of all waste in the European Union is waste from mining and quarrying, most of which is major mineral waste. In order to contribute to the reduction of the amount of mineral waste stored in heaps, the possibility of their recycling should be sought. Ray et al. [16] added waste marble powder to the epoxy resin to improve the thermal conductivity of the epoxy resin. Sharma et al. [17] used waste granite powder and improved hardness, impact strength and wear resistance. Krzywiński et al. [18, 19] used fine aggregate from demolition waste, which improved the pull-off strength of epoxy coatings.

Chowaniec et. al. Have previously carried out research on the possibility of using waste quartz and limestone powders as fillers for epoxy coatings. These tests [1, 20] showed that all waste quartz powders in the amount of up to 29% by weight of the coating mass do not deteriorate the pull-off strength, reduce the toxicity and costs of epoxy coating.

In the work [21], the influence of waste limestone powder on the epoxy coating was investigated, where 85% of the powder particles had a size of 0.1 to 1.2 mm. This limestone powder sedimented a lot. The addition of limestone powder did not affect the hardness of the coating, it caused a decrease in tensile strength of the coating by 60–62%, flexural strength by 10–59% and a decrease in pull-off strength of coating by 1–7%.

2   Research Goals and Significance

The motivation of this work was the desire to extend the research on the possibility of using waste limestone powders as fillers for epoxy coatings. These activities support: the idea of a circular economy (recycling of mineral waste), the natural environment (reducing the demand for harmful epoxy resin) and economy (lowering the cost of the floor). As mentioned in the introduction, the adhesion tests of the epoxy coating with waste limestone powders were previously carried out [21]. As the next stage of research, it was decided to analyze epoxy coatings with waste limestone powder, but with a much smaller particle size. The specific objectives of the research are:

- characterization of the effect of waste limestone powder on the hardness, tensile strength and flexural strength of epoxy coating by making standard samples,
- broadening the knowledge about the adhesion of the epoxy coating with waste limestone powders by making fragments of the floor,
- the comparison of the research results with the standard requirements for epoxy floors will allow to verify the applicability of the developed solution in real civil engineering areas.

3   Materials

3.1   Materials

The epoxy coating was obtained by mixing two components: epoxy resin and phenalkamine curing agent. Waste mineral powder (limestone powder) was selected as filler. The chemical composition and the particle size distribution of the powder was determined by spectronometric analysis using the SEM JEOL model JSM-6610A. All particles were smaller than 0.03 mm. The results of both analyses are presented in Table 1 and Fig. 1a.

Table 1. Chemical composition of the waste limestone powder.

Compound	CaCO$_3$	MgO	SiO$_2$	Fe$_2$O$_3$	Al$_2$O$_3$
%	97.12	1.00	1.50	0.08	0.30

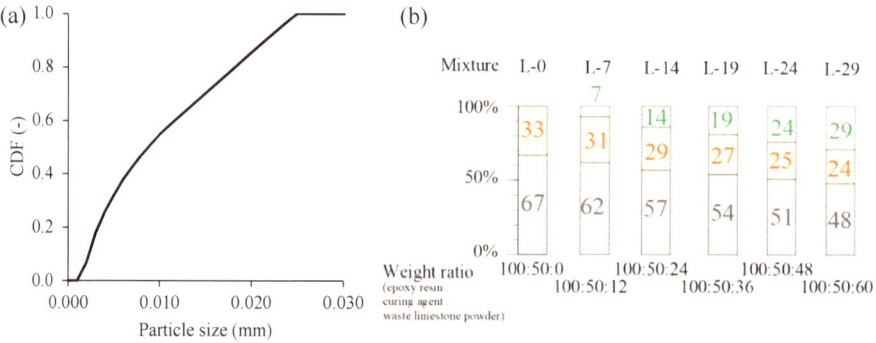

Fig. 1. (a) Particle size distribution curves for waste limestone powder, (b) Mixing ratio of epoxy coating (ration in g/100g of the mixture)

As in the previous tests [21], five mixtures of epoxy resin with a curing agent and with limestone powder were prepared. One reference mixture was also prepared without powder. In each case, the weight ratio between the epoxy resin and the curing agent was the same (100: 50). The composition of each mixture in grams per 100 g of the mixture was shown in Fig. 1b.

3.2   Preparation of the Samples

Samples for testing hardness, tensile and flexural strength.
The dimensions of the samples were determined on the basis of the PN-EN ISO 527–1:2020–01 standard for tensile samples and according to PN-EN ISO 178: 2019–06 for flexural samples (see Fig. 2). The dimensions of the samples and the process of their preparation were prepared in the same way as in the last tests [21]. Five samples of the mixtures were made for each test. The samples were made at room temperature.

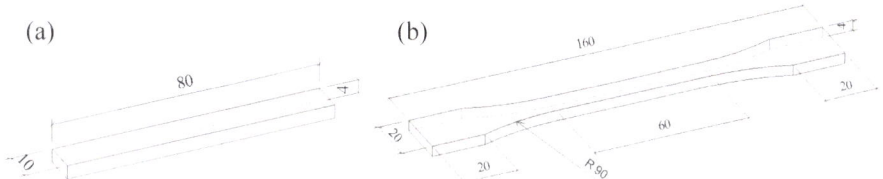

Fig. 2. Shapes of the samples (in mm) for: (a) flexural, and (b) tensile strength tests

Samples for the pull-off strength tests.
The epoxy coating for the pull-off strength tests was made on the same substrate and in the same way as in the previous tests [21]. The thickness of the bonding agent layer was about 0.1–0.2 mm, the epoxy coating was 1.4 mm and the gray finish was 0.5 mm (see Fig. 3).

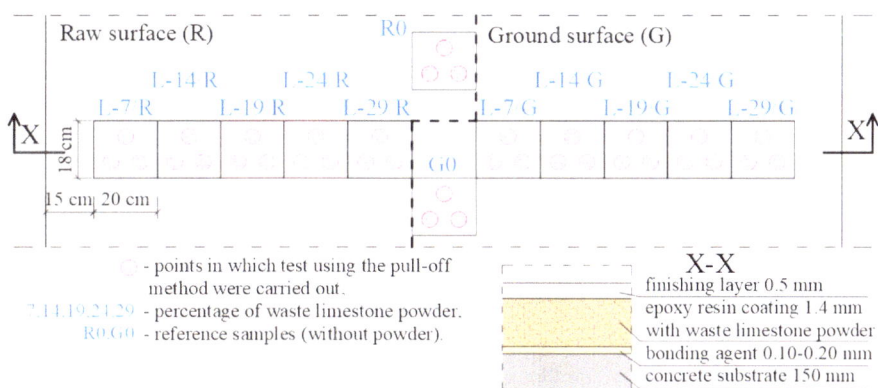

Fig. 3. View of the substrate

4   Methods

The hardness test procedure was carried out in accordance with ISO 868: 2003, using a Shore D durometer. The tensile strength tests were carried out in accordance with PN-EN ISO 527–1: 2020–01, using a universal testing machine with a load range of up to 2 kN.

The rate of application of the axial force was 2 mm/min. The flexural strength tests were carried out in accordance with the PN-EN ISO 178: 2019–06, using the same machine as in the tensile tests. The rate of force application was 1 mm/min, and the support spacing was 64 mm. For each test, the average of the measurements was calculated from five samples.

The pull-off strength tests were performed in accordance with ASTM D4541 using a DY-216 machine (Proceq, Switzerland). The rate of force application was 0.050 MPa/s. The average pull-off strength was calculated from the 3 results for each area as in Fig. 4.

5   Results

5.1   Hardness

Studies with previous waste limestone powder [21] showed that limestone powder sediment in the epoxy coating. This creates a layer on the bottom of the samples with a higher powder content than on the remaining height of the samples, which affects the hardness results. Therefore, hardness tests were performed on the top and bottom of each sample.

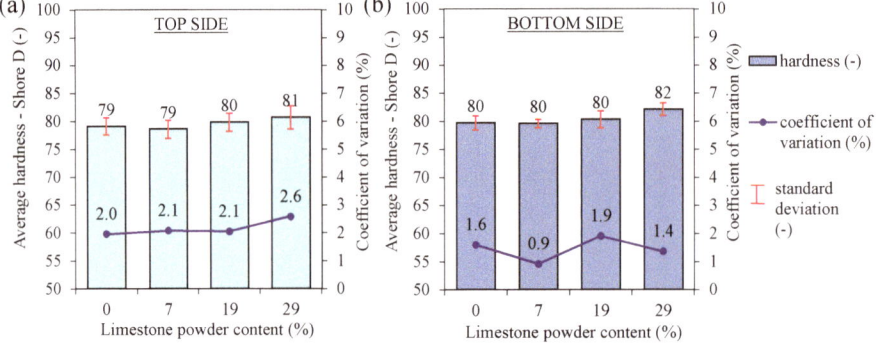

Fig. 4. The dependence of the hardness of the epoxy coating on the content of waste limestone powder for: (a) top side, (b) bottom side

Analyzing the results for the top side (see Fig. 4a) it can be seen that the waste limestone powders had little effect on the hardness of the epoxy coating. All results are between 79 and 81. Standard deviation (SD) for all results ranged from 1.6 to 2.1. The coefficient of variation (CV) ranged from 2.0 to 2.6. The bottom side of the samples gave slightly better results than the top side (on average by 1). Epoxy coating hardness results ranged from 80 to 82. SD for all scores ranged from 0.7 to 1.5, CV ranged from 0.9 to 1.9.

According to the PN-EN 1504–2 standard, the minimum Shore D hardness of the floor coating is 60. Analysing the product data sheets of other epoxy floor coatings available on the European market, their Shore D hardness is usually in the range of 65 to 80. Thus, the tested epoxy coating with waste limestone powders has a hardness well above the required value.

5.2   Tensile and Flexural Strength

The waste limestone powder at all percentages reduced the tensile and flexural strength of the epoxy coating (see Fig. 5). The reduction in tensile strength ranged from 28% to 47% compared to the reference sample. The reduction in flexural strength ranged from 36% to 39% compared to the reference sample. SD and CV significantly worsened. The deterioration of the SD and CV parameters may result from the accidental distribution of the powder particles in the samples and hence the lower reproducibility of the results.

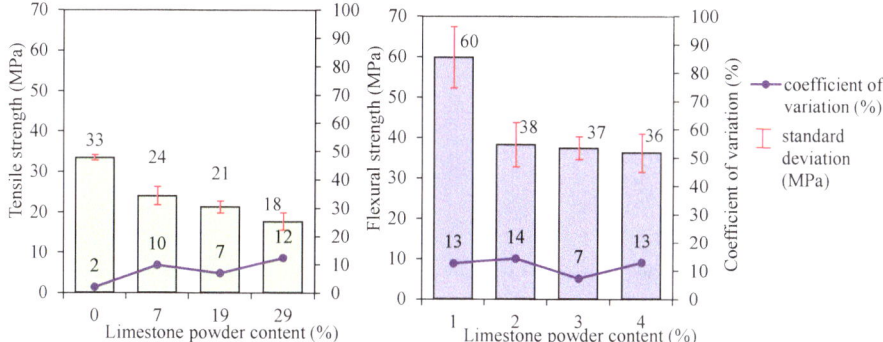

Fig. 5. The dependence of strength of the epoxy coating on the content of waste limestone powder for: (a) tensile strength, (b) flexural strength

No standard specifies the tensile and flexural strength values required for floor coatings. Therefore, the decision to use the developed solution depends on the designer and the expected loads on the floor. However, it can be assumed that the tensile strength of the epoxy coating must be the same or greater than the tensile strength of the substrate material. Fiber-reinforced concrete is most often used as the substrate for the floor. According to Marcalikova et al. [22] the tensile strength of the fiber-reinforced concrete substrate usually does not exceed 13 MPa. Therefore, it is likely that the reduction in tensile strength should not be a barrier for most floors.

Some European companies provide the declared values of the flexural strength of the epoxy coating in the product sheets. On the basis of these product sheets, it was estimated that usually epoxy coatings of floors have a flexural strength in the range of 20–60 MPa. Thus, the obtained values of flexural strength for epoxy coating with the addition of limestone powders are within the limits of standard epoxy coatings found in Europe.

5.3   Pull-Off Strength

The pull-off strength results are shown in Fig. 6. As the content of limestone powder increased, the pull-off strength decreased or increased slightly. Thus, no unequivocal downward or upward trend was observed. For the reference coating without powder, the pull-off strength was 2.9 ± 0.2 MPa for the raw surface and 2.9 ± 0.3 MPa for the ground surface. The pull-off strength results for the coating with the addition of limestone powder ranged from 2.3 MPa to 3.2 MPa for the raw surface and from 2.5 MPa to 3.1 MPa for the ground surface.

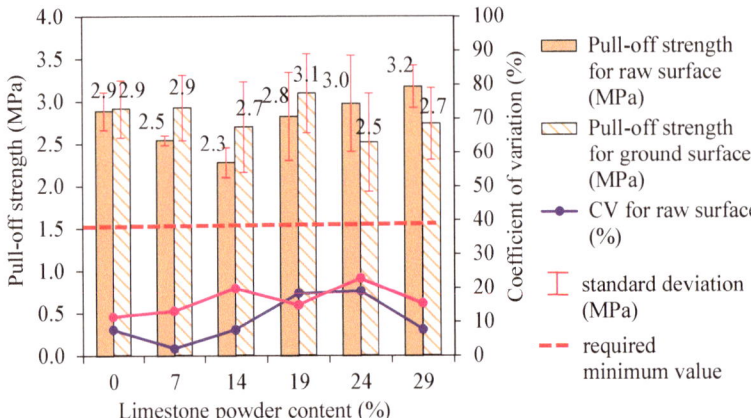

Fig. 6. The dependence of the pull-off strength of the epoxy coating on the content of waste limestone powder

The PN-EN 1504–2 standard characterizes that the floor coating should have an average pull-off strength of min. 1.5 MPa. All the obtained results are significantly above the standard requirements. In summary, the addition of waste limestone powders does not generally deteriorate the pull-off strength of the epoxy coating.

6   Conclusions

Based on the research, the following conclusions were made:

- The addition of waste limestone powder slightly improves the hardness of the epoxy coating (by about 1–2). In previous studies, in which limestone powder with larger particles was used, the improvement was greater (by about 1–8), but only for the bottom side of the samples. The use of a powder with a smaller particle size reduced the difference between the results for the top and bottom sides.
- The addition of limestone powder significantly deteriorated the tensile and flexural strength of the epoxy coating. However, the powder with smaller particles showed higher results compared to the previously used powder with larger particles [21]. The tensile strength of the powder with smaller particles was 18–24 MPa, and that of the powder from previous tests was 13–14 MPa. The flexural strength for the powder with smaller particles was 36–38 MPa, and for the powder from previous tests it was 25–54 MPa.
- The addition of waste limestone powder in an amount up to 29% generally does not deteriorate the pull-off strength of the epoxy coating. All samples obtained results above the standard required value of 1.5 MPa.

Funding.   The authors received funding from the project supported by the National Science Centre (NCN), Poland [grant no. 2020/37/N/ST8/03601] "Experimental evaluation of the properties of epoxy resin coatings modified with waste mineral powders (ANSWER). The scholarship holder, A. Chowaniec-Michalak, was supported by the Foundation for Polish Science (FNP)."

References

1. Chowaniec, A., et al.: Environ. Sci. Pollut. Res. (2022). https://doi.org/10.1007/s11356-022-19772-0
2. Miroslav, M., et al.: Lecture Notes in Mech Engin (2019). https://doi.org/10.1007/978-3-319-99353-9_6
3. Ojha, S., et al.: Environ. Sci. Eng. (2022). https://doi.org/10.1007/978-3-030-96554-9_32
4. Bartoli, M., et al.: J. Mater. Sci. (2022). https://doi.org/10.1007/s10853-022-07550-9
5. Omah, E., et al.: Int J Adv Manuf (2022). https://doi.org/10.1007/s00170-022-09593-3
6. Naik, P., et al.: J Indian Acad Wood Sci (2021). https://doi.org/10.1007/s13196-020-00272-y
7. Jadhav, A., Jadhav, C.: Iran. Polym. J. (2022). https://doi.org/10.1007/s13726-022-01034-y
8. Behin, J., et al.: Korean J. Chem. Eng. (2018). https://doi.org/10.1007/s11814-017-0301-0
9. Salasinska, K., et al.: Polym. Bull. (2018). https://doi.org/10.1007/s00289-017-2163-3
10. Kocaman, S., Ahmetli, G.: J. Polym. Environ. (2020). https://doi.org/10.1007/s10924-020-01675-1
11. Sogancioglu, M., et al.: Environ. Sci. Pollut. Res. (2020). https://doi.org/10.1007/s11356-019-07028-3
12. Sheel, A., Pant, D.: Int. J. Environ. Sci. Technol. (2022). https://doi.org/10.1007/s13762-022-04365-8
13. Sharma, V., et al.: Fibers Polym (2021). https://doi.org/10.1007/s12221-021-0145-4
14. https://ec.europa.eu/eurostat/statistics-explained/index.php?title=Waste_statistics (11. 2023)
15. Yeih, W., et al.: Cem. Concr. Compos. (2004). https://doi.org/10.1016/S0958-9465(02)00142-7
16. Ray, S., et al.: Recent Advances in Mechanical Engi (2023). https://doi.org/10.1007/978-981-16-9057-0_62
17. Sharma, A., Gautam, V.: Mechanical and wear charact. of epoxy resin-based functionally graded Mat. for Sustainable utilization of stone industry waste. Adv in Manuf Syst (2021)
18. Krzywiński, K., Sadowski, Ł.: Proc of the 3rd RILEM Spring (2021). https://doi.org/10.1007/978-3-030-76543-9_23
19. Krzywiński, K., et al.: Sci. Eng. Compos. Mater. (2021). https://doi.org/10.1515/secm-2021-0029
20. Chowaniec, A., et al.: Int. J. Adhes. Adhes. (2022). https://doi.org/10.1016/j.ijadhadh.2021.103009
21. Chowaniec-Michalak, et al. J Cleaner Prod (2022). https://doi.org/10.1016/j.jclepro.2022.133828
22. Marcalikova, Z., et al.: Procedia Struct Integrity (2020). https://doi.org/10.1016/j.prostr.2020.11.068

Alkaline Hydrogels—Multifunctional Materials for Concrete Rehabilitation

Tim Mrohs, Andre Jung, and Oliver Weichold[✉]

Institute of Building Materials Research, RWTH Aachen University, Schinkelstraße 3, 52062 Aachen, Germany
Weichold@ibac.rwth-aachen.de

Abstract. The most important factor for the protection of steel reinforcement in cementitious materials such as concrete is the alkalinity. As well as slowing down the penetration of atmospheric carbon dioxide, it delays to a certain extent the action of chloride ions. Both act at the molecular level in the form of discrete, individually mobile objects that can trigger steel corrosion. Therefore, maintenance materials designed to address these problems at the molecular level benefit from their own high pH value.

To accomplish this an alkaline hydrogel based on diallyldimethylammonium hydroxide was developed which proved to be a multitool for modern building maintenance. The gel structure can be modified in order to tune macroscopic properties such as viscosity and stickiness relevant for applications. These are e. g. the restoration of the alkaline buffer of carbonated concrete, coupling material for the electrochemical chloride extraction, and crack injection, where the gel performs three functions simultaneously.

Keywords: DADMAOH · Hydrogel · swelling · rehabilitation

1 Introduction

Steel reinforced concrete forms a good composite material, as the properties of the individual components complement each other ideally. The steel reinforcement is very ductile and, thus, can strengthen the concrete with its tensile properties. The cement, in turn, protects the steel by passivating the surface, which is due to the high alkaline pH of the concrete [1]. These excellent properties of the composite and the good availability of the compounds leads to the fact that steel reinforced concrete is the most used material in the world after water [2]. The typical lifespan of a steel-reinforced structure is estimated to be about 50–60 years. The main reason for the shortening of this lifetime is the corrosion of the steel reinforcement by two environmental factors. One of them is the carbonation of the concrete [3] and the other is the ingress of chloride ions down to the layer of the steel reinforcement [4]. In the carbonation process, the corrosion is occurring on a large area following a chemical reaction of the concrete with atmospheric carbon dioxide, which reduces the pH value of the cementitious matrix and dissolves the passive layer of the steel [5]. The ingress of chloride is a phenomenon triggered mainly by de-icing salts or sea spray. The dissolved chloride ions diffuse through the porous matrix of

© The Author(s) 2025
L. Czarnecki et al. (Eds.): ICPIC 2023, 61, pp. 365–373, 2025.
https://doi.org/10.1007/978-3-031-72955-3_37

the concrete to the steel reinforcement [6]. In this process, the protective passive layer is dissolved only locally, resulting in pitting corrosion. Consequently, the cross-section of the reinforcing steel is reduced, leading to a global failure of the composite [7]. Additionally, cracks can further reduce the lifespan of the affected buildings by increase the ingress of carbon dioxide and chlorides.

Prevention of corrosion and maintenance of existing structures is of major interest in the context of the climate change. To maintain the carbonated structure, the pH value of the concrete is raised over a certain threshold again to prevent the corrosion. This process is called realkalisation. Yet, the drawbacks of these methods are either their high invasive procedure, permanent application, or the quick loss of the alkaline environment after the treatment [8]. For the chloride ingress maintenance, the most commonly used method is the electrochemical chloride extraction (ECE) [9]. In this method, an external electrode is coupled to the steel reinforcement and placed on the concrete surface. Then the steel is cathodically polarized, and chloride is extracted from the surface by the driving voltage of 40–50 V. The high voltage is a result of the poor bonding of the currently used coupling material to the concrete surface. Further disadvantages of this method are that chloride gas generated at the anode can be poorly captured, and the coupling material must be frequently re-watered by hand to maintain conductivity. This leads to high personnel costs for such treatments.

Therefore, in the presented work, a new type of organic compound will be presented that opens novel pathways for structure maintenance. A highly alkaline diallyldimethylammonium hydroxide (DADMAOH) hydrogel consisting of a monomer, a comonomer, and a crosslinker will be demonstrated. It will be shown that there are several methods for producing the DADMAOH monomer and what advantages this charged molecule brings for the maintenance of concrete. In addition, it will be presented how the use of different comonomers and crosslinkers can influence the properties of the resulting gels in order to tune macroscopic properties relevant for applications such as viscosity and stickiness, as well as the four major advantages for use in building rehabilitation: 1. The ability to realkalise carbonated concrete [10], 2. The passivating effect on reinforcing steel, 3. The sealing of water-bearing cracks in concrete [11] and 4. The high conductivity, which makes the material an ideal coupling material for the electrochemical chloride extraction [12]. The latter is demonstrated with a practical field test on chloride-infested car-park columns.

2   Results and Discussion

The monomer DADMAOH polymerizes radically under ring closure and, due to the cationic charge of the ammonium group, provides a high charge density and thus also a high alkali density in the resulting polymer. To date, there are two working methods to prepare the highly alkaline hydroxide form of the DADMA hydrogel: The first on is ion exchange starting from the commercially available diallyldimethylammonium chloride (DADMAC) solution [10]. However, exchange of the monomer solution without ion exchange resin in a NaOH solution leads to Hoffmann-like decomposition of the ammonium compound to the amine, resulting in loss of cationic charges of the resulting gels. On the other hand, completely polymerized DADMAC gel can be converted to the hydroxide form by equilibration with NaOH solution, as the gel is significantly more stable against hydrolytic decomposition than the monomer [13] (Fig. 1).

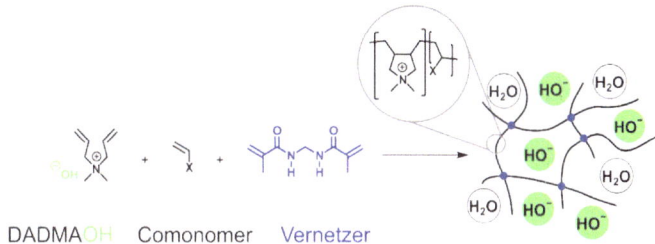

DADMAOH   Comonomer   Vernetzer

Fig. 1. Schematic composition of a DADMAOH hydrogel

A crosslinker is required to bridge the polymer chains to form a swellable gel network. In this context, the gels using N,N-methylenebisacrylamide (BIS) as crosslinker are well documented. However, BIS comes with the disadvantages of low solubility and low alkali resistance. By using tetraallylammonium-based crosslinkers such as tetraallylammonium bromide (TAAB) or tetraallyltrimethylenedipiperidine (TAMPB), much more homogeneous networks can be obtained, due to the similarity of the chemical structure to DADMAOH. This leads to stable gel formation while achieving higher degrees of swelling, even with lower amounts of crosslinker [14]. Furthermore, it can be shown that DADMAOH hydrogels with tetraallylammoinium-based crosslinkers are significantly more resistant to hydrolysis in the alkaline milieu than those with BIS [15]. This is of particular importance for subsequent applications in the construction research.

The third element in the gel formulation is the comonomer. The choice of molecule can be used to control the macroscopic and rheological properties of the resulting gels, such as degree of swelling, adhesion, stiffness, or curing time of the polymerization solution. The changes in properties are mainly due to intramolecular interactions. For example, the use of methylacrylamide allows the formation of hydrogen bonds, while the use of acrylic acid leads to ionic interactions, which additionally physically crosslink the gel network and thus lead to a stiffening of the structure [11].

In contrast to the small hydroxide ions, the polycationic backbone of the hydrogel is immobile, so that it is able to exchange anions with the surrounding medium. This property can be used in the context of maintenance of steel concrete structures. The concrete of these buildings increasingly loses its alkali buffer due to the action of carbon dioxide and water forming calcium carbonate. As a result, the pH of the concrete decreases and the passive layer of the embedded reinforcing steel is destroyed. Common repair methods in this case are the removal of the old concrete and the application of alkaline fresh concrete. In contrast, the DADMAOH hydrogel can be applied to the surface of the carbonated concrete, as shown in Fig. 2a.

The cationic polymer backbone remains on the surface as a stationary phase, while the hydroxide ions can penetrate the concrete by diffusion driven by the concentration gradient. The water required for diffusion is provided by the gel itself, which also fills the pore spaces of the concrete, so that only concretes with a sufficient pore volume are suitable for this application. The carbonate ions of the concrete migrate into the

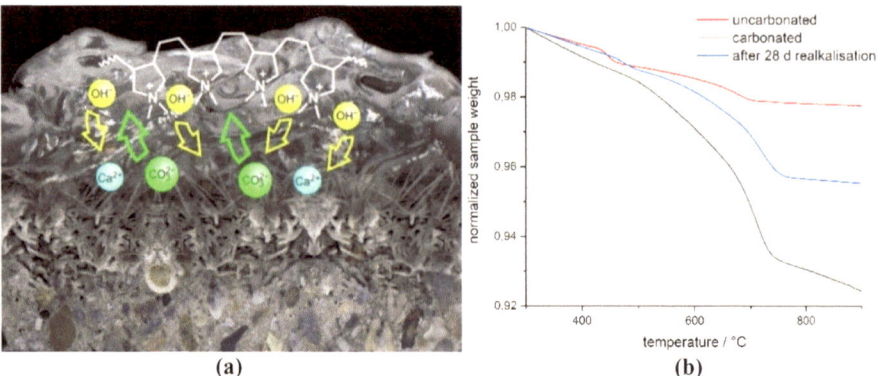

Fig. 2. a) DADMAOH hydrogel on a concrete surface [10]. **b)** Thermogravimetric analysis of uncarbonated, carbonated and realkalized concrete specimens normalized to weight at 300 °C.

hydrogel according to the gradient. An additional driving force for this is achieved by the high flexibility of the polymer chains, since these are capable to complex the bivalent carbonate ions and thus bind them considerably stronger than the monovalent hydroxide ions. In order to verify this thesis, concrete samples were made from a CEM I with a *w/c* ratio of 0.6. After 28 days of storage, the samples were transferred to a climate chamber with elevated CO_2 pressure and were fully carbonated after 6 months. The verification of complete carbonation was performed using phenolphthalein according to DIN EN 14630:2007-01. Subsequently, the samples were loaded with gel on the surface and sealed with foil to protect them from the influence of air. After 28 days, a realkalization depth of approx. 1 cm was observed by phenolphthalein test. Samples for thermogravimetric analysis of the concrete were taken from the fresh sample after storage (Fig. 2b red), after carbonation (Fig. 2b black) and after realkalization (Fig. 2b blue). The thermogravimetric analysis was performed between 30 °C and 900 °C under N_2 and the resulting curves were normalized to the value at 300 °C in order to neglect the weight loss of the previously evaporated water. Figure 2b shows a decrease in mass at about 500 °C for the uncarbonated fresh concrete, which is due to the decomposition of calcium hydroxide to calcium oxide. Between 650 °C and 700 °C the mass decrease of calcium carbonate can be observed. After carbonation, the shoulder of calcium hydroxide disappears, whereas significantly more calcium carbonate is found in the sample. By realkalisation with DADMAOH hydrogel, the calcium hydroxide fraction was lowered and a calcium hydroxide content was detected. This allows the conclusion that the gel is able to restore a partial amount of the alkali buffer.

Since high pH values are capable of passivating steel, the DADMAOH hydrogel should also be able to achieve this effect in the case of direct contact, such as cracks in concrete structures. To verify this assumption, an experiment in accordance with DIN EN 480-14 was prepared. Technically, a polarisation experiment with 500 mV for 24 h should be performed in a corrosive environment. This experiment is crucial in civil engineering application, as it shows, whether the used material could potentially corrode the steel reinforcement. Therefore, it was necessary to perform this experiment on the poly(DADMAOH-*co*-acrylic acid) gels and thereby verify its non-corroding nature. A technical drawing of the experimental setup is shown in Fig. 3a. Mortar cubes with the dimension of $100 \times 100 \times 100$ mm³ were prepared from CEM I 42.5 with a *w/c* ratio of 0.5. On one side of the samples a mixed-metal oxide coated titanium mesh (MMO) was embedded in a distance of 10 mm to the surface. After curing the cubes for 28 days, a 20 mm hole was drilled in the middle of the sample and filled with Gel. For the potentiostatic experiments, the entire sample was first placed in a bucket of water, since the conductivity of dry mortar is not sufficient. Furthermore, a 3-electrode setup was used. The reference electrode consisted of a MMO mesh, which was positioned in the water outside of the sample (reference electrode and water bath not shown in Fig. 3a). The embedded MMO mesh was used as the counter electrode and an 8 mm diameter steel pin placed in the drill hole and surrounded in DADMAOH gel served as the working electrode. Lastly, the surface of the Gel was covered with silicon oil to prevent moisture exchange with the surrounding air. An open circuit delay was measured to determine the resting potential between the electrodes before the potentiostatic experiments were performed. The results of the potentiostatic measurement can be seen in Fig. 3b [11].

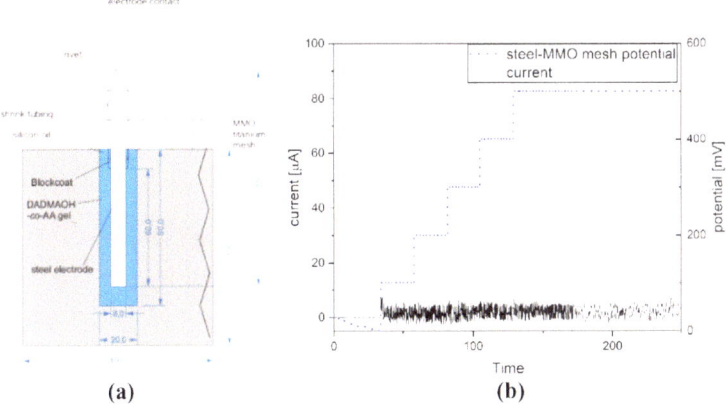

(a)   (b)

Fig. 3. a) technical drawing of a polarisation setup (DIN EN 480-14) **b)** stepwise increase of potentiostatic polarisation (from 0 to 500 mV) with a steel electrode as an anode. Test derived from DIN EN 480-14 [11].

To observe possible corrosion, it was chosen to increase the potential between the electrodes stepwise. Therefore, 100 mV/day steps were used until the potential reached 500 mV. The resulting current must not reach a value above $10 \mu A$ to fullfill the requirements of the DIN EN 480-14. During the whole polarization procedure described, no significant current outside the signal-to-noise ratio was detected. Even after 5 days of maximum polarization of 500 mV, all values remain below $5\mu A$. Therefore, it appears that even under these electrochemical conditions, the alkaline environment of the gel prevents the steel from depassivation. This measurement also shows that the remaining chloride ions, which may have been left in the Hydrogel by the ion exchange process of gel production ($< 5\%$), do not corrode steel-reinforced concrete [11].

In addition to the strongly alkaline pH value, the high charge density also offers further advantages: the cationic charges interact very well with surfaces containing silicates, i.e. mainly negatively charged surfaces. The gel is therefore able to adhere to concrete. Furthermore, the gel matrix swells when water is added and could therefore be well suited to seal water-bearing cracks. This was tested according to WTA-Merkblatt 4–6 "sealing of structural elements in contact with soil at a later stage" under moderate impact of pressing water (exposure class W2.1-E) on mortar prisms. For this application mortar prisms were used with dimensions of $160 \times 40 \times 40$ mm^3. In order to test for the above mentioned realkalisation feature, the mortar prisms were fully carbonated. To obtain a defined crack, the samples were broken in a 3-point bending test and the two halves were brought back together with a screw rack. The crack was adjusted to a typical water-bearing crack width in cementitious materials of 0.3 ± 0.03 mm. An unpolymerized DADMAOH reaction solution was applied into the crack by using an injection packer and cured there for 7 days. Next, a hose coupling was attached to the samples and then connected to a water reservoir through a tube. Further, the reservoir was put in a height of 5 m to ensure a water pressure difference of around 0.5 bar. The schematic presentation is shown in Fig. 4 [11].

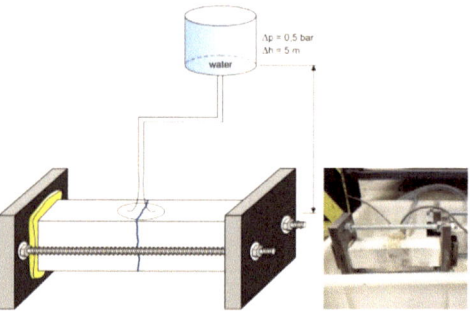

Fig. 4 Schematic presentation of the water leakage test leaned on WTA test W2.1-E [11].

According to WTA procedure the sealing test was performed for 28 days and the rear side of the samples had been observed to check upon the occurrence of leakages. During the whole time, no leakage has occurred for all samples. This observation could be correlated with the gel-blocking effect, which would seal the water leaking crack. After this successful test of the material, the sealed mortar samples were broken and the mortar surface then cautiously wiped off to remove residues of the gel. A phenolphthalein test was performed to examine the mortar for realkalisation. It was obtained that the previously carbonated samples were realkalised for the most part, leaving only the edges untouched, due to the pressureless filling method of the curing solution. If the carbonated sample would have had a steel reinforcement, the injection of the gel, as a maintenance procedure, would be an effective 3-in-1-system to seal, realkalise and passivate the steel reinforcement [11].

The fourth major advantage of DADMAOH gel is its high conductivity. This, together with the good bonding to concrete, enables the gel to serve as an effective coupling material in the electrochemical chloride extraction, by which chloride ions are extracted from the pore system of the concrete using an externally applied voltage. In preliminary laboratory tests, it was shown that with a DADMAOH-*co*-methacrylamide gel voltages of 1 V are sufficient to obtain extraction of chloride ions with this coupling material, whereas previous commercial methods used 40–50 V. Furthermore, any chlorine gas that is formed at the anode is immediately trapped and bound by the highly alkaline gel, preventing it from leaking into the environment. As a practical application example, a field test was performed on four chloride-infested car-park columns using this novel coupling system. For economic reasons, the gel was mixed with water-saturated hydroxyethyl cellulose in a 1:2 ratio. The gel was applied to the columns with an MMO mesh and then wrapped with PE film to protect it from drying out (Fig. 5a). The extraction was planned in cycles comprising 21 days of ECE and 7 days of rest. To observe the evolution of the chloride concentration, samples were taken from three depths during the resting periods. In the first cycle the voltage was set to 5 V, to achieve a current density of 1 A/m^2. As the concrete resistance increased with the application time, due to the continued extraction of charge carriers, the voltage had to be increased to 8 V in the second cycle to maintain the desired current density. After two of these ECE Cycles three of four treated columns were successfully reduced by more than 95% of the original chloride content. The fourth column required an additional 21-day period, in which the voltage was raised to 15 V, potentially due to a lower porosity and/or lower moisture content. The results of this procedure are shown in Fig. 5b [12].

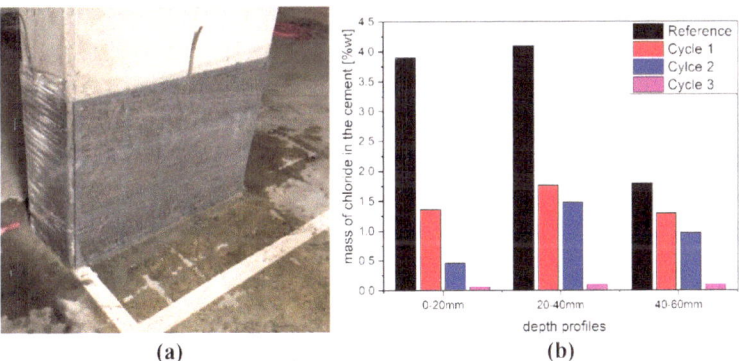

(a)   (b)

Fig. 5. **a**) Prepared installation of the MMO mesh anode with gel/hydroxy cellulose as electrolyte wrapped with PE foil to prevent drying processes. **b**) Evolution of the chloride content in the column requiring three ECE cycles [12].

As already observed during the mortar experiments, the chloride ions close to the surface were first extracted from the concrete during the ECE application. In the first cycle, about 70% of the chloride close to the surface was extracted within 21 days. This was also observed regarding the 20–40 mm profile where the decrease was slightly lower. In the depths close to the reinforcement, the decrease was significantly lower than for the other profiles. Hence, the chloride close to the surface could be extracted more efficiently. In the second cycle, the chloride at the surface was again extracted more efficiently than from deeper layers. Only a small change could be observed for the second profile. In the third cycle, it is clearly seen that the procedure is more efficient than a standard approach. More than 95% of the chloride could be extracted from the columns. Finalizing, the mixture of the gel/hydroxy cellulose could be used in the ECE maintenance while applying voltages of about 1/10 of typical application. Additionally, moistening has not been needed during the process, which significantly reduces the personnel cost [12].

3   Conclusion

The presented highly alkaline diallyldimethylammonium hydrogel proves to be a true multitool in the field of maintenance. The material was shown to have four fundamental advantages: a) realkalizing carbonated concrete, b) repassivating exposed reinforcing steel, c) sealing water-bearing cracks, and d) electrolytic coupling to concrete surface in electrochemical chloride extraction. Furthermore, due to the chemical structure, a tailor-made material for the respective application can be produced by a clever choice of comonomers such as acrylic acid or methacrylamide. The above-mentioned application possibilities of the material were verified by water leakage test (WTA W2.1-E) and polarization test according to DIN EN 480-14. In addition, electrochemical chloride extraction with the new gel-containing coupling system was successfully performed in a field test on chloride-contaminated parking garage columns. The operating period for these highly alkaline hydrogels in maintenance depends on the area of application. For realkalization, it is recommended to perform the treatment as soon as the covering 1 cm of concrete is completely carbonated. Concrete with a compressive strength class C20/25 reaches this carbonation depth after approx. 7 years, whereas a concrete C30/37 will take approx. 30 years. The crack injection solution should be applied in the case of water-bearing crack formation, while the ECE should be performed in the presence of elevated chloride contents of approx. 4 mass percent.

References

1. El-Reedy, M.: Steel-Reinforced Concrete Structures: Assessment and Repair of Corrosion. CRC Press (2007)
2. Gagg, C.R.: Cement and concrete as an engineering material: an historic appraisal and case study analysis. Eng. Fail. Anal. **40**, 114–140 (2014)
3. Leemann, A., Moro, F.: Carbonation of concrete: the role of CO2 concentration, relative humidity and CO2 buffer capacity. Mater. Struct. **50**, 30 (2016)
4. Montemor, M.F., Simões, A.M.P., Ferreira, M.G.S.: Chloride-induced corrosion on reinforcing steel: from the fundamentals to the monitoring techniques. Cement Concr. Compos. **25**(4), 491–502 (2003)
5. Possan, E., Thomaz, W.A., Aleandri, G.A., Felix, E.F., dos Santos, A.C.P.: CO2 uptake potential due to concrete carbonation: a case study. Case Stud. Constr. Mater. **6**, 147–161 (2017)
6. Wuman, Z.: Chloride diffusion coefficient and service life prediction of concrete subjected to repeated loadings. Mag. Concr. Res. **65**, 185–192 (2013)
7. Taha, N.A., Morsy, M.: Study of the behavior of corroded steel bar and convenient method of repairing. HBRC J. **12**(2), 107–113 (2016)

8. Raupach, M., Orlowsky, J.: Schutz und Instandsetzung von Betontragwerken. Bau + Technik-Verlag (2008)
9. Elsener, B., Angst, U.: Mechanism of electrochemical chloride removal. Corros. Sci. **49**(12), 4504–4522 (2007)
10. Jung, A., Weichold, O.: Preparation and characterisation of highly alkaline hydrogels for the re-alkalisation of carbonated cementitious materials. Soft Matter **14**(40), 8105–8111 (2018)
11. Jung, A., Weichold, O.: A 3-in-1 alkaline gel for the crack injection in cement-based materials with simultaneous corrosion protection and re-passivation of crack-crossing steel rebars. Constr. Build Mater. **344** (2022)
12. Jung, A., Faulhaber, A., Weichold, O.: Alkaline hydrogels as ion-conducting coupling material for electrochemical chloride extraction. Mater. Corros. Werkst. Korros. (2021)
13. Olsson, J.S., Pham, T.H., Jannasch, P.: Poly(N, N-diallylazacycloalkane)s for anion-exchange membranes functionalized with n-spirocyclic quaternary ammonium cations. Macromolecules **50**(7), 2784–2793 (2017)
14. Mrohs, T.B., Weichold, O.: Multivalent allylammonium-based cross-linkers for the synthesis of homogeneous, highly swelling diallyldimethylammonium chloride hydrogels. Gels **8**(2) (2022)
15. Mrohs, T.B., Weichold, O.: Hydrolytic stability of crosslinked, highly alkaline diallyldimethy-lammonium hydroxide hydrogels. Gels **8**(10) (2022)

The Use of Polymer Concrete as a Cost-Effective and Durable Alternative for Rapid Pothole Repair in Asphalt Surfaces

Frans Willem van Zyl[✉] and Deon Kruger

Faculty of Engineering and the Built Environment, University of Johannesburg, Auckland Park Campus, Johannesburg, South Africa
fransvanzyl98@gmail.com

Abstract. The paper describes an investigating into the application of a polymer concrete using a recyclable material such as polypropylene and chopped tire rubber as aggregate material to provide a durable and cost-effective alternative to the existing cold mix asphalt (CMA) pothole repair material. Three different aggregate mix designs, including traditional crusher- sand and -stone, rubber crumbs, and plastic chips, in combination with Vinyl Ester, Polyester, Polyurethane and Furan resin were combined to create 12 possible polymer concrete mixes to be used as a rapid pothole repair material. These polymer concrete mix designs were tested for common properties and characteristics typically experienced on road surfaces. These include characteristics such as compressive strength, flexural strength, and abrasion resistance. All results obtained from the various tests performed indicated that the polymer concrete mix designs exhibit enhanced physical properties compared with CMA and would thereby increase the durability and lifespan of repaired potholes when using this material.

Keywords: Rapid Pothole Repair · Polymer Concrete · Recycled Aggregates

1 Introduction

Roadways are an extremely important and expensive part of a country's infrastructure. In 2019, the South African budget estimate for national expenditure allowed for a total of 900 000 km^2 of blacktop patching to repair potholes. In 2015, an estimated area of 1 497 281 km^2 was done [3]. Potholes in roadways are a worldwide obstacle although the most common method of repair offers a short-term solution [1].

Potholes are predominantly caused by the infiltration and erosion of water through cracks in the asphalt layer. Cold-mix asphalt, a bituminous mixture containing medium to fine aggregates, is typically the most common material used for repairing potholes. Cold patches typically have a one-year lifespan before a more permanent method such as re-layering or resurfacing can be implemented. Polymer concrete, however, could serve as a viable replacement for cold-patch material [1, 2].

The use of polymer concrete to repair potholes in asphalt surfaces was a concept developed to rapidly repair runways of bomb craters [4]. The need arose for a rapidly

© The Author(s) 2025
L. Czarnecki et al. (Eds.): ICPIC 2023, 61, pp. 374–383, 2025.
https://doi.org/10.1007/978-3-031-72955-3_38

curing material that could easily be installed using a small crew and little to no specialized equipment. To date, advancements have been made to repair potholes using polymer concrete, although the concept has not widely been accepted [5].

This study deals with the design and assessment of various polymer concrete (PC) mixes using recycled material such as unrefined polypropylene shards, rubber crumbs refined from tires and crusher-sand and -stone. The three different aggregate mixes include; recycled plastic, rubber crumbs, and crusher-sand and -stone. The polymer binders used in conjunction with the three different aggregates mixes include: Vinyl Ester, Polyester, Polyurethane, and Furan resin. The various PC mix designs were all tested for the properties and characteristics typically required for use on road surfaces, and which may affect the road surface durability and service life of the repair. These include characteristics such as compressive strength, flexural strength, and abrasion resistance. Each of the PC mix designs were tested and compared to a cold mix asphalt material typically used for pothole repair to compare the suitability of the PC mix as a rapid pothole repair material.

2 Material and Methods

2.1 Materials

The selection of the polymer binder was primarily based on the expected physical characteristics, such as compressive and flexural strength, of the composite material created using the polymer as binder. The polymer's UV resistance was the second criteria in the selection process because of the prolonged UV exposure that the repair material would experience. Polyester (PE), Vinyl Ester (VE), Polyurethane (PU), and Furan resin (FU) were selected for evaluation as binders. These resins were also considered suitable for polymer concrete pothole repair material due to the unit cost and availability of local suppliers in Johannesburg. The polyurethane resin obtained contained a bitumen emulsion used to improve flexibility and adhesion with in-situ bitumen material.

The aggregates were selected based on cost-effectiveness and individual physical properties that may aid in creating a more durable composite material. The following material was selected:

- 7.1 mm crusher stone SANS 1083: 2006
- Unwashed crusher sand SANS 1083: 2006
- Unrefined recycled polypropylene plastic shards (2.00–10.00 mm)
- Rubber crumbs (0.2–2.00 mm) recycled from used motor vehicle tyres

The more traditional aggregates (crusher-sand and -stone) were selected; to reduce the cost of the final repair material, wide availability, and its proven record of providing superior mechanical properties to the composite material. Recycled materials such as the plastic shards and rubber crumbs were used to potentially impart greater flexibility and a higher resistance to impact loading cracking onto the composite material.

2.2 Specimen Preparation Procedures

Various specimen sizes were required for the different testing procedures although each of test specimens manufactured were compacted in a similar method using an electrical rotary hammer as a compactor. This procedure was used to proportionally simulate the on-site installation of pothole repair material using either a plate or roller compactor. The rotary hammer drill, set to the hammer setting, was applied to each layer with a downward force of 0.6 kN for a duration of 30 s using a $45 \times 45 \times 10$ mm square high tensile steel plate. The material was compacted twice for each layer. All moulds were coated with a release agent to decrease adhesion of the material to the mould.

2.3 Optimal Mix Design

The optimal mix design was derived from a trial-and-error method whereby small specimens were created and evaluated based on void content, surface texture, and proportions of aggregate to polymer ratios. Various factors such as workability, void content, cost-effectiveness, and ease of use when mixing and installing were taken into consideration when examining the small specimens created. From the observations made, the mix design was adjusted accordingly, and different aggregates were tested to observe the behaviour of the final composite material. All tests conducted were compared to a CMA baseline specimen that was prepared similar to the polymer concrete specimens.

2.4 Compressive Strength

Compressive strength tests were conducted following the SANS 5863: 2006 standard for compressive strength testing of hardened concrete using 100 mm cubes. The cube moulds were filled in three layers and compacted as per the SANS 5863: 2006 standard. The specimens were removed from the moulds 24 h after casting and were then stored in a climate-controlled room at a temperature of 25 °C for 5 days prior to testing. The compressive strength of a 15 MPa Portland Cement concrete mix design and CMA specimens were also determined to be used as a baseline for comparison. The concrete specimens were prepared in accordance with SANS 5861–3:2006 and cubes were also cured at 25 °C and submerged in water, every day for a period of two hours, to simulate precipitation or dew conditions. The compressive strength testing apparatus applied a constant load of 0.25 kN/second and was set to cease further loading once the applied load had decreased more than 15% of the peak applied load. The maximum applied loading was recorded, and photographic evidence of the tested cubes was taken to evaluate how the specimen failed.

2.5 Flexural Strength

The flexural strength of the polymer concrete was determined following the SANS 5864 code on flexural strength testing for hardened concrete. The beams were cast using 40x40x160 mm moulds that were placed and compacted in three separate layers. A simple three-point, centre loading test was conducted to determine the flexural strength following the standard mentioned. The machine increased the load applied from the centre point onto the beam at a constant rate of 3.03 kN/min and recorded the maximum load once the beam had failed either in excessive yielding or total beam failure as per the SANS 5864 code requirement.

2.6 Abrasion Resistance

As no applicable asphalt abrasion resistance standards or procedures were found in the literature, a comparative test was developed to investigate the abrasion resistance of the polymer repair material. A towable sledge was developed to facilitate the exposure of several 100 mm cube specimens to a constant wearing process by being dragged over an asphalt surface of 0.61 mm texture depth. A constant load of 3kg were applied vertically to each of the specimens during the process. The sledge with attached specimens were dragged at a constant speed of 5–7 km/h over 400 m intervals whereafter each specimen was cleaned of any loose debris and the mass loss due to the abrasive wear was recorded.

3 Results

3.1 Optimal Mix Design

Using the crusher stone and crusher sand as common base aggregate, three different PC mix designs were selected via a trial-and-error method based on its unique physical properties and proportions. These three PC mix designs are shown in Table 1. The three different aggregate mix designs (N - Normal, P - Plastic and R - Rubber) were tested in more detail using the four different polymers as binders selected to evaluate each specific mix design for the suitability for use as a rapid road repair material. As can be seen from Table 1, the trial-and-error method led to a final, optimal mix design for each specific aggregate. Using only the plastic and rubber aggregates, respectively, a mix design with acceptable mechanical properties was not achieved. Therefore, crusher-stone and -sand were added to the plastic and rubber aggregate mix designs to provide additional mechanical strength. It was hypothesized that both the recycled aggregates will impart characteristics such as flexibility to the final composite material. Unrefined polypropylene shards, as aggregate, were selected due to its increased adhesion properties and its reduced cost compared to refined polypropylene pellets. It should be noted that the increased polymer binder content of the "R" design was required due to the rubber crumb aggregate consisting of small particles thus increasing the surface area compared to the larger plastic particles. Table 2 provides illustrates the properties of the polymers used as binders.

The optimal mix designs were determined by systematically increasing the fines content of the mixes to decrease the overall void content resulting in minimizing the polymer binder content requirement to decrease the overall costs of the mix designs.

3.2 Compressive Strength

Five 100 mm cube specimens of each mix design were randomly selected from the specimen batch for testing 5 days after casting and curing in a climate-controlled room at 25 °C. Figure 1 indicates the average compressive strengths obtained for each of the various mix designs. Although the South African road authorities, such as SANRAL, do not recommend the use of Portland Cement concrete to repair potholes on asphalt pavements, it is common for communities to fill potholes using Portland Cement concrete in urban roads. As such, the compressive strength determined of a standard 15 MPa

Table 1: Optimal aggregate mix designs selected by a trial-and-error method.

Design code	Ratio (1 = 150ml)	Material	Void Content %
-N	0.5	Crusher stone	2.13
	1	Crusher sand	
	0.2	Polymer binder	
-P	0.33	Plastic shards	3.08
	0.33	Crusher stone	
	1	Crusher sand	
	0.2	Polymer binder	
-R	0.33	Rubber crumb	3.87
	0.33	Crusher stone	
	1	Crusher sand	
	0.24	Polymer binder	

Table 2: Polymer resin used in conjunction with Table 1.

| | Resin | | | |
Properties	Polyester (PE)	Polyurethane (PU)	Vinyl Ester (VE)	Furan (FU)
Catalyst to resin % volume	1–3	17	1.2	22
Geltime @ 25 °C (minutes)	9–13	30–40	120–240	*
Tensile strength, MPa	64	>15	15–95	70–80
Tensile Elongation @ break, %	3.3	>40	6	3
Flexural Strength, MPa	120	>15	115	60–80
Barcol hardness	46	60	36	>45

*Gel time dependant on type and volume of catalyst used.

Portland Cement concrete mix and standard CMA specimen under the trial testing and curing regime were used as a baseline for comparison.

All the PC mixes outperformed the traditional CMA baseline significantly. The increased compressive strength indicates that the PC mix designs have higher compressive resistance to heavy loads experienced on roadways and using the PC mixes could thus benefit the lifespan of pothole repairs compared to the traditional cold patch material.

The difference in compressive strengths between the N and R mix designs decrease by 91.9% for VE and 80.93% for PE respectively. The general decline in compressive

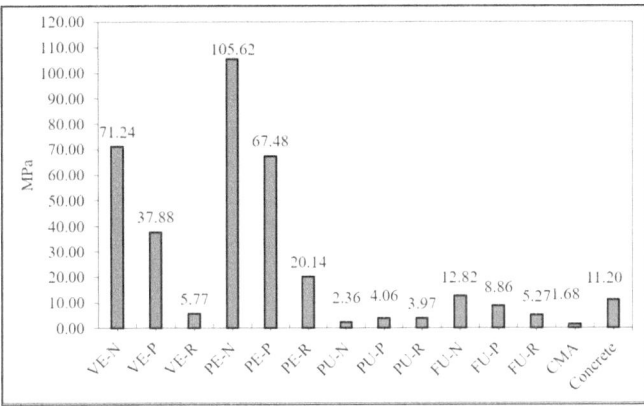

Fig. 1: Compressive strength of various mix designs.

strength from the N to P to R mixes, with the sole exception of PU binder, indicate a reduction of compressive resistance caused by the more compressible plastic shard and rubber crumb aggregates. Typically, when a compressive load is applied to a material, the particles within the material distribute the load uniformly throughout the material to minimise the stress-stain on the individual particles. As a result of the lower stiffness of the plastic and rubber aggregates, these particles deform in the direction of the applied load resulting in the surrounding natural aggregate and binder matrix experiencing higher stress-strain loadings. This condition causes a decrease in compressive resistance of the material and hence a lower recorder strength.

A substantial decrease in compressive strength is observed when comparing the PU and FU mixes with VE and PE, due to the rigid 3D structure formed after polymerization of both the VE and PE polymers. The addition of the bitumen emulsion to the Polyurethane leads to an increase in flexibility that resulted in a decrease in rigidity and therefore a reduced compressive strength. It is noteworthy that all the PU resin mix designs, although failing under the compressive load in accordance with standard SANS 5863: 2006, only showed signs of deformation whereas the VEN and PEN specimens failed completely in an hourglass shape.

3.3 Flexural Strength

The flexural strength of a pothole repair material, specifically in terms of a flexible asphalt pavement, is a crucial element as the material needs to withstand cyclic vehicular induced loading without developing flexural failure. An increase in the flexural strength of the repair material will allow the repaired area to withstand loading more effectively, especially if deterioration occurs in the base layer. Figure 2 represent the results obtained of the flexural strength tests performed on the various mix designs.

The CMA mix exhibited an average of 0.81MPa with the singular highest value reached being 1.2 MPa. All PC mix designs outperformed the CMA test group significantly with the VE, PE, and FU based mixes exhibiting increased flexural strength results of up to 4950% than that of the CMA mix.

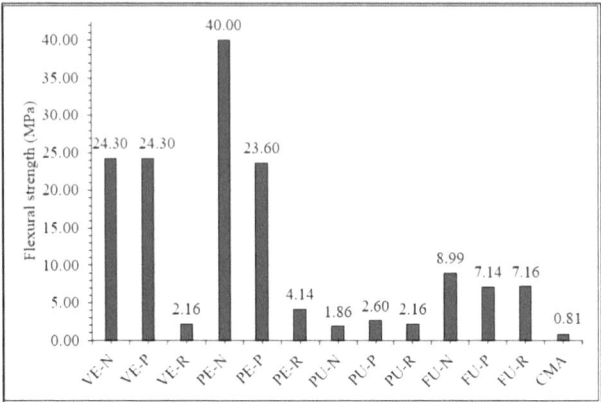

Fig. 2: Flexural strength of various mix designs.

Although the PU specimens recorded substantially lower results compared to other binder, the PU specimens did not break when failing under the applied load but deformed significantly with the largest deformation of 19 mm recorded at the centre of the test beam. In addition, the PU specimens, were re-tested after being rotated 90 degrees. All PU specimens achieved a minimum of four rotations before small surface cracks were observed. When considering the different aggregate mix designs, a decrease in flexural strength is observed for the R mixes compared to that of the P and N mixes, as shown in Fig. 2.

It is notable that specimens containing the plastic aggregate, did not fail by breaking into two sections as the elongated plastic shards served as tensile reinforcement within the aggregate matrix. When studying Fig. 2, the PE and VE mixes outperformed the other binder mixes. The PEN maximum flexural strength could not be measured as the specimen did not fail at the testing machine's maximum load capacity of 40 MPa. The FU specimens outperformed the CMA baseline mix, and the FUR mix exhibited the largest flexural strength when compared with all the other binder mixes containing rubber as aggregate. In addition, it is noteworthy that the FUR mix exhibited the most uniform flexural strength independent of the aggregate type used in the mix.

3.4 Abrasion Resistance

The abrasion resistance of the various PC mix designs were tested to compare the wearability and possible lifespan of the various mix designs. Material used for road surfacing requires an acceptable level of abrasion resistance caused by regular traffic interaction. Abrasive force is applied when the tire surface exerts forces (horizontal, vertical, acceleration and de-acceleration) upon the road surface during vehicular motion. On roads with inclines or declines, the need for increased traction also increases the abrasive force exerted onto the road. Figure 3 presents the results of the comparative abrasion resistance.

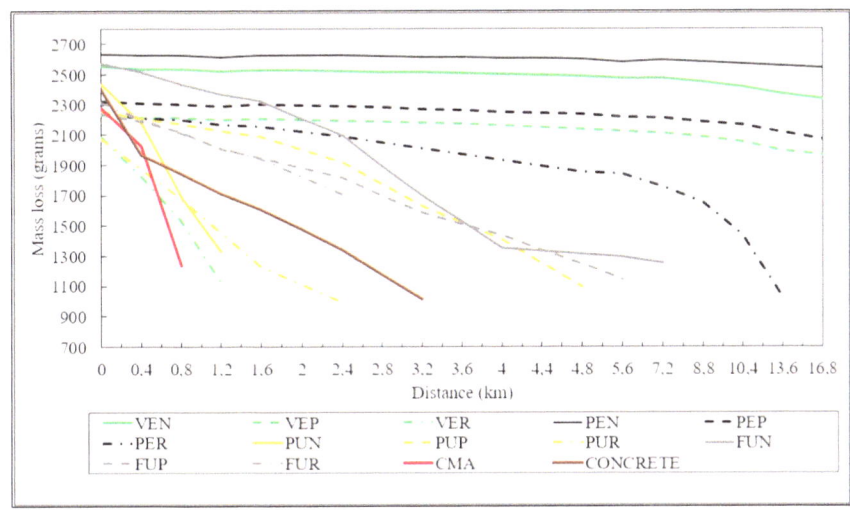

Fig. 3: Abrasion resistance illustrated by mass loss vs distance dragged over asphalt surface.

When observing Fig. 3, the difference in abrasion resistance between the various PC mix designs are significant. The mass loss of each specimen was recorded until the 100 mm cube specimens were abraded down to a height of 40 mm.

When observing Fig. 3, the CMA specimen exhibited the lowest abrasion resistance and failed at a dragging distance of 0.8 km at which point the specimens weight decreased by 45.7%. All PC mixes outperformed the traditional CMA baseline mix and thus exhibited higher abrasion resistance. After the specimens were dragged for 4.8 km, all of the specimens excluding the PEN, VEN, PEP and VEP specimens, reached the height of 40 mm and were thus considered failed. When considering the rate of mass loss of the PEN, VEN, PEP, and VEP specimens during the 16.8 km dragging exposure, it is estimated that in order for the respective PC mix designs to reach a similar 45.7% decrease in mass loss, as compared with the CMA mix which reached this value after 0.8 km of dragging, a dragging distance of 224, 93, 70, and 64 km, respectively, needed to be applied. Therefore, an increase of 280, 116, 87 and 80 times the dragging distance when comparing the CMA mix with those of the PEN, VEN, PEP, and VEP mixes respectively. In addition to these findings, no cracks were visible on the exposed surfaces of the PC mix specimens after the extended abrasion exposure.

It was observed that the VER and PER specimens underperformed when compared with their plastic and normal aggregate alternatives. When physically handling the PER specimens, an increase in specimen temperature was observed when compared to the PEN and PEP specimens which were all simultaneously tested. At the 13.6 km dragging interval, the PER specimen had a recorded temperature of 37.4 °C on the exposed surface, whereas 32.4 °C and 33.1 °C were recorded for PEN and PEP respectively. The increase in surface temperature caused by friction forces, softened the rubber aggregates resulting in decreased adhesion between aggregates and the binder in the PER specimen.

Similarly, to the PE and VE specimens, FUN and FUP specimens outperformed their rubber aggregate counterpart. Whilst FUR failed after 2.4 km travelled, FUP and FUN failed after 5.6 and 7.2 km, respectively. The PU mix designs exhibited increased mass loss rates compared to the other specimens and, when examining the exposed surface, was evident that the PU specimens exhibited lower surface rigidity than that of other polymer binder mixes. The exposed surfaces of all PU mixes exhibited clear signs of small material particles being ripped from the cube. All other polymer mixes illustrated a smooth and intact surface during testing.

In terms of abrasion resistance, PEN, VEN, PEP, and VEP outperformed all other mix designs including CMA by significant margins, followed by PER, VER, FUN, FUP, and PUP, respectively.

4 Conclusion

The results obtained during this investigation indicated that polymer concretes manu-factured using specific polymeric binders and specific aggregates, can be used as a rapid pothole repair material. The compressive and flexural strength, in addition to the abrasion resistance of these selected polymer concretes, exceeded that of the traditionally used CMA material. Given the mechanical strength tests conducted, the PEN, VEN, PEP, and VEP mix designs outperformed the alternative mix designs. The use of the polyurethane with a bitumen emulsion (PU mixes) as a pothole repair material exhibited a lower durability compared to alternative polymers when exposed to abrasion forces. Since both the traditional CMA mixes and the proposed PC mixes require similar preparation, placement, and compaction methods, only an increased initial material cost is applicable when using the polymer concrete mixes. Using the known typical CMA lifespan and strength characteristics as basis for comparison, the increased durability and lifespan of the PEN, VEN, PEP, and VEP polymer concrete mixes tested, it can be considered that these mixes incur a lower Equivalent Uniform Annual Cost (EUAC) as compared to the CMA material resulting in a more cost-effective pothole repair material. As per Van Zyl et al. (2021) it was shown that the PEN, VEN, PEP, and VEP, may have an EUAC of as low as 8% that of the CMA material [6]. In addition to the cost effectiveness of the repair material, the use of polymer concrete with increased durability will allow road construction workers to repair potholes less frequently before re-layering or resurfacing of the road is required compared to using the traditional CMA material.

References

1. Somasundaram, J., et al.: Pothole formation and occurrence in black vertisols of central and western India. Agric. Res. **3**(1), 87–91 (2014)
2. Siew, E.F., Ireland-Hay, T., Stephens, G.T., Chen, J.J., Taylor, M.P.: A study of the fundamentals of pothole formation. Light Metals **2005**, 763–769 (2005)
3. Treasury.gov.za: Estimates of National Expenditure 2019 [online] (2019). Avail-able at: http://www.treasury.gov.za/documents/national%20budget/2019/enebooklets/Vote%2035%20Transport.pdf. Accessed 23 February 2021
4. Kruger D.: Recent developments in the use of polymer concrete. Mater. Soc. 9(3) (1985)
5. Jung, K.C., Roh, I.T., Chang, S.H.: Evaluation of mechanical properties of polymer concretes for the rapid repair of runways. Compos. B Eng. **58**, 352–360 (2014)

6. Van Zyl, F.W., Kruger, D.: The use of polymer concrete as a cost-effective and durable alternative for rapid pothole repair in asphalt surfaces. Unpublished M.Eng thesis. Department Of Civil Engineering Science, University of Johannesburg. South Africa (2021)

Reinforcement and Strengthening

Fatigue Behaviour of Patch-Repaired and CFRP Strengthened Reinforced Concrete Beams

Valontino James and Pilate Moyo[✉]

Department of Civil Engineering, University of Cape Town, Cape Town, South Africa
pilate.moyo@uct.ac.za

Abstract. The service life of corrosion-damaged reinforced concrete (RC) infrastructure can be improved through patch repairs and structural strengthening. However, the effect of varying corrosion damage and patch repair extent has not yet been clearly pronounced. The fatigue performance of corrosion-damaged RC beams that have been patch repaired to varying lengths and subsequently strengthened with carbon fibre-reinforced polymer (CFRP) laminates was evaluated by conducting four-point bending tests and cyclic load tests on simply supported beams. Three criteria were identified to evaluate performance: fatigue life, crack development and stiffness degradation. Various data acquisition techniques, such as neutral axis DEMEC strain targets, strain gauges, linear variable differential transducers (LVDT) and digital image correlation (DIC) were employed to investigate these performance criteria. The experimental results indicated that an increase in corrosion damage and patch repair extent lowered the ultimate static failure load and increased fatigue life. An increase in specimen stiffness was observed for the specimens with the longer damage extent compared to the specimens with the shorter damage extent, where stiffness was gauged in terms of midspan deflection, composite material strain and neutral axis shift. Moreover, the results yielded through the DIC process showed potential to identify potential failure locations, quite early in the specimen fatigue life by comparison of tangential strain, peak vertical deflection and the eventual failure location.

Keywords: Reinforced concrete · fatigue behaviour · carbon fibre reinforced polymer strengthening · patch repair · digital image correlation

1 Introduction

Reinforced concrete (RC) infrastructure deterioration is a growing global concern, whether premature failure is caused by higher operating loads that influence structural capacity or due to substandard durability design. For example, in the South African heavy-haul railway industry, there has been a drive to use longer trains which would invariably lead to higher fatigue loads [1]. In addition, reinforced concrete structures in chloride and carbon-dioxide-rich environments are susceptible to reinforcement corrosion.

© The Author(s) 2025
L. Czarnecki et al. (Eds.): ICPIC 2023, 61, pp. 387–398, 2025.
https://doi.org/10.1007/978-3-031-72955-3_39

A common approach to repair corrosion-damaged RC structures involves the removal of damaged concrete and corrosion reaction products, followed by applying a cementitious repair mortar. Where additional capacity is required, structural strengthening may be considered. Patch repairs restore the durability of concrete elements, and their success relies on compatibility with the concrete substrate. Fibre-reinforced polymer (FRP) strengthening has become a favourable structural strengthening method, given its high strength-to-weight ratio and ease of application [2]. Moreover, FRP has been proven effective in not only restoring the structural performance of corrosion-damaged structures in terms of immediate ultimate limit state (ULS) capacity as well as long-term serviceability limit state (SLS) fatigue performance [3].

While the fatigue behaviour of FRP-strengthened RC structures has been extensively reported in the literature [4–6], there is a dearth of information on the fatigue behaviour of corrosion-damaged, patch-repaired and FRP-strengthened RC elements with varying degrees of damage [5]. Considered the effect of both externally bonded reinforcement (EBR) and near-surface mounted (NSM) fibre-reinforced strengthening on quasi-full-scale RC beams under fatigue loading. The study found that although EBR specimens performed better than NSM-reinforced specimens under cyclic loading, both FRP strengthening methods improved the fatigue performance of RC beams [6]. Considered the long-term behaviour of FRP-strengthened RC beams that have been corrosion damaged and found that an increase reduced fatigue life in the degree of corrosion. However, neither of the studies above considered the effect of the patch repair component [7]. Considered this combined effect for the same damage extent [8, 9]. Considered the effect of varying damage and patch repair extent but only reviewed its performance under monotonic loading [10]. Studied the behaviour of patch-repaired beams with varying damage lengths but under impact loading. This research was focused on the fatigue behaviour corrosion damaged RC beams that have been patch repaired to varying lengths and subsequently CFRP strengthened under two different cyclic load stress ranges.

2   Specimen Details and Material Properties

The experimental programme comprised fifteen (15) quasi-full-scale $155 \times 254 \times 2000$ mm RC beams with tensile reinforcement that either remained uncorroded or was subjected to accelerated corrosion and subsequently patch repaired. The dimensions of the beam were selected to conform to a series of full-scale tests previously conducted with the same size [9–11] as shown in Fig. 1.

The concrete used in this experiment was designed yield a compressive strength of 40 MPa. The mechanical properties of the concrete are presented in Table 1.

Reinforcement corrosion was induced electrochemically in a controlled laboratory environment. The electrochemical cell comprised of the tensile steel reinforcement, which served as anode and a 12 mm stainless-steel rod, submerged in the sodium chloride solution, which served as the cathode. The corrosion extent was varied to obtain damage lengths of 450 mm, 800 mm, 1300 mm and 1800 mm; however, over each damage extent, a uniform 10% degree of corrosion was maintained, which equates to 5% corrosion per tensile reinforcing bar. A corrosion pond containing 5% sodium chloride solution was assembled on the beam tensile face, as shown in Fig. 2(b). The beams were connected in

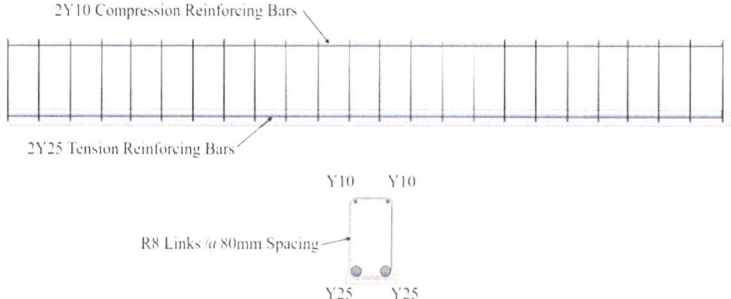

Fig. 1. Beam specimen reinforcement layout

Table 1. Concrete 40 MPa mix material properties

	28 days	Standard Deviation
Compressive Strength (MPa)	43.5	±1.1
Tensile Strength (MPa)	2.7	±0.2
Modulus of Elasticity (GPa)	39.6	±1.2

series to a DC power supply to achieve uniform corrosion damage. The time required to induce 10% corrosion in the RC beams was calculated using Faraday's law. For a constant corrosion current of 1 ampere (A), the desired degree of corrosion can be obtained in approximately 30.8 days.

Fig. 2. Sustained load beams setup to induce a partial cracking moment

All 12 beams were corroded under sustained loading through an inverse 4-point bending system where the loads applied at the beam ends, inducing a bending moment equivalent to 60% of the beam cracking moment. Pinned supports were created at the loading points to allow the beams to deflect during corrosion.

The patch repair process involved the removal of damaged cover concrete in varying lengths of either 450 mm, 800 mm, 1300 mm or 1800 mm to a depth of at least 20 mm below the corroded reinforcement and 50 mm beyond the damage extent. This research used a locally sourced cementitious grout as the patch repair mortar. The mechanical properties of the patch repair mortar are presented in Table 2.

Table 2. Cementitious mortar material properties

	28 days	Standard Deviation
Compressive Strength (MPa)	79.2	±1.2
Tensile Strength (MPa)	2.8	±0.4
Modulus of Elasticity (GPa)	35.6	±2.6
Tensile bond strength (MPa)	2.5	±0.5

The FRP strengthening for the RC beam was designed for flexure following recommendations in [2]. The RC beams were designed to resist a ULS capacity of 62.3 kNm; this capacity was reduced 10% by accelerated corrosion. The difference in performance capacity was used to calculate the FRP strengthening required. The induced capacity reduction of 6.22 kNm required a $9.71mm^2$ of FRP reinforcing. Two CFRP laminates, each with a cross-sectional area of $60mm^2$ were used. Locally sourced epoxy adhesive Sikadur 30 was used to bond CFRP strips and the patch repair mortar or concrete surface once the substrate had reached a minimum tensile strength of 1.5 MPa. The mechanical properties of the various FRP and epoxy materials used in this experiment are shown in Table 3.

Table 3. Sika CFRP structural strengthening material properties

	FRP Laminate	FRP Wrap	Epoxy resin for repair mortar	Epoxy resin for FRP laminate	Epoxy resin for FRP wrap
Compressive Strength (MPa)	-	-	60–70	70–80	-
Tensile Strength (MPa)	3100	4900	18–20	24–27	30
Modulus of Elasticity (GPa)	165	230	-	11.2	3.8
Thickness (mm)	1.2	0.127	-	2	-

Each test specimen in this study was assigned a label to identify and track it during experimental testing. These labels are listed under the identity column in Table 4.

Table 4. Test specimen notation and details

Identity	Corrosion level (%)	Patch Repair Length	FRP strengthening	Test Regime
S_CNTRL 1	0	No patch repair	Strengthened	Monotonic testing
S_CNTRL 2	0	No patch repair	Strengthened	Fatigue 40% stress range
S_CNTRL 3	0	No patch repair	Strengthened	Fatigue 60% stress range
S_450 mm 1	10	450mm	Strengthened	Monotonic testing
S_450 mm 2	10	450mm	Strengthened	Fatigue 40% stress range
S_450 mm 3	10	450mm	Strengthened	Fatigue 60% stress range
S_800 mm 1	10	800mm	Strengthened	Monotonic testing
S_800 mm 2	10	800mm	Strengthened	Fatigue 40% stress range
S_800 mm 3	10	800mm	Strengthened	Fatigue 60% stress range
S_1300 mm 1	10	1300mm	Strengthened	Monotonic testing
S_1300 mm 2	10% corrosion	1300mm	Strengthened	Fatigue 40% stress range
S_1300 mm 3	10% corrosion	1300mm	Strengthened	Fatigue 60% stress range
S_1800 mm 1	10% corrosion	1800mm	Strengthened	Monotonic testing
S_1800 mm 2	10% corrosion	1800mm	Strengthened	Fatigue 40% stress range
S_1800 mm 3	10% corrosion	1800mm	Strengthened	Fatigue 60% stress range

3   Instrumentation

Each test specimen was instrumented with four strain gauges. One strain gauge was placed on each of the following surfaces: the compression surface, tension steel surface, tension concrete (or patch repair) surface and the CFRP laminate surface. DEMEC strain targets were placed on one side of each specimen to track the neutral axis migration. An extensometer was used to measure the relative movement of the DEMEC targets.

Crack behaviour was monitored using two different techniques. The first method entailed visual monitoring, where crack patterns were tracked using a permanent marker at pre-determined load intervals. The second method to used monitor crack development was DIC. This method involved spraying and painting a stochastic matt black pattern on the test surface. A high-resolution 5-megapixel (MP) monochrome Basler digital camera captured the test surface from 2m away from the calibrated test specimen. The images were then post-processed using Dantec software specimen.

4   Test Setup and Procedure

Monotonic and fatigue tests were conducted under a four-point bending simply supported configuration. One specimen from each patch repair extent was tested under monotonic loading to establish and verify ULS failure loads. The remaining two specimens of each damage extent were subjected to cyclic loading under different stress ranges, where a minimum load of 6kN was chosen to avoid impact loads and the maximum loads of either 40% or 60% of the beam ULS capacity.

Two-point loads 450mm apart were applied on the compression surface using an Instron actuator through a spreader beam. The sinusoidal loads were applied at a frequency of 4Hz which is relatively high, but an acceptable test frequency for fatigue testing of RC concrete [2, 5, 7, 12–14].

5   Experimental Results

5.1   Accelerated Corrosion

Subsequent to fatigue testing tension steel was retrieved to examine the extent of corrosion damage in terms of mass loss, type of corrosion as well as the locality of the corrosion damage. Table 5 presents findings from the post-fatigue assessment of corrosion damaged tension steel.

5.2   Monotonic Behaviour

The static test results indicated that the specimens with the shortest damage length outperformed those with longer damage lengths. The 450 mm specimen had an 18.6% higher failure load than the 0 mm (control) specimen, whereas 1800 mm specimen only had a 5.8% higher failure load than the control specimen. The results further show that the ULS capacity of the 450mm damage length specimen was 8.3%, 10.9% and 12.1% higher than the 800 mm, 1300 mm and 1800 mm specimens, respectively.

Table 5. Accelerated corrosion results

Identity	Corrosion Type	Measured Mass Loss (g/m)	Equivalent Uniform Depth (mm/bar)	Percentage Mass Loss (%)	Standard Deviation (%)
S_450 mm 1	*Pitting*	189.96	0.31	9.87	±0.30
S_800 mm 1	*Pitting*	136.85	0.22	7.11	±1.46
S_1300 mm 1	*Pitting*	136.97	0.22	7.11	±0.95
S_1800 mm 1	*Pitting*	109.33	0.18	5,68	±0.26

5.3 Fatigue Life

Table 6 presents the relative fatigue life cycles of specimens tested under 40% and 60% stress ranges.

Figure 7 shows the predicted fatigue life cycles based on the Helgason and Hanson model.

Overall, the experimental results suggest that as the stress range is reduced by 20% from medium cycle fatigue stress (60%) to low cycles fatigue stress (40%), the fatigue life can be increased by 5 to 8 times. Under both stress range conditions, as the damage extent was increased from 450mm to 1800mm, the fatigue life was extended by as much as 76.7%.

5.4 Crack Development and Failure Mode

All specimens tested under fatigue loading exhibited a similar crack propagation and failure mode, as summarized in Table 7. A few unique stages can characterize this process:

1. A crack pattern with predominantly flexural cracks and shear and flexural-shear cracks was clearly defined during the first load cycle. This crack pattern remained relatively unchanged throughout the test.
2. Crack propagation remained low until the rupture of the tension steel. The rupture of tension steel caused a rapid increase in crack propagation in terms of crack heights and densities.
3. After the rupture of steel, FRP laminates started to delaminate at the position of the main cracks. Shortly after FRP debonding, compression concrete would crush, leading to the ultimate failure of the section.

Figure 3 shows the crack pattern sketched from an actual specimen at ultimate failure as well as a DIC crack pattern of the specimen at maximum load during the first load cycle. The comparison of these two different images was done intentionally for two reasons. The first reason being that it was found overall that the crack patterns did not change significantly after the first load cycle. The second reason was to evaluate the possibility of using DIC to identify possible failure locations early in the structural

Table 6. Monotonic and fatigue loading test results

Identity	Load Range (kN)	Total No. Cycles	Ultimate Static Load (kN)	Fatigue Failure Mode	Static Failure Mode	Static Load after Fatigue Testing (kN)
S_CNTRL 1	Static	-	274	-	CC & FD	-
S_CNTRL 2	6–109.6	1000000	-	No Failure		276
S_CNTRL 3	6–164.4	256000	-	SR, FD & CC	-	-
S_450 mm 1	Static	-	325	-	CC & FD	-
S_450 mm 2	6–130	788303	-	SR (2 bars), FD & CC		-
S_450 mm 3	6–195	119716	-	SR (1 bar), FD & CC		-
S_800 mm 1	Static	-	300	-	CC & FD	-
S_800 mm 2	6–120	1150000	-	No Failure		300
S_800 mm 3	6–180	102750	-	SR, FD & CC		-
S_1300 mm 1	Static	-	293	-	CC & FD	-
S_1300 mm 2	6–117.2	1083935	-	SR, FD & CC		-
S_1300 mm 3	6–175.8	201450	-	SR, FD & CC		-
S_1800 mm 1	Static	-	290	-	CC	-
S_1800 mm 2	6–116	2000000	-	No Fatigue, SR		216
S_1800 mm 3	6–174	247000	-	SR, FD & CC		-

[*]IC = inconclusive results, CC = concrete crushing, FD = carbon fibre debonding, SR = steel rupture

service life. The DIC crack patterns were obtained by plotting tangential strain in the x-direction, where positive strain concentrations indicate areas where cracks were likely to form, and conversely negative strain concentrations indicate areas where concrete crushing was likely.

The DIC crack pattern correlates with the actual specimen's crack pattern, as shown in Fig. 8 above. It does not show the shear cracks quite as accurately as the flexural

Table 7. Crack behaviour and failure mode results

Identity	Average Crack Spacing (mm)	Average Crack Height (mm)	Predominant Crack Type	Fatigue Failure Mode	Failure Position
S_CNTRL 2	58.06	63.81	flexural	No Fatigue Failure	Centre
S_CNTRL 3	85.71	74.20	flexural	SR, FD & CC	Right Pin Load
S_450 mm 2	69.23	59.14	flexural	SR,FD & CC	Centre
S_450 mm 3	75.00	65.70	flexural	SR,FD & CC	Left Pin Load
S_800 mm 2	48.65	54.22	flexural	No Fatigue Failure	Centre
S_800 mm 3	75.00	59.81	flexural	SR (1 bar), FD & CC	Right Pin Load
S_1300 mm 2	46.15	63.66	flexural	SR (2 bars), FD & CC	Centre
S_1300 mm 3	69.23	52.61	flexural	SR (2 bars), FD & CC	Centre
S_1800 mm 2	45.00	75.08	flexural	No Fatigue, SR (1 bar)	Centre
S_1800 mm 3	66.67	47.72	flexural	SR (2 bars), FD & CC	Right Pin Load

[*]IC = inconclusive results, CC = concrete crushing, FD = carbon fibre debonding, SR = steel rupture

cracks. This may be due to the fact that the correlation algorithm considered only relative movement in the x-direction when it computed tangential x-strain.

The average crack spacing of the specimens tested under the 40% and 60% stress ranges followed similar patterns. For the 40% stress range specimens, as the damage extent increased from 450 mm to 1800 mm there was a 53.9% reduction in the average crack spacing. In comparison to the 60% stress range test specimens the 40% stress range test specimens had a 47,6% lower average crack spacing as well as a lower average crack spacing reduction as the damage extent was increased from 0 mm to 1800 mm.

If one considers the overall performance of the 40% stress range specimen, the reduction of average crack spacing indicates an increase in the total number of cracks as the damage extent was increased. The increase in average crack height with the damage extent indicates that there may be a stiffness reduction as the damage extent is increased.

5.5 LVDT vs DIC Deflections

A summary of the maximum deflection results obtained from LVDT measurements and theoretical design calculations is presented in Table 8. The results are presented

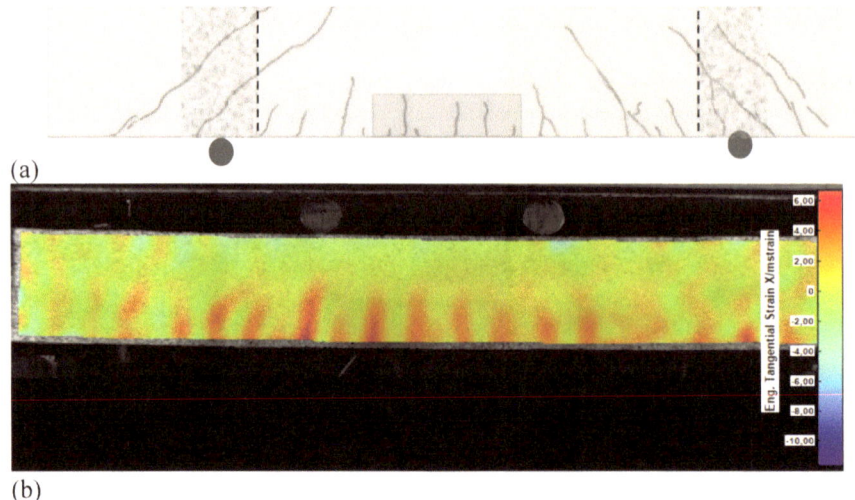

(a)

(b)

Fig. 3. 450 mm specimen: Final actual crack pattern versus DIC crack pattern at 1 load cycle (60% ULS load)

for midspan span deflection measurements after 100 000th load cycle. There is a good correlation between theoretical and measured deflections.

Table 8. Summary of midspan deflection results from LVDT measurements and theoretical calculations

Identity	Maximum Test Load P_{max} (kN)	Calculated Deflection (mm)	LVDT Measurement 100 000th load cycle (mm)
S_CNTRL 2	109.6	2.57	2.77
S_CNTRL 3	164.4	3.86	2.46
S_450 mm 2	130	6.56	2.74
S_450 mm 3	195	9.84	2.26
S_800 mm 2	120	6.06	2.58
S_800 mm 3	180	9.09	7.15
S_1300 mm 2	117.2	5.92	3.11
S_1300 mm 3	175.8	8.87	2.42
S_1800 mm 2	116	5.85	3.44
S_1800 mm 3	174	8.78	8.09

6 Concluding Remarks

This paper presented an experimental study of the fatigue behaviour of corrosion-damaged RC beams that were patch-repaired and CFRP-strengthened. Fatigue performance was assessed in terms of fatigue life, crack development, failure mode and stiffness degradation. The experimental results indicated that under low cyclic stress, the fatigue life of rehabilitated beams was up to eight times higher than the high-stress range. Under both stress range conditions, as the damage extent was increased from 450 mm to 1800 mm, the fatigue life increased by as much as 76.7%. As the damage extent was increased from 450 mm to 1800 mm, average crack spacing and average crack height were reduced, culminating in an overall stiffer section. Moreover, crack densities were found to increase under the lower stress range as those specimens experienced a longer fatigue life. Crack densities tended to increase at the location of steel rupture. The location of steel rupture often coincided with the points of maximum.

References

1. Busatta, F., Moyo, P.: How testing and monitoring can support heavy haul railway bridge management: the experience gained in South Africa. In Fröhling, R.D., Gräbe, P.J. (eds.) 11th International Heavy Haul Association Conference 2017, Cape Town (2017)
2. Täljsten, B.: FRP strengthening of existing concrete structures: design guidelines, 4th edn. Division of Structural Engineering, Lulea University of Technology, Sweden, Lulea (2006)
3. Dong, J.F., Wang, Q.Y., Guan, Z.W.: Structural behaviour of RC beams externally strengthened with FRP sheets under fatigue and monotonic loading. Eng. Struct. **41**, 24–33 (2012)
4. Aidoo, J., Harries, K. A., Petrou, M. F. (2004). Fatigue behavior of carbon fiber reinforced polymer-strengthened reinforced concrete bridge girders. ASCE J. Compos. Constr. **8**, 501–509 (2004)
5. Mahal, M., Blanksvärd, T., Täljsten, B., Sas, G.: Using digital image correlation to evaluate fatigue behavior of strengthened reinforced concrete beams. Eng. Struct. **105**, 277–288 (2015)
6. Song, L., Yu, Z.: Fatigue performance of corroded reinforced concrete beams strengthened with CFRP sheets. Constr. Build. Mater. **90**, 99–109 (2015)
7. Gregan, S.: The fatigue performance assessment of corrosion-damaged RC beams, patch repaired and externally strengthened using CFRP. University of Cape Town (2012)
8. Dladla, T.: The behaviour of patch repaired and FRP strengthened RC beams. University of Cape Town (2014)
9. Mundeli, S.: Behavior of RC beams patch repaired and strengthened with FRP composites: a numerical study (2014)
10. Habimana, P.: Behaviour of FRP strengthened RC beams with patch repairs subjected to impact loading. University of Cape Town (2017)
11. Tigeli, M.: Effect of structural repair and strengthening on stiffness and ultimate capacity of corrosion-damaged RC beams. University of Cape Town (2014)
12. Charalambidi, B.G., Rousakis, T.C., Karabinis, A.I.: Analysis of the fatigue behavior of reinforced concrete beams strengthened in flexure with fiber-reinforced polymer laminates. Compos. B Eng. **96**, 69–78 (2016)

Experimental and Numerical Investigation of Patch-Repaired and CFRP-Strengthened Beams

Pilate Moyo[1]([⊠]) and Salathiel Mundeli[2]

[1] Department of Civil Engineering, University of Cape Town, P Bag X3, Cape Town, South Africa
pilate.moyo@uct.ac.za
[2] Department of Civil, Environment and Geomatics Engineering, College of Science and Technology, University of Rwanda, Kigali, Rwanda

Abstract. Carbon fibre reinforced polymers (CFRP) have emerged as an effective material for strengthening reinforced concrete structures. While many studies have been reported on CFRP strengthening of reinforced concrete elements subject to corrosion, there is a dearth of information on strengthening patch-repaired concrete elements. This paper reports on the experimental and numerical investigation of the behaviour of corrosion-damaged RC beams that have been patch repaired and strengthened with CFRP. Fifteen beams were cast; twelve of these were subjected to simulated corrosion of 5%, and the remaining three were used as control beams. The damage length was varied from 450 mm, 800 mm, 1300 mm and 1800 mm while keeping the depth of patch repair at 105 mm. All the CFRP-strengthened beams failed by intermidiate crack (IC) debonding. The control beams had a lower average crack density than the retrofitted beams. There was an increase in the average crack density for increasing damage lengths. Patch repair combined with CFRP strengthening retrofit method restored the damaged beams load carrying capacity but reduced the beams' ductility compared to the control beams. The yield loads for patch repaired and strengthened (RS) results increased by 10%, 17% and 20% for beams with 450 mm, 800 mm and 1800 mm damage lengths, respectively. It was also observed that for increasing damage lengths; 450 mm, 800 mm and 1800 mm the peak loads increased by 13%, 20%, 20%, respectively. The experimental results correlated well with the finite element modelling (FEM) results.

Keywords: CFRP strengthening · CFRP multi-layer modelling · Patch-repairs

1 Introduction

Corrosion of steel reinforcement is the most common cause of the deterioration of reinforced concrete elements. It leads to the reduction of the area of steel, cracking of concrete and loss of structural capacity. The typical maintenance approach for corroded reinforced concrete elements is to; 1) remove the concrete in the corrosion-affected

L. Czarnecki et al. (Eds.): ICPIC 2023, 61, pp. 399–406, 2025.
https://doi.org/10.1007/978-3-031-72955-3_40

regions, 2) clean and protect the corroded steel, 3) repair the concrete element using an appropriately designed concrete mix or cementitious grout, and 4) strengthen the element to restore the lost structural capacity. Reinforced fibre polymer plates (FRP) have emerged as a viable solution for strengthening reinforced concrete elements due to their inherent advantages such as corrosion resistance, lightweight and high strength. The behaviour of concrete elements that have been repaired and strengthened using FRP plates is complex due to their multiple-layer nature, including the old concrete layer, the new concrete layer, the epoxy layer and the CFRP layer. A good understanding of the interaction between these components is essential to understand the system's failure modes and for the development of design procedures for such systems. This study focuses on the flexural behaviour of corrosion-damaged reinforced concrete beams repaired using concrete and strengthened using carbon fibre polymer (CFRP) plates.

Malumbela et al. [1] investigated the load-carrying capacity of patch-repaired and CFRP-strengthened RC beams. They concluded that combined patch repair and strengthening is effective for retrofitting corrosion-damaged beams. They achieved up to 50% increase in load-carrying capacity. Xie and Hu [2] achieved an ultimate load-carrying capacity increase as high as 93.8% for corrosion levels between 15% and 50% compared to the control pristine beam. The above studies focused on corrosion damage localised near the mid-span of the beams.

Numerical approaches by [3–6] have been proposed to model the behaviour of CFRP-strengthened beams. However, these existing models do not consider the effect of the patch repair material, which introduces an additional material law to the complex multi-layer system. Such complexity for concrete is due to the nonlinear load-deformation response of concrete and difficulty in forming suitable constitutive relationships under combined stresses, progressive cracking of concrete under increasing load and the complexity in the formulation of the failure behaviour for various stress states, the consideration of steel and its interaction with concrete and time-dependent effects such as creep and shrinkage of concrete [7].

2   Experimental Study

2.1   Experimental Program

Five types of beams were cast and replicated three times. The five categories of beams cast were control beams, set 1 beams (450 mm damage), set 2 (1800 mm) beams, set 3 (1300 mm) beams and set 4 (800 mm) beams. The set numbering has to do with the order in which the beams were cast. The damaged beams were subjected to simulated corrosion of 5%. Simulated corrosion was completed by means of milling the steel perpendicular to the cross section to the longitudinal tensile reinforcement. All the damaged beams were patch repaired with the same mortar mix and strengthened with the same CFRP plate with dimensions 1700 mm × 50 mm × 1.2 mm in flexure. Identical anchorage was also provided at the CFRP plate ends with FRP wrap. After sufficient curing, all beams were tested under four-point bending. The CFRP plate used was 50 mm wide and 1.2 mm thick. The FRP wrap was 300 mm wide.

2.2   Test Beams

All test beams had identical cross-section dimensions. The reinforcement layout was also identical except for the tensile longitudinal reinforcement milled in the maximum bending regions to simulate the various damage lengths. The beam section and reinforcements for a typical beam are shown in Fig. 1.

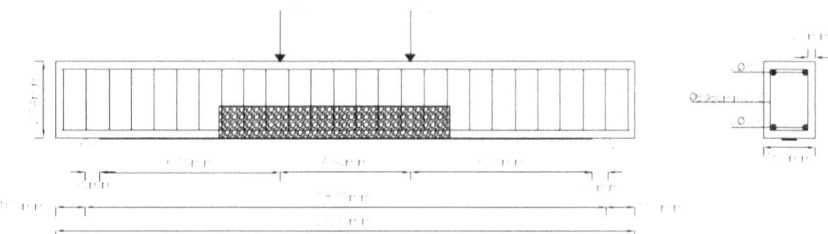

Fig. 1. Reinforcement layout, patch repair and CFRP plate for typical 800 mm patch

2.3   Material Characterisation

The materials used for this study included concrete, steel, epoxy, CFRP plate and wrap and their mechanical properties are summarised in Table 1.

Table 1. Mechanical properties of materials

Material	Compressive strength (MPa)	Tensile strength (MPa)	Modulus of Elasticity (GPa)
CFRP Plate	-	3100	165
CFRP Wrap	-	4900	230
Steel bar (tension and compression)	-	630	200
Steel bar (Stirrups)	-	300	200
Epoxy/FRP plate	70–80	24–27	11.2
Repair mortar	70	5.5	
Concrete	50		

3   Finite Element Modelling

The finite element analysis performed in this study consisted of modelling the nonlinear behaviour of the same reinforced concrete beams of the experimental studies, which are patch repaired and strengthened with CFRP bonded to their tension face to investigate the behaviour of such RC beams under four points bending. The commercial finite element package ABAQUS software was used.

3.1 Material Properties and Constitutive Models

Concrete

Concrete was modelled using the concrete damaged plasticity model in Abaqus to capture both inelastic deformation and stiffness degradation that concrete undergoes at low confining pressure. The material properties specified in Table 1 were used for modelling. The elastic parameters necessary for establishing the first part of the model were the secant modulus of elasticity E_{cm} and mean axial tensile strength, f_{ctm} and were calculated according to [8]. The post-peak behaviour in tension was represented with tension stiffening to simulate the effects of concrete/steel effects such as bond slip and dowel action. The nonlinear uniaxial compression stress-strain curve was constructed based on the expression proposed by [8]:

$$\frac{\sigma_c}{f_{cm}} = \frac{k\eta - \eta^2}{1 + (k - 2)} \tag{1}$$

where $\eta = \frac{\varepsilon_c}{\varepsilon_{c1}}$, $k = 1.05\frac{|\varepsilon_{c1}|}{f_{cm}}$ and $\varepsilon_{c1} = 0.7f_{cm}^{0.31} \leq 2.8‰$, f_{cm} is the mean value of concrete cylinder compressive strength derived from concrete cube strength and σ_c is the compressive stress in concrete.

Steel Reinforcement

Tension, compression and shear reinforcements were assumed to behave in an elastic perfectly plastic manner in both compression and tension.

Fibre Reinforced Polymers Material and Adhesive-Concrete/CFRP Interface

CFRP material behaves in a linear elastic manner up to failure. For flexural strengthening, the elastic modulus in fiber direction is of most importance. The properties presented in Table 1 were assigned to CFRP during modelling. The layer of adhesive between concrete and CFRP of 1 mm thick was modeled using cohesive zone model and the values suggested by [9] were assigned.

3.2 Numerical Analysis

In this study, different parts of the complete model were assembled in Abaqus/Standard in 3D modelling space. A typical model for 800 mm patch is shown in Fig. 2.

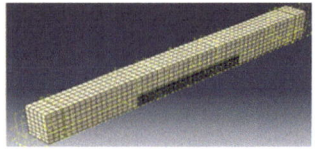

Fig. 2. Finite element model for patch-repaired and CFRP strengthened RC beam (800 mm-Patch)

4   Results and Discussions

As stated earlier, the main objective of this study was to investigate the behavior of reinforced concrete beams patch repaired and strengthened with FRP composites under static loading. Such behavior was studied in terms of crack initiation and propagation, load-deflection relationships, and failure mechanisms. A comparison with experimental results obtained from the same beams is done for validation.

4.1   Cracking Initiation and Evolution

Cracking was initiated whenever the maximum principal stress was greater than the tensile strength of concrete which was 3.5 MPa for concrete and 4.3 MPa for repair material (Figs. 3 and 4).

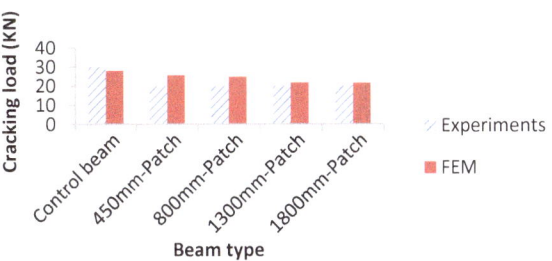

Fig. 3. Cracking loads

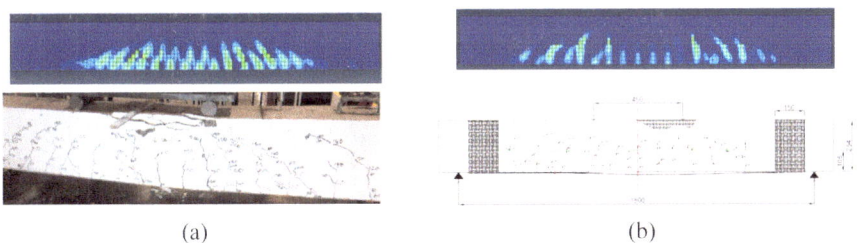

Fig. 4. Structural crack pattern: (a) Control beam, (b) 1800 mm patched beam

4.2   Load Deflection Relationships

Load deflection curves obtained from control beam and four patch repaired and strengthened beams are shown in Fig. 5 in comparison to experimental results. As it can be seen, there is a close agreement between numerical results and experimental findings and it is seen that patch repair and FRP strengthening increases load carrying capacity of damaged reinforced concrete beams.

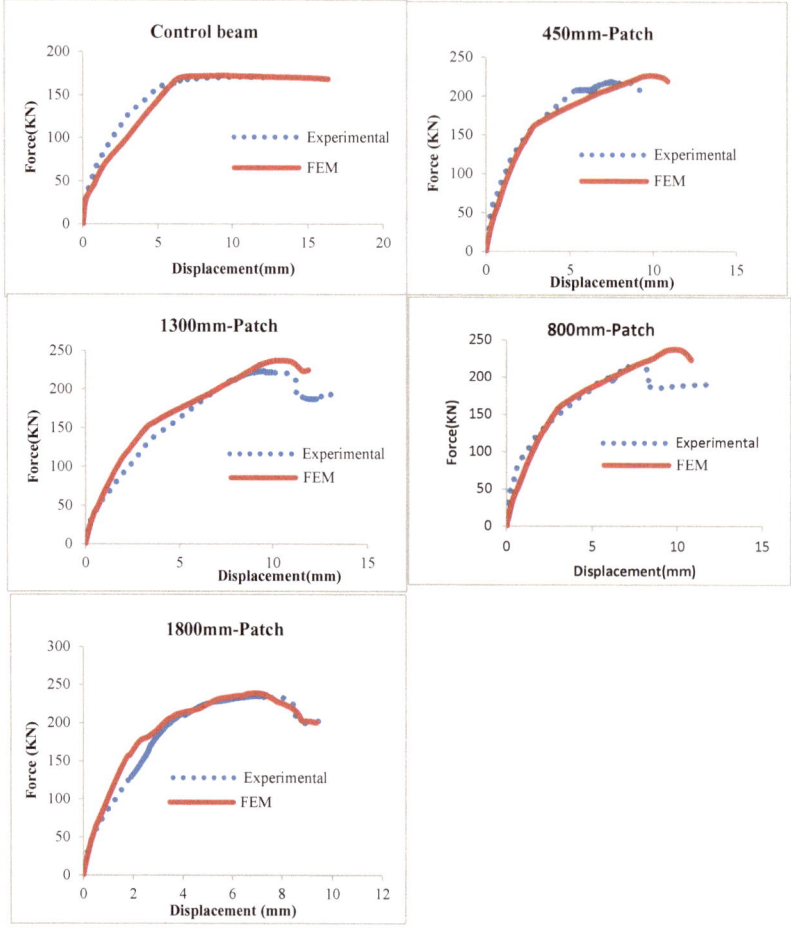

Fig. 5. Load-deflection relationships

4.3 Failure Mechanism

From both numerical and experimental studies, the failure mode was intermediate crack induced debonding followed by concrete crushing as shown in Fig. 6.

5 Conclusions

In this study, a finite element model was developed for analysis of reinforced concrete beams patch repaired and strengthened with FRP composites by varying the length of the patch. Results from finite elements analysis agreed with experimental findings regarding cracking load, structural crack distribution, load deflection relationships and failure mechanism. Repair and strengthening of corrosion-damaged reinforced concrete beams was found to increase the load-carrying capacity. Despite good results obtained from

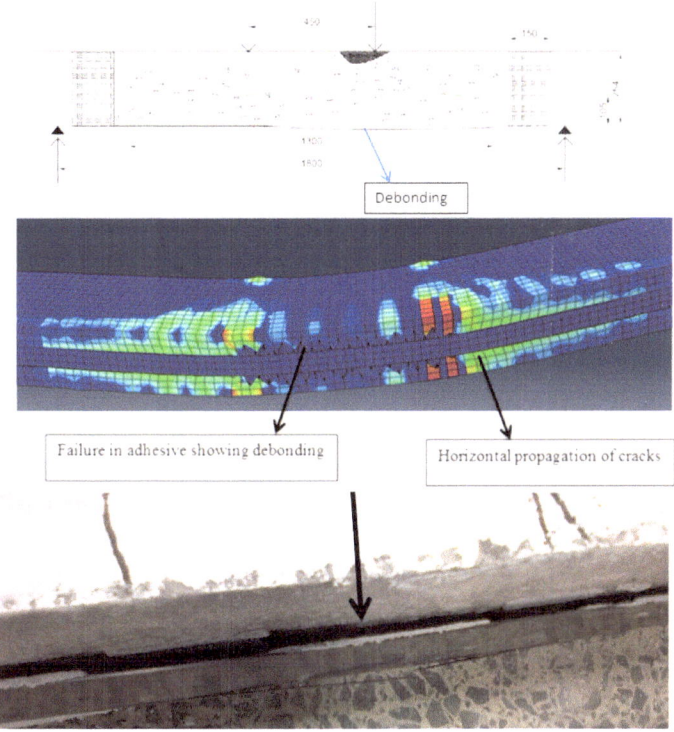

Fig. 6. Intermediate-crack induced debonding.

this study, future researches are necessary putting much effort in the energy approaches to study the behavior of patch-repaired and strengthened reinforced concrete beams, particularly the debonding failure.

References

1. Malumbela, G.: Measurable parameters for performance of corroded and repaired RC beams under load. PhD Thesis. University of Cape Town (2010)
2. Xie, J.-H., Hu, R.-L.: Experimental study on rehabilitation of corrosion-damaged reinforced concrete beams with carbon fiber reinforced polymer. Constr. Build. Mater. **38**, 708–716 (2013)
3. Camata, G., Spacone, E., Zarnic, R.: Experimental and nonlinear finite element studies of RC beams strengthened with FRP plates. J. Compos.: Part B **38**, 277–288 (2007)
4. Chen, G.M., Teng, J.G., Chen, J.F., Rosenboom, O.A.: Finite element model for intermediate crack debonding in RC beams strengthened with externally bonded FRP reinforcement. In: Proceedings of the Fourth International Conference on FRP Composites in Civil Engineering. Zurich, Switzerland (2008)
5. Supaviriyakit, T., Pornpongsaro, P., Pimanmas, A.: Finite element analysis of FRP-strengthened RC beams. Songklanakarin J. Sci. Technol. **26**(4), 497–507 (2004)
6. Kwak, H.-G., Filippou, F.C.: Finite element analysis of reinforced concrete structures under monotonic loads (Report N^0 UCB/SEMM-90/14). University of Calfornia: Structural Engineering, Mechanics and Materials (1990)

7. Simonelli, G.: Finite element analysis of RC beams retrofitted with fibre reinforced polymers. PhD Thesis. Università degli Studi di Napoli Federico II, London (2005)
8. Eurocode 2.: Design of concrete structures. Part 1-1: general rules and rules for buildings (2004)
9. Dassault Système Simulia Corp.: Abaqus/CAE User's Manual. Providence: Dassault Système. (2010)

Bond Characteristics of BFRP and GFRP Bars in Concrete with Additives—Results from a Beam Test Study

Marek Urbański[✉], Elżbieta Szmigiera, Grzegorz Adamczewski,
Piotr P. Woyciechowski, and Kostiantyn Protchenko

Warsaw University of Technology, Warsaw, Poland
marek.urbanski@pw.edu.pl

Abstract. This article presents a comparative analysis of the bond behavior of steel bars in concrete and bars made of basalt fiber-reinforced polymer (BFRP) and glass fiber-reinforced polymer (GFRP) in modified concrete. While steel bars have been the conventional choice for reinforcement in concrete structures, their bonding properties are well established. In contrast, FRP bars possess distinct mechanical and physical properties, which can lead to different bonding behavior in concrete. The study investigated the effects of concrete properties and bar characteristics on the bond behavior of GFRP and BFRP bars. Specifically, the study analyzed the relationships between bond stress-slip, modes and mechanisms of failure, and changes in bond strength of concrete with the addition of zeolite and metakaolin, with the presence of GFRP, BFRP, and steel bars. The findings of the study reveal that the adhesion of composite bars to modified concrete is enhanced to varying degrees. The bond stress of GFRP bars to concrete with metakaolin addition was found to be 50% higher than to normal concrete, while the bond stress to concrete with zeolite was similar. On the other hand, BFRP bars exhibited an increase in bond stress of 7% in the presence of concrete with metakaolin. Moreover, BFRP bars displayed a greater bond to steel reinforcement that underwent plasticization or rupture. The study also noted that the change in bond strength of GFRP and BFRP bars due to their linear deformability was gradual, characterized by a several times greater slip range compared to steel bars.

Keywords: BFRP · GFRP · Beam test · Bond behavior · Zeolite · Metakaolin · Slip

1 Introduction

The maintenance and repair of infrastructure remains one of the major challenges of civilization [1], with the global cost of infrastructure repair and maintenance estimated at over €100 billion [2]. Corrosion of the steel reinforcement of concrete structures as a result of carbonation of the concrete cover, the use of de-icing salts, as well as a combination of moisture, temperature and chlorides reducing the alkalinity of concrete are the most common causes of damage to concrete structures during their operation

© The Author(s) 2025
L. Czarnecki et al. (Eds.): ICPIC 2023, 61, pp. 407–421, 2025.
https://doi.org/10.1007/978-3-031-72955-3_41

[3], while road surfaces, bridges, viaducts, tunnels, and underground garages are the main problems [4]. The cost of road and bridge corrosion is estimated at US$276 billion annually (approximately 3.1% of GDP) for US industry and government agencies [5].

To address this issue, fiber-reinforced polymers (FRP) have emerged as a promising alternative to traditional steel reinforcement [6]. Among the different types of FRP bars, basalt fiber reinforced polymer (BFRP) and glass fiber reinforced polymer (GFRP) bars are gaining increasing attention due to their high strength-to-weight ratio and excellent corrosion resistance. However, the bond behavior of FRP bars in concrete remains a subject of ongoing research.

GFRP bars are the most commonly used FRP structural composites due to their cost effectiveness, with BFRP bars being a relatively new type of rebar that can provide an economical alternative to GFRP [7–11]. BFRP bars composed of basalt fibers and epoxy matrix have been tested and demonstrated excellent resistance to environmental conditions. Test data for beams with BFRP bars confirmed their usefulness in this respect. It should be noted that both GFRP and BFRP bars exhibit different mechanical properties depending on their components, and the design of FRP-RC elements requires different considerations and measures than conventional reinforced concrete structures [3, 12–15].

1.1 Zeolite and Metakaolin Properties

One of the key mechanical properties of rebar is its bond to concrete. The interaction between concrete and reinforcement is made possible by bond behavior—i.e. the ability to transfer forces between two building materials. On the one hand, the type of ribbing affects the improvement of the bond of the reinforcement, and on the other hand, the compressive and tensile strength of the concrete. Substitution of part of the cement with mineral additives in the form of zeolite and/or metakaolin increases the parameters of concrete and thus its bond behavior.

It should be noted that the use of mineral additives to concrete in the form of zeolite and metakaolin significantly reduces the energy consumption of RC structures. At the same time, it is a factor that significantly affects sustainable development due to the low energy-intensive mineral components that are a favorable alternative to some cement. The production of each ton of Portland cement requires the consumption of about 1.2 tons of limestone and 0.11 tons of standard coal. This results in the emission of about 0.85–0.92 tons of CO_2 and a significant amount of NO_x [16–18]. A cement substitute in the form of zeolite or metakaolin without high-temperature calcination or sintering can reduce CO_2 emissions by about 70% during production and use [19].

Zeolites are porous aluminosilicates containing large amounts of reactive SiO_2 and Al_2O_3. Preliminary studies confirmed the beneficial effect of the modifier in the form of zeolite on the increase in compressive and flexural strength, but only with the share of this modifier below 15% of the cement mass. The addition of zeolite also increases the durability of conventional concrete not only by reducing the permeability of concrete, but above all by improving the resistance to the reaction of alkaline aggregate.

Metakaolin, a highly reactive aluminosilicate material, has been found to exhibit high pozzolanic activity, making it a suitable substitute for cement in concrete, and an additional component for improving concrete's tightness [20]. The addition of metakaolin to concrete has been found to result in a 10% increase in compressive strength and a

50% increase in flexural tensile strength. In all cases, the use of metakaolin has shown a more favorable effect on tensile strength [21, 22]. The utilization of metakaolin as a partial replacement for cement in concrete is significant in the construction of sustainable development and environmental protection related to building structures.

1.2 Bond Behavior Research

Research on the bond behavior of GFRP bars to concrete has been conducted for several decades, with an emphasis on understanding the mechanisms governing the behavior of the FRP-concrete bond. Two commonly used experimental methods to test the bond of FRP bars in concrete are the pull out test and the beam test [23, 24].

In pull out tests, a single FRP bar is embedded in a cubic sample of concrete and a pull-out force is applied to the bar in a direction perpendicular to the concrete surface. The test provides information about the bond strength of the FRP bar to the concrete and the distribution of bonding stress along the bar length.

In order to evaluate the bonding behavior of fiber-reinforced polymer (FRP) bars in concrete, beam tests are commonly employed. In these tests, a concrete beam is reinforced with FRP bars and subjected to a bending load. The bonding behavior is then evaluated by measuring bond slip, which is the relative displacement between the FRP bar and the concrete under an applied load. Compared to other methods, such as pullout tests, beam tests provide a more realistic representation of the bonding behavior of FRP members in real structures because they take into account the effects of concrete compressive stresses and bar curvature [25, 26].

Studies have shown that the bond properties of FRP bars in concrete are affected by several factors, including concrete strength, surface roughness of FRP bars, surface treatment of FRP bars, and bar diameter [27, 28].

In general, the studies have shown that the bond strength of FRP bars in concrete is lower than that of steel bars, but the bonding slippage of FRP bars is much lower, indicating a more plastic bond behavior of FRP-concrete. In addition, studies have shown that the bonding properties of FRP bars in concrete can be improved by using surface treatments such as sandblasting or acid etching to increase the surface roughness of FRP bars and increase the bond strength [29, 30].

GFRP bar tests by Baena et al. [31] showed that the bond behavior of the bars does not depend on the strength of the concrete, but basically on the surface properties of the GFRP bars. The reason is the damage to the GFRP bars occurring mainly in the resin layer between the layers covered with sand concrete bars with a strength of about 50 MPa.

Achilledes and Pilakoutas [32] showed that the bond strength between GFRP bars and normal strength concrete mainly depends on the strength of the concrete when the strength is less than 15 MPa. However, in the case of concrete with a strength greater than 30 MPa, the destruction of adhesion occurs partly on the surface of the GFRP bars [33]. This was due to the control of the bond strength by the interlayer shear strength of the resin layer.

Tighiouart et al. after performing adhesion tests of FRP bars, they found that the bond strength between the GFRP bars and the surrounding concrete does not increase with increasing concrete strength [34].

Research results by Dai et al. suggest that an increase in temperature and humidity can lead to a decrease in bond strength, while aging can have a significant effect on the bonding behavior of FRP bars in concrete. The study also showed that the type of FRP used and the type of resin used to bond the FRP-to-the concrete can also affect the bonding behavior of the FRP rods in the concrete [35].

Research results by Wang et al. and Xiong et al. [36, 37] showed that BFRP and GFRP bars have excellent bonding properties in concrete, and their bond strength is comparable or even higher than that of steel bars. The bond strength of FRP bars in concrete is influenced by several factors, including the type of FRP used, the type of resin used to bond the FRP to the concrete, the concrete mix and the surface roughness of the FRP bars.

These findings suggest that FRP bars could potentially be a suitable alternative to steel bars as a reinforcement material in concrete structures. However, it should be noted that specific results may vary depending on the type of FRP used, concrete mix and testing conditions.

2   Research Program

2.1   Purpose and Course of the Research

In the present investigation, the effect of additives in the form of zeolite and metakaolin to concrete on the bond stress and slip behavior of BFRP and GFRP bars was compared and, for comparison, steel bars. The use of steel bars as reinforcement in concrete structures is a well-established practice, and their bonding properties are well documented. However, due to the distinct mechanical and physical properties of FRP bars, their bonding behavior in concrete may differ from that of steel bars. The results of this study will provide an understanding of the distinctive bonding properties of FRP bars to concrete, and compare them to the conventional bonding of steel bars to concrete. Such information will be of great value to engineers and researchers in comprehending the potential advantages and constraints of using FRP as reinforcement in concrete structures.

Three types of concrete were subjected to bond strength tests utilizing BFRP and GFRP composite reinforcement, as well as steel reinforcement, which served as the reference reinforcement in the beam test. The research was conducted to assess the feasibility of utilizing unconventional modifiers, in the form of mineral additives, to increase the adhesion of composite reinforcement made of FRP bars in selected concrete elements used in building infrastructure facilities. The optimal concrete mix was developed by selecting the ingredients to increase adhesion to FRP bars. Two modified concrete mixes were chosen, with 10% addition of zeolite (Z) and 10% addition of metakaolin (K), respectively, in relation to the ordinary concrete mix (C).

2.2   Strength Characteristics of Concrete and Reinforcement

The concrete mixes used in the beam test were composed of CEM I 42.5R cement provided by Lafarge. The density of the concrete was 2270 kg/m^3, and the water-to-cement (w/c) ratio was 0.45. The consistency of the concrete was of class S3, as specified

Table 1. Summary of concrete mix compositions

Mix proportion	C (ordinary)	Z (zeolit)	K (metakaolin)
Cement	360	324	324
Water	162	162	162
Gravel 2/8	574	574	574
Gravel 2/16	631	631	631
Sand 0/2	708	708	708
Zeolit	–	36	–
Metakaolin	–	–	36

by [38]. The detailed compositions of the concrete mixes, including the proportions of cement, water, aggregates, and mineral additives, are presented in Table 1.

In order to determine the strength parameters of the concrete, 150 mm cubic samples and 150×300 mm cylindrical samples were prepared. Three types of strength tests were performed on these samples, which enabled the determination of the compressive and splitting tensile strength, as well as the modulus of elasticity of the concrete. The test results are presented in Table 2.

Table 2. Strength properties of concrete types

	C (ordinary)			Z (zeolit)			K (metakaolin)		
	average	SD	COV	average	SD	COV	average	SD	COV
	MPa	MPa	%	MPa	MPa	%	MPa	MPa	%
f_c	46,36	1,01	2,19	48,37	0,98	2,02	49,55	0,93	1,87
f_{ct}	2,70	0,35	12,84	3,22	0,33	10,39	2,96	0,43	14,68
E_c	30017	340	1,13	31958	1138	3,56	36614	1339	3,66

Note f_c—cube compressive strength, f_{ct}—splitting tensile strength, E_L—modulus of elasticity, SD—standard deviation, COV—coefficient of variation

Table 3 presents the mechanical properties of GFRP, BFRP, and steel bars used in the beam tests. The properties include the tensile strength, yield strength (for steel bars only), and modulus of elasticity. The standard deviation and coefficient of variation are also provided to indicate the variability in the test results.

2.3 Test Procedure

The bond test, utilizing the beam test, was conducted following the guidelines of the PN-EN 10080 standard [39]. The beam test methodology is used to determine the bonding characteristics of reinforcing bars to concrete. The test configuration consists of two

Table 3. Mechanical properties of bar types.

	GFRP			Steel			BFRP		
	average	SD	COV	average	SD	COV	average	SD	COV
	MPa	MPa	%	MPa	Mpa	%	MPa	MPa	%
f_y	–	–	–	519,7	3,50	0,67	–	–	–
f_t	1033,3	30,49	2,95	615,8	3,64	0,59	1024,7	69,65	6,80
E_L	41892	3410	8,14	227293	25783	11,32	41470	3570	8,62

Note f_y yield strength, f_t—tensile strength, E_L—modulus of elasticity, SD—standard deviation, COV—coefficient of variation

concrete beams, each with dimensions of 80 mm × 160 mm × 375 mm, linked in the tension zone by the tested bar and in the compression zone by a steel joint, which is in the form of a cylinder with a diameter of 30 mm (as illustrated in Fig. 1).

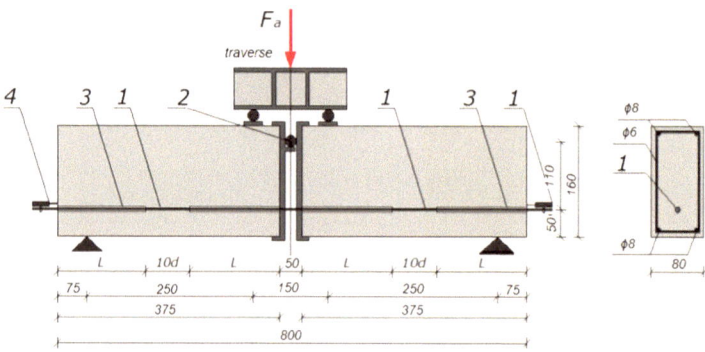

Fig. 1. Set-up for the test: 1—tested bar, 2—steel joint Ø30 mm, 3—PVC pipe, 4—slip measurement sensor

The bond between the reinforcing bar and the concrete was evaluated at a location in the middle of the 10d beams, where d represents the diameter of the bar. The remainder of the bar was enclosed in PVC pipes and had no adhesion with the concrete of the beam. The bond test was conducted under load using two concentrated forces, with the displacement of the tested bars at the ends of the beams being recorded during the test, as illustrated in Fig. 1. The test beam, supported by two rotating roller bearings, was loaded with two equal forces applied symmetrically about the center of the span. The concrete age of the tested beam was required to be within the range of 21 to 35 days. The load was applied incrementally corresponding to the stress σ in the bar, starting from 0 MPa and increasing in successive increments of 80 MPa and up to 240 MPa. For each increment, the total force applied to the set of beams was recorded.

$$F_a = \frac{2 \cdot A_n \cdot \sigma \cdot z}{a} \tag{1}$$

where: A_n is the nominal cross-sectional area of the bar, σ is the tensile stress in the test bar, z is the distance from the center of the hinge to the center of the test bar, and a is the shear distance.

The test shall continue until complete loss of adhesion in both beams. The adhesion stresses τ_b for a given value of the slip value for the force F_a are:

$$\tau_b = \frac{\sigma}{40} \tag{2}$$

where: σ is the stress in the tested bar.

The bond stress must be computed for four measured slip values: $\tau_{0.1}$—bond stress at 0.01 mm slip, $\tau_{0.1}$—bond stress at 0.1 mm slip, τ_1—bond stress at 1 mm slip, and τ_{max}—bond stress at maximum force.

A total of 27 sets of beams were prepared to measure the bond behavior of GFRP and BFRP bars with a nominal diameter of 12 mm, and for comparison, steel bars with a nominal diameter of 12 mm. The dimensions and reinforcement of all the beams in the test sets were identical. Particular attention was paid to the proper placement of the tested rods and the appropriate length of the rod ends to enable the installation of IL065 laser gauges for slip measurement. After casting the beam sets, the formwork elements were left in place to avoid stresses in the bars that could arise during the transfer of the element to the test stand.

Moreover, two clamps made of 15 mm thick steel sheet were fabricated to transfer force from the steel pin to the concrete, connecting two beams with the tested rod. The load was transferred symmetrically through the traverse onto the beam set at two points 150 mm apart, as shown in Fig. 2.

Fig. 2. A example of the beam test—the bond testing of BFRP bar

For slip measurement, a CMOS laser sensor IL065 with a measuring range of 50 mm and a measuring accuracy of 2 μm was installed at both ends of the tested bars. To immobilize the tested bars against the PVC sheaths and enable their axial displacement, a special assembly foam was used at the ends of the beams.

To measure the slip, a CMOS multi-function analog laser sensor, specifically the Keyence IL065, was installed at both ends of the tested bars. The sensor had a measuring range of 50 mm and a high measuring accuracy of 2 μm. Each of the tested rods was immobilized at the ends of the beams against the PVC sheaths using special assembly foam, which enabled their axial displacement for accurate measurements.

For the beam tests, three types of concrete were used with BFRP and GFRP composite reinforcement, as well as steel reinforcement which served as the reference reinforcement in the beam test. The load was transmitted through an actuator with a load range of 200 kN, mounted on an articulated joint in the ZD20 testing device. Throughout the test, the slippage of the tension bar at both ends of the beam set was continuously recorded. The results were recorded every second throughout the study period, as the load increased monotonically until maximum slip was reached.

3   Results and Discussion

Table 4 presents the bond stress for successive slip values: 0.01 mm, 0.1 mm, 1 mm and the maximum τ_{max} and average τ_m bond stress.

The interaction between concrete and reinforcement is facilitated by bond—the ability to transfer forces between two building materials. The formation of cracks in the concrete is necessary to activate the bond effect and make the reinforced concrete (e.g., FRP bars, steel bars) useful. Bond analysis was conducted for GFRP, steel, and BFRP bars in normal concrete, concrete with the addition of zeolite, and concrete with the addition of metakaolin. GFRP and BFRP bars with braid ribbing were compared to ribbed steel bars with similar equivalent diameter. As the composite bars differed in equivalent diameter due to technological conditions, it was decided to compare the bond behavior using the ratio of equivalent to nominal diameter. Since the bars had similar diameters, the bond stress could be accurately assessed. According to EN 1992–1 [40], sufficient bond stress are ensured if the average bond stress $\tau_m \geq 6.42$ MPa and the maximum bond stress $\tau_{max} \geq 10.51$ MPa for a bar diameter of 12 mm. All tested bars met the requirements of the above standard. The tests were conducted under standard conditions for both average and maximum bond stress.

For ordinary concrete, the average slip at maximum bond stress for GFRP bars was 2.20 mm, which was higher than that for steel (0.35 mm) and BFRP (0.61 mm). In concrete with the addition of zeolite, the average slip for GFRP bars at maximum bond stress was 0.82 mm, which was higher than the average slip for steel (0.11 mm) and BFRP (0.37 mm).

In concrete with the addition of metakaolin, the average slip for GFRP bars at maximum bond stress was 0.83 mm, which was higher than the slip for steel (0.35 mm) and greater than that for BFRP (0.67 mm).

Figure 3 presents the maximum bond stress for GFRP, steel, and BFRP bars in ordinary concrete, concrete with zeolite addition, and concrete with metakaolin addition.

The results showed that for ordinary concrete, the maximum bond stress τ_{max} for GFRP bars was 11.86 MPa, which was lower than the bond strength of other types of bars. Specifically, for steel bars, τ_{max} was 15.44 MPa, which was 30.2% larger than GFRP bars. On the other hand, the maximum bond stress was observed for BFRP bars, with a value of 21.27 MPa, which was higher than GFRP bars by 79.3% and steel bars by 37.8%. It is noteworthy that the steel bars were plasticized, which resulted in the inhibition of slip due to plastic deformation of the middle section of the bar (Fig. 3a).

Table 4. Bond stresses of bars to concrete for successive slip values in MPa.

Type	Side	$\tau_{0,01}$	$\tau_{0,1}$	$\tau_{1,0}$	τ_{max}	τ_m
BG-1	Left	7,27	9,15	10,57	13,28	8,99
	Right	5,32	7,63	9,16		7,37
BG-2	Left	6,66	8,05	9,69	10,52	8,13
	Right	5,77	6,49	9,08		7,11
BG-3	Left	5,73	8,30	11,60	11,77	8,54
	Right	7,73	8,95	10,71		9,13
ZG-1	Left	7,22	9,45	-	11,76	–
	Right	3,21	8,47	9,76		7,15
ZG-2	Left	7,16	9,96	10,84	12,70	9,32
	Right	7,29	10,49	7,68		8,49
ZG-3	Left	5,72	7,62	9,98	10,87	7,77
	Right	1,54	7,7	10,02		6,43
MG-1	Left	13,24	13,74	–	14,31	–
	Right	10,26	12,20	12,20		11,55
MG-2	Left	15,15	15,74	20,13	21,93	17,01
	Right	14,07	14,70	17,37		15,38
MG-3	Left	12,50	13,46	16,73	16,79	14,23
	Right	12,93	13,82	14,96		13,90
BS-1	Left	10,85	13,23	–	16,14	–
	Right	11,08	13,38	–		–
BS-2	Left	0,14	11,88	–	15,18	–
	Right	6,31	10,33	15,10		10,58
BS-3	Left	8,75	11,93	11,93	14,99	10,87
	Right	6,11	9,90	–		–
ZS-1	Left	15,35	-	–	16,03	–
	Right	12,45	15,33	–		–
ZS-2	Left	10,34	14,60	–	16,00	–
	Right	10,31	14,18	–		–
ZS-3	Left	12,02	13,81	–	16,08	–
	Right	9,45	15,14	–		–
MS-1	Left	10,97	14,58	–	16,26	–

(*continued*)

In the case of concrete with the addition of zeolite, the maximum bond stress τ_{max} for GFRP bars was 11.78 MPa, which was lower than other types of bars. For steel

Table 4. (*continued*)

Type	Side	$\tau_{0,01}$	$\tau_{0,1}$	$\tau_{1,0}$	τ_{max}	τ_m
	Right	9,77	12,68	16,16		12,87
MS-2	Left	9,79	11,02	–	16,86	–
	Right	8,80	12,13	–		–
MS-3	Left	7,79	10,28	16,68	16,73	11,58
	Right	7,90	11,73	–		–
BB-1	Left	13,30	14,77	–	22,12	–
	Right	15,66	19,25	21,54		18,82
BB-2	Left	15,57	16,48	20,92	21,28	17,66
	Right	12,82	14,50	–		–
BB-3	Left	13,93	17,00	20,29	20,42	17,07
	Right	12,73	17,54	–		–
ZB-1	Left	18,79	18,80	20,87	21,80	19,49
	Right	16,86	19,79	–		–
ZB-2	Left	21,49	22,75	–	23,67	–
	Right	11,55	18,35	23,59		17,83
ZB-3	Left	20,21	21,18	–	22,73	–
	Right	15,36	18,65	22,47		18,83
MB-1	Left	15,84	18,97	22,17	22,91	18,99
	Right	14,52	15,88	19,20		16,54
MB-2	Left	14,67	17,72	–	19,51	–
	Right	13,17	14,29	17,85		15,10
MB-3	Left	14,34	16,53	–	18,43	–
	Right	12,50	14,73	–		–

Note The initiation of slip of the bars relative to the surrounding concrete was observed first on one side of the sample beam system due to the slight variation in the concrete mix, and it is equally important to consider the side of the beam system where slippage occurred later. This is crucial because it affects the course of bar slip after exceeding the slip for maximum bond stress

bars, τ_{max} was 16.04 MPa, which was 36.2% larger than GFRP bars. The highest bond strength was again observed for BFRP bars, with a value of $\tau_{max} = 20.64$ MPa, which was higher than GFRP bars by 75.2% and steel bars by 28.7% (Fig. 3b).

Moreover, for concrete with the addition of metakaolin, the maximum bond stress for GFRP bars was 17.67 MPa, which was lower than other types of bars. For steel bars, τ_{max} was 16.61 MPa, which was smaller than GFRP bars by 6%. Once again, the highest bond strength was noted for BFRP bars, with a value of $\tau_{max} = 21.21$ MPa, which was higher than GFRP bars by 20.0% and steel bars by 27.7% (Fig. 3c).

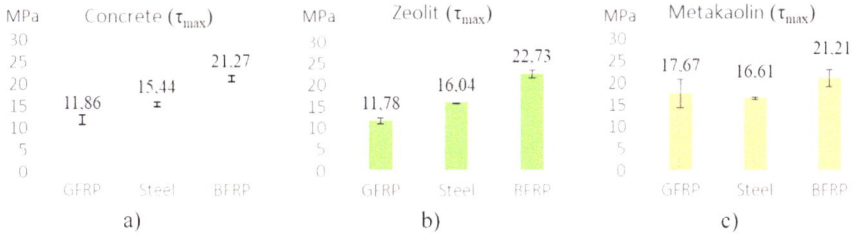

Fig. 3. Relationship of bond stress of GFRP, steel and BFRP bars to the type of concrete (a) ordinary concrete, (b) concrete with zeolite, (c) concrete with metakaolin

The changes in the average bond stress τ_m, depending on the influence of zeolite and metakaolin additions to the concrete of GFRP, steel, and BFRP bars, are presented in Fig. 4.

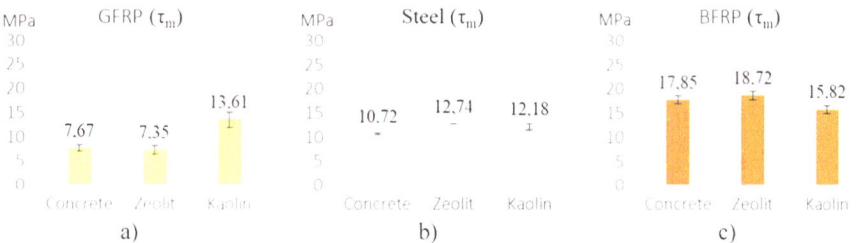

Fig. 4. Comparison of the average bond stress of bars for the samples with ordinary concrete, with the addition of zeolite and metakaolin: (a) GFRP bars, (b) Steel bars, (c) BFRP bars

The results of the study revealed that the average bond stress for GFRP bars decreased by 4% in concrete with the addition of zeolite. In contrast, in the presence of metakaolin, a significant increase in the average bond stress of 77.4% was observed compared to ordinary concrete (Fig. 4a).

For steel bars, an increase in the average bond stress by 19% was observed for concrete with the addition of zeolite. However, for concrete with the addition of metakaolin, a modest increase of 14% was noted compared to normal concrete (Fig. 4b).

In the case of BFRP bars, the average bond stress increased by 5% in the presence of zeolite addition to concrete. However, in the case of metakaolin addition, a decrease of 11% in the average bond stress was observed compared to normal concrete (Fig. 4c).

4   Conclusions

This research study was designed to investigate the bond behavior of BFRP and GFRP bars in modified concrete using beam tests. The effects of concrete properties and bar characteristics on the bond behavior of FRP bars were thoroughly analyzed and compared. The findings of this study provide important information for the design and implementation of FRP reinforced concrete structures and contribute to the development of improved models of FRP bar bond in concrete.

1. It was ensured that standard conditions of average and maximum bond stress were met in all tests.
2. For the tested types of bars, the maximum bond stress (τ_{max}) and the average bond stress (τ_m) determined for the slip of 0.01 mm, 0.1 mm, and 1 mm were sufficiently high and met or exceeded those required by EN 1992–1-1.
3. The results showed that the highest maximum bond stress ($\tau_{max} = 22.73$ MPa) was recorded for BFRP bars in concrete with zeolite addition, which were 75.2% and 28.7% higher than GFRP and steel bars, respectively. However, in the presence of metakaolin addition in the concrete, the maximum bond stress of BFRP bars was $\tau_{max} = 21.21$ MPa, which was higher by 27.7% and 20.0% compared to steel and GFRP bars, respectively.
4. In terms of average bond stress, the highest values were recorded for BFRP bars in concrete with zeolite addition ($\tau_m = 18.72$ MPa), which were 47% and 154% higher than steel and GFRP bars, respectively. However, in the presence of metakaolin in the concrete, the average bond stress of BFRP bars was $\tau_m = 15.82$ MPa, which was higher by 24.2% and 16.2% compared to steel and GFRP bars, respectively.
5. The addition of zeolite and metakaolin to concrete reduced the slippage at maximum bond stress in the case of GFRP bars by more than two and a half times. For BFRP and steel bars, reduced slip was observed in the presence of concrete with the addition of zeolite.
6. A diametrically different behavior in the case of GFRP and BFRP bars was observed after reaching the stress peak. Bond behavior on the "retarded slip" side decreased gradually and remained at the level of over 80% with the slip several times higher than the slip at maximum bond stress.

In conclusion, the bonding behavior of FRP bars in concrete is a complex and ongoing area of research, and further research is necessary to fully comprehend and optimize the bonding of FRP bars in concrete structures.

Acknowledgements. The authors gratefully acknowledge Astra company for generously providing GFRP and BFRP bars, as well as the concrete additives, zeolite and metakaolin, for the purpose of this study. This research was conducted under the Grant 35.2022 RND from Warsaw University of Technology, for which the authors express their deep appreciation.

References

1. Czarnecki, L.; Emmons, P.: Repair and protection of concrete structures, Polski Cement, Kraków, 2002
2. fib Bulletin 40: FRP reinforcement in RC structures. Technical report. International Federation for Structural Concrete: Lausanne—Switzerland, 2007, 160
3. Garbacz, A., Urbański, M., Łapko, A.: BFRP bars as an alternative reinforcement of concrete structures—compatibility and adhesion issues. Adv. Mater. Res. **1129**, 233–241 (2015)
4. ACI 440.3R-04, „Guide test methods for fiber-reinforced polymers (FRPs) for reinforcing or strengthening concrete structures
5. Koch, G.H., Brongers, M.P.H., Thompson, N.G., Virmani, Y.P., Payer, J.H.:Corrosion costs and preventive strategies in the United States, Report No FHWA-RD-01–156, US. DoT, FHWA, Washington, D.C. (2002)

6. ASTM A615 / A615M - 09b Standard Specification for Deformed and Plain Carbon-Steel Bars for Concrete Reinforcement. Americaa

7. Garbacz, A., Szmigiera, E., Urbański, M., Protchenko, K., Kubas, M.: On research on FRP hybrid reinforcement for infrastructure RC structures. Eng. Construction, Warsaw 2017; 8/2017: 428–432 (2017) (in polish)

8. Szmigiera, E., Protchenko, K., Urbański, M., Garbacz, A.: Mechanical properties of hybrid FRP Bars and Nano-Hybrid FRP Bars. Arch. of Civ. Eng. **65**(1), 97–110 (2019). https://doi. org/10.2478/ace-2019-0007

9. Protchenko, K., Szmigiera, E.D., Urbański, M., Garbacz, A.: Development of Innovative HFRP Bars. MATEC Web of Conf. **196**, 1–6 (2018). https://doi.org/10.1051/matecconf/201 819604087

10. Protchenko, K., Dobosz, J., Urbański, M., Garbacz, A.: Wpływ substytucji włókien bazaltowych przez włókna węglowe na właściwości mechaniczne prętów B/CFRP (HFRP). Czasopismo Inżynierii Lądowej, Środowiska i Architektury. JCEEA **63**, 1/1:149–156 (2016). http://doi.prz.edu.pl/pl/pdf/biis/454

11. Urbanski, M.: Compressive strength of modified FRP hybrid bars. Materials **13**(8), 17 (1898). https://doi.org/10.3390/ma13081898

12. Protchenko, K., Szmigiera, E.D.: Post-fire characteristics of concrete beams reinforced with hybrid FRP Bars. Materials **13**(5), 1–15 (2020). https://doi.org/10.3390/ma13051248

13. Protchenko, K.: Residual fire resistance testing of Basalt- and hybrid-FRP reinforced concrete beams. Materials **15**, 1–18 (2022). https://doi.org/10.3390/ma15041509

14. Protchenko, K., Zayoud, F., Urbański, M., Szmigiera, E.: Tensile and shear testing of basalt fiber reinforced polymer (BFRP) and hybrid basalt/Carbon fiber reinforced polymer (HFRP) Bars. Mater. **13**, 1–16 (2020). https://doi.org/10.3390/ma13245839

15. Protchenko, K., Leśniak, P., Szmigiera, E.D., Urbański, M.: New model for analytical predictions on the bending capacity of concrete elements reinforced with FRP bars. Mater. **14**, 1–17 (2021). https://doi.org/10.3390/ma14030693

16. Shobeiri, V., Bennett, B., Xie, T., Visintin, P.: A comprehensive assessment of the global warming potential of geopolymer concrete. J. Cleaner Prod. 297. (2021). https://doi.org/10. 1016/j.jclepro.2021.126669

17. Groves, M.C.E., Sasonow, A.: Uhde EnviNOx® technology for NOX and N2O abatement: a contribution to reducing emissions from nitric acid plants. J. Integr. Environ. Sci. **7**, 211–222 (2010). https://doi.org/10.1080/19438151003621334

18. Elizondo-Martínez, E.-J., Andrés-Valeri, V.-C., Jato-Espino, D., Rodriguez-Hernandez, J.: Review of porous concrete as multifunctional and sustainable pavement. J. Build. Eng. **27** (2020). https://doi.org/10.1016/j.jobe.2019.100967

19. Naqi, A., Jang, J.: Recent progress in green cement technology utilizing low-carbon emission fuels and raw materials: a review. Sustainability **11**(2) (2019) https://doi.org/10.3390/su1102 0537

20. Rashad, A.M.: Metakaolin as cementitious material: history, scours, production and composition—a comprehensive overview. Constr. Build. Mater. **41**, 303–318 (2012). https://doi.org/ 10.1016/j.conbuildmat.2012.12.001

21. Sudagar, A., Andrejkovičová, S., Patinhaa, C., Velosa, A., McAdam, A., Ferreira da Silva, E., Rocha, F.: A novel study on the influence of cork waste residue on metakaolin-zeolite based geopolymers. Appl. Clay Sci. **152**, 196–210 (2018). https://doi.org/10.1016/j.clay.2017. 11.013

22. Bakera, A.T., Alexander, M.G.: Use of metakaolin as a supplementary cementitious material in concrete, with a focus on durability properties. RILEM Tech. Lett. **4**, 89–102 (2019). https:// doi.org/10.21809/rilemtechlett.2019.94

23. Hao, Q., Wang, Y., He, Zheng H.; Jinping O.: Bond strength of glass fiber reinforced polymer ribbed rebars in normal strength concrete. Constr. Build. Mater. **23**(2), 865–871 (2009). https://doi.org/10.1016/j.conbuildmat.2008.04.011

24. Weichen, X., Qiaowen, Z., Yu, Y., Zhiqing, F.: Bond behavior of sand-coated deformed glass fiber reinforced polymer rebars. J. Reinforced. Plastics. Composit. **33**(10), 895–910 (2014). https://doi.org/10.1177/0731684413520263

25. Zemour, N., Asadian, A., Ahmed, E.A., Khayat, K.H., Benmokrane, B.: Experimental study on the bond behavior of GFRP bars in normal and self-consolidating concrete. Constr. Build. Mater. **189**, 869–881 (2018)

26. Hossain, K.M.A., Ametrano, D., Lachemi, M.: Bond strength of standard and high-modulus GFRP bars in high-strength concrete. J. Mater. Civ. Eng. **26**, 449–456 (2014)

27. Zhang, P., et al.: Influence of rib parameters on mechanical properties and bond behavior in concrete of fiber-reinforced polymer rebar. Adv. Struct. Eng. **24**, 196–208 (2021)

28. Hao, Q., Wang, Y., He, Z., Ou, J.: Bond strength of glass fiber reinforced polymer ribbed rebars in normal strength concrete. Constr. Build. Mater. **23**, 865–871 (2009)

29. Park, C., Won, J., Cha, S.: Bond properties of CFRP rebar in fiber reinforced high strength concrete with surface treatment methods of reinforcing fibers. J. Korea Concr. Inst. **21**, 275–282 (2009)

30. Kang, J., Kim, B., Park, J., Lee, J.: Influence evaluation of fiber on the bond behavior of GFRP bars embedded in fiber reinforced concrete. J. Korea Concr. Inst. **24**, 79–86 (2012)

31. Baena, M., Torres, L.L., Turon, A., Barris, C.: Experimental study of bond behavior between concrete and FRP bars using a pull-out test. Compos: Part B **40**: 784–797 (2009)

32. Achillides, Z., Pilakoutas, K.: Bond behaviour of fiber reinforced polymer bars under direct pullout conditions. J. Compos. Constr. **8**, 173–181 (2004)

33. Tepfers, R.: Bond clause proposals for FRP bars/rods in concrete based on CEB/FIP Model Code 90. Part 1: Design bond stress for FRP reinforcing bars. Structural Concrete **7**(2), 47–55 (2006)

34. Tighiouart, B., Benmokrane, B., Gao, D.: Investigation of bond in concrete member with fiber reinforced polymer (FRP) bars. Constr. Build. Mater. **12**, 453–462 (1998)

35. Dai, Jian-Guo, Yokota, H., Iwanami, M., Kato, E.: Experimental investigation of the influence of moisture on the bond behavior of FRP to concrete interface. J Composit. Constr. **14**(6), 834–844. https://doi.org/10.1061/(ASCE)CC.1943-5614.0000142

36. Wang, Z.K., et al.: Long-term durability of basalt- and glass-fibre reinforced polymer(BFRP/GFRP) bars in seawater and sea sand concrete environment. Constr. Build. Mater. **139**, 467–489 (2017). https://doi.org/10.1016/j.conbuildmat.2017.02.038

37. Xiong, Z., Wei, W., Liu, F., Cui, C.Y., Li, L.J., Zou, R., Zeng, Y.: Bond behaviour of recycled aggregate concrete with basalt fibre-reinforced polymer bars. Compos. Struct. 256 (2021). https://doi.org/10.1016/j.compstruct.2010.11.30.78

38. European Standard EN 12350–2:2019. Testing fresh concrete - Part 2: Slump test. CEN-CENELEC Management Centre: Rue de la Science 23, B-1040 Brussels

39. European Standard EN 10080:2007. Steel for the reinforcement of concrete - Weldable reinforcing steel – General. CEN-CENELEC Management Centre: Rue de la Science 23, B-1040 Brussels

40. European Standard EN 1992–1–1:2008. Eurocode 2: Design of concrete structures - Part 1–1 : General rules and rules for buildings. CEN-CENELEC Management Centre: Rue de la Science 23, B-1040 Brussels

Effect of Hybridization of BFRP Bars on Their Microstructure and Mechanical Properties

Karolina Ogrodowska$^{(\boxtimes)}$, Marek Urbański, and Andrzej Garbacz

Faculty of Civil Engineering, Warsaw University of Technology, Warsaw, Poland
`karolina.ogrodowska@pw.edu.pl`

Abstract. The FRP (Fiber Reinforced Polymer) bars are increasingly used as the main reinforcement of concrete structures, replacing traditional steel reinforcement. In this paper results of Basalt Fiber Reinforced Polymer bars (BFRP) modification by partial replacement of basalt fibers with carbon fibers were presented. The analysis of an effect of hybridization on a microstructure and mechanical properties of BFRP bars were performed. This analysis was thought out based on tests performed: tensile strength and shear strength and a microstructure observation with scanning electron microscope. The results obtained indicate that the hybridization effectively increases elasticity modulus compared to unmodified BFRP and the tensile strength and shear strength increase in lower extant. The nonhomogeneous distribution of carbon fiber in the cross-section of HFRP bars has relatively small effect of mechanical properties and their scattering.

Keywords: Fiber · Reinforced · Polymer · Bars · FRP · Hybridization · Microstructure

1 Introduction

The development of building materials market and increasing importance of use of high-performance materials construction is conducive to the increasing use of FRP composites in Civil Engineering [1–3]. The main advantages of FRP composites are: high tensile strength to weight ratio, corrosion resistance, electromagnetic transparency, fatigue resistance, easy production of unique shapes, flexible aesthetics and a low carbon footprint in the perspective of the life cycle [4]. Currently, there are more and more applications regarding use of FRP composites as the main reinforcement in concrete structures [5]. Among others, basalt fibers seem to be a compromise, between new plant fibers and the most used in construction glass and carbon fibers. Basalt Fiber Reinforced Polymer (BFRP) bars characterize very good mechanical properties and are more stable chemical resistance compared with Glass Fiber Reinforced Polymer (GFRP) bars. BFRP also demonstrates a wider range of working temperatures and provide significantly better cost-effectiveness compared to Carbon Fiber Reinforced Polymer (CFRP) bars [6–8]. The main drawback of BFRP is lower stiffness, which affects the fulfillment of the SLS conditions of the structure, such as deflection and cracking. The increasing importance

L. Czarnecki et al. (Eds.): ICPIC 2023, 61, pp. 422–430, 2025.
https://doi.org/10.1007/978-3-031-72955-3_42

of rational material usage promotes materials modification. Combining fibers with different properties allows to use the advantages of individual fibers, their mechanical and physical properties, and allows for a more rational use of the material in terms of cost [9–11]. To obtained high properties of Hybrid Fiber Reinforced Polymer (HFRP) bars many factors should be control during manufacturing; among others, the fiber placements in cross-sectional, which can affect their properties.

In this paper the effect of hybridization of BFRP bars by substitution of the part of basalt fibers with carbon fiber on microstructure and mechanical properties was discussed.

2   General Concept

In order to improve the mechanical parameters (especially stiffness), the configurations of HFRP bars consisting of basalt or glass and carbon fibers were parametrically analyzed and optimized using rules of mixture (ROM). Initially, two types of hybrids were considered: carbon-basalt or carbon-glass in the fiber volume fraction ratio: 1:1; 1:2; 1:3, 1:4 or 1:9. [12, 13]. Then, it was decided to use a mix of carbon and basalt fibers, due to the better mechanical properties of basalt fibers compared to glass fibers. Carbon fibers are much stiffer than basalt fibers but at the same time several times more expensive. Moreover, the tensile strain of the carbon fibers should be close to the strain of the basalt fibers to avoid shear lag.

The location of the fibers in the cross-section of the bar is also important. The analyzes carried out indicated that it was more advantageous to arrange carbon fibers in the surface layer of the bar and basalt fibers in its core (Fig. 1a) due to the increase in stiffness. After starting the trial production of HFRP bars, it was noticed that in the surface layer of the HFRP bar, rovings consisting of carbon fibers were degraded as a result of scorching caused by excessive temperature (Fig. 1b).

a)   b)

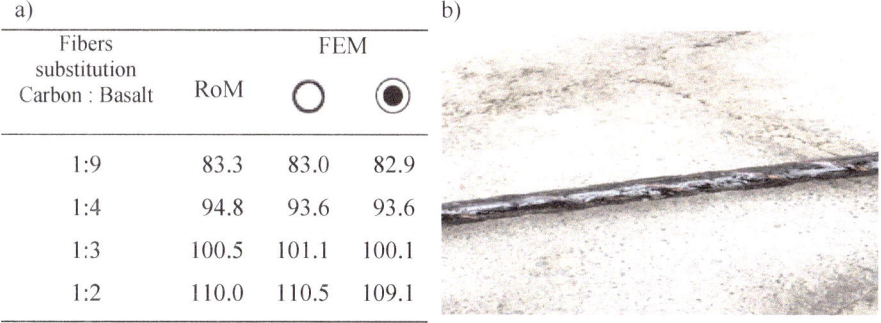

Fibers substitution Carbon : Basalt	RoM	FEM	
		◯	◉
1:9	83.3	83.0	82.9
1:4	94.8	93.6	93.6
1:3	100.5	101.1	100.1
1:2	110.0	110.5	109.1

Fig. 1. a) The modulus of elasticity of the HFRP bars calculated using RoM (Rule of Mixtures) and obtained by FEM (Finite Element Method) simulation for at two different ideal arrangement of fibers in bar; b) visible burn of the carbon roving of the HFRP bar

Therefore, the concept of carbon fiber distribution was changed. It was decided to place the carbon fibers in the core of the bar, and the basalt fibers near its surface.

Additional stiffness analyzes of the bar modified in this way indicated a minimal decrease in the stiffness of the bar in relation to its original configuration, which was compensated by an increase in the share of carbon fibers. At the same time, the phenomenon of carbonization of carbon fibers, which can only occur under ventilation conditions, has been eliminated. In the pultrusion process, some fiber properties influenced the final architecture of the bar structure. The problem with uneven distribution was caused by the difficulty of keeping the carbon fibers in the core of the bar during the pultrusion process - the tendency of the carbon fiber to float in the consolidation part. Some carbon fibers were also burned during the pultrusion process.

3   Materials and Methods

Table 1 presents properties of individual ingredients, collected on the basis of manufacturers data: Toho Tenax Europe GmbH for carbon fibers, Kamenny Vek for basalt fibers and Ciech Sarzyna S.A. company for resin matrix (four-component 1300 System®).

Table 1. Properties of constituents used for preparing BFRP and HFRP bars.

Property	Epoxy resin	Carbon fiber	Basalt fiber
Density ρ (g/cm^3)	1.16	1.79	2.89
Filament diameter (μm)	–	7.00	17.17
Young's Modulus E_{11} (GPa)	3.45	242.10	89.00
Poisson Number ν_{12} (-)	0.35	0.28	0.26
Shear Modulus G_{12} (MPa)	1.28	24.00	21.70
Tensile Strength σ_{11} (MPa)	55	4240.8	3000

Taking into account results of computer simulation and the trail production the compositions of BFRP and HFRP were selected (Tab.2). After an in-depth analysis of possible configurations, HFRP bars consisting of carbon and basalt fibers were recommended for production in a volume ratio of 1:4 in relation to the total volume of fibers. Thus, the weight ratio of carbon to basalt fibers in the HFRP bar was 1:3.

Table 2. Types of tested FRP bars.

Bar type	In the total fibers by weight		In the total bar by weight	
	Basalt Fiber	Carbon fiber	Basalt fiber	Carbon fiber
BFRP	100%	–	70.3%	–
HFRP	75.3%	24.7%	49.4%	16.2%

In this research project two type of tests of mechanical properties were conducted in room temperature: transverse shear and longitudinal tensile (performed in accordance with the ACI 440.3R [14] guidelines and ASTM methods ASTM D7617/D7617M-11(2017) [15] and ASTM D7205/D7205M-21(2021) [16], respectively. Five specimens were tested for each measurement conditions. The observation with SEM electron microscopy were conducted on cross-section of samples before mechanical tests and taken nearby of failure place after shear test. Details about the tests carried out in the author's previous research works are in the literature [17, 18].

4   Results and Discussion

4.1   Mechanical Properties

The test results of longitudinal tensile test and transverse shear test of samples of BFRP and HFRP bars are presented the Table 3.

Table 3. Results of longitudinal tensile test and transverse shear test

Property	BFRP		HFRP	
	Mean value	COV, %	Mean value	COV, %
Tensile strength, MPa	1103.33	2.07	1277.92	4.34
Tensile strain, %	2.52	2.09	1.73	4.33
Modulus of elasticity, GPa	43.87	1.95	73.89	4.15
Shear strength, MPa	205.37	5.63	229.44	1.77

Note: COV – coefficient of variation (ratio of standard deviation to mean value) in %

The results of mechanical tests indicates that hybridization with carbon fibers effectively increases elasticity modulus ~70% compared to unmodified BFRP. The tensile strength and transverse shear strength increase in lower extant: ~16% and ~11% respectively. The hybridization increase two times value of COV in the case of tensile properties, but they were still low below 5%. In the case of shear strength the COV was even three times lower.

4.2   SEM Observation of Microstructure

The examples of SEM microstructures of BFRP and HFRP bars (dimeter of 8mm) at magnification 60× are presented in Fig. 2. They confirmed nonhomogeneous distribution of fibers in the case BFRP and HFRP. The question is how it influences mechanical properties of both types of FRP bars. In the case of HFRP bars carbon fibers were located in the core of the bars. However, the SEM micrographs of the samples after shear tests showed uneven carbon fibers distribution in the HFRP bar, where they were located outside the core zone (Fig. 3). It was also observed that the distribution of the fibers in one bar varied along its entire length. The supervision of the constant distribution of the fibers in the hybridization process, can be a great challenge in the production process. The analysis of the SEM images also indicates that there is no one way of cracks propagation trough HFRP bar during shear tests. It seems that even voids in bar were not preferable way of cracks propagation. It was observed that some cracks were blocked on voids and interfaces between basalt fibers zone and carbon fibers one.

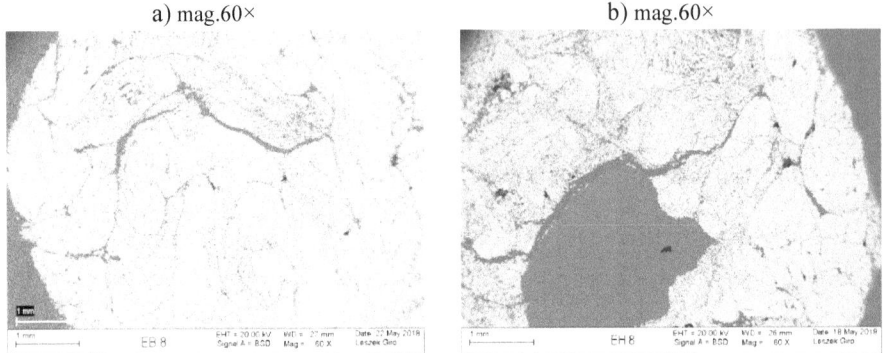

a) mag.60×   b) mag.60×

Fig. 2. Examples of microstructures BFRP (a) and HFRP (b) bars of 8 mm diameters

4.3 Experimental Results vs Theoretical Estimation of Mechanical Properties

Based on the parameters provided for each of the components by the manufacturer, the values were calculated based on theoretical consideration. The composite will usually break when the stresses in the fibers reach their strength f_{fu}. After the fibers break, the matrix is unable to carry to load. Therefore, the composite strain to failure ε_{cu} is equal to the fiber strain to failure ε_{fu}. At this strain level, the matrix has not failed yet because it is more compliant and can sustain larger strains. Under these conditions, it can be assumed that the longitudinal tensile force is controlled by the fiber tensile stress and is represented by Eq. (1):

$$F_{1t} = \sigma_f V_f + \sigma_m V_m = \sigma_f V_f + \sigma_f \frac{E_m}{E_f}\left(1 - V_f\right) \tag{1}$$

where: σ_f – tensile stress in the fiber; σ_m – tensile stress in the matrix; V_f – fiber volume in composite; V_m – matrix volume in composite; E_m – longitudinal tensile modulus of elasticity for matrix; E_f – longitudinal tensile modulus of elasticity for fibres.

This equation assumes that the strain in the matrix and the fibers are the same, which is true if the fiber-matrix bond is perfect. The ultimate strain or stress of the matrix is not realized, because the fibers are more brittle (fail at a lower strain). The underlying assumption is that once the fibers break, the matrix is not capable of sustaining the load and the composite fails. The shear stress-strain law of the composite was assumed to be linear Eq. (2):

$$\tau = G_{12} \cdot \gamma \tag{2}$$

a) mag. 42×   b) mag.65×

c) mag.150×   d) mag. 300×

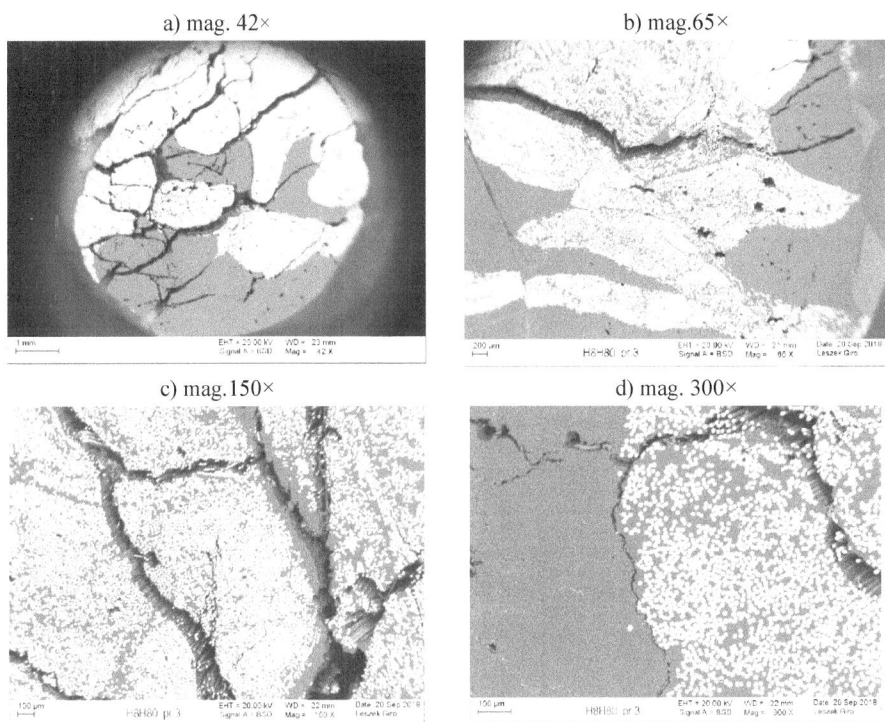

Fig. 3. The SEM micrographs of the HFRP bars cross-section previously subjected to a shear test at different magnification of the bar.

where: τ – shear stress; G_{12} - shear modulus; γ- shear strain

The maximum-shear-strain criterion limited the strains in the tension-compression quadrants to account for shear failure of the fibres. The shear strain is computed as Eq. (3):

$$\gamma = (1 + \nu_{12})\varepsilon_{1t}/2 \tag{3}$$

where: ε_{1t} – strain in the fiber at failure of the unidirectional composite in tension;

ν_{12} – fibre's Poisson coefficient.

The performed calculations overestimate the measured values by about 25% (Tab.4).

The predictions are higher than measured values. This is due to defects exist in the composites, which translates into an imperfect fiber-matrix bond. This shows that with ideal production process (reduced imperfection effect) the performance of composites can be higher. The measured to calculated shear strength ratio (87.6%) is relatively closer to the test values compared to the analogous tensile strength ratio (82.5%) for HFRP composites. In addition, test to calculated ratios are greater for HFRP compared to BFRP for both tensile strength and transverse shear strength. This indicates a better technological processing of HFRP bars compared to homogeneous BFRP bars. In the analysis of this case should be underlined that the above forms do not differentiate the

Table 4. Calculated values vs measured values of tensile and transverse shear strength.

	Calculated values		Mean of measured values		Ratio of measured value/calculated value	
Bar type	Tensile strength F_{1t} [MPa]	Shear strength τ [MPa]	Tensile strength [MPa]	Shear strength [MPa]	Tensile strength [%]	Shear strength [%]
BFRP	1368.7	267.3	1103.3	205.4	80.6	76.8
HFRP	1549.5	261.8	1277.9	229.4	82.5	87.6

distribution of fibers in the bars in any way, but only include the content of individual fractions.

5 Conclusions

On the basis of the results obtained the following main conclusions on hybridization of BFRP bars with carbon fibers can be drawn:

- the hybridization effectively increases elasticity modulus ~70% compared to unmodified BFRP. The tensile strength and shear strength increase in lower extant: ~16% and ~11% respectively;
- the hybridization increase two times value of coefficient of variation in the case of tensile properties, but they were still low below 5%. In the case of shear strength the COV was even three times lower;
- the nonhomogeneous distribution of fibers in the cross-section of HFRP bars has relatively small effect of mechanical properties. The analysis of the SEM images also indicates that there is no one preferable way for cracks propagation trough HFRP bar during shear tests.

Additionally, it can be concluded that better situation is if the carbon fibers are more randomly dispersed across in the HFRP cross-section because of the fibers with higher elongation can safely withstand higher loads.

Acknowledgements. The article was prepared within the framework of the Internal Grant of the Faculty of Civil Engineering at Warsaw University of Technology no. 504/04740/1080/44.000000. The results of the project "Innovative Hybrid – FRP composites for infrastructure design with high durability" NCBR: PBS3/A2/20/2015 were partially used during preparation of this paper.

References

1. Hollaway, L.C.: A review of the present and future utilisation of FRP composites in the civil infrastructure with reference to their important in-service properties. Constr. Build. Mater. **24**, 2419–2445 (2010)

2. Siwowski, T., Rajchel, M., Kaleta, D., Własak, L.: FRP bridges in Poland: state of practice. Arch. Civ. Eng. **67**(3), 5–27 (2021)
3. Kotynia, R., Szczech, D., Kaszubska, M.: Bond behavior of GRFP bars to concrete in beam test. Procedia Eng. **193**, 401–408 (2017)
4. Karbhari, V.M., et al.: Durability gap analysis for fiber-reinforced polymer composites in civil infrastructure. J. Compos. Constr. **7**(3), 238–247 (2003)
5. Zhou, L., Zheng, Y., Taylor, S.E.: Finite-element investigation of the structural behavior of basalt fiber reinforced polymer (BFRP)-reinforced self-compacting concrete (SCC) decks slabs in Thompson bridge. Polymers **10**(6), 678 (2018)
6. Preinstorfer, P., Reichenbach, S., Huber, T., Kromoser, B.: Potential fields of application for CFRP reinforcement in concrete infrastructure engineering: Material availability, application areas and static parametric study with consideration of the GWP. Concrete Structures, New Trends for Eco-Efficiency and Performance
7. Fiore, V., Scalici, T., Di Bella, G., Valenza, A.: A review on basalt fibre and its composites. Composites. Part B, Eng. **74**, 74–94 (2015)
8. Elgabbas, F., Ahmed, E.A., Benmokrane, B.: Physical and mechanical characteristics of new basalt-FRP bars for reinforcing concrete structures. Constr. Build. Mater. **95**, 623–635 (2015)
9. Subagia, I.A., Kim, Y., Tijing, L.D., Kim, C.S., Shon, H.K.: Effect of stacking sequence on the flexural properties of hybrid. Composites Tom B **58**, 251–258 (2014)
10. Swolfs, Y., Gorbatikh, L., Verpoest, I.: Fibre hybridisation in polymer composites: a review. Composites. Part A, Appl. Sci. Manuf. **67**, 181–200 (2014)
11. Ogrodowska, K., Łuszcz, K., Garbacz, A.: The effect of temperature on mechanical properties of hybrid FRP bars applicable for reinforcing of concrete structures. MATEC Web Conf. **322**, 01029 (2020)
12. Protchenko, K., Młodzik, K., Urbański, M., Szmigiera, E., Garbacz, A.: Numerical estimation of concrete beams reinforced with FRP bars. In: MATEC Web of Conferences, vol. 86, p. 02011. EDP Sciences (2016)
13. Szmigiera, E.D., Protchenko, K., Urbański, M., Garbacz, A.: Mechanical properties of hybrid FRP bars and nano-hybrid FRP bars. Arch. Civil Eng. **65**(1), 97–110. https://doi.org/10.2478/ace-2019-0007 (2019)
14. American Concrete Institute ACI 440.3R-04. Guide Test Methods for Fiber-Reinforced Polymers (FRPs) for Reinforcing or Strengthening Concrete Structures; American Concrete Institute: Farmington Hills, MI, USA (2004)
15. ASTM: Standard test method for transverse shear strength of fiber-reinforced polymer matrix composite bars. ASTM D7617/D7617M-11. ASTM, West Conshohocken, PA (2017)
16. ASTM: Standard test method for tensile properties of fiber reinforced polymer matrix composite bars. D7205/D7205M-21. ASTM, West Conshohocken, PA (2021)
17. Ogrodowska, K., Łuszcz, K., Garbacz, A.: Nanomodification, hybridization and temperature impact on shear strength of basalt fiber-reinforced polymer bars. Polymers **13**(16), 2585 (2021)
18. Ogrodowska, K., Urbański, M.: Nanosilica modification of epoxy matrix in hybrid basalt-carbon FRP bars—impact on microstructure and mechanical properties. Materials **16**(5), 1912 (2023)

Crack Propagation Analysis of Model Concrete Columns with BFRP Reinforcement Bars

Małgorzata Wydra[1]([envelope]), Grzegorz Sadowski[1], Piotr Dolny[1], and Jadwiga Fangrat[2]

[1] Faculty of Civil Engineering, Mechanics and Petrochemistry, Warsaw University of Technology, Warsaw, Poland
malgorzata.wydra@pw.edu.pl
[2] Building Research Institute (ITB), Warsaw, Poland

Abstract. Available studies on concrete structural parts with FRP reinforcement bars concern mostly investigations on bent elements (beams, slabs) [1, 2]. There are also available a few theoretical analyses on columns [3–5]. Though, there is still little experimental data concerning concrete columns with FRP bars [6–8], especially subjected to eccentric load, as also underlined in the review article [9]. This research aims to fulfill this research gap. Also, basalt FRP bars were chosen as relatively new type of non-metallic bars with low ecological impact [1].

A total of eight columns with the height of either 750 mm or 1500 mm having 150 mm x 150 mm rectangular cross section were examined under axial or eccentric mechanical load up to 290 kN. Columns were reinforced with four BFRP main bars with the diameter of either 8 or 10 mm, and 8 mm steel stirrups in each case. The results on the thermal and mechanical properties' investigations on BFRP bars were presented in [10]; the compressive strength values of the used BFRP bars were in the range of 441.2–466.8 MPa and elasticity modulus at compression values were equal to 31.0–38.4 GPa. Tested compressive strength of concrete, from which all columns were made (in one concrete pouring) were equal to 33.8 MPa. Each column was loaded in three cycles of loading-unloading, increasing the eccentricity, from 0 to 2 cm, and finally to 4 cm. DIC (Digital Image Correlation) method was used for the analysis of crack propagation (as in earlier research of bent elements [11]), but also unexpectedly there were visualised intensification areas of compression micro-damages. Failure was noted for two elements - B075_8_2 at the eccentricity of 4 cm (failure load – 290 kN after 60 s of sustained load) and B150_10_2 at the eccentricity of 4 cm (280 kN). Other specimens did not fail under load up to 290 kN. Maps from DIC method were also compared with results from numerical modelling (in Abaqus software) with good resemblance.

Keywords: Basalt Fibre Reinforced Polymer · reinforcement bars · Digital Image Correlation · eccentric load · concrete columns

1 Introduction

Available studies on concrete structural parts with FRP reinforcement bars concern mostly investigations on bent elements (beams, slabs) [1, 2]. There are also available a few theoretical analyses on columns [3–5]. Though, there is still little experimental

© The Author(s) 2025
L. Czarnecki et al. (Eds.): ICPIC 2023, 61, pp. 431–439, 2025.
https://doi.org/10.1007/978-3-031-72955-3_43

data concerning concrete columns with FRP bars [6–8], especially subjected to eccentric load, as also underlined in the review article [9]. This research aims to fulfill this research gap. Also, basalt FRP bars were chosen as relatively new type of non-metallic bars with low ecological impact [1].

In this study 8 concrete columns with BFRP main reinforcement and steel transverse reinforcement were investigated experimentally and numerically. Each column (150 mm x 150 mm x 750 mm or 1500 mm) was loaded in three cycles of loading-unloading, increasing the eccentricity, from 0 cm to 2 cm, and finally to 4 cm. DIC (Digital Image Correlation) method was used for the analysis of crack propagation (as in earlier research of bent elements [11]), but also unexpectedly there were visualised intensification areas of compression micro-damages.

2   Specimens

A total of eight columns with the height of either 750 mm or 1500 mm having 150 mm x 150 mm rectangular cross section were examined. The scheme of the column is presented in Fig. 1, while types of analyzed columns in Table 1.

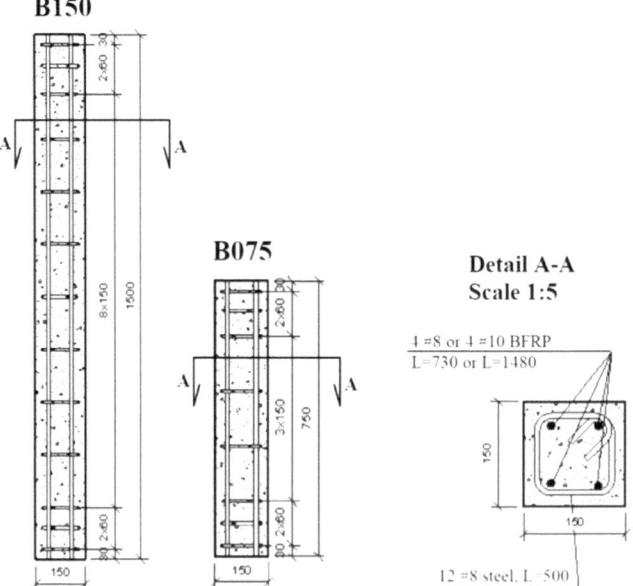

Fig. 1. Scheme of the reinforcement.

Mechanical properties of the BFRP bars were subject of the earlier authors' investigation [10] (Table 2). The concrete used for columns had *w/c* ratio equal to 0.57. Total water amount was 170 kg per 1 m^3 of concrete. Type of used cement was CEM I 42.5N-NA. Consistency class was S1 according to [12]. The tested tensile strength of the steel stirrups was equal to 611.2 MPa, while the offset yield strength 0.2% was equal to 568.0 MPa (Table 3).

Table 1. Parameters of half-scale columns tested at room temperature.

Designation	Column height [mm]	Diameter of the main bar [mm]
B075_8_1	750	8
B075_8_2	750	8
B075_10_1	750	10
B075_10_2	750	10
B150_8_1	1500	8
B150_8_2	1500	8
B150_10_1	1500	10
B150_10_2	1500	10

Table 2. Properties of the BFRP bars (mean values) [10].

Diameter of the bar [mm]	Compressive strength [MPa]	Elasticity modulus at compression [GPa]
8	483	38
10	467	31

3 Methods

3.1 Experiment

Each column was loaded in three cycles of loading-unloading, increasing the eccentricity, from 0 to 2 cm, and finally to 4 cm. DIC (Digital Image Correlation) method was used for the analysis of crack propagation. Each column was photographed from two perpendicular directions (Fig. 2).

3.2 Numerical Model

The numerical model consisted of:

- two rigid bodies – upper and lower (r3d4 mesh elements), at which boundary conditions for the displacement were defined (lower: no rotation and movement, upper: enabled movement at Y axis and rotation against
- Z axis);
- one concrete part (modelled with the use of 10 mm c3d8r finite elements);
- reinforcement modelled by truss (t3d2) elements with the length of 10 mm;
- two steel elements (by which the force was transferred from the hydraulic press to the element), modelled with the use of c3d8r 10 mm elements.

The embedded region function was used for modelling the interaction between concrete and reinforcement. Calculations were performed in one step (static general), in

Table 3. Properties of the used concrete.

Parameter	Mean value	Standard deviation	Specimens	Methods
Compressive strength after 7 days (demoulding day)	24.41 MPa	0.12 MPa	3, cube 150	[13]
Compressive strength after 28 days (stored in water)	33.82 MPa	0.52 MPa	6, cube 150	
Compressive strength after 317 days[1] (stored with columns)	36.89 MPa	0.45 MPa	3, cube 150	
Compressive strength after 360 days[2] (stored with columns)	33.40 MPa	4.59 MPa	3, cube 150	
Stabilised secant elasticity modulus after 318 days	29.0 GPa	1.5 GPa	2, cylinder Φ150x300	[14]
Initial secant elasticity modulus after 361 days	27.7 GPa	-	1, cylinder Φ150x300	
Stabilised secant elasticity modulus after 361 days	31.0 GPa	-		

[1]in that day columns with height of 750 mm tested; [2]in that day columns with the height of 1500 mm tested

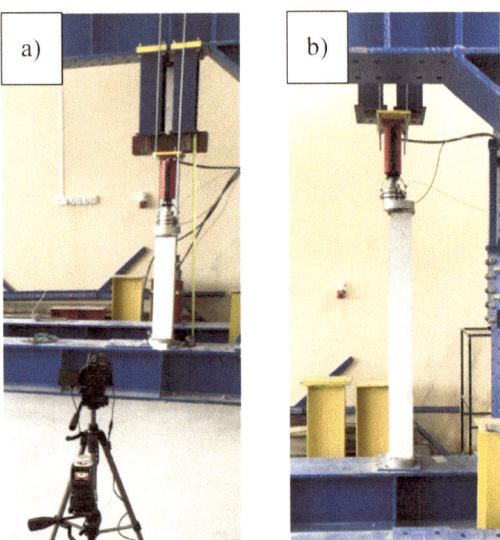

Fig. 2. Test stand - columns with the height of: a) 750 mm; b) 1500 mm.

which vertical displacement of upper rigid body was forced and reaction in the reference point in lower rigid body was measured in order to register a maximum value, after which decrease of reaction was noted (as a failure force value). Similar method was used with satisfactory agreement to experimental results in [6]. The representation of numerical model meshes are shown in Fig. 3.

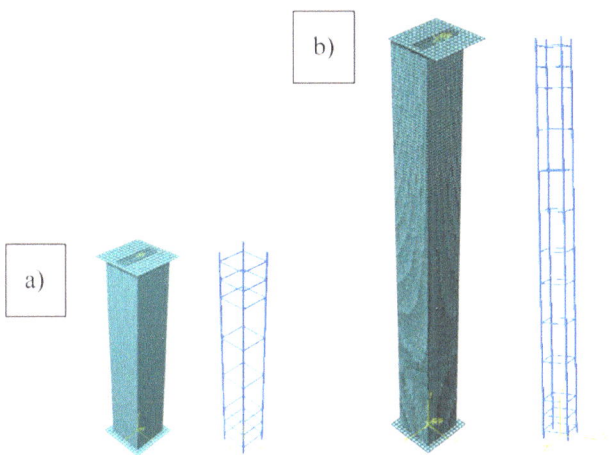

Fig. 3. Numerical model - columns with the height of: a) 750 mm; b) 1500 mm.

Concrete was modelled with the use of Concrete Damaged Plasticity. The general assumptions for CDP (which is modification of Drucker-Prager model) model are described in [15–17].

Parameters for Concrete Damaged Plasticity was following: dilatation angle - Ψ: 36, eccentricity - $\in$: 0.1, f_{b0}/f_{c0}: 1.16, κ: 0.667 and viscosity parameter: 5E-05 [4]. Poisson's ratio is assumed as 0.2. Damage parameters d_c and d_t were calibrated followingly:

$$\sigma_c = (1 - d_c)E_0\left(\varepsilon_c - \varepsilon_c^{pl}\right) \tag{1}$$

$$\sigma_t = (1 - d_t)E_0\left(\varepsilon_t - \varepsilon_t^{pl}\right) \tag{2}$$

where:

σ_c – compressive stress, MPa

σ_t – tensile stress, MPa

E_0 – undamaged modulus of deformation, GPa

ε_c – compression strain, ‰

ε_t – tensile strain, ‰

ε_c^{pl} – plastic compression strain, ‰

ε_t^{pl} – plastic tensile strain, ‰.

Tensile strength at room temperature value was assumed as per [18] and calculated with the following equation:

$$f_{ct} = 0.3f_c^{2/3} \tag{3}$$

where:

f_{ct} – tensile strength at room temperature, MPa

f_c – compressive strength (room temperature), MPa.

The relation between stress and stress-related strains for concrete and steel was based on the European standard [19] for steel-reinforced concrete elements at high temperatures (as the further aims of these calculations were related with the analysis of such structures at high temperatures).

The material parameters were assumed as follows: concrete compressive strength = 35.0 MPa; steel yield strength = 550 MPa, elasticity modulus = 210 GPa for steel, and 35 GPa for BFRP.

4 Results and Discussion

DIC method have been proved useful in determination of crack propagation (as in earlier research of bent elements [11]), but also enabled analysis of compression micro-damages. Observation of compression zones of damages in the experimental part was especially interesting (example shown in Fig. 4b). Vertical cracks at the top of the columns were visible in both – DIC maps (Fig. 4a) and numerical considerations (Fig. 5a,b). At failure, these cracks resulted in detachment of part of the concrete. Also, good resemblance of the damaged areas of the specimen at failure (Fig. 5b) to maps of damages from numerical part of the study (d_c and d_t values) was noted (Fig. 5a).

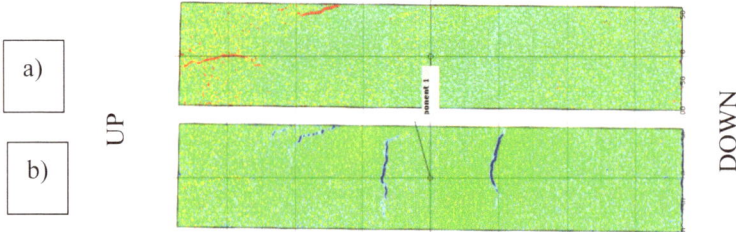

Fig. 4. Observed damaged zones (B075_8_2 column, 4 cm eccentricity): a) major strains (visible cracks); b) minor strains (visible damages in compression).

Failure was noted for two elements in the experiment- B075_8_2 at the eccentricity of 4 cm (failure load – 290 kN after 60 s of sustained load) and B150_10_2 at the eccentricity of 4 cm (280 kN). Other specimens did not fail under the mechanic load up to 290 kN with eccentricity varying from 0 cm to 4 cm.

The predictions on failure force from numerical part of the study were higher than loads applied during the experimental part and are given in Table 4. The values of maximum forced were in the range from 303.1 kN to 355.9 kN. In most cases the numerically predicted failure force was lower with the eccentricity increase. Better utilisation of the BFRP bars (higher stresses) at failure of the column was noted for higher eccentricity (97.8–120.7 MPa for 0 cm; 126.1–133.6 MPa for 2 cm and 221.3–249.7 MPa for 4 cm), but in each case they were much lower than the experimentally determined compressive strength. Lower values of failure force (by 3.5–9.1%) was noted for higher columns.

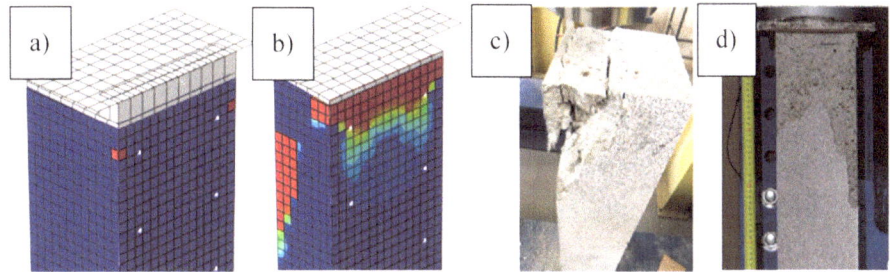

Fig. 5. a) Damages of the column (d_t) noted in numerical considerations (cross section view of 1500 mm column with 10 mm BFRP reinforcement at 4 cm eccentricity) 100 kN; b) at failure; c) column after failure – B075_8_2; d) column after failure - B150_10_2.

Table 4. Maximum force values observed in numerical model along (values in brackets – stresses in the BFRP bars).

Type of column	Eccentricity [cm]		
	0	2	4
Column height: 75 cm, main reinforcement: 8 mm	355.9 kN (118.1 MPa)	327.5 kN (126.1 MPa)	330.8 kN (246.2 MPa)
Column height: 75 cm, main reinforcement: 10 mm	354.2 kN (97.8 MPa)	326.6 kN (128.2 MPa)	341.0 kN (221.3 MPa)
Column height: 150 cm, main reinforcement: 8 mm	340.6 kN (120.7 MPa)	311.7 kN (124.6 MPa)	303.1 kN (249.7 MPa)
Column height: 150 cm, main reinforcement: 10 mm	340.0 kN (110.7 MPa)	315.3 kN (133.6 MPa)	314.6 kN (245.6 MPa)

5   Conclusions

The following conclusions can be drawn from this study:

1. Most of the columns (6 out of 8) did not fail under the applied load (290 kN), which is in line with prediction on the values of failure forces in numerical model (303.1–355.9 kN).
2. In the case of two columns that failed at mechanical load at the level of 280 kN (B150_10_2) and 290 kN (B075_8_2), both at the eccentricity of 4 cm, the experimentally determined failure load was lower than numerical by 11% and 12%, respectively. Also, the column B075_8_2 did not fail immediately, but after 60 s of sustained load.
3. The usefulness of DIC method in location of crack propagation in concrete columns has been confirmed. Also, compressive damaged zones were registered with the use of that method.
4. Failure mode had a good resemblance in damage (dc and d_t) parameters location in numerical model.

References

1. Inman, M., Thorhallsson, E.R., Azrague, K.: A mechanical and environmental assessment and comparison of Basalt Fibre Reinforced Polymer (BFRP) rebar and steel rebar in concrete beams. Energy Procedia **111**, 31–40 (2017)
2. Urbanski, M., Lapko, A., Garbacz, A.: Investigation on concrete beams reinforced with basalt rebars as an effective alternative of conventional R/C structures. Procedia Eng. **57**, 1183–1191 (2013)
3. Korentz, J.: Nośność mimośrodowo ściskanych słupów betonowych ze zbrojeniem niemetalicznym. Builder **4**(297) (2022)
4. Hamze, A.A., Al-Taher, R., Taji, A., Yazbak, D., Abed, F.: Developing interaction diagram for BFRP-RC short columns using FEA. In: 2019 8th International Conference on Modeling Simulation and Applied Optimization, IEEE, Manama, Bahrain, Bahrain (2019)
5. Zadeh, H.J., Nanni, A.: Design of RC columns using glass FRP reinforcement. J. Compos. Constr. **17**, 294–304 (2013)
6. Wydra, M., Włodarczyk, M., Fangrat, J.: Nonlinear analysis of compressed concrete elements reinforced with frp bars. Materials **13**, 1–16 (2020)
7. Włodarczyk, M., Trofimczuk, D.: Prediction of ultimate capacity of FRP reinforced concrete compression members. In: Concrete Innovations in Materials, Design and Structures B. Abstract 2019 Fib International Symposium, May 27–29, Kraków (2019)
8. AlAjarmeh, O.S., Manalo, A.C., Benmokrane, B., Karunasena, W., Mendis, P., Nguyen, K.T.Q.: Compressive behavior of axially loaded circular hollow concrete columns reinforced with GFRP bars and spirals. Constr. Build. Mater. **194**, 12–23 (2019)
9. Elmessalami, N., El Refai, A., Abed, F.: Fiber-reinforced polymers bars for compression reinforcement: a promising alternative to steel bars. Constr. Build. Mater. **209**, 725–737 (2019)
10. Wydra, M., Dolny, P., Sadowski, G., Grochowska, N., Turkowski, P., Fangrat, J.: Analysis of thermal and mechanical parameters of the BFRP bars. Mater. Proc. **13**(1), 1–9 (2023)
11. Sadowski, G., Wydra, M.: Comparison of methods applied to analysis of crack propagation in reinforced concrete composite beam. Porównanie metod badawczych stosowanych w analizie procesu zarysowania belki zespolonej (in Polish). ACTA SCIENTIARUM POLONORUM - Architectura Budownictwom vol. 18, pp. 3–12 (2019)
12. EN 12350-5 : 2011 Testing fresh concrete – Part 5 : Flow table test (2011)
13. EN 12390-3:2019 Testing hardened concrete - Part 3: Compressive strength of test specimens (2019)
14. EN 12390-13:2014 Testing hardened concrete - Part 13: Determination of secant modulus of elasticity in compression (2014)
15. Lubliner, J., Oliver, J., Oller, S., Onate, E.: A plastic-damage model for concrete. Int. J. Rock Mech. Mining Sci. Geomech. Abst. **26**, 252 (1989)
16. Alfarah, B., López-Almansa, F., Oller, S.: New methodology for calculating damage variables evolution in Plastic Damage Model for RC structures. Eng. Struct. **132**, 70–86 (2017)
17. Wosatko, A., Winnicki, A., Polak, M.A., Pamin, J.: Role of dilatancy angle in plasticity-based models of concrete. Arch. Civ. Mech. Eng. **19**, 1268–1283 (2019)
18. EN 1992-1-1:2008 Eurocode 2: Design of concrete structures - Part 1-1 general rules and rules for buildings, n.d
19. EN 1992-1-2 (2004): Eurocode 2: Design of concrete structures - Part 1-2: General rules - Structural fire design (2004)

Serviceability Limit State of Fiber Reinforced Concrete Beams with BFPB Bars and Stirrups

Julita Krassowska[✉] and Marta Kosior-Kazberuk

Department of Building Structures and Mechanical Structures, Bialystok University of Technology, 15-351 Bialystok, Poland
j.krassowska@pb.edu.pl

Abstract. The work analyzes the serviceability ultimate limit state of 4.5 m long fiber-reinforced concrete beams with basalt bars and stirrups (BFRP). On the basis of previous tests, deformations in beams with composite reinforcement are above acceptable values. Beams were made of concrete with basalt fibers to improve deformability, cracks resistance and deflection. The tests showed that the load capacity of beams reinforced with BFRP bars was lower than that of beams with steel reinforcement, resulting from different failure mechanisms of both beams. The failure of beams with BFRP reinforcement was rapid. Deformations in the concrete were reduced by using basalt fibers in the concrete. Increasing the stiffness of the structure with reinforcement with BFRP bars and stirrups using concrete with basalt fibers can meet the SLS requirements for limiting the deflection and cracking of concrete elements reinforced with them.

Keywords: BFRP · BFRP · concrete deformation · composite reinforcement · basalt fibers · fiber-reinforced concrete

1 Introduction

Over the years, various attempts have been made to reduce the effects of corrosion in reinforced concrete structures. The durability of reinforced concrete structures holds utmost significance when considering industrial facilities and various elements exposed to coastal environments, bridge pavements, ground contact, thin-walled components, or situations where achieving high-quality concrete is impractical [1–3]. The solution may be to use non-metallic reinforcement. An important feature is the good corrosion resistance of FRP fibers [4]. The most commonly used composite materials currently include: composites with glass fibers GFRP (Glass Fiber Reinforced Polymer), with carbon fibers CFRP (Carbon Fiber Reinforced Polymer) and with basalt fibers BFRP (Basalt Fiber Reinforced Polymer) [5].

The research carried out so far shows that the modulus of elasticity of composite bars is about five times smaller than the modulus of elasticity of steel reinforcement. The results presented in [4, 6] shown a much greater reduction of the cross-section stiffness of the FRP reinforced concrete element after its cracking than in the case

© The Author(s) 2025
L. Czarnecki et al. (Eds.): ICPIC 2023, 61, pp. 440–448, 2025.
https://doi.org/10.1007/978-3-031-72955-3_44

of a concrete element with steel reinforcement. The moment of inertia of the cross-section after cracking in beams with composite reinforcement is about four times lower than in beams with steel reinforcement [7]. As a result, in the serviceability limit state (SLS), much greater values of deflections and crack widths are observed, comparing to reinforced concrete elements (beams and slabs). Tests of concrete with basalt fibers used in two-span beams show the possibility of reducing the deflection and deformation values on the concrete surface [8].

The tests presented in the paper aimed to evaluate the behavior of concrete beams with BFRP bars and stirrups made of concrete with basalt fibers. Beams were subjected to bending. During the tests, the development of cracking and deformation of beam elements was analyzed.

2   Materials and Research Methods

The concrete was made of Portland cement CEM I 42.5R and natural aggregate with a grain size of 0-16mm. The cement content was 320 kg/m^3. Concretes with w/c = 0.5 ratio were selected for the test. Basalt fibers were added to the concrete in an amount of 8 kg/m^3. Basalt fibers are made of thin chopped basalt fibers with a fiber elementary diameter of 20 μm, tensile strength of 750 MPa and Young's modulus of 89 GPa. The concrete recipe is given in Table 1.

The concrete had an average compressive strength f_{ck} = 43,78 MPa, tensile strength f_{ctm} = 5,55 MPa and a modulus of elasticity E_{cm} = 40,64 GPa. Basalt fiber reinforced concrete, respectively: f_{ck} = 44,52 MPa, f_{ctm} = 6,11 MPa and E_{cm} = 42,02 GPa.

The subject of the tests were beams with dimensions of 120x300x4500 mm. The reinforcement of the beam was steel and basalt bars and stirrups. Ribbed bars with a diameter of Ø6mm and Ø14mm, made of BSt500s steel with a yield strength of f_{yk} = 500MPa. BFRP reinforcement with a diameter of Ø6 mm and Ø14 mm and guaranteed tensile strength equal to $f_{u,ave}$ = 1180 MPa, guaranteed modulus of elasticity E_f = 47,6 GPa and guaranteed deformation at break ε^*_{fu} = 2,0%.

Two series of test elements were made, differentiated due to the longitudinal reinforcement A - steel and B - basalt. In series A, beams of concrete with basalt fibers in the amount of 8 kg/m3 (WB) and reference beams of concrete without fibers were also made (W0). In series B, concrete with basalt fibers in the amount of 8 kg/m3 was used in all beams (WB).

In the program of testing single-span beams, three series were assumed, differing in the distribution of transverse reinforcement. Series I beams were shear reinforced with stirrups with spacing determined according to PN-EN 1992-1-1 [9] due to the maximum spacing. In series II, the stirrups had a spacing twice as large as that established in PN-EN 1992-1-1 [9]. Series III beams have not stirrups. The Fig. 1 shows a diagram of the research program.

Single-span beams were loaded in a the 4-point bending with a span length of l_{eff} = 4500mm. The support and load diagram of the beams is shown in Fig. 2. The distances between the supports were 4200 mm. The measured values were the displacements of the beams on the supports (Fig. 3) and in the span, deformations at various points of the height of the support section and the measurement of span deformations using the DIC

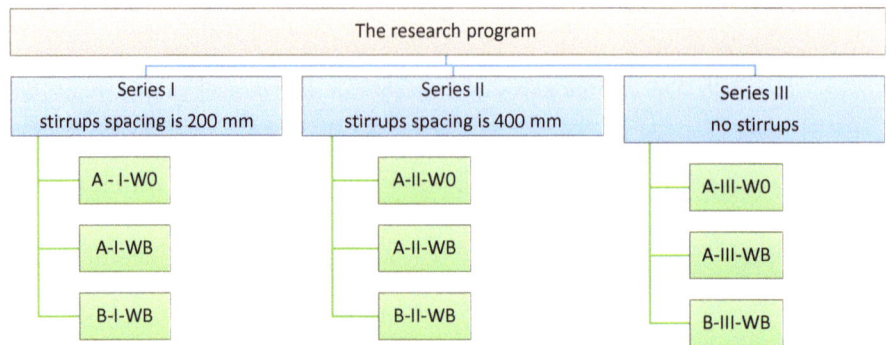

Fig. 1. The research program

method. The test load increment was 5.0 kN. At each stage, the values of deflections and deformations were recorded. At the same time, crack propagation was mapped. Deformation measurement was made using a contact extensometer with a measurement base of 250 mm and an accuracy of 0.001 mm. The deflections were measured using inductive sensors with a measurement base of 50 mm and an accuracy of 0.01 mm. The deflections were measured using inductive sensors with a measurement base of 50 mm and an accuracy of 0.01 mm.

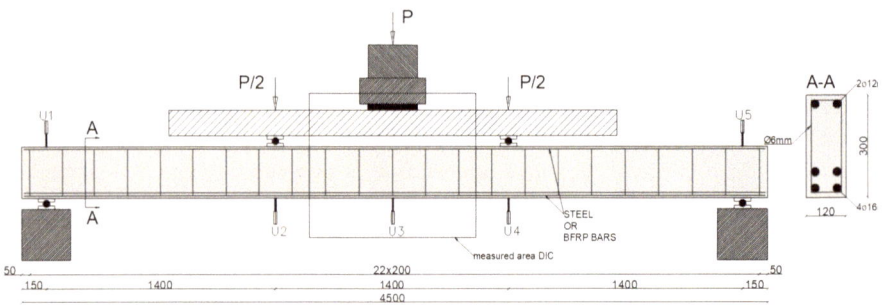

P- force; U1-5 - deflection;

Fig. 2. Geometry, details of reinforcement, loads and instrumentation

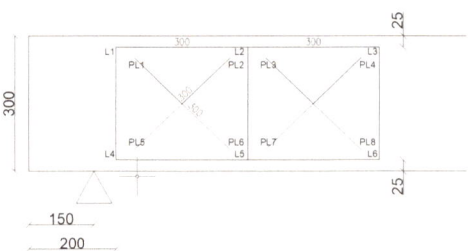

Fig. 3. Scheme of the distribution of deformation measurement points on the beam surface

3   Analysis of the Serviceability Limit State of BFRP Beams

3.1   The State of Deformation

Figure 4 shows comparative diagrams of deformation values on the concrete surface in the support zones in particular series in the failure phase.

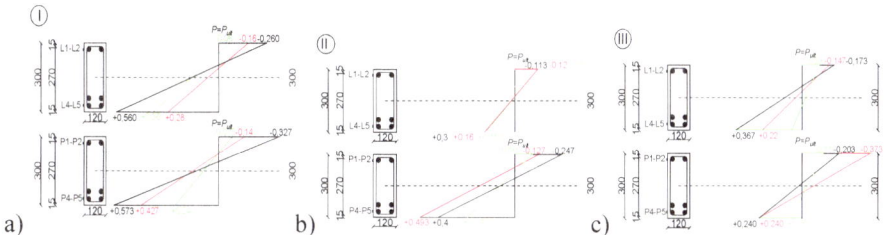

Fig. 4. Comparative diagram of the deformation values on the concrete surface in the support zones in the series a)I b II) c)III (black-A-W0; red-A-WB; green B-WB)

The extent of the compression zone was significantly reduced in the case of beams with composite reinforcement. The highest tensile stress values were also recorded for them.

The different character of deformations of cracked beams reinforced with BFRP bars, compared to reinforced concrete beams, results from the way of deformation of both types of reinforcement. Steel bars deform elastically until they reach their elastic limit. BFRP rods, on the other hand, deform elastically until they break. The addition of basalt fibers in the amount of 8 kg/m^3 significantly reduced the deformation of the concrete element. Higher deformation values are typical for elements with BFRP reinforcement, which results from the similar values of the modulus of elasticity of the composite reinforcement and concrete.

Figure 5 shows the maps of the main deformations, obtained using DIC (Digital Image Correlation System) and diagrams of deformations on the concrete surface at the level of the longitudinal compression and tension reinforcement. The analysis of the deformation maps of the individual series corresponds to the recorded cracking, as shown in Fig. 5.

Beams reinforced with BFRP bars and stirrups were characterized by smaller ranges of the compressed zone in the middle of the span than the beams with steel reinforcement. In Fig. 6, one can observe the change in the position of the neutral axis of the cross-section in individual series and the difference in the height of the perpendicular cracks. The height of the compressed zone decreased with the use of composite reinforcement. The multiplicity of tensile stresses in all series turned out to be the highest in BFRP beams. In the case of the addition of basalt fibers, the neutral axis of the cross-section and the size of the deformations in the failure phase were smaller with the deformations in the reference beams.

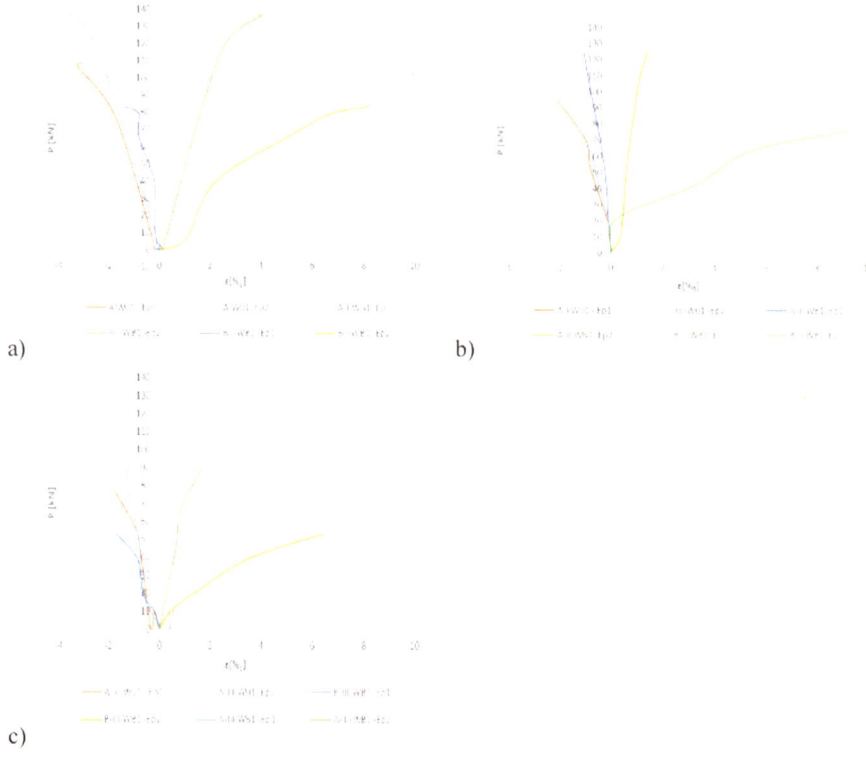

a)

b)

c)

Fig. 5. Strains in concrete for series a) I, b) II c) III

3.2 Failure Mode

Figure 7 shows the image of failure of the example beams of the series with composite reinforcement.

In series I and II with steel reinforcement, the beams failed by bending. Most of the cracks were perpendicular to the axis of the element, single diagonal cracks appeared at the effort of approx. 75% P_{ult}. In beams with composite reinforcement, the failure was determined by the transverse force, causing failure by shear in the support sections. By far the largest number of cracks with the largest crack width were characteristic of beams with composite reinforcement. The failure of the BFRP beams was sudden, caused by brittle fracture of the stirrups. In the case of beams with steel reinforcement, the failure process was much slower. The number of cracks in the BFRP beams was greater, mainly perpendicular cracks in the middle part of the element caused by bending.

The failure of series III beams resulted from the shearing of the support zone in all series. The destructive crack started from the place where the load was applied and ended at the support. In the beams of the A-III-W0 and A-III-WB series, cracks along the reinforcement caused by slippage were observed. The steel reinforcement under destructive loads was subject to local adhesion loss, while the composite reinforcement was subject to brittle fracture.

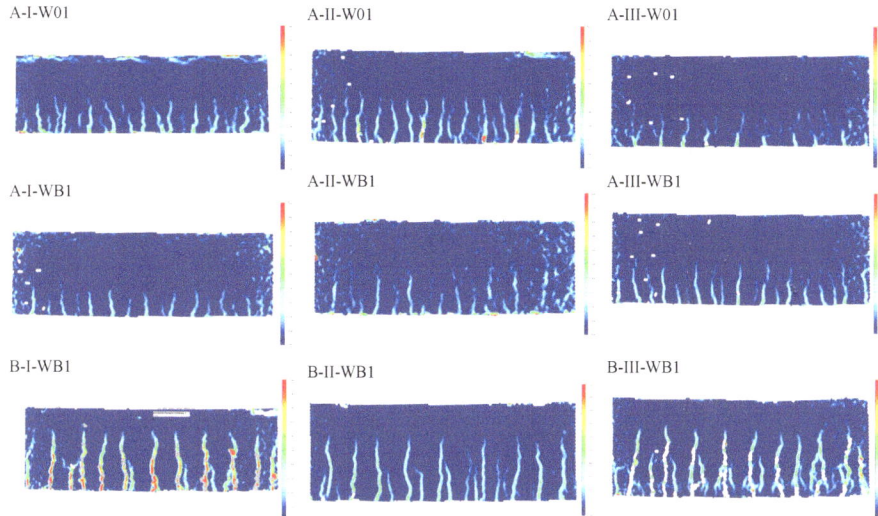

Fig. 6. Strains recorded using DIC in span area

Figure 8 shows a model of failure by brittle cracking of the longitudinal or transverse reinforcement typical for BFRP structures. The BFRP reinforcement fractured suddenly and unexpectedly, or in the case of beams without transverse reinforcement, the longitudinal members buckled and completely failed the compression zone. The destruction of the stirrups BY high shear force caused the stirrups to open or break in the vicinity of the bend.

4   Conclusions

Based on the tests, it was found that the deformability of the tested beams reinforced with BFRP bars was lower than that of beams with traditional reinforcement, which resulted from different failure mechanisms of both types of beams.

Beams reinforced with BFRP experienced rapid failure. In both Series I and II, BFRP stirrups exhibited brittle cracking. Beams without transverse reinforcement showed signs of slippage along the reinforcement.

Thanks to the use of basalt fibers in concrete, the size of deformations at the same levels of the beams was reduced without the addition of fibers. Increasing the stiffness of the structure with reinforcement with BFRP bars and stirrups using concrete with basalt fibers can meet the SLS requirements for limiting deflection and cracking of concrete elements reinforced with them.

Fig. 7. Model of failure of beams of series a) A-I-W0 b) A-I-WB c) B-I-WB

Fig. 8. Selected details of failure of beams with composite reinforcement

Acknowledgements. The work was carried out at the Białystok University of Technology as part of a research project financed by the National Center for Research and Development entitled "Innovative hybrid FRP reinforcement for infrastructural structures with increased durability" project number PBS3 / A2 / 20/2015 (ID 245084) and as part of financing by the Ministry of Science and Higher Education of the Republic of Poland; project number WZ/WB-IIL/ 4/2020.

References

1. Abed, F., El Refai, A., Abdalla, S.: Experimental and finite element investigation of the shear performance of BFRP-RC short beams. Structures **20**, 689–701 (2019). https://doi.org/10.1016/j.istruc.2019.06.019
2. Lapko, A., Urbański, M.: Experimental and theoretical analysis of deflections of concrete beams reinforced with basalt rebar. Arch. Civil Mech. Eng. **15**(1) (2015). https://doi.org/10.1016/j.acme.2014.03.008
3. Zhao, Y., Wanga, L., Lei, Z., Han, X., Shi, J.: Study on bending damage and failure of basalt fiber reinforced concrete under freeze-thaw cycles. Const. Build. Mat. **163**, 460–470 (2018)
4. Hollaway, L.C.: A review of the present and future utilisation of FRP composites in the civil infrastructure with reference to their important in-service properties. Const. Build. Mat. **24**, 2419–2445 (2010)
5. Alshannag, M., Alshmalani, M., Alsaif, A., Higazey, M.: Flexural performance of high-strength lightweight concrete beams made with hybrid fibers. Case Stud. Const. Mater. **18**, e01861 (2023). https://doi.org/10.1016/j.cscm.2023.e01861
6. Urbanski, M., Lapko, A., Garbacz, A.: Investigation on concrete beams reinforced with Basalt Rebars as an effective alternative of conventional R/C structures. Procedia Eng. **57**, 1183–1191 (2013). https://doi.org/10.1016/j.proeng.2013.04.149
7. Zou, X., Lin, H., Feng, P., Bao, Y., Wang, J.: A review on FRP-concrete hybrid sections for bridge applications. Compos. Struct. 113336 (2020). https://doi.org/10.1016/j.compstruct.2020.113336
8. Krassowska, J., Kosior-Kazberuk, M.: The effect of steel and Basalt Fibers on the Shear behavior of double-span fiber reinforced concrete beams. Materials **14**(20), Art. nr 20, paź. 2021. https://doi.org/10.3390/ma14206090
9. EN 1992–2:2005 - Eurocode 2. Design of concrete structures. Concrete bridges. Design and detailing rules

Shear Deformability of Reinforced Concrete Beams Strengthened with the FRCM System

Pavlo Vegera[1]([✉]), Iryna Grynyova[2], Zinoviy Blikharskyy[1], Roman Khmil[1], and Oksana Korobko[2]

[1] Department of Building Constructions and Bridges, Lviv Polytechnic National University, Lviv, Ukraine
`Pavlo.I.Vehera@lpnu.ua`
[2] Odessa State Academy of Civil Engineering and Architecture, Odesa, Ukraine

Abstract. The article presents the results of experimental studies of reinforced concrete beams on the shear without transverse reinforcement strengthened by the FRCM system. For the implementation of the research, four experimental samples were designed and manufactured, with cross-sectional dimensions of 200x100 mm and a length of 2100 mm. The beams are designed in such a way that even after strengthening the support areas, the failure occurs due to the shear force. None of the samples is destroyed by the bending moment. The tests were carried out according to the authors' improved methodology, by testing each sample twice. The samples were strengthened by the FRCM composite system at load levels of 0, 0.3, and 0.5 of the bearing capacity of the control samples. Reinforced concrete beams were strengthened by gluing P.B.O. fabrics in the form of vertical strips with a width of 70 mm, for the possibility of fixing the concrete strains in the support areas. Samples strengthened by the FRCM system are destroyed more smoothly and plastically than unstrengthened beams, and there is no mass fallout of concrete particles. According to the obtained data, graphs of the strain distribution in support area and the isofield of their distribution were constructed. In accordance with the results of the research, the maximum effect of the composite system use for the shear reinforcement was established by 26…57%. With increasing the load level at which the sample is strengthened, the effect of the strengthening decreased.

Keywords: shear · strengthening RC beams · FRCM system

1 Introduction

Research measuring the residual bearing capacity of reinforced concrete elements has been developed in response to growing tendencies in cost minimization during reconstruction. This is particularly true of industrial structures, where reinforced concrete components are frequently utilized and whose application varies depending on market demands. Therefore, it is necessary to modify existing buildings and structures in order to reduce expenses while implementing new jobs.

In the article [1], the necessity of evaluating the technical state of building structures and figuring out the remaining bearing capacity of damaged flexural reinforcing concrete

© The Author(s) 2025
L. Czarnecki et al. (Eds.): ICPIC 2023, 61, pp. 449–457, 2025.
https://doi.org/10.1007/978-3-031-72955-3_45

elements is taken into consideration. It is emphasized how difficult it is to calculate damaged components as a result of numerous flaws and damage and how composite reinforced concrete is. The article [2] provides a case study of the cost-effectiveness of various combinations of concrete class, rebar type, and reinforcing ratio in terms of the dependability of prestressed reinforced concrete structures according to the shear bearing capacity.

A non-destructive method of acoustic emission [4] may be used to examine such structures in non-laboratory conditions. Another example is the combination of a traditional reinforcing frame and dispersed reinforcement of concrete with steel fiber with the variation of the volumetric distribution of steel fiber [3]. In the work [5], the fundamental rules for calculating the flexural elements made of fiber concrete and reinforced concrete in relation to bending moments are discussed.

The article [6] discusses the significance of researching how damage affects the bearing capacity of reinforcing concrete parts in the building sector. The example study in the paper [7] illustrates how the unpredictability of the vibration load impacts bearing capacity and deformability. The analysis entails assessing structural damage brought on by machine operation as well as measuring the actual vibration level.

The article [8] presents the findings of an experimental study of reinforced concrete beams damaged in the compression zone. The main focus of attention is on how concrete degradation affects the beams' capacity to support themselves.

The impact of varying reinforcement damage on bent parts of rectangular sections is a topic covered in the article [9]. In these studies, reinforced concrete beams with tensile reinforcement damage are experimentally examined to determine the effects of introducing an initial load. The distinctive qualities of the reinforcement's construction must also be considered when analyzing the consequences of corrosion; in work [10], this effect is taken into account when using thermally enhanced reinforcement. The examination into T-shaped reinforced concrete beam damage is described in [11]. The shear span, the ratio between the thickness of the overhang and the working height of the beam section, the reinforcement ratio for shear reinforcement, and the degree of pre-stress in the tensile rebar were all taken into consideration. In accordance with the physics of reinforced concrete, it is advisable to think of the grid as a continuous reinforced concrete beam [12]. Consider the work in [13], which illustrates the use of FRCM reinforced with a composite system, as a method for reinforcing weak beams. Although prestressed bending sections also have a similar vital importance as discussed in the article [14], the issue of high-quality suitable work of diverse materials is particularly crucial for reinforcements built of composite materials. Due to the fact that reinforced concrete is a composite material, it is crucial to take into account both its material properties and crack resistance, especially when reinforcement is utilized. The relationship between crack resistance and load type was looked at in [15].

2   Aim of the Research

The purpose of this paper is to research parameters of the strength and deformability of reinforced concrete beams without transverse reinforcement, strengthened by the FRCM system under different loading levels.

3   The Results of Experimental Research

To realize the aim, four experimental samples were designed and manufactured, with cross-sectional dimensions of 200x100 mm and a length of 2100 mm. The working tensile reinforcement is adopted class A400C Ø18 mm, compressed reinforcement - A400C Ø10 mm. There is no transverse reinforcement in the zone of action of the shear force. The estimated beam span is 1900 mm.

Beams are labeled, according to the following type: BC - control beam or BSC - beam strengthened with composite material; the first digit is the serial number, the second digit is the test sample number, and the third digit is the tested support area. For example, BC 1.2–2 means that the second support area of the second beam from the first series was tested. The index 0...0.5 means the level at which strengthening was performed, taken from the obtained destructive one, from control beams data.

The beams are designed in such a way that even after strengthening the support area, the failure occurs due to the shear force. None of the samples is destroyed by the bending moment. During the research, each sample was tested twice - each support area separately [13].

According to the research program, reinforced concrete beams were reinforced by gluing P.B.O. fabrics in the form of vertical strips with a width of 70 mm, for the possibility of observing the concrete strains in the support area. Samples BSC 1.1-0 were reinforced without initial load; beams BSC 1.2-0.3 and BSC 1.3-0.5 were strengthened at the level of the initial load equal to 0.3 and 0.5 from the destructive one determined by experimental testing of control samples. The criterion for the loss bearing capacity was adopted similar to that for unreinforced samples: the exhaustion of the bearing capacity on the shear was equated to the physical destruction of the compressed concrete zone above the top of the diagonal crack. The destruction of a reinforced concrete beam on the shear strengthened with a composite system occurred in the following sequence:

– opening of the diagonal crack of maximum width ($w_k = 0.4$ mm) on the concrete surface;
– the spread of the diagonal crack to the concrete compressed zone and the appearance of cracks branching with an opening width of $w_k = 0.05...0.2$ mm on the surface of the strengthening system;
– destruction of the concrete above the top of the diagonal crack in the zone of action of the main tensile stresses and detachment of the reinforcement in this zone;
– plastic deformation of the rebars of the reinforcing frame and destruction of the concrete of the compressed zone, significant deformations of the fabric of the strengthening system, which can be seen due to the violation of the protective layer.

When the load is further increased, the ends of the fabric are completely peeled off and its anchoring is disturbed.

Exhaustion of the bearing capacity occurred at the moment of exfoliation of the concrete compressed zone, together with a sharp elongation of the fabric tape and damage to the protective layer of the FRCM system in the area of propagation of the diagonal crack (Fig. 1).

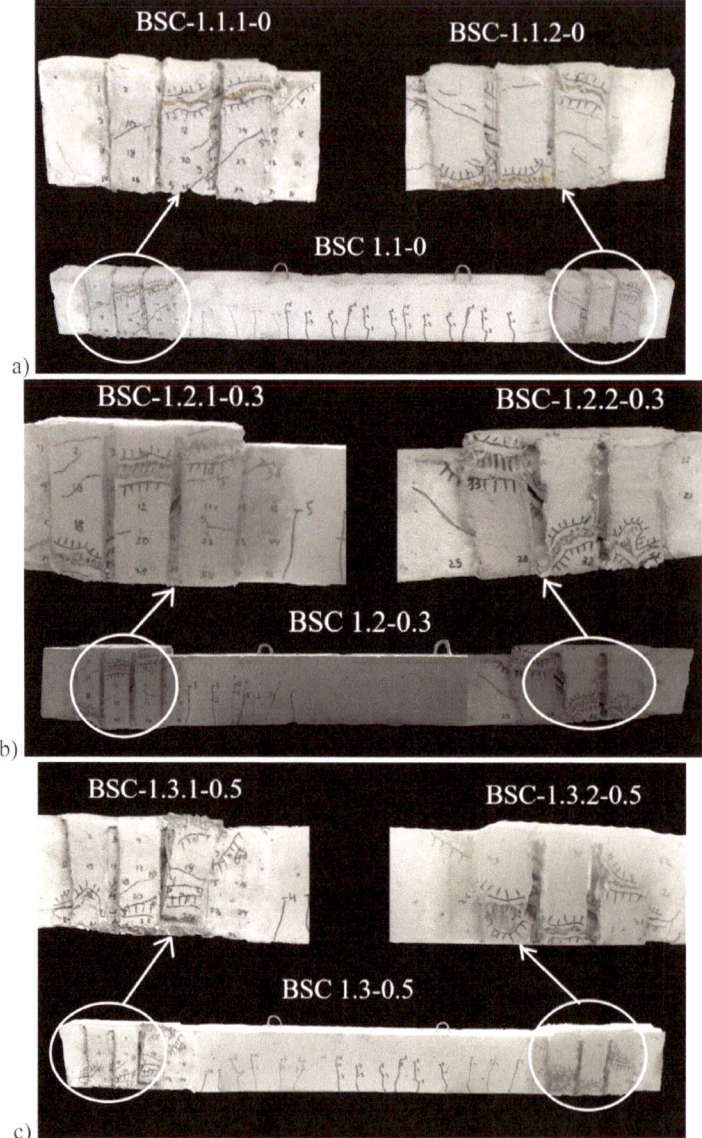

Fig. 1. Tested experimental samples: a) BSC 1.1; b) BSC 1.2-0.3; c) BSC 1.3-0.5

The bearing capacity for the action of the transverse force was: for the sample BSC 1.1 – $V_{Ed} = 137.5 \, kN$, for the beam BSC 1.2-0.3; $V_{Ed} = 120 \, kN$ and for BOD 1.3-0.5 - $V_{Ed} = 110 \, kN$.

At the same time, the nature of the failure has changed for the strengthened samples: the beam loses its bearing capacity more plastically, there is no fallout of concrete particles and no visible plastic deformation of the reinforcing frame. The deformation distribution, which is shown on the isopoles (Fig. 2), indicates the distribution of tensile forces over a larger area of the support area.

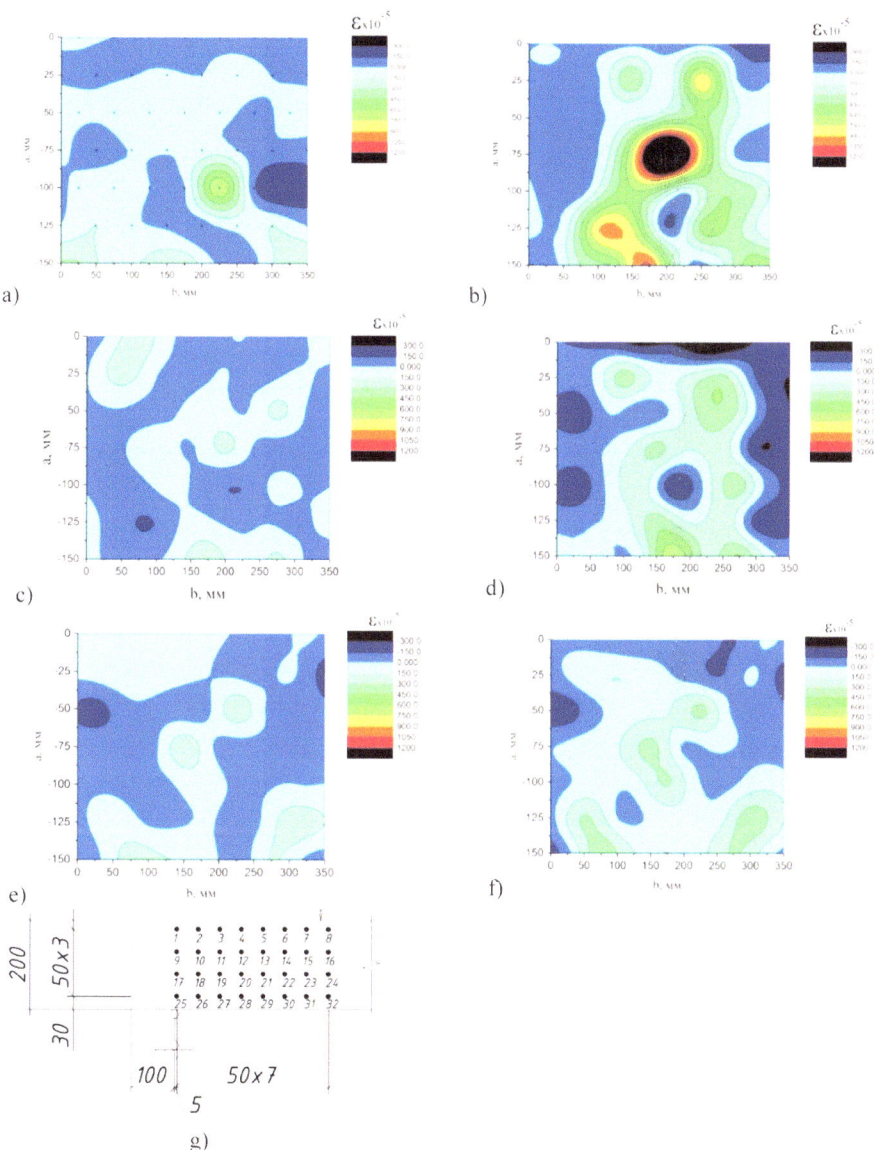

Fig. 2. Isofields - deformations distribution in the support areas: isofields before the opening of the diagonal crack for a beam: a) BSC 1.1-0; c) BSC 1.2-0.3; e) BSC 1.3-0.5; isofield before the beam's bearing capacity is exhausted: b) BSC 1.1-0; d) BSC 1.2-0.3; f) BSC 1.3-0.5; g) a scheme for placing the benchmarks of the comparator

For the sample strengthened without initial load, stress concentration occurred at the half height of the support area, as in the unstrengthened sample. Therefore, this is not characteristic of the strengthened samples under the action of the load. This deformations distribution is caused by a more effective inclusion of the tape in the work, during strengthening without the action of the load. For other samples, strengthening was performed in the presence of significant tensile deformations in the element, which led to a change in the distribution of forces in the section.

Concrete tensile deformations, together with the opening width of diagonal cracks, are similar in nature to those of the control samples but reach significantly higher values (Fig. 3).

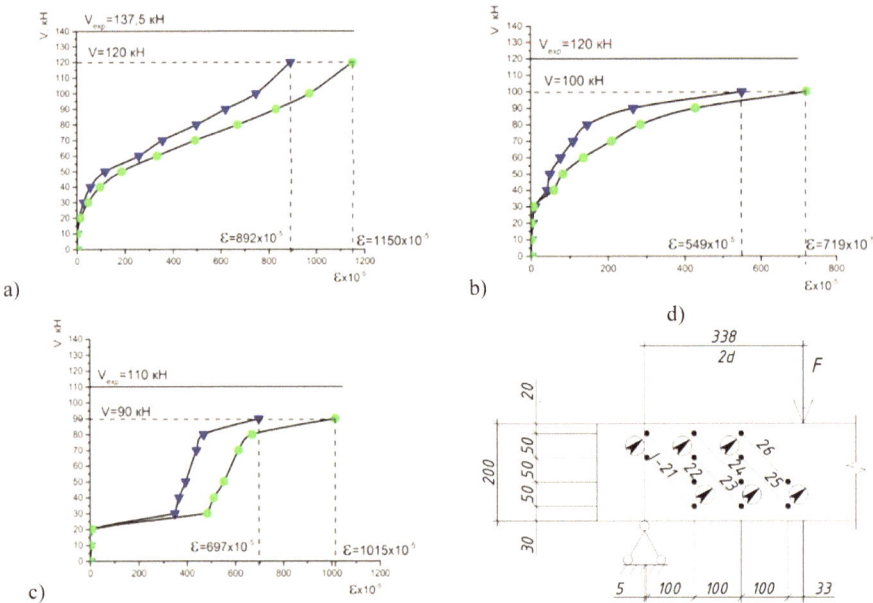

Fig. 3. Concrete tensile deformation of the support area of the beam: a) BSC 1.1-0; b) BSC 1.2-0.3; c) BSC 1.3-0.5; d) sections of strain measurement

For the beam BSC 1.3-0.5, strengthening was performed already after the opening of the diagonal crack, as evidenced by the rapid increase in deformations without increasing the load. Tensile deformations reach their maximum values for the beam BSC 1.1-0 and decrease by the decrease in the bearing capacity on the shear.

The deformations of the strengthening system were measured in the longitudinal direction - the direction of placement of working fibers. The strain graph of the fibers of the reinforcement tape is shown in Fig. 4.

The maximum deformations reach, is 57% of the ultimate elongation, for the BSC 1.1-0 beam. This is a very high indicator of the use of reinforcing tape. With a change in the load level, the maximum deformations of the reinforcing tape also change and amount to 26% for the beam BSC 1.2-0.3 and 43% for the beam BSC 1.3-0.5. Strains of the strengthening element of the beam BSC 1.2-0.3 showed the lowest values, which is associated with the strengthened at the onset of the ultimate tensile deformations of concrete, and the inclusion of the strengthened element in the operation before the opening of the diagonal crack. During the exhaustion of the bearing capacity on the shear, the strengthening fabric received significant deformations, which led to the loss of its initial length, but the fabric rupture was not observed.

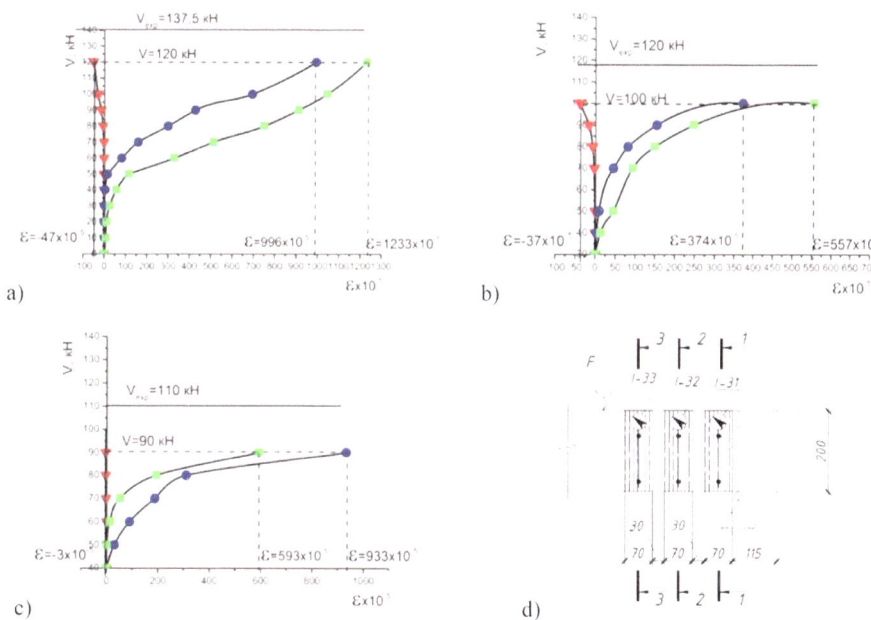

Fig. 4. Strengthening tape strains: a) BSC 1.1-0; b) BSC 1.2-0.3; c) BSC 1.3-0.5; d) sections of strain measurement

According to [16], the use of strengthening tape is recommended to be designed at the level of 40% of its ultimate elongation strain, based on which we can conclude - according to experimental data, this type of strengthening is also an effective way of using high physical and mechanical characteristics of the composite material when reinforced beams on the shear.

4   Conclusions

Based on the above, the following conclusions can be drawn:

- Samples strengthened by the FRCM system are destroyed more slowly and plastically, and there is no massive fallout of concrete particles;
- The deformations of the beam support area are found in the same range for all samples, and are smaller than the tensile deformations recorded on the control samples;
- The maximum deformations of the elongation of the strengthening tape are 57% of the ultimate elongation, which is a high indicator;

- With an increase in the load level at which strengthening is performed, the effect of using the tape changes and depends on the presence of cracks in the support area.

References

1. Lobodanov, M., Vegera, P., Blikharskyy, Z.: Influence analysis of the main types of defects and damages on bearing capacity in reinforced concrete elements and their research methods. Prod. Eng. Arch. **22**(22), 24–29 (2019)
2. Ahaieva, O., Vegera, P., Karpiuk, V., Posternak, O.: Design reliability of the bearing capacity of the reinforced concrete structures on the shear. In: Blikharskyy, Z. Proceedings of EcoComfort 2022, Lecture Notes in Civil Engineering, vol. 290, pp. 1–15. Springer, Heidelberg (2023)
3. Babych, Y.M., Savitskiy, V.V., Andriichuk, O.V., Ninichuk, M.V., Kysliuk, D.Y.: Results of experimental research of deformability and crack-resistance of two span continuous reinforced concrete beams with combined reinforcement. In: Panchenko, S. (eds.) TRANSBUD-2019, IOP Conference Series: Materials Science and Engineering, vol. 708, No. 1, p. 012043. IOP Publishing Ltd (2019)
4. Adamczak-Bugno, A., Lipiec, S., Vavruš, M., Koteš, P.: Non-destructive methods and numerical analysis used for monitoring and analysis of fibre concrete deformations. Materials **15**(20), 7268 (2022)
5. Kochkarev, D., Galinska, T., Tkachuk, O.: Normal sections calculation of bending reinforced concrete and fiber concrete element. Int. J. Eng. Technol. **7**(3), 176–182 (2018)
6. Blikharskyy, Z., Lobodanov, M., Vegera, P.: Investigation of defective reinforced concrete beams with obtained damage of compressed area of concrete. Prod. Eng. Arch. **28**(3), 225–232 (2022)
7. Ilnytskyy, B.M., Kramarchuk, A.P., Bula, S.S., Bobalo, T.V.: Study of the vibration influence on load-bearing floor structures in case of machinery operation. In: Panchenko, S. (eds.) TRANSBUD-2019, IOP Conference Series: Materials Science and Engineering, vol.708, No. 1, p. 012052. IOP Publishing Ltd (2019)
8. Lobodanov, M., Vegera, P., Khmil, R., Blikharskyy, Z.: Influence of damages in the compressed zone on bearing capacity of reinforced concrete beams. In: Blikharskyy, Z. Proceedings of EcoComfort 2020, Lecture Notes in Civil Engineering, vol. 100, pp. 260–267. Springer, Heidelberg (2021)
9. Kopiika, N., Vegera, P., Vashkevych, R., Blikharskyy, Z.: Stress-strain state of damaged reinforced concrete bended elements at operational load level. Prod. Eng. Arch. **27**(4), 242–247 (2021)
10. Khmil, R., Blikharskyy, Z., Vegera, P., Kopiika, N.: Bearing capacity of reinforced concrete beams with and without damages of rebar. Prod. Eng. Arch. **29**(3), 298–303 (2023)
11. Karpiuk, V., Somina, Y., Karpiuk, F., Karpiuk, I.: Peculiar aspects of cracking in prestressed reinforced concrete T-beams. Acta Polytechnica **61**(5), 633–643 (2021)
12. Kos, Z., Klymenko, Y., Karpiuk, I., Grynyova, I.: Bearing capacity near support areas of continuous reinforced concrete beams and high grillages. Appl. Sci. **12**(2), 685 (2022)
13. Kos, Z., Blikharskyi, Z., Vegera, P., Grynyova, I.: A calculation model for determining the bearing capacity of strengthened reinforced concrete beams on the shear. Appl. Sci. **13**(8), 4658 (2023)
14. Borzovič, V., Laco, J., Pecník, M., Pažma, P.: The crack development mechanism of prestressed girder influenced by different bond between pre-stressed tendons and concrete. Key Eng. Mater. **691**, 309–320 (2016)

15. Golewski, G.L.: The Phenomenon of cracking in cement concretes and reinforced concrete structures: the mechanism of cracks formation, causes of their initiation, types and places of occurrence, and methods of detection—a review. Buildings **13**(3), 765 (2023)
16. Externally bonded FRP reinforcement for RC structures. Technical report / [T. Triantafillou, S. Matthys, K. Audenaert, G. Balázs, and oth]. – St.: International Federation for Structural Concrete (fib)., 130p. (2001)

Fast Tannic Acid Surface Modification for Improving PE Fiber-Cement Matrix Bonding Performances

Ali Bashiri Rezaie[✉], Marco Liebscher, Mahsa Mohammadi, and Viktor Mechtcherine

Faculty of Civil Engineering, Institute of Construction Materials, Technische Universität Dresden, Dresden, Germany
ali.bashiri_rezaie@tu-dresden.de

Abstract. In cementitious composites, an application of various fibers can contribute to endow a controlled crack propagation, moderated brittle failure, superior tensile strength and higher energy absorption capacity. Fiber-matrix bonding properties play a key role in fiber strengthening efficiency and the final mechanical performances of the reinforced matrices. This is true specifically for high-performance polyethylene (PE) fibers which yield very high tensile strength and modulus of elasticity, but do not interact properly with cementitious matrix due to their inert hydrophobic surface lacking functional groups.

In the presented work, PE fibers are functionalized by using fast tannic acid modification technique to enhance the bonding properties between a cementitious matrix and the fibers. Environmental scanning electron microscopy (ESEM) confirmed the presence of polymer coating layers on the fiber surfaces. Micromechanical tests indicated that the modified fibers considerably improved the maximum fiber pullout force, interfacial shear strength and pullout work in comparison with the reference fibers. This enhancement in bonding properties could be traced back to the created functional layer on the PE surface triggering a better interaction with cement hydrates as well as a rougher surface enhancing fiber-matrix mechanical interlocking at interfaces. Overall, the introduced approach can be applied for different fibers to promote their bonding behavior with cementitious matrices resulting in an enhanced fiber reinforcing effect in composites.

Keywords: Cementitious Composites · Fiber Reinforced Concrete · Polyethylene Fibers · Interfacial Properties · Fast Tannic Acid Modification

1 Introduction

Cementitious materials are characterized by high compressive strengths, while possessing poor tensile properties, low strain capacity and meager resistance to cracking with noticeable brittleness [1]. As a consequence, different non-polymeric and polymeric fibers have been embedded into cementitious matrices to produce fiber reinforced cementitious composite (FRCC) with enhanced tensile strength, controlled crack propagation, mitigated brittle failure and higher energy absorption capacity [2]. Ultra-high

L. Czarnecki et al. (Eds.): ICPIC 2023, 61, pp. 458–465, 2025.
https://doi.org/10.1007/978-3-031-72955-3_46

molecular weight polyethylene (UHMWPE) fibers, a kind of special polyethylene (PE) fibers, represent superb modulus of elasticity and tensile strength along with outstanding stability in alkaline/acidic environments, making them as an appropriate candidate for FRCC among the other polymeric fibers [3]. Hence, they have also caused considerable attention to be used in strain-hardening cementitious composites (SHCC) [4]. Nonetheless, PE fibers possess hydrophobic property with no chemical active groups on their surface which yields weak adhesion properties between these fibers and their water-based cementitious matrices. Such low fiber-matric interaction is usually undesired [4].

As a composite material, it is well-known that the effectiveness of load transfer between the constituents, i.e. fibers and surrounding matrix, depends on fiber-matrix interfacial properties, particularly determining the degree of fiber reinforcing power in FRCC [5]. In general, it is preferable that the fibers tolerate applied forces through a gradual delamination process followed by full pullout from the matrix instead of either simple fiber pullout or pure fiber rupture. An optimized fiber-matrix bonding performance, hence, is required to exploit full advantages of utilization of fibers in FRCC/SHCC [6].

As stated above, PE fibers do not have any functional groups and are therefore unable to adhere properly to cementitious matrices. On this basis, researchers have rendered a variety of fiber modification approaches comprising chemical and physical treatments [7]. In recent years, application of tannic acid (TA) as a cost-effective and plant-based polyphenol has significantly attracted huge interest to be used for surface modification of diverse substrates [8]. In comparison to dopamine (DA), which is widely applied for the same goal, TA has an analogous chemical structure but at a considerably lower price, and similarly shows an ability to form a polymeric layer having numerous hydroxyl groups on different surfaces under alkaline conditions [5]. There are frequent studies reported on the use of a 24-h TA modification or a 12-h TA modification on various substances to make them hydrophilic to be exerted for different applications [9, 10]. However, such long treatment durations hinder practical uses so far. To overcome this issue, sodium periodate (SP) has been introduced to accelerate the TA-based surface modification of different substrates [11, 12]. Since an oxidation step is necessary to form the TA hydrophilic layer, SP as a strong oxidizing can make TA self-polymerization easier and faster, shortening duration of the modification [11]. To the best of the author's knowledge, no paper can be found in the literature stating rapid TA functionalization of PE fibers via using SP to be incorporated in cementitious matrices. Therefore, the current research suggests a simple and rapid technique to activate the PE fiber surface for improving interfacial adhesion properties in cementitious matrices for the first time.

2 Experimental

2.1 Materials

A kind of special PE fiber i.e. UHMWPE fiber (SK62) was provided from Dyneema® (DSM, The Netherlands) with an average diameter of 18 μm according to microscopic measurements. TA (ACS reagent, CAS-Number: 1401–55-4), SP (NaIO$_4$, ACS reagent, $\geq$ 99.8%) and ethanolamine (EA, $\geq$ 98%, CAS Number: 141-43-5) were provided by Sigma-Aldrich (St. Louis, MO, USA). The fibers were washed in deionized water/isopropanol solution inside a sonication bath for 1 h and the washed fibers were then dried in ambient temperature for later use.

2.2 Rapid Functionalization of PE Fibers by TA, EA and SP

To rapidly modify PE fiber surface, the washed PE fibers (0.10 g) were placed into an aqueous 200 mL-solution containing a certain amount of 4 g/L of TA. After that, 4 mL of EA was added to the above solution followed by addition of a constant SP amount of 8 g/L. The final solution was kept under mild shaking at ambient temperature for various durations. Ultimately, the modified fibers were taken out from the solution, washed several times in deionized water, dried in room temperature and put inside a storage bag for the tests. The samples with modification durations of 30 min, 1 h and 3 h were labeled as PE-TA-EA-SP-30 min. PE-TA-EA-SP-1 h and PE-TA-EA-SP-3 h, respectively (one-step method).

In addition to the mentioned one-step process, a two-step modification was further performed. The same TA solution including 0.10 g of PE fibers was prepared and then 8 g/L of SP was added to the solution. The provided solution was kept under mild shaking at room temperature for 1 h. Afterwards, 4 mL of EA was inserted and the experiment was proceeded for 2 h. The modified fibers were finally taken out from the solution, washed several times in deionized water, dried in room temperature and kept in a storage bag for the tests This sample was named as PE-TA-SP/EA-3 h (two-step method).

2.3 Fabrication of Specimens for Single Fiber Pullout Tests

A high-strength cementitious matrix with a water-to-binder ratio of 0.18 was used according to the literature regarding SHCC with PE fibers [13]. By means of a hand mixer containing a planetary rotating blade, the dry components including 296.25 g of cement powder (CEM I 52,5 R-SR3/NA, Holcim Technology Ltd., Zurich, Switzerland), 57.22 g of silica fume (Elkem Microsilica 971, Elkem ASA Silicon Materials Company, Oslo, Norway) and 29.42 g of quartz sand (0.06–0.2 mm, Quarzwerke Strobel, Germany), were mixed as the first step. Low, medium and high speeds were used for 30 s for each in sequence. The mixing process was proceeded by the addition of 7.10 g of superplasticizer (Glenium ACE 460, BASF, Germany) and 63.92 g of water to the solid powders for an additional 5 min under high speed. A specifically designed mold as reported in Ref [14] was used to produce specimens for the single fiber pullout test with a fiber embedded length of 2 mm. The prepared mortar mixture was cast into the mold in which the casting process was conducted under vibration for 40 s to remove air bubbles. The specimens were subsequently demolded after one day and stored at standard climate conditions for 14 days to be tested.

2.4 Characterization Methods

In order to evaluate surface changes after rapid TA functionalization of PE fibers, an environmental scanning electron microscope (ESEM) Quanta 250 FEG by the FEI Company (Hillsboro, OR, the Netherlands) was used.

Investigation of micro-mechanical properties i.e., single fiber pullout tests was done by means of a Zwick line testing machine (ZwickRoell GmbH, Ulm, Germany), utilizing a 10-N load cell and a constant displacement rate of 0.01 mm/s. Furthermore, single fiber pullout specimens were prepared with a 2-mm free length according to [14]. At least eight samples were considered to be tested for each modification series. Besides, interfacial shear strengths (τ) were calculated using Eq. (1) with regard to P_{max}, d, and L as the maximum pullout force, diameter of the PE fiber and the fiber length embedded in the matrix, respectively [3]:

$$\tau = \frac{P_{max}}{\pi \, dL} \tag{1}$$

3   Results and Discussion

3.1   Investigation of PE Fiber Surface Changes After Rapid TA Modification

The surface changes before and after rapid TA modification of PE fibers were evaluated by means of ESEM analysis and the acquired images are shown in Fig. 1. For the pristine sample, long grooves over the fiber length along with slight micro-pits can be seen on the fiber surfaces implying relatively smooth surface. In contrast, all the rapid TA-modified PE fibers clearly indicate a TA coating layer on their surfaces which reveals that the PE functionalization process was successfully performed. Additionally, it can be obviously seen that the surfaces of the treated fibers become rougher through deposition of TA on the fibers' surfaces.

According to Fig. 1, it can be further realized that TA modification durations and order of EA addition to the synthesis solution can affect the surface morphologies of the treated PE fibers. An uneven surface with partially distributed TA coatings was observed in the modified sample with the lowest treatment time, i.e. sample PE-TA-EA-SP-30 min, while a relatively smoother coated surface with more TA coating was found for the modified sample with the longest modification time, i.e. sample PE-TA-EA-SP-3 h. Indeed, a longer modification duration can trigger more polymerization of TA in the presence of EA deposited on the PE fiber surface as reported in Ref [5] as well. Moreover, it seems that sample PE-TA-SP/EA-3h showed an inhomogeneous surface with the highest surface roughness among the modified samples. This can be likely explained by a better chance of Schiff-base and/or Michael-type addition reactions [5] between the EA as an amine compound and formed quinones of TA in presence of SP when EA is added after 1 h of addition of SP compared to simultaneous addition of SP and EA. It is worthwhile to note that this surface roughness seen in the treated fibers will directly influence fiber-matrix interfaces, reflecting on interfacial properties between fibers and their matrix.

In addition to formation of surface roughness as a physical effect of fast TA functionalization of PE fibers, the chemical impact of such modification can be described as follows:

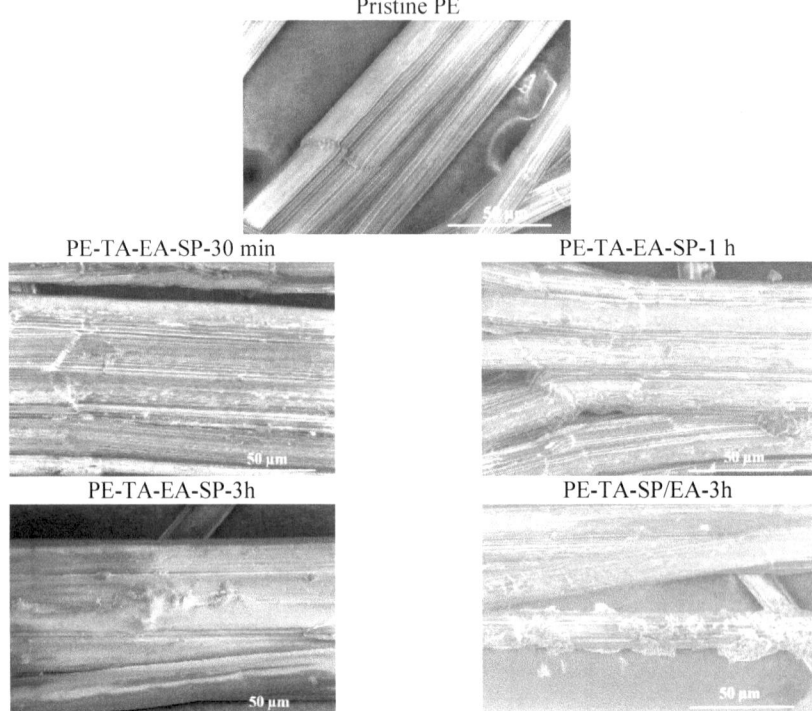

Pristine PE

PE-TA-EA-SP-30 min   PE-TA-EA-SP-1 h

PE-TA-EA-SP-3h   PE-TA-SP/EA-3h

Fig. 1. ESEM images of pristine PE fibers and various rapid TA-modified PE fibers

a) Highly reactive quinones can be generated via oxidation of numerous phenolic hydroxyl groups of TA in presence of SP as a strong oxidizing agent which subsequently follows a self-crosslinking reaction between the aryl rings [5]. More reactivity of the radicals produced by SP namely superoxide and hydroxyl radicals than O_2 (conventional TA treatment in presence of air) can result in forming carboxyl groups in addition to hydroxyl groups through oxidative cleavage of TA aromatic units during the modification process [11].

b) A Michael addition reaction and/or a Schiff base reaction can take place by addition of EA as an amine compound leading to formation of amino groups on the fiber surface as well. Finally, a thin TA polymeric layer containing active polar groups is generated on the PE fiber surfaces [5, 6].

3.2   Micromechanical Assessment of Rapid TA-Modified PE Fibers/cementitious Matrix Interactions

Interfacial shear strength values of the specimens containing pristine and modified PE fibers were achieved from single fiber pullout tests and the calculated results are shown in Fig. 2(a). A poor fiber-matrix interaction is seen for the pristine PE fibers with an interfacial shear strength of 0.69 MPa. In contrast, the rapid TA-modified PE fibers showed improved fiber-matrix adhesion properties with interfacial shear strengths from

1.29 MPa to 1.73 MPa, expressing a minimum improvement of 87% compared to the pristine samples. The fast TA modification, in fact, appropriately activates PE fiber surfaces through forming hydroxyl, carboxyl and amino groups [5, 11] leading to more reactions between these polar groups and cement hydrates which cause superior interfacial properties. With regard to modification time for one-step modified samples, sample PE-TA-EA-SP-30 min had the highest interfacial shear strength. It can be traced back to the shortest modification time for this sample which led to an uneven and partially covered fiber surface by TA layer as observed in ESEM images which is beneficial for fiber-matrix interfacial properties. With increasing modification duration up to 1 h, the shear strength decreased which was followed by an increase for 3-h modification. This can be explained by the smoothening/roughening effect of reproducing a TA layer on the fiber surface over the time. For the sample using the two-step method, i.e. sample PE-TA-SP/EA-3h, the maximum shear strength was gained among all the modified samples. A better chance of reaction between the reactive TA layer and EA, and a higher surface roughness as seen in ESEM images resulted in enhanced fiber-matrix interactions for this sample.

Moreover, pullout energy values of the pristine and modified samples are represented in Fig. 2(b). The pullout energy is 0.062 N*mm for the pristine PE fibers indicating a weak interfacial performance with the cementitious matrix. Contrarily, the modified samples demonstrated a promoted pullout energy ranging from 0.068 N*mm to 0.146 N*mm. Deposition of a reactive TA layer along with created roughness on the PE fiber surface, indeed, not only improved the fiber-matrix interactions chemically but also strengthened the mechanical interlocking of the fiber-matrix interface. It can be seen that the longest modification duration, i.e. 3 h, led to maximum pullout energy values owing to completion of TA polymerization and TA/EA reaction in a longer synthesis duration. Furthermore, the one-step modification method showed greater pullout energy than the two-step modification, while the trend was reversed for interfacial shear strength. This suggests that the physicochemical surface phenomena at fiber-matrix interfaces induced by these two different methods are dissimilar affecting interfacial performances of the specimens.

4   Conclusions

The current research proposes a simple method to enhance interfacial performances between PE fibers and a cementitious matrix through functionalization of PE fiber by rapid TA surface modification. The findings are as follows:

- The PE fiber surface became rougher after rapid TA surface modification as shown by ESEM images. It was further observed that the two-step modification method resulted in more surface roughness and more TA deposition on the fiber surface.
- Both interfacial shear strength and pullout energy of the modified samples were improved by rapid TA modification due to forming polar active groups on the fiber surface which can react with cement hydrates as well as creating surface roughness for better mechanical interlocking at fiber-matrix interfaces.

(a) **(b)**

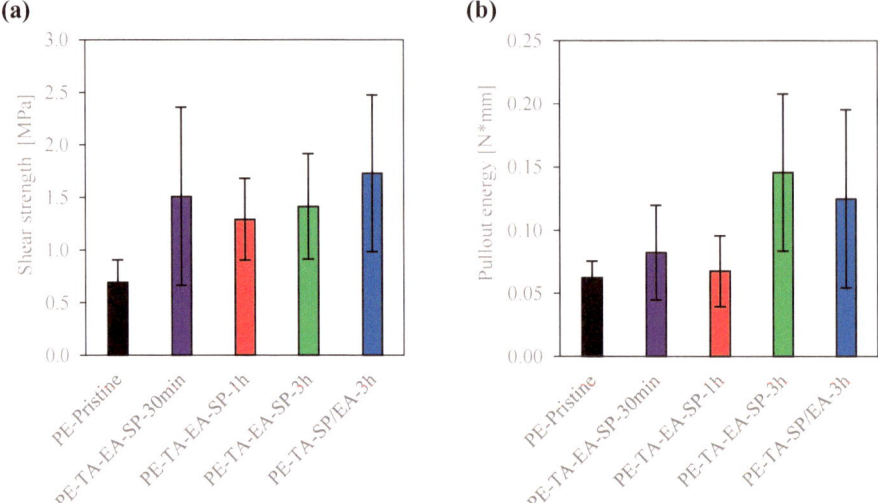

Fig. 2. Interfacial shear strength and fiber pullout energy values of the pristine and rapid TA-modified PE fibers

- It was seen that even the 30-min TA modification was effective for enhancing PE fiber-cementitious matrix interaction; nonetheless, the 3-h TA modification yielded more pronounced improvement in both interfacial shear strength and pullout energy parameters.

Acknowledgements. The authors express their gratitude to the German Research Foundation (Deutsche Forschungsgemeinschaft - DFG) for the financial support of the project 455631638.

References

1. Dai, J.G., Huang, B.T., Shah, S.P.: Recent advances in strain-hardening UHPC with synthetic fibers. J. Compos. Sci. **5**(10), 283 (2021)
2. Li, H., Schamel, E., Liebscher, M., Zhang, Y., Fan, Q., Schlachter, H., Köberle, T., Mechtcherine, V., Wehnert, G., Söthje, D.: Recycled carbon fibers in cement-based composites: Influence of epoxy matrix depolymerization degree on interfacial interactions. J. Clean. Prod. 137235 (2023)
3. Bashiri Rezaie, A., et al.: Enhancing the interfacial bonding between PE fibers and cementitious matrices through polydopamine surface modification. Compos. B Eng. **217**, 108817 (2021)
4. Lu, Z., Yin, R., Yao, J., Leung, C.K.: Surface modification of polyethylene fiber by ozonation and its influence on the mechanical properties of strain-hardening cementitious composites. Compos. B Eng. **177**, 107446 (2019)
5. Bashiri Rezaie, A., Liebscher, M., Drechsler, A., Synytska, A., Mechtcherine, V.: Tannic acid/ethanolamine modification of PE fiber surfaces for improved interactions with cementitious matrices. Cement Concr. Compos. **131**, 104573 (2022)

6. de Souza Castoldi, R., Bashiri Rezaie, A., Liebscher, M., de Souza, L.M.S., Mechtcherine, V., de Andrade Silva, F.: Effect of surface modification of sisal fibers with polyphenols on the mechanical properties, interfacial adhesion and durability in cement-based matrices. Cellulose, 1–22 (2023)
7. Lu, Z., Yao, J., Leung, C.K.: Using graphene oxide to strengthen the bond between PE fiber and matrix to improve the strain hardening behavior of SHCC. Cem. Concr. Res. **126**, 105899 (2019)
8. Sathishkumar, G., Gopinath, K., Zhang, K., Kang, E.T., Xu, L., Yu, Y.: Recent progress in tannic acid-driven antibacterial/antifouling surface coating strategies. J. Mater. Chem. B **10**(14), 2296–2315 (2022)
9. Qian, X., Yang, H., Wang, J., Fang, Y., Li, M.: Eco-friendly treatment of carbon nanofibers in cementitious materials for better performance. Case Stud. Constr. Mater. **16**, e01126 (2022)
10. Pawłowska, A., Stepczyńska, M., Walczak, M.: Flax fibres modified with a natural plant agent used as a reinforcement for the polylactide-based biocomposites. Ind. Crops Prod. **184**, 115061 (2022)
11. Ong, C., et al.: Tannin-inspired robust fabrication of superwettability membranes for highly efficient separation of oil-in-water emulsions and immiscible oil/water mixtures. Sep. Purif. Technol. **227**, 115657 (2019)
12. Li, R., et al.: A facile method to modify polypropylene membrane by polydopamine coating via inkjet printing technique for superior performance. J. Colloid Interface Sci. **552**, 719–727 (2019)
13. Curosu, I., Liebscher, M., Mechtcherine, V., Bellmann, C., Michel, S.: Tensile behavior of high-strength strain-hardening cement-based composites (HS-SHCC) made with high-performance polyethylene, aramid and PBO fibers. Cem. Concr. Res. **98**, 71–81 (2017)
14. Ranjbarian, M., Mechtcherine, V.: A novel test setup for the characterization of bridging behaviour of single microfibres embedded in a mineral-based matrix. Cement Concr. Compos. **92**, 92–101 (2018)

Foamed Concrete Reinforced
with Polypropylene Fibers and Geotextile
in Geotechnical Applications

Marta Kadela[1][(✉)] and Marian Drusa[2]

[1] Building Research Institute, Filtrowa 1, 00-611 Warsaw, Poland
m.kadela@itb.pl
[2] Department of Geotechnics, Faculty of Civil Engineering, University of Žilina, Univerzitná
8215/1, 01026 Zilina, Slovakia

Abstract. Foamed concrete is known as lightweight or cellular concrete. It is commonly defined as a cementitious material with a minimum of 20% (by volume) mechanically entrained foam in the mortar mix where air-pores are entrapped in the matrix by means of a suitable foaming agent. Although the foamed concrete has been patented in 1923, it is mainly used as a filling or leveling material. The use of foamed concrete has been limited e.g. for backfilling retaining walls. As a material used in contact with the ground, it is a relatively new material. In order to use foamed concrete in road construction as a replacement for hydraulically bound mixtures, the improvement of foamed concrete was considered. The article presents the two different type of polypropylene reinforced of foamed concrete. Polypropylene fibers with content of 0.3 kg/m^3 and 6.37 kg/m^3 were used. For foamed concrete samples with addition of 6.37 kg/m^3 PP fibers, the splitting tensile strength increased. In second case, the polypropylene geotextiles with weight 150 g/m^2, 200 g/m^2 and 500 g/m^2 were used. It can be observed that flexural strength for foamed concrete reinforced with geotextile samples was higher compared to the base (unreinforced) foamed concrete sample. On this basis, the suitability of using reinforced foamed concrete in the road pavement-subsoil system was determined.

Keywords: Foaming Agent · Lightweight Concrete · Polypropylene Fiber · Polypropylene Geotextile · Road Pavement · Tensile Splitting Strength · Flexural Strength

1 Introduction

Foamed concrete is known as lightweight or cellular concrete. It is commonly defined as a cementitious material with a minimum of 20% (by volume) mechanically entrained foam in the mortar mix where air-pores are entrapped in the matrix by means of a suitable foaming agent (protein or synthetic). The most important features of foamed concrete include low density (from approximately 100 kg/m^3 to 1800 kg/m^3 [1–3]), good insulation parameters (thermal conductivity coefficient λ ranging from approximately

© The Author(s) 2025
L. Czarnecki et al. (Eds.): ICPIC 2023, 61, pp. 466–474, 2025.
https://doi.org/10.1007/978-3-031-72955-3_47

0.05 W/(m·K) to 1 W/(m·K) K) [4–8]), speed and ease of implementation and low cost of production. However, although the foamed concrete has been patented in 1923, it is mainly used as a filling or leveling material. The use of foamed concrete has been limited e.g. for backfilling retaining walls. As a material used in contact with the ground, it is a relatively new material (Figs. 1 and 2). In addition, the use of a structure or building element in contact with the ground, as is the case, among others, in road pavement or floors on the ground, requires specific material and strength properties related to the soil and water conditions occurring in a given area [9–14].

Fig. 1. Foamed concrete used in running track [15]

(a)   (b)

Fig. 2. Applications of foamed concrete in the construction of a parking lot in Trencin, Slovakia: in section (a) and in view (b) [16]

In this purpose special test stand was created at the Faculty of Civil Engineering of the University of Žilina (FCE Uniza) [17, 18]. In order to use a foamed concrete layer in the lower subbase layer, it is necessary to meet the requirements for tensile stresses at the bottom of this layer. The reinforced foamed concrete is rarely presented in the literature. The aim of the article is to attempt to improve foamed concrete through various types of reinforcement: by using polypropylene fibre (first solution (I)) and polypropylene geotextile (solution II).

2 Materials and Methods

2.1 Case I—Foamed Concrete Reinforced with PP Fiber

The tests in case I have been performed on foamed concrete with density of 800 kg/m^3. The materials used in this study were Portland cement, water and synthetic foaming agent. The industrial cement was Portland cement CEM I 42.5 R, according to PN-EN 197–1:2011 [19].

Polypropylene fiber with a length of 13 mm was used (Fig. 3). The fiber content was 0.30 kg/m^3 and 6.37 kg/m^3.

Fig. 3. Used polypropylene fiber

The splitting tensile strength of foamed concrete reinforced with PP fiber was performed according to PN-EN 12390–6 [20], but loading speed was equal 0,05 ± 0,01 MPa/s.

2.2 Case II—Foamed Concrete Reinforced with PP Geotextile

The tests in case II have been performed on foamed concrete with density of 500 kg/m^3. The materials used in this study were cement, ground granulated blast-furnace slag (GGBS), aggregate with fraction of 0/2 mm, clean water without chemical residues and protein foaming agent. The Portland cement CEM I 42.5 R according to EN 197–6:2023 [21] was used.

Used reinforcement was polypropylene geotextile with a tree different weight: 150 g/m^2, 200 g/m^2 and 500 g/m^2, see Fig. 4. The parameters of geotextile were presented in Table 1 and in Fig. 5. The geotextile is made of cut polypropylene fibers with density of 0.91 kg/dm^3 and melting temperature of 165 °C connected in the needling process.

The flexural strength of foamed concrete reinforced with PP geotextile was determined in a four-point flexural test in accordance with the EN ISO 12390–5 standard [22]. The reinforced foamed concrete beam with dimensions of 100 mm x 100 mm x 400 mm was tested.

(a)   (b)

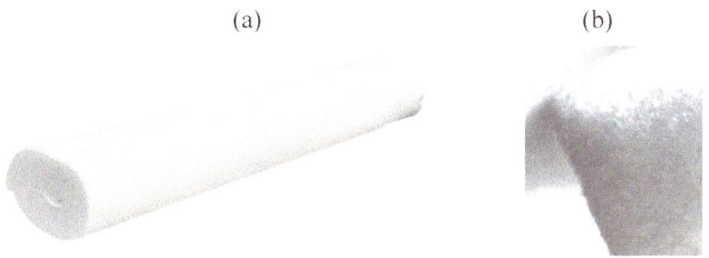

Fig. 4. Used geotextile Filtek 500 non-woven PP geotextile ironed (a) and detail (b)

Table 1. Parameters of geotextiles

Parameter	Geotextile with weight of:		
	150 g/m^2	200 g/m^2	500 g/m^2
tensile strength (kN/m$^-$)	3.4 ± 0.4	12.0 ± 1.0	33 ± 2
• longitudinal direction	9.5 ± 1.0	7.5 ± 1.0	19 ± 2
• transversal direction			
ductility (%)	120 ± 35	75 ± 15	70 ± 20
• longitudinal direction	80 ± 20	115 ± 15	115 ± 15
• transversal direction			
dynamic puncture resistance (mm)	19 ± 4	14 ± 2	6 ± 2
static puncture resistance (N)	850 ± 150	1400 ± 200	3600 ± 300

(a)   (b)

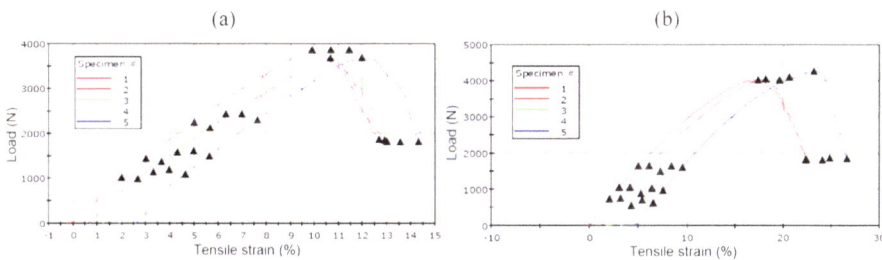

Fig. 5. Load-Tensile strain curves of geotextile with weight 500 kg/m^3 for longitudinal (a) and transversal direction (b)

3   Results and Conclusions

3.1   Case I – Foamed Concrete Reinforced with PP Fiber

The effect of the polypropylene fiber content on the splitting tensile strength for foamed concrete (without the addition of sand) is shown in Fig. 6. For a density of 800 kg/m^3, the addition of polypropylene fibers in the amount of 0.30 kg/m^3 (approximately 0.1% of the cement mass) does not improve the splitting tensile strength of foamed concrete. However, the addition of polypropylene fibers in the amount of 6.37 kg/m^3 (approximately

2% of the cement mass) is characterized by approximately 30% higher splitting tensile strength compared to the value obtained for foamed concrete without the addition of fibers. Similar results were obtained by Bing et al. [23], who showed an increase in splitting tensile strength by 31.7% by using the addition of polypropylene fibers compared to foamed concrete without fibers.

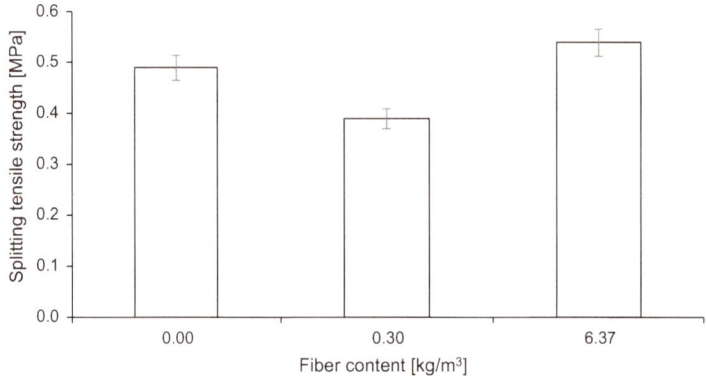

Fig. 6. The effect of polypropylene fiber content on splitting tensile strength for foamed concrete with a density of 800 kg/m^3

Moreover, the use of fibers in foamed concrete, in addition to strengthening it, also changes the nature of the degradation of foamed concrete samples, from brittle to ductile (Fig. 7).

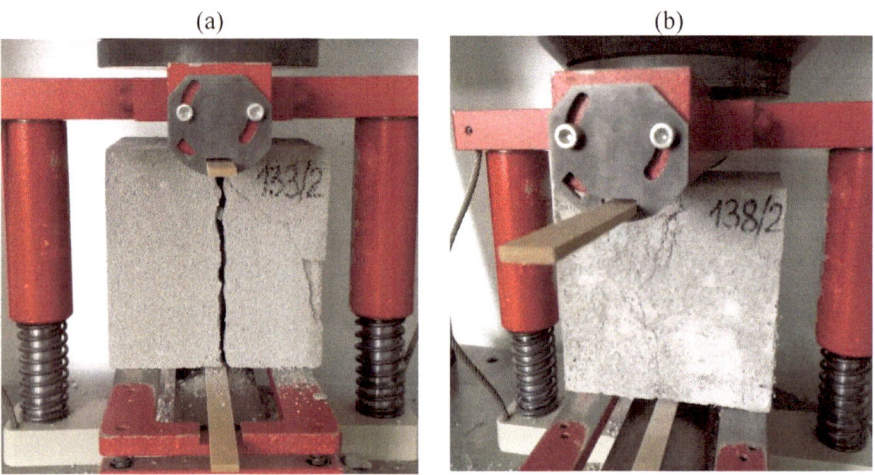

Fig. 7. Degradation of samples in the splitting tensile strength test for foamed concrete with a density of approximately 750 kg/m^3 without the addition of fibers (a) and with the addition of polypropylene fibers in the amount of 6.37 kg/m^3 (b)

3.2   Case II—Foamed Concrete Reinforced with PP Geotextile

Figure 8 presents result of four-point flexural test of foamed concrete with density of 500 kg/m^3. It can be observed that flexure strength increased for foamed concrete with geotextile. Increase in flexural strength of foamed concrete with geotextile is proportional to weight of the geotextile. Flexural strength increased by about 11%, 20% and 56% compared to the unreinforced foamed concrete sample for the geotextile with weight of 150 g/m^2, 200 g/m^2 and 500 g/m^2, respectively.

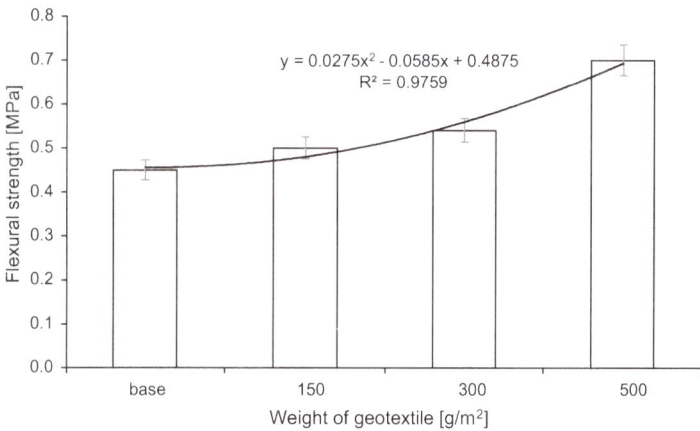

Fig. 8. Flexural strength of foamed concrete with density of 500 kg/m^3 without and with the PP geotextile with different weight

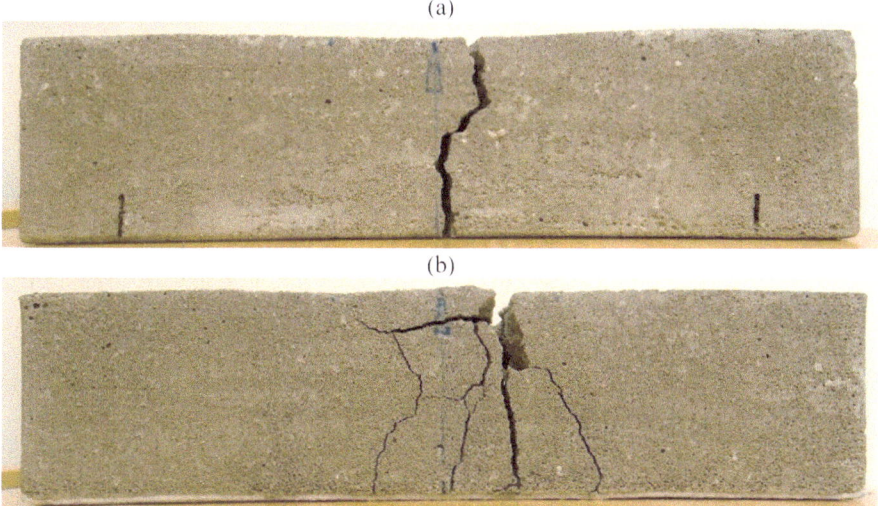

Fig. 9. Degradation of samples in the flexural test for foamed concrete with a density of approximately 500 kg/m^3 without (a) and with the geotextile with weight of 500 g/m^3 (b)

A different nature of degradation of foamed concrete with geotextile compared to the base sample (without reinforcement) was demonstrated, see Fig. 9.

4 Conclusions

The research aimed to assess the possibility of using propylene reinforcement as an improvement for foamed concrete. For this purpose, two different types of PP reinforcement were assumed. Propylene fibers with a length of 13 mm and a content of 0.3 kg/m^3 and 6.37 kg/m^3 were used (Case I). In addition, geotextiles of three different weights (150 g/m^2, 200 g/m^2 and 500 g/m^2) were used (Case II). The positive effect of polypropylene reinforcement in improvement foamed concrete has been demonstrated. Each time, the tensile strength was higher, and the nature of failure changed from brittle to ductile. The following conclusions from the research described above can be drawn:

- for foamed concrete samples with addition of 6.37 kg/m^3 PP fibers, the splitting tensile strength increased,
- increase in flexure strength of foamed concrete with geotextile is proportional to weight of the geotextile.

On this basis, the suitability of using reinforced foamed concrete in the road pavement-subsoil system was determined. Moreover, the article is part of a wider research project aimed at assessing reinforced foamed concrete.

Acknowledgements. The results presented in this article are the result of the scientific internship and to conduct classes in the frame of the Personal exchange program for students and scientists as part of bilateral cooperation—mobility offer for the academic year 2021/2022 financed by the NAWA (Polish National Agency for Academic Exchange) No. PPN/BIL/2020/1/00318/U/01 and the result of the research project No. VEGA 1/0484/20 'Experimental and numerical analysis of base layers of foamed concrete reinforced with geosynthetics'.

References

1. Nambiar, E.K., Ramamurthy, K.: Air-void characterization of foam concrete. Cem. Concr. Res. **37**, 221–230 (2007). https://doi.org/10.1016/j.cemconres.2006.10.009
2. Raj, A., Sathyan, D., Mini, K.M.: Physical and functional characteristics of foam concrete: a review. Constr. Build. Mater. **221**, 787–799 (2019). https://doi.org/10.1016/j.conbuildmat.2019.06.052
3. Fu, Y., Wang, X., Wang, L., Li, Y.: Foam concrete: a state-of-the-art and state-of-the-practice review. Adv. Mater. Sci. Eng. **2020**(6153602), (2020). https://doi.org/10.1155/2020/6153602
4. Kadela, M., Winkler-Skalna, A., Łoboda, B., Kukiełka, A.: PIANOBETON – charakterystyka materiałowa oraz możliwości zastosowania. Materiały budowlane 2015, Nr 7, s. 108–110. https://doi.org/10.15199/33.2015.07.30
5. Mohd Zahari, N., Abdul Rahman, I., Zaidi, A., Mujahid, A.: Foamed concrete: potential application in thermal insulation [in:] Proceedings of Malaysian Technical Universities Conference on Engineering and Technology (MUCEET), MS Garden, Kuantan, Pahang, Malaysia, 2009
6. Newman, J., Owens, P.: Properties of lightweight concrete. In: Newman, J., Choo, B.S. (eds.) Advanced Concrete Technology, pp. 3–29. Butterworth-Heinemann, Oxford (2003)

7. Proshin, A., Beregovoi, V.A., Beregovoi, A.M., Eremkin, I.A.: Unautoclaved foam concrete and its constructions, adapted to the regional conditions. Use of foamed concrete in construction, pp. 113–120. Thomas Telford, London, 2005
8. Gołaszewski, J., et al.: Effect of foaming agent, binder and density on the compressive strength and thermal conductivity of ultra-light foam concrete. Buildings **12**, 1176 (2022). https://doi.org/10.3390/buildings12081176
9. Kozłowski, M., Kadela, M.: Mechanical characterization of lightweight foamed concrete. Adv. Mater. Sci. Eng. **2018**(6801258), (2018). https://doi.org/10.1155/2018/6801258
10. Kadela, M.; Kozłowski, M.; Kukiełka, A.: Application of foamed concrete in road pavement–weak soil system. In Proceedings of the Procedia Engineering, International Conference on Analytical Models and New Concepts in Concrete and Masonry Structures AMCM'2017, 2017, Vol. 193, pp. 439–446. https://doi.org/10.1016/j.proeng.2017.06.235
11. Kadela, M., Babiak, B.: Pianobeton w budownictwie komunikacyjnym. Materiały Budowlane 2018, nr 3, s. 80–81. https://doi.org/10.15199/33.2018.03.32
12. Kadela, M., Kozłowski, M., Kukiełka, A.: Application of foamed concrete in road pavement – weak soil system. Procedia Engineering **193**, 439–446 (2017). https://doi.org/10.1016/j.proeng.2017.06.235
13. Hulimka, J., Krzywoń, R., Knoppik-Wróbel, A.: Use of foamed concrete in the structure of passive house foundation slab. Proc. 7th International Conference on Analytical Models and New Concepts in Concrete and Masonry Structures AMCM2011, Wyd. Pol. Krakowskiej, 2011, s. 221–222+CD
14. Kadela, M., Kozłowski, M.: Foamed concrete layer as sub-structure of industrial concrete floor. Procedia Engineering, Tom **161**, 468–476 (2016). https://doi.org/10.1016/j.proeng.2016.08.663
15. http://www.provoton.com/
16. Drusa, M., Fedorowicz, L., Kadela, M., Scherfel, W.: Application of geotechnical models in the description of composite foamed concrete used in contact layer with the subsoil Proc. 10th Slovak Geotechnical Conference Geotechnical problems of engineering constructions. Vysokie Tatry (2011)
17. Hajek, M., Decky, M., Drusa, M., Orininová, L., Scherfel, W.: Elasticity modulus and flexural strength assessment of foam concrete layer of Poroflow. IOP Conf. Series: Earth Environ. Sci. **44**(022021),2016https://doi.org/10.1007/978-3-319-21602-7_8
18. Decký, M., Drusa, M., Zgútová, K., Blaško, M., Hájek, M., Scherfel, W.: Foam concrete as new material in road constructions. Procedia Eng., Proceeding WMCAUS **161**(2016), 428–433 (2016). https://doi.org/10.1016/j.proeng.2016.08.585
19. PN-EN 197–1:2011 Cement - Część 1: Skład, wymagania i kryteria zgodności dotyczące cementów powszechnego użytku
20. PN-EN 12390–6:2011 Badania betonu - Część 6: Wytrzymałość na rozciąganie przy rozłupywaniu próbek do badań
21. EN 197–6:2023 Cement - Part 6: Cement with recycled building materials
22. EN ISO 12390–5 Testing hardened concrete. Flexural strength of test specimens, 2009
23. Bing, C., Zhen, W., Ning, L.: Experimental research on properties of high-strength foamed concrete. J. Mater. Civ. Eng. **24**(1), 113–118 (2012)

Polypropylene Fiber Reinforced - Latex Modified Mortar for Installation of Granite Paving Blocks on Various Road Sections

Jakob Šušteršič[1]($\boxtimes$), Rok Ercegovič[1], Sandi Drolc[1], and Naser Kabashi[2]

[1] IRMA Institute for Research in Materials and Application, Ljubljana, Slovenia
`jakob.sustersic@irma.si`
[2] Faculty of Civil Engineering and Architecture, University of Pristina, Pristina, Kosovo

Abstract. The paper discusses the use of Polypropylene Fiber Reinforced - Latex Modified Mortar (PFR-LMM) for installation of granite paving blocks (GPBs) on various road sections in Slovenia. The following four examples are considered: two inner rings of roundabouts on both sides of the bridge over the Sava River on the bypass near Krško, a roundabout in front of the entrance to the Šoštanj thermal power plant and one part of a street in Ljubljana. GPBs were installed on a slab made with Steel Fiber Reinforced Concrete (SFRC). The workability of the fresh PFR-LMM had to be such that it filled the joints between the GPBs. Hardened PFR-LMM, however, must provide a good bond between the GPBs to ensure resistance to traffic loads and resistance to constantly changing weather and temperature influences. The age of the subject applications is between 5 and 11 years.

Keywords: polypropylene fiber reinforced latex-modified mortar · granite paving block · workability · compressive strength · flexural strength · bond

1 Introduction

Larger granite paving blocks (GPBs) are generally used on roadway sections where higher traffic loads are expected and where the visual appearance of the surface needs to be improved. Examples of such use have been considered for the projects presented in this paper. In all these projects, the GPBs were placed on top of a steel fiber reinforced concrete (SFRC) slab using Polypropylene Fiber Reinforced - Latex Modified Mortar (PFR-LMM). The thickness of the PFR-LMM layer between the GPBs and the SFRC slab was 3 cm. The width of the joints between the GPBs was 2 cm. The subgrade of SFRC slab was constructed as elastic, isotropic and homogeneous body with modulus of deformation $E_{v2} \geq 120$ MPa.

L. Czarnecki et al. (Eds.): ICPIC 2023, 61, pp. 475–484, 2025.
https://doi.org/10.1007/978-3-031-72955-3_48

2 Brief Description of Applications on Different Road Sections

2.1 Two Inner Rings of Roundabouts on Both Sides of the Bridge Over the Sava River on the Bypass Near Krško

The roundabout on the left bank of the river was built in 2012 and the one on the right bank a little later (see Fig. 1a and b).

(a)

(b)

Fig. 1. (**a**) Location of roundabouts on both sides of the bridge over the Sava River (Source: https://www.google.com/maps/). (**b**) Inner ring of the roundabout on the left side of the Sava River (Source: https://www.google.com/maps/).

This bypass passes the Krško Nuclear Power Plant. Due to the occasional heavy traffic for the Krško NPP, it was required that the inner ring of both roundabouts should also be able to withstand the heavy traffic loads. The $20 \times 20 \times 20$ cm GPBs were installed on a 24 cm thick SFRC slab.

PFR-LMM was used for the installation of the GPBs with the following mix-proportion: Portland cement CEM I 52,5 R (480 kg/m^3), silica fume (7,5 w. % to cement content), styrene-butadiene copolymer latex (solid particles 3,5 w. % to cement content), effective water-binder ratio $(w/b)_{eff} = 0,33$, high-range superplasticizer (0,6 w. % to cement content), admixture for expansion (0,5 w. % to cement content), polypropylene fibers with length of 10 mm and with diameter of 30 to 40 m (1,0 kg/m^3), crushed gravel aggregate with $D_{max} = 4$ mm.

The mix-proportion of the SFRC was as follows: Portland cement CEM I 52,5 R (420 kg/m^3), silica fume (2,5 w. % to cement content), effective water-binder ratio $(w/b)_{eff} = 0,37$, high-range superplasticizer (0,75 w. % to cement content), hooked steel fibers with a length of 16 mm and a diameter of 0,5 mm (0,38 vol. %) and wavy steel fibers with a length of 39 mm and a diameter of 1,25 mm (0,64 vol. %), natural gravel and crushed aggregate with $D_{max} = 32$ mm.

2.2 Roundabout in Front of the Entrance to the Šoštanj Thermal Power Plant

This roundabout was built during the construction of Šoštanj TPP Block 6 in 2012. It is in front of the entrance to the TPP, next to the cooling tower of Block 4 (see Fig. 2).

Fig. 2. Location of the roundabout in front of the TPP entrance, next to the cooling tower of Block 4 (Source: https://www.google.com/maps/).

During the construction of Block 6, very large and heavy steel structural elements and components of the TPP, such as the turbine (200 t) and the transformer (60 t), were transported. Therefore, most of the central part, as well as the inner ring of the roundabout, was built in the same composition as that used for the two roundabouts on both sides of the bridge over the Sava River - the 20 × 20 × 20 cm GPBs were installed on a 24 cm thick SFRC slab (see Sect. 2.1). In this way, large and heavy vehicles were able to drive through the roundabout to the entrance of the TPP (see Fig. 3).

The mix-proportions of the PFR-LMM and SFRC was also like that of the roundabouts discussed above (see Sect. 2.1).

Fig. 3. The inner ring and most of the central part of the roundabout are lined with GPBs with dimensions 20 × 20 × 20 cm (Source: https://www.google.com/maps/).

2.3   Part of a Street in Ljubljana

During the renovation of Gosposvetska Street in Ljubljana in 2018, the architect (Med-prostor - architectural atelier) designed one part of the street in front of the church to be made of decorative GPBs (see Fig. 4 a and b). The dimensions of the granite paving block are length 30 cm, width 20 cm and thickness 10 cm. The floor layout also included the insertion of smaller parts from the basic module of granite paving block (see Fig. 4 b).

The thicknesses of the upper pavement bearing layers are (from top to bottom): GPBs 10 cm, PFR-LMM binder 3 cm, SFRC slab 21 to 24 cm, subbase asphalt 6 cm and tampon 20 cm.

The mix-proportion of the PFR-LMM used for the installation of the granite paving blocks on the SFRC slab was slightly modified from the mix-proportion of the PFR-LMM in Sect. 2.1 and was: Portland cement CEM I 42,5 R (480 kg/m3), silica fume (7,5 w. % to cement content), styrene-butadiene copolymer latex (solid particles 3,5 w. % to cement content), effective water-binder ratio $(w/b)_{eff} = 0,39$, high-range superplasticizer (1,4 w. % to binder content), admixture for shrinkage reduction (1,6 w. % to binder content), polypropylene fibers with length of 10 mm and with diameter of 30 to 40 m (1,5 kg/m^3), crushed gravel aggregate with Dmax = 4 mm.

The mix-proportion of the SFRC was approximately the same as that used in the construction of the roundabouts described in Sects. 2.1 and 2.2. The required compressive strength class of the SFRC was C35/45 in accordance with SIST EN 206:2013 + A1:2016 [1].

(a)

(b)

Fig. 4. (**a**) The part of Gosposvetska Street in front of the church is lined with decorative GPBs (Source: https://www.google.com/maps/). (**b**) Floor plan combination of GPBs 30 × 20 cm and smaller parts of the base module 20 × 20 cm. (Source: https://www.medprostor.si/projekti/projekt-gosposvetska-cesta/).

3 Investigation of the PFR-LMM

This chapter discusses the properties of PFR-LMM based on the results obtained from laboratory investigation prior to the construction of a part of Gosposvetska Street in Ljubljana, which was carried out as part of the reconstruction of this street. The mix-proportion of the PFR-LMM given in Sect. 2.3 is like the PFR-LMMs used for the construction of the roundabouts discussed in Sects. 2.1 and 2.2. Therefore, the performance of all PFR-LMMs is similar, as shown by the results of the permanent controls carried out during the construction of all the road sections discussed in Sect. 2.

3.1 Fresh PFR-LMM

The consistency of the fresh PFR-LMM, measured by the slump - flow test method given in the SIST EN 12350-8:2019 [2], was SF $= 500 \pm 100$ mm. This means that PFR-LMM was a self-compacting mortar that was able to fill the joints between the GPBs without additional compaction (see Fig. 5).

Fig. 5. PFR-LMM completely fills the joints between GPBs without additional compaction.

The volume change of the PFR-LMM was measured according to the method given in SIST EN 445:2008 [3]. After 24 h, a volume change of $+ 0{,}1\%$ was obtained. So PRF-LMM expanded very little, with no inhomogeneity and no water bleeding detected. In addition, the effect of polypropylene fibers on the reduction of autogenous as well as total shrinkage of concrete is well known [4, 5].

3.2 Hardened PFR-LMM

3.2.1 Compressive Strength

Compressive strength tests were carried out on the halves of the prisms with dimensions $40 \times 40 \times 160$ mm in accordance with SIST EN 13892 – 2:2003 [6]. The age of the PFR-LMM was 20 days.

The results obtained are: average compressive strength $R_{c,aver.} = 66{,}3$ MPa; minimum compressive strength $R_{c,min.} = 57{,}3$ MPa; standard deviation $s_c = 4{,}5$ MPa; $n = 12$.

3.2.2 Flexural Strength

The flexural test was not carried out according to SIST EN 13892–2:2003, which specifies a three-point bending, but a four-point bending configuration was used to determine the flexural strength of the PFR-LMM. ACI Committee 544 mentions that both methods can be used to determine the flexural strength [7]. Since a larger proportion of the total volume is loaded in the four-point bending configuration, the flexural strength is lower compared to the three-point bending configuration [8]. The flexural strength obtained with the three-point bending configuration test is about 22% higher than that obtained with the four-point bending configuration test [9]. A slightly smaller difference was obtained when high-performance SFRCs with very high fiber content were investigated [10].

The flexural strength of 20 day-old PFR-LMM was tested on prisms with dimensions $40 \times 40 \times 160$ mm, applying a four-point bending configuration. The span length was 120 mm.

The results obtained are: average flexural strength $R_{f,aver.} = 8,6$ MPa; minimum flexural strength $R_{f,min.} = 7,9$ MPa; standard deviation $s_f = 0,5$ MPa; n = 6.

3.3 Evaluation of the Bond Between SFRC and GPBs Made from PFR-LMM

The ability to bond the GPBs to the SFRC slab with PFR-LMM was evaluated based on the results of the flexural test. The test was carried out on prisms with dimensions $100 \times 100 \times 400$ mm, applying a four-point bending configuration. The span length was 300 mm (see Fig. 6 a). The age of the PFR-LMM was 20 days.

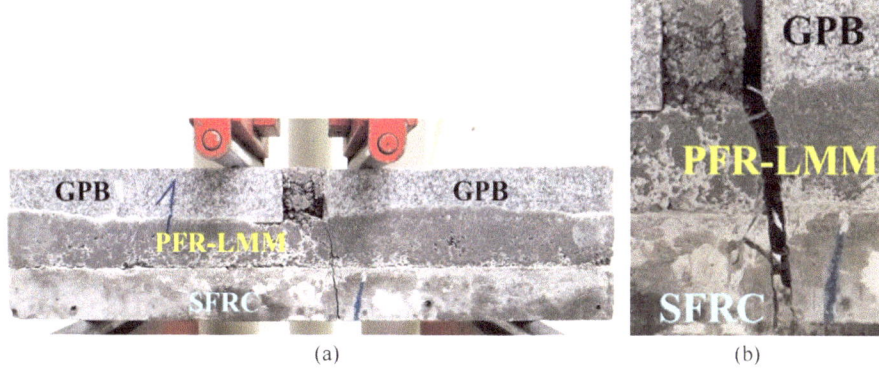

Fig. 6. (**a**) A prism made of three layers with dimensions $100 \times 100 \times 400$ mm during a flexural test with a four-point bending configuration. (**b**) Progress of the vertical crack through all layers after the test.

The prisms were made of three layers: the lower layer with a thickness of 35 mm was made with SFRC, the middle layer with a thickness of 30 mm was made with PFR-LMM and the upper layer with a thickness of 35 mm was composed of two GPBs with a joint between them with a width of 20 mm made with PFR-LMM (see Fig. 6 a). In Fig. 6 a, a vertical crack can be seen running through the whole thickness of the lower SFRC layer, which has already extended into the second PFR-LMM layer and is oriented at the joint angle between the GPBs.

Figure 6 b shows the final progression of the vertical crack, which completely split the prism in two through all the layers. The steel fibers still bridge the crack in the bottom layer, while the polypropylene fibers bridge the crack in the middle layer. In the upper layer, there is a crack at the interface between the GPB and the PFR-LMM, which fills the joint between the two GPBs.

One very important finding is that all the prisms tested never developed horizontal cracks at the interfaces between the layers. This means that the PFR-PMM provided a good bond to the SFRC slab and the GPBs.

During the test, the deflections were measured continuously as a function of the load. A typical load - deflection diagram is given in Fig. 7.

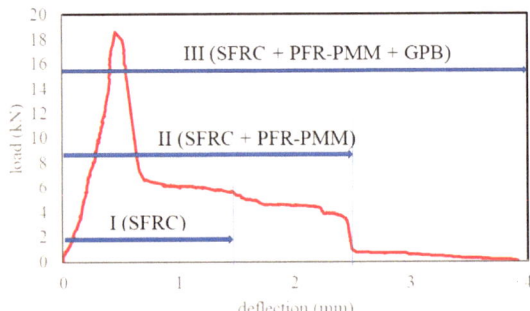

Fig. 7. Typical load - deflection diagram.

From the load-deflection diagram, the maximum flexural strength R_{fmax} and the toughness as measure of absorption energy G are calculated. The results of the test on the three prisms are given in Table 1. The absorbed energy, or toughness, is obtained by calculating the area up to a certain deflection.

Table 1. Maximum flexural strength R_{fmax} and the toughness GI, GII and GIII.

prism designation	$R_{f,max}$ (MPa)	toughness		
		GI (J)	GII (J)	GIII (J)
A	6,1	9,8	11,1	12,0
B	5,6	11,3	13,0	13,2
C	5,8	11,3	15,8	16,4
average	5,8	10,8	13,3	13,9
minimum	5,6	9,8	11,1	12,0
s	0,25	0,70	1,95	1,83

At the end of the elastic zone, a crack is formed which propagates in the first layer of SFRC as the external load continues to be applied. At the point of maximum load, the stress concentration is highest and the relatively greatest reduction in absorbed energy occurs. Due to the presence of the steel fibers, the reduction in absorbed energy is moderated up to the point where the crack reaches the horizontal interface between the SFRC layer and the PFR-LMM layer. In the diagram in Fig. 7, this is the end point of zone I, when the crack passes into the second layer of the PFR-LMM. The reduction in absorbed energy is moderate because the crack propagation is restrained by the polypropylene fibers in the PFR-LMM and still by the steel fibers in the SFRC.

However, when the crack reaches the interface between the PFR-LMM layer and the GPB (this is the end point of zone II in the load - deflection diagram), the absorbed energy decreases rapidly, and a prism fracture occurs at the deflection of 4 mm. The crack runs along the vertical interface between the GPB and the PFR-LMM, which fills the joint between the two GPBs. The poorer adhesion between the PFR-LMM and the vertical face of the GPB was expected because this surface was sawn and therefore smooth. In practice, GPBs with rough side surfaces have been used, so their bonding to PFR-LMM is much better.

4 Conclusions

The findings of the permanent observation of the road sections in question, where polypropylene fiber reinforced - latex modified mortar (PFR-LMM) was used for the installation of granite paving blocks (GPBs), show that no changes affecting traffic safety and the durability of these sections have occurred during several years of use (between 5 and 11 years). During construction, it was found that fresh PFR-LMM was very easy to cast without additional compaction and despite this, good installation of the GPBs was ensured. Laboratory tests have shown that PFR-LMM provides good bond to steel fiber reinforced concrete (SFRC) base slab and GPBs.

References

1. SIST EN 206:2013+A1:2016 Beton – Specifikacija, lastnosti, proizvodnja in skladnost (Concrete – Specification, performance, production and conformity)
2. SIST EN 12350-8:2019 Preskušanje svežega betona – 8. del: Samozgoščevalni beton – Poskus razleza s posedom (Testing fresh concrete – Part 8: Self-compacting concrete – Slump-flow test)
3. SIST EN 445:2008 Injekcijska masa za prednapete kable – Preskusne metode (Grout for prestressing tendons – Test methods)
4. Saje, D., Bandelj, B., Šušteršič, J., Lopatič, J., Saje, F.: Shrinkage of polypropylene fibre reinforced high performance concrete. J. Mater. Civ. Eng. **23**(7), 941–952 (2011)
5. Saje, D., Bandelj, B., Šušteršič, J., Lopatič, J., Saje, F.: Autogenous and Drying Shrinkage of Fibre Reinforced High-Performance Concrete. J. Adv. Concr. Technol.Concr. Technol. **10**(2), 59–73 (2012)
6. SIST EN 13892 – 2:2003 Metode preskušanje za mešanice za estrih – 2. del: Določanje upogibne in tlačne trdnosti (Methods of test for screed materials – Part 2: Determination of flexural and compressive strength)
7. ACI Committee 544: Report on measuring mechanical properties of hardened fiber-reinforced concrete. In: ACI Collection of Concrete Codes, Specifications and Practice. ACI 544.9R-17. 2022 Edition
8. Neville, A.M.: Hardened concrete: physical and mechanical aspects. ACI, Detroit and The Iowa State University Press, Ames (1971)
9. Wright, P.J.F.: The flexural strength of plain concrete – its measurement and use in designing concrete mixes. Road Research Technical Paper No. 67, H. M. S. O., London (1964)
10. Šušteršič, J., Leskovar, I., Ercegovič, R., Korla, J.: Behavior of concrete with very high fiber content. Fibre reinforced concrete: design and applications BEFIB 2008. pp. 183–190. Indian Institute of Technology, Madras (2008)

Underwater Abrasion Resistance of Fibre Reinforced-Latex Modified Concrete with Granulated Rubber

Mateja Klun[1], Jakob Šušteršič[2], Rok Ercegovič[2], Matjaž Mikoš[1], and Andrej Kryžanowski[1(✉)]

[1] University of Ljubljana, Ljubljana, Slovenia
andrej.kryzanowski@fgg.uni-lj.si
[2] IRMA Institute for Research in Materials and Application, Trzin, Slovenia

Abstract. In this paper, we present and discuss the initial results of a large-scale research project involving laboratory and field investigations of abrasion resistance of different types of concrete. The decision to study in more detail the abrasion resistance of fibre-reinforced concrete with granulated rubber was based on the results of previous research projects, as well as on observations on the behaviour of concretes placed in the spillways of hydro power plants loaded with water and water-borne particles. Gravel aggregate, steel, polypropylene fibres, and granulated rubber were used to prepare the concrete. In the fibre-reinforced concretes without granulated rubber, the binding component consisted of cement and silica fume, but when granulated rubber was added, the binding component consisted of cement and a dry proportion of SBR latex. The results obtained by now, at an age of 90 days show that fibre-reinforced concretes with granulated rubber have an improved resistance to underwater abrasion, compared to fibre-reinforced concretes without granulated rubber.

Keywords: Abrasion resistance · Concrete · Polymer binder

1 Introduction

In hydraulic structures the term 'abrasion' means the process of disintegration of exposed concrete surfaces, resulting from loads arising from sediment transport (Kryžanowski 2009). The disintegration rate of the concrete surface largely depends on the transport capacity of water and the manner of transport of solid matter (Kryžanowski 2009) and (Mikoš 1993). The protection of structures against abrasion damage is provided by protective linings made of abrasion-resistant materials, together with the appropriate structural solutions. The development trend in this area is moving towards finding the appropriate technical solutions and material analyses to determine the parameters for an abrasion resistant material, taking into consideration the following criteria: (1) high resistance to physical processes and chemical action in the water stream, (2) availability of the materials used, (3) feasibility and economy of the project, (4) minimal maintenance costs, and (5) durability of the design solutions. The latter is of vital importance from

© The Author(s) 2025
L. Czarnecki et al. (Eds.): ICPIC 2023, 61, pp. 485–494, 2025.
https://doi.org/10.1007/978-3-031-72955-3_49

the scope of reducing maintenance costs and ensuring good structural condition during normal operation (Kryžanowski et al. 2009) and (Jakobs et al. 2004).

The main challenge in studying the abrasion resistance of concrete arises from the inability to create proper hydraulic conditions in the laboratory to mimic the fully developed underwater abrasive action. The methods which enable modelling of wear mechanisms of the water current with bed load are the ones that come closest to the conditions present in the natural environment. Common to the majority of the methods investigating abrasion resistance is that they provide only qualitative comparisons between the tested specimens, based on a proportional loss of mass or an input of the abrasive medium during the investigation. The validation of results and applicability of the methods for forecasting the behaviour of concretes in natural conditions can only be achieved by performing the test under the conditions similar to those in the actual ambient (operational) conditions as the natural environment of the designed structure, including the monitoring of all relevant hydraulic and hydrological parameters. As part of the dam construction project on the Lower Sava River, the adequacy of the ASTM C 1138 laboratory method for the assessment of abrasion resistance of concretes in hydraulic structures, by performing a comparison between the laboratory measurements and measurements in the natural environment, was analysed. A good correlation of the abrasion resistance results according to the ASTM C 1138 laboratory method and measurements in the natural environment was achieved. At the same time, high abrasion resistance of concrete was achieved with a special concrete mixture with the addition of granulated rubber and polymer binder (Kryžanowski et al. 2009).

In 2022, a large-scale research project involving laboratory and field investigations of abrasion resistance of different types of concrete has initiated. Based on the experience from previous studies, as well as from observations of the behaviour of concretes placed in the spillways of hydro power plants on the Lower Sava River, we decided to perform a detailed study on the abrasion resistance of fibre-reinforced – latex modified concrete with granulated rubber.

2 Concrete Compositions

2.1 Determining the Concrete Compositions

In determining the appropriate concrete composition, we relied on previous research, where we designed suitable concrete compositions to meet the requirements to be installed in the spillways of the dam structures on the Lower Sava River, considering the following findings:

- The use of polymer-modified concrete demonstrates excellent workability, water-resistance, and improved compressive strength. The addition of a polymer binding improves the adhesion of abrasion-resistant concrete linings to the massive concrete base in dam structures.
- The addition of fibres improves the strength characteristics and contributes to the increase of the abrasion resistance of concrete, taking into account that steel fibres contribute to the resistance of concrete at the macro level, and polypropylene fibres to improve the cement paste.

• As the literature suggests substituting gravel aggregate with a rubber granulate additive in concrete increases the abrasion resistance of the concrete's surface, therefore, concrete compositions with rubber granulate aggregate were included in the research. The rubber granulates used in the research is a waste product from car tyre recycling. (Toutanji 1996; Eldin and Senouci 1993; Šušteršič et al. 2004; Kryžanowski et al. 2012; Kozjek et al. 2015).

2.2   Preparation of Concrete Compositions

Concrete compositions were prepared by using Portland cement with a minimum 80% proportion of clinker and a mixed addition of limestone and slag, type: CEM II/A-M(LL-S) 42.5 R, which is in accordance with the SIST EN 197–19 standard. The aggregate was obtained by separation of the natural crushed gravel from the alluvial quaternary deposit of the Sava River on the site. Fractions 0–4, 4–8, and 8–16 mm were used. Four samples of different concrete composition were intended for test purposes (Table 1): The ABR-1 composition is adopted as control composition, which is basically the same as the composition of abrasion resistant concrete built in the spillways of the first hydropower plant in the cascade on the Lower Sava River. In ABR-2 composition and all further modifications the nominal maximum gravel of 8 mm was adopted. The ABR-2 composite with smaller modifications was used in concretes on the spillway of the second hydropower plant in the cascade on the Lower Sava River.

In the following compositions, the proportion of fine fraction mineral aggregate (0–4 mm) was partly replaced by rubber aggregate. The rubber aggregate used in this study is the end product of recycling scrap for vehicle tyres with a characteristic cubic grain shape, similar to the usual crushed mineral aggregate. Rubber aggregate of different fractions (from 0 to 3,5 mm), mostly of uniform composition, were used in the study. With the ABR-3 composition the mineral additive ($SiO_2 > 90\%$) was replaced by polymeric binder (styrene-butadiene copolymer latex with dry portion in dispersion 48%); the proportion of steel fibres was replaced by doubling polypropylene fibres (L = 10 mm, $\varnothing 30 \sim 40 \ \mu m$); the proportion of the finest fraction (0–4 mm) was partially replaced by rubber aggregate. The ABR-4 composition represents a minor modification of the ABR-3 composition, in which steel fibres were added in a doubled amount compared to the control composition. The value of the w/c ratio in the composites did not vary considerably.

The mixtures of concretes were prepared in the laboratory mixer with a vertical shaft and with a volume of 75 dm^3. Right after the mixing the fresh concrete properties, such as: temperature (SIST EN 12350–1:2019), slump (SIST EN 12350–2:2019), air content (SIST EN 12350–7:2019) and density (SIST EN 12350–6:2019) were determined, following the standard procedures (Table 2). The average values of investigation results of the hardened concrete are given in Table 3: (1) Compressive strength and density were performed at the ages of 28, 56, and 90 days (SIST EN 12390–3:2019 and 12390–7:2019), respectively, on cubicles of dimensions of 15 cm, by taking three samples of each composition; (2) The static modulus of elasticity of the concrete was defined at the ages of 28, 56, and 90 days on prisms of 10/10/40 cm, taking three samples per each composition (DIN 1048, Part 5); (3) Abrasive resistance test was performed at the ages

Table 1. Concrete mixture propositions.

Parameter				Designation of concrete composition			
				ABR-1	ABR-2	ABR-3	ABR-4
Binder	Cement		kg	440	450	450	450
	Silica fume		kg	20	22.5	-	-
	SBL-solid particles		kg	-	-	43.2	43.2
Water-to-binder ratio (effective)			-	0.40	0.41	0.38	0.36
Superplasticizer			kg	1.84	2.22	2.12	2.48
Steel fibres			kg	40	40	-	80
Polypropylene fibres			kg	0.5	0.5	1.0	1.0
Fraction of gravel aggregate (mm)	natural	0–4	kg	553	643	426	405
	crushed	0–4	kg	490	565	372	353
	natural	0–4	kg	138	474	309	293
	natural	0–4	kg	560	-	-	-
Fraction of granulated rubber (mm)		8–16	kg	-	-	60	57

(continued)

Table 1. *(continued)*

Parameter				Designation of concrete composition			
		0–0.5	kg	-	-	27	26
		2–3.5	kg	-	-	49	47

of 90 and 180 days, on cylinders of Ø30/10 cm, taking one sample per composition (ASTM C 1138).

3 Programme of the Research Work

3.1 Testing the Wear of Concrete Samples

The research work was performed in accordance with standard ASTM C 1138 method for the reason that a number of comparisons and results exist in the literature on the basis of which the results of our research work could be evaluated. The test method can only be used to determine the relative resistance of the material to the abrasion action of waterborne particles. The standard procedure of the investigation provides for the measurement of the specimen surface wear at 12-h intervals, while the total investigation time is 72 h. The result of the test is the average depth of wear expressed by the average wear volume on the surface of the specimen for the duration of the test (Liu 1981).

It can be seen from the Figs. 1 and 2 that all concretes achieve adequate abrasion resistance, while concretes with the addition of rubber aggregate show a significantly

Table 2. Fresh concrete test results.

Types of test			Average values			
			ABR-1	ABR-2	ABR-3	ABR-4
Air temperature		°C	22.3	26.8	27.1	24.8
Fresh concrete temperature		°C	23.3	26.9	26.0	5.0
Slump (SIST EN 12350–2:2019	immediately after mixing	mm	180	140	220	200
	30 min after mixing	mm	160	100	220	150
Air content (SIST EN 12350–7:2019/AC:2022, Chapter 5)	immediately after mixing	%	3.8	4.7	13.3	13.3
	30 min after mixing	%	3.7	4.8	n.d	10.0
(w/b)$_{eff}$ (SIST 1026:2016. Appendix NC)		-	0.38	0.40	n.d	n.d
Density (SIST EN 12350–6;2019)		Kg/m^3	2,358	2,333	1,882	1,924

Table 3. Test results for hardened concrete.

	Density (SIST EN 12390–7)			Compressive strength (SIST EN 12390–3)			Modulus of elasticity (DIN 1048. Part 5)			Average depth of abrasion (ASTM C1138–19)	
	(kg/m^3)			(MPa)			(GPa)			(mm)	
	28 days	56	90	28 days	56	90	28 days	56	90	90 days	180
ABR-1	2,397	2,410	2,413	69.97	71.17	7.13	34.77	36.33	37.77	2.77	2.79
ABR-2	2,357	2,350	2,390	63.77	64.83	72.50	33.27	34.83	36.73	3.18	3.72
ABR-3	1,953	1,953	1,893	14.60	14.63	16.27	16.27	13.67	15.30	1.57	1.20
ABR-4	1,993	1,987	1,983	15.10	16.33	18.73	13.90	15.30	16.20	1.16	1.03

higher abrasion resistance than conventional compositions. A comparison between compositions with rubber aggregate show an improvement in the abrasion resistance of the ABR-4 composition with the addition of steel fibres. The abrasion resistance of the conventional compositions has not increased with age, moreover it has even slightly decreased in the case of ABR-2 composition. However, generally we can confirm that also the concrete samples with the addition of rubber aggregate show an improved resistance towards abrasion with the increasing age of the sample.

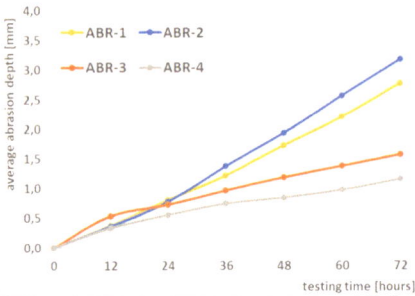

Fig. 1. Wear of concrete samples at 90 days according to ASTM C1138 test method

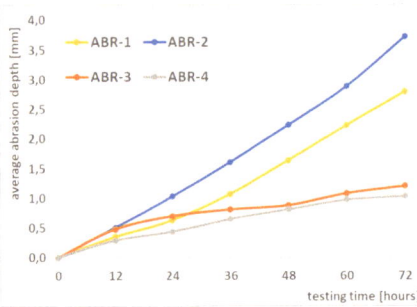

Fig. 2. Wear of concrete samples at 180 days according to ASTM C1138 test method

The dynamics of wear progression was also analysed, where the test sample ABR-3 at 90 days of age have the highest initial wear increment, while all other compositions show similar initial wear increment and ABR-4 having the lowest one (Fig. 3). As expected, in the case of compositions with a rubber aggregate after 24 h of test time a decrease in the dynamic of wear progression is detected and the wear progression remains the same until the end of the test, while on the contrary, with conventional concrete compositions, the wear progression increased slightly throughout the duration of the test.

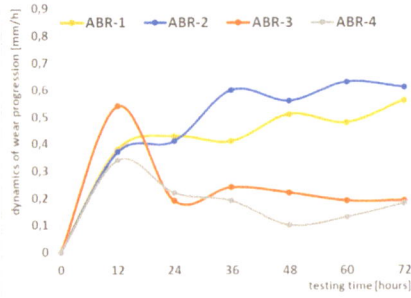

Fig. 3. Dynamics of wear progression – 90 days

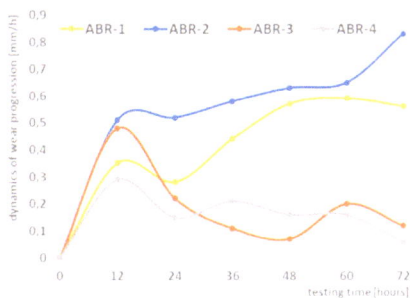

Fig. 4. Dynamics of wear progression – 180 days

At the age of 180 days the initial wear increment after 12 h of testing is lower for all compositions than for those at the age of 90 days (Figs. 3 and 4). For rubber aggregate compositions the wear progression decreases with the duration of the test and is lower than for the cases at the age of 90 days The ABR-4 composition shows a very uniform wear progression throughout the test and a lower one compared to the ABR-3 composition. However, the conventional compositions show a different wear dynamics pattern to that observed at 90 days. After an initial wear increment a slight decrease is observed after 24 h of testing after which the wear starts to continuously increase. The more pronounced is the wear dynamics of the ABR-2 composition, which is consistently higher compared to others, while the wear progression of the ABR-1 composition remains of comparable magnitude.

3.2   Wear Measurements in Natural Condition

Given that the previous field measurements showed a very good agreement between the field wear results and the ASTM C1138 abrasion resistance results, we decided to repeat the abrasion resistance studies on a large-scale field model in natural conditions. The opportunity to set up a field model was demonstrated by the restoration work on a small torrent in the highlands. As part of the remediation works, it is planned to establish extensive monitoring on the watercourse, in the scope of monitoring hydrological-hydraulic parameters and sediment transport along the watercourse. The field model is designed as a trough spillway in a watercourse bed with reinforced banks that direct the water flow to the spillway chute, where the test plots are placed at the bottom of the chute.

In 2022, we placed 8 test plots, 0.5/0.5 m in size, 10 cm thick, at a distance of 0.5 m from each other on the chute of the spillway. The test plots were prepared at the same time as the test specimens for the laboratory tests, 2 test plots for each composition. The test plots were prepared by pouring the concrete in special wooden moulds, using a vibration pin, and afterwards they were manually finished. Then the moulds with test plots were kept covered with plastic foil in controlled climate conditions at 20 °C. After one day the test plots were taken out of the moulds and kept in controlled climate condition until installation in the field. Prior to installation in the field, the test plots were measured under laboratory conditions using a photogrammetric method with a data capture accuracy between 0.3 and 0.4 mm. The test plots were framed with 2.5 cm thick wooden slats before installation and the bottom was covered with a foil to prevent the test plot from sticking to the concrete base and to allow the test plot to be removed after the investigation is completed. A network of geodetic points is embedded in the concrete base of the spillway chute to allow for periodic geodetic surveys to be carried out during the duration of the investigation, which is expected to last at least 2 years (Fig. 5). After the investigation is completed, the test plots will be removed from the field model and a final geodetic survey will be carried out under laboratory conditions, which will also serve as a reference for the wear rate in natural condition. After the completion of field measurements, we will carry out an abrasion resistance test according to the ASTM C1138 method and perform comparisons between the results of laboratory measurements in the entire duration of the investigation and measurements in test plots. The purpose of this comparison is to confirm the suitability of the ASTM C1138 method for predicting the development of abrasion resistance of concrete on water structures.

Fig. 5. Field model with test plots after installation (red dots represent geodetic points)

4   Conclusions

This paper presents the preliminary results of a research project involving laboratory and field measurements of the abrasion resistance of concretes of different compositions. The following has been established:

- All the concrete compositions used in the research, which are based on the those used in the dam constructions on the Lower Sava River, demonstrate adequate abrasion resistance.
- Concretes with the addition of rubber aggregate and a binding component consisting of cement and a dry proportion of SBR latex show comparatively significantly higher abrasion resistance than conventional compositions with cement binder and gravel aggregate.
- Fibre reinforced concretes with granulated rubber have better abrasion resistance compared to fibre-reinforced concretes with gravel aggregate, as well as concretes with granulated rubber without added fibres.

- Given that previous field investigations showed a very good agreement between the field wear results and the ASTM C1138 abrasion resistance results, it was decided to repeat the abrasion resistance investigations on a large-scale field model in natural conditions in this research with the aim of obtaining additional arguments in favour of confirming the suitability of the ASTM C1138 method for predicting the development of abrasion resistance of concretes on water structures.

Acknowledgements. The authors would like to thank Hidrotehnik, d.d., which financed the implementation of the field model. This research received funds from the Slovenian Research Agency (research core funding No. P2–0180).

References

Eldin, N.N., Senouci, A.B.: Rubber-tire particles as concrete aggregate. Cement Concrete Aggregates. **15**(1), 74–84 (1993)

Jakobs, F. & Winkler, K. & Hunkeler, F. & Volkart, P. Betonabrasion im Wasserbau, VAW, 168. Zürich: ETH. (2001)

Kryžanowski, A., Mikoš, M., Šušteršič, J., Planinc, I.: Abrasion Resistance of Concrete in Hydraulic Structures. ACI mater. j. **106**(4), 349–356 (2009)

Kryžanowski, A. Abrasion Resistance of Concrete on Hydraulic. Structures PhD thesis. University of Ljubljana (2009)

Kryžanowski, A., Mikoš, M., Šušteršič, J., Ukrainczyk, V., Planinc, I.: Testing of Concrete Abrasion Resistance in Hydraulic Structures on the Lower Sava River. Strojniški vestnik-Journal of Mechanical Engineering **58**(4), 245–254 (2012)

Kozjek, D., Pavlovčič, U., Kryžanowski, A., Šušteršič, J., Jezeršek, M.: Three-dimensional characterization of concrete's abrasion resistance using laser profilometry. Strojniški vestnik—J. Mech. Eng.J. Mech. Eng. **61**(5). pp. 311–318. (2015)

Liu, T.C.: Abrasion resistance of concrete. Journal Proceedings – ACI, vol. 78, no. 5, p. 641–350 (1981)

Mikoš, M. Fluvial abrasion of gravel sediments, Acta hydrotechnical. vol. 11. p. 10 107. University of Ljubljana (1993)

Šušteršič, J., Kryžanowski, A., Planinc, I., Zajc, A., Dobnikar, V., Leskovar, I., Ercegovič, R.: Technical report: Performance of concrete exposed to underwater abrasion loading (in Slovenian). Ljubljana: IRMA. (2004)

Toutanji, A.H.: The use of rubber tire particles in concrete to replace mineral aggregates. Cement Concr. Compos. **18**, 135–139 (1996)

Fast Cured Mineral-Impregnated Carbon-Fiber (MCF) Reinforcements Made of Geopolymer as a Promising Alternative to Conventional Fiber Reinforced Polymer (FRP) Systems

Jitong Zhao, Marco Liebscher[(✉)], Golrokh Airom, and Viktor Mechtcherine

Technische Universität Dresden, Institute of Construction Materials, 01062 Dresden, Germany
marco.liebscher@tu-dresden.de

Abstract. This study introduces the design and realization of a fast-setting technology for an efficient industrial production of a novel mineral-impregnated carbon-fiber (MCF) reinforcements for the building sector. By employing mineral-based matrices for carbon fiber (CF) reinforcements, numerous advantages can be achieved, including high temperature resistance, cost-effectiveness, reliable bonding with concrete substrates, and enhanced flexibility in automated processing.

This study focuses on the impact of different thermal curing regimes for the forming process of the MCF composite. The fabrication process involves commercially available raw materials and the utilization of a continuous pultrusion line, followed by oven heating at temperatures of 50 °C and 75 °C for short durations. The purposefully designed impregnation suspension allowed a sufficient long-lasting processing window at the early age. Extensive experimental investigations have been conducted to examine the development of the resulting MCF performance and the implementation of the MCF as reinforcement in GP concrete at varying temperature levels.

Keywords: Carbon-Fiber Composite · Mineral Impregnation · Geopolymer · Reinforcement · Automated Processing

1 Introduction

Amidst depleting energy reserves and increasing environmental pollution, carbon fiber reinforced polymer (CFRP) composites have gained prominence in modern society. By utilizing carbon fiber (CF) reinforcement, these composites enable the construction of lightweight structures and the strengthening of traditional building materials, resulting in reduced fuel consumption and harmful emissions [1]. For the available CF composites, commonly polymeric matrices are applied to secure their shape stability, inner stress-transfer, and reinforcing ability to the concrete matrix [2]. But, insufficient fire resistance and poor compatibility with the concrete matrices greatly restrict their broad application [3]. To tackle these challenges, a promising alternative to traditional steel or

© The Author(s) 2025
L. Czarnecki et al. (Eds.): ICPIC 2023, 61, pp. 495–502, 2025.
https://doi.org/10.1007/978-3-031-72955-3_50

polymer-based reinforcements mineral-impregnated carbon fiber (MCF) reinforcements has emerged. This innovative impregnation technology involves currently utilizing minerals, specifically hydraulic micro-cements [1], silica fume [4] and alternative binders, i.e., aluminosilicate [5, 6] or calcium silicate cement [7]. MCF reinforcements, with their unique composition and profiles, enable comparable load-bearing capacity to FRP while offering improved durability, fire resistance, and compatibility with concrete. Additionally, their high geometrical flexibility during the fresh and forming stages unlocks vast potential for automation and digitalization [8]. Amongst abundant variants, geopolymer (GP) impregnating suspensions offer a promising solution in terms of the long-lasting processing window and reliable impregnation quality during the early stages for efficient industrial production [9]. The syntheses of GP via moderate thermal activation at temperatures below 100 °C facilitate rapid hardening and strength evolution, addressing the concept development of the fast-setting inorganic matrix composites as advanced construction reinforcement materials, akin to thermosetting resins.

The stimulus of the present study leans upon developing fast-setting forming process for GP-based MCF composite via brief, controlled thermal treatment. A highly automated inline MCF production was employed. To identify the optimal combination, the investigating parameters involve curing temperatures of 50 °C and 75 °C for varying durations, ranging from 2 h to 8 h. The resulting MCF composites were qualified regarding their physicochemical and mechanical behavior at the early age and 28 days and implemented in a GP concrete matrix. Load transfer capacity of MCF was validated at various temperature levels from 20 °C to 200 °C.

2 Experimental Program

2.1 Materials and Manufacturing of Mineral-Impregnated Carbon-Fiber

To manufacture MCF, a commercial carbon tow, SIGRAFIL® C T50–4.4/255-E100, manufactured by SGL Group, Germany, was selected. This tow consisted of 50,000 individual filaments, each with a diameter of 6.9 μm and tensile strength of 4,400 MPa, treated with epoxy sizing. The impregnating suspension and corresponding mixing procedure were specifically tailored to attain a complete impregnation of the yarn, following a previous study [10]. The suspension comprised metakaolin (MK) ($Al_2O_3 \cdot 2SiO_2$) from BASF, Germany, a superplasticizer (SP) Sapetin D27 from Woellner, Germany, and a commercial potassium silicate activator Geosil® 14517 from Woellner, Germany. The mixing was performed intensively using a T 50 digital ULTRA-TURRAX® at a speed of 7000 rpm for 7 min. The designed GP impregnation matrix, made from metakaolin, revealed a delayed setting time up to approximately 14 h under room temperature condition. The manufacture of the unidirectional MCF element was achieved via an automated, continuous pultrusion process, as detailed in [5]. The carbon rovings were continuously pulled at a velocity of 6 m/min under constant tension. Pre-wetting was carried out using a kiss-coater, followed by impregnation in a bath equipped with five-roller-foulard. The finally shaped MCF was obtained using a conical nozzle with an inner opening diameter of 4.1 mm. After initial impregnation, the semi-finished composites were thermally cured under sealed conditions in oven. Curing parameters involved the temperature conditions of 50 °C and 75 °C for varying durations, ranging from 2 h to 8 h. Subsequently, the cured MCF specimens were stored at 20 °C and a relative humidity of 65% until testing. Figure 1 shows the used CF roving and the obtained MCF bar. The made reinforcement possessed a nearly circular cross-section after solidification with a fiber volume fraction of approximately 16 vol.-%.

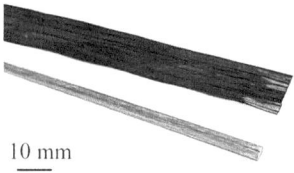

10 mm

Fig. 1. Ready-made bar of GP-based MCF (below) and used CF roving (top).

2.2 Characterization of Impregnating Matrix and MCF

To assess the impact of heating treatment on the flexural performance of MCF composites, three-point bending tests were conducted using a Zwick-Roell testing machine (model Z 1445) equipped with a load cell (capacity: 1 kN) on the young and 28 day age specimens. The tests were performed with a span of 100 mm, following the procedure outlined in [11]. The displacement rate during the tests was set at 5 mm/min. The bond-slip relationships between the yarn and GP concrete substrate was characterized by one-sided in-situ pullout tests with an Instron machine (model 8501), load cell and Instron climate chamber at a displacement rate of 1 mm/min at room temperature (~ 20 °C), 100 °C and 200 °C following a previously established testing setup [12]. The considered GP concrete matrix comprised of MK, the same potassium silicate activator (Geosil® 14517), quartz sand and rough sand, associated with 28 d compressive strength of 62.8 MPa, as described in [13]. A commercial FRP, GRID Q85/85 – CCE – 21 from Solidian (Albstadt, Germany), impregnated with epoxy (EP) resin and with tensile strength of 3300 MPa was used as reference. The pullout data has been as well published previously in [13]. The fiber matrix distribution and fracture surface of the composites were observed using an environmental scanning electron microscopy (ESEM) with a Quanta 250 FEG instrument manufactured by FEI (The Netherlands).

3   Results and Discussion

3.1   Characterization of Impregnating Matrix

Figure 2 shows representative ESEM images of the impregnation matrices fractured after curing at 50 °C and 75 °C for 2 h and 8 h, respectively. The matrices treated at 50 °C for 2 h indicated an amorphous binder phase with more unreacted or partially reacted MK particles. When extended for 8 h, less unreacted particles and more built aluminosilicate gel were visible, declaring the superior mechanical performance for the MCFs with the extended curing regime. At 75 °C after 2 h of curing, the size and number of unreacted MK particles were apparently reduced, while both the GP gel and void contents stemming from dissolved MK particles increase slightly in the structure. With extending time to 8 h, a more porous and less ordered microstructure was particularly built.

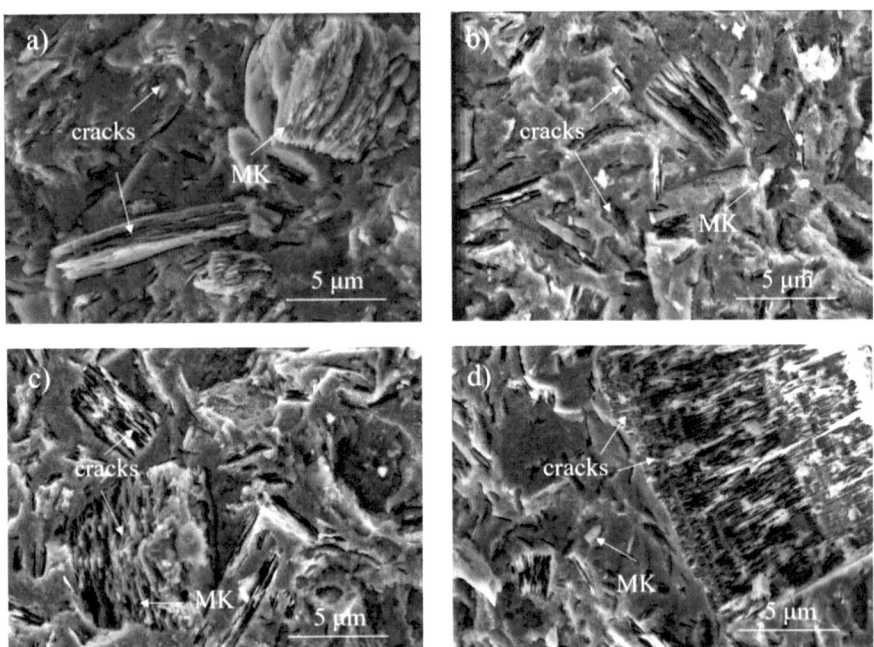

Fig. 2. ESEM images of the GP matrices cured at 50 °C and 75 °C (a, b) for 2 h and (c, d) for 8 h, respectively.

The morphological characteristics of the homogeneous filament-matrix distribution provide compelling evidence in support of the high quality of the designed GP suspension; as determined using microscopic image of the cross-section of MCF; cf. Figure 3a. The specimen was cured at 50 °C for 8 h and these images accurately represent specimens produced with varying curing conditions. The black circles denote the locations of individual filaments. Instances of impregnation matrix accumulation without embedded filaments were observed infrequently. An even incorporation of carbon filaments within the matrix yields excellent interfacial stress transfer between components, thereby contributing to the exceptional mechanical performance of the composite. In Fig. 3b, a consistent and continuous embedding of the fiber without discernible gaps was evident, suggesting a favorable physical interaction between the components.

Differences obtained in the morphological evolution correlated well with considerable changes in the flexural performance of MCF composites; see Fig. 4. Since the thermal curing effectively accelerated the geopolymerization reaction and therewith a rapid development in matrix strength in the early stage of the forming process, outstanding flexural performance was attained within few hours. With rising temperature, here from 50 and 75 °C, a faster strength gain both under flexural load can be seen merely before 4 h, relating to the advanced geopolymerization [14], whereas further constantly heating yielded in inferior flexural strength. For the thermally cured specimens for 8 h at 50 °C, the early-age flexural strength reached the highest value of 590 MPa, 30% greater than 75 °C. This decreasing trend was attributed to the increased matrices' porosity, seen

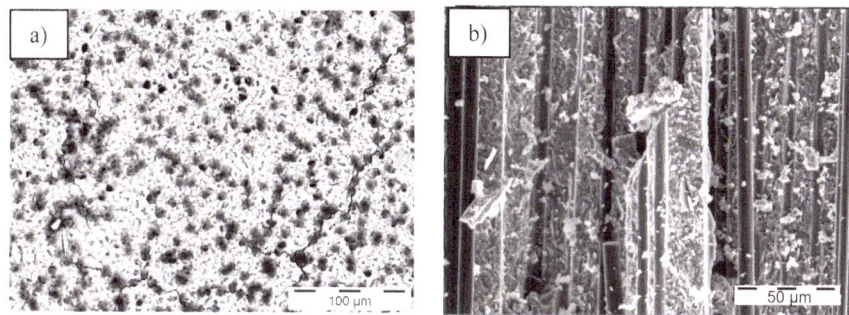

Fig. 3. ESEM images showing in a) the cross-section and in b) fracture surface of MCF cured at 50 °C for 8 h.

in Figs. 2 and 3. At both temperatures, the extension of curing, here from 2 h to 8 h, resulted in a gradual increase in flexural strength, triggered by the increased degree of geopolymerization [5]. After storing for additional 28 days at 20 °C, the flexural strength exhibited similar values to the young age except for 2 h-cured composite counterparts and generated also an increasing trend with extended curing. Consequently, the above findings suggested an extended curing duration at a moderate temperature, e.g., 50 °C for 16 h, for the post-treatment of MCF, which was further used for the resulting MCF in the yarn pullout tests with GP concrete.

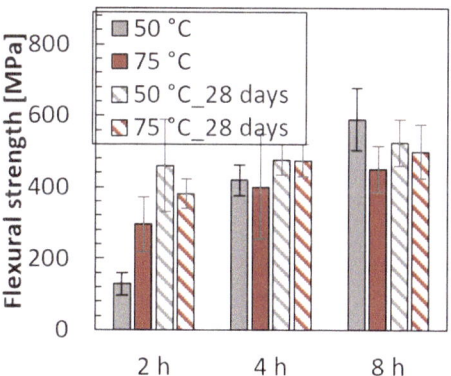

Fig. 4. (a) Flexural strength and (b) tensile strength of MCFs tested immediately after curing at 50 °C and 75 °C as well as at the age of 28 days.

Finally, pullout results highlighted comparable average shear bond strength in GP concrete to the EP yarn for the developed MCF reinforcement at room temperature; see Fig. 5a. Whereas, distinct pullout curve profiles were seen between them; see Fig. 5b. The MCF bars revealed a fast ascending branch in the initial loading stage until reaching the peak debonding force at a small displacement and with a higher shear modulus followed by a sudden drop to a moderate pullout force level. The particular curve profile emphasized good chemical compatibility of MCF towards GP concrete matrices. The reference EP yarn was characterized by a slowly and nonlinearly ascending branch and afterwards a gradual decay in frictional force. The bond of EP yarn was dominated by the mechanical interlock from the surface deformation, rather than the chemical adhesion between components. With rising environmental temperature, a significant downward trend was accompanied by reductions of 21% and 95% at 100 °C and 200 °C, respectively. The significantly diminished bond observed in the EP yarns was triggered by the substantial softening of the viscoelastic epoxy and the degradation of the shear modulus upon surpassing its glass transition temperature [13]. However, considerable enhancement in bonding behavior at elevated temperatures levels were depicted with GP impregnation, maintaining approx. 60% bond strength at 200 °C.

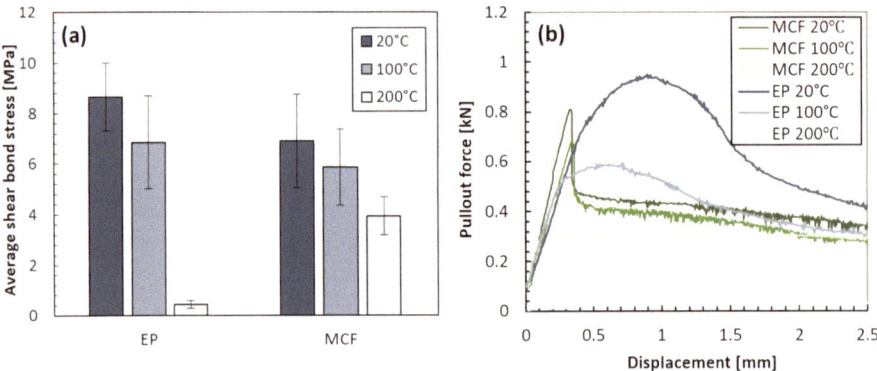

Fig. 5. (a) Shear bond strengths and (b) representative pullout force-displacement curves at different temperature levels for the epoxy yarns and the MCFs, adapted from [13].

4   Summary and Conclusions

This paper highlights the potential of automated produced and fast-setting MCF as promising fiber reinforcement system for the construction industry. Efficient rapid MCF setting was achieved via a purposefully designed GP impregnation matrix based on MK. A sufficiently long-lasting processing window allows an easy usage of the mineral suspension as well as a high flexibility for industrial applications.

By applying targeted heating at 50 °C and 75 °C, a rapid setting and early strength gain of the derived MCF prototype was realized at different rates and almost completed merely within the initial several hours, similar to the forming process of FRP with thermosets. The produced MCF achieved a maximal flexural strength of 590 MPa after 8 h of curing at 50 °C, being comparable to conventional FRP. The temperature treatment with prolonged duration indicated the advancement of geopolymerisation of the impregnation matrix and enhanced mechanical performance of resulting MCFs. However, negative impacts of the heat curing promoted the formation of a more porous and less ordered microstructure. For this reason, longer curing duration at relatively low temperatures, e.g., 50 °C, is expected to be beneficial for durable MCF.

Finally, the novel type of reinforcement delivered equivalent bond strength to commercial CF yarns with epoxy coating at room temperature and evident improvement in bond strength at elevated temperatures, as proven by yarn pullout tests.

References

1. Mechtcherine, V., Michel, A., Liebscher, M., et al.: Mineral-impregnated carbon fiber composites as novel reinforcement for concrete construction: Material and automation perspectives. Autom. Constr. **110**, 103002 (2020)
2. Dvorkin, D., Poursaee, A., Peled, A., Weiss, W.J.: Influence of bundle coating on the tensile behavior, bonding, cracking and fluid transport of fabric cement-based composites. Cem. Concr. Compos. **42**, 9–19 (2013)
3. Bisby, L.A., Green, M.F., Kodur, V.K.R.: Response to fire of concrete structures that incorporate FRP. Prog. Struct. Mat. Eng. **7**, 136–149 (2005)
4. Nadiv, R., Peled, A., Mechtcherine, V., et al.: Micro- and nanoparticle mineral coating for enhanced properties of carbon multifilament yarn cement-based composites. Compos. B Eng. **111**, 179–189 (2017)
5. Zhao, J., Liebscher, M., Michel, A., et al.: Development and testing of fast cured mineral impregnated carbon fiber (MCF) reinforcements based on geopolymers from metakaolin. Cem. Concr. Compos. **116**, 103898 (2021)
6. Zhao, J., Liebscher, M., Köberle, T., et al.: Mineral-impregnated carbon-fiber (MCF) composites made with differently sized fly-ash geopolymers for durable light weight and high temperature applications. Cem. Concr. Compos. 104950 (2023)
7. Strauss Rambo, D.A., De Andrade Silva, F., Toledo Filho, R.D., et al.: Tensile strength of a calcium-aluminate cementitious composite reinforced with basalt textile in a high-temperature environment. Cem. Concr. Compos. **70**, 183–193 (2016)
8. Mechtcherine, V., Michel, A., Liebscher, M., Schmeier, T.: Extrusion-based additive manufacturing with carbon reinforced concrete: Concept and feasibility study. Materials **13**, 2568 (2020)
9. Zhao, J., Trindade, A.C.C., Liebscher, M., et al.: A review of the role of elevated temperatures on the mechanical properties of fiber-reinforced geopolymer (FRG) composites. Cem. Concr. Compos. **137**, 104885 (2023)
10. Zhao, J., Liebscher, M., Tzounis, L., Mechtcherine, V.: Role of sizing agent on the microstructure morphology and mechanical properties of mineral-impregnated carbon-fiber (MCF) reinforcement made with geopolymers. Appl. Surf. Sci. **567**, 150740 (2021)
11. Liebscher, M., Zhao, J., Wilms, G., et al.: Influence of roller configuration on the fiber-matrix distribution and mechanical properties of continuously produced, Mineral-Impregnated Carbon Fibers (MCFs). Fibers **10**, 42 (2022)
12. Zhao, J.: Marco Liebscher, Schneider K, et al Effect of surface profiling on the mechanical properties and bond behaviour of mineral-impregnated, carbon-fibre (MCF) reinforcement based on geopolymer. Construc. Build. Mater. **367**, 130199 (2023)
13. Zhao, J., Zhao, D., Liebscher., M., et al.: Temperature-dependent pullout behavior of geopolymer concrete reinforced with polymer- or mineral impregnated carbon fiber composites: an experimental and numerical study. ACS Sustain. Chem. Eng. in press (2023)
14. Rovnaník, P.: Effect of curing temperature on the development of hard structure of metakaolin-based geopolymer. Constr. Build. Mater. **24**, 1176–1183 (2010)

Influences of the Effectiveness of a Column Confinement with Textile Reinforced Concrete (TRC)

Wladislaw Polienko[✉] and Klaus Holschemacher

Structural Concrete Institute, Leipzig University of Applied Sciences, Leipzig, Germany
{wladislaw.polienko,klaus.holschemacher}@htwk-leipzig.de

Abstract. In the present paper the results of uniaxial compression tests conducted on textile reinforced concrete (TRC)—confined reinforced concrete (RC) columns are reported. By confining the column with TRC, the lateral expansion of the concrete can be impeded. The resulting multiaxial compressive stress state allows to enhance the components axial capacity. Due to the corrosion-resistant textile, the usual concrete cover in reinforced concrete construction is reduced, which allows slender but at the same time highly load-bearing components to be created. Consequently TRC provides a sustainable, environmentally friendly and lighter option for column reinforcement due to the material savings. The aim of this study is the investigation of various influences on the achievable strengthening effect. The impact of ratio of textile reinforcement and the applied fine grain concrete jacket is evaluated. In addition, the influence of the concrete strength of the strengthened component on the overall increase in load-bearing capacity was investigated. With the help of experiments on TRC reinforced RC columns with a circular cross-section mechanical property and constraint mechanism under uniaxial compression were documented and analyzed. Based on the test results, stress distribution and failure mechanisms of the reinforced specimens is studied. Furthermore, stress-strain relationship of strengthened members is investigated. The results show an increasing in both strength and ductility related to the unstrengthened reference columns. The specimens with a lower compressive strength can achieve a higher degree of reinforcement. A high ductility of the reinforced columns could also be observed.

Keywords: Concrete columns · Textile reinforced concrete · Confinement · Strength · Reinforced concrete · Experimental tests · Strain-Stress

1 Introduction

1.1 Research Significance

The reuse of existing buildings is becoming increasingly relevant due to progressive urbanization and the resulting demand for residential and commercial space in urban areas. This approach is now being understood as the key to climate protection in the

© The Author(s) 2025
L. Czarnecki et al. (Eds.): ICPIC 2023, 61, pp. 503–510, 2025.
https://doi.org/10.1007/978-3-031-72955-3_51

construction industry. In 2014, for example, investments in existing buildings in Germany alone amounted to around €173 billion. This includes monitoring, inspection, maintenance and redesign of structures. Changed and increased requirements, a stricter normative framework, but also damage that has occurred during the service life of the building can lead to the need to improve the load-bearing capacity, serviceability or durability of individual components or of an entire structural system [1].

The standards available to the designer name and classify different retrofitting measures. These essentially include cross-section additions, modifications to the structural system, injections and prestressing technologies. Cross-section additions can be realized by shotcrete or in-situ concrete, while reinforced and unreinforced concrete are possible. In addition, there are newer methods such as confinement with TRC.

TRC is an innovative building material which has been intensively researched in the last decade. Due to the use of non-metallic, textile reinforcement the concrete cover required to protect the reinforcement from corrosion can be reduced and allows the production of slender, durable components with high load-bearing capacity. Compared to classic reinforced concrete, much lower layer thicknesses and thus lower dead loads can be achieved with even higher load-bearing capacity and lower material consumption.

In order to enable the new type of reinforcement to enter the market, a generally applicable and reliable analytical design model must be created based on certain relevant structural parameters. This must cover the design of load-bearing capacity, serviceability and durability. For this purpose, the behavior of the composite material under different loading situations must be investigated.

1.2 Column Retrofitting

RC columns are an indispensable structural element for transferring vertical loads for the entire structural system. By restoring or increasing the load-bearing capacity of the columns, the overall load-bearing capacity of a building can be effectively strengthened.

In practice, shotcrete is increasingly being used for retrofitting columns. The increase in ultimate load is achieved by a combination of subsequently applied additional layer of concrete and steel reinforcement. Sufficient concrete cover must be ensured to comply durability and protect the applied reinforcement from corrosion. This material application, which is usually not statically relevant, contributes to total layer thicknesses of up to 10 cm, resulting in considerable increases in cross-section and significantly higher dead loads that must be taken into account in the structural analysis. The increase in load-bearing capacity results here from the enlargement of the cross-sectional dimensions and the additional longitudinal and transverse steel reinforcement.

By confining with TRC systems the mode of action results from the transverse strain restraining effect of the textile reinforcement. The carbon fiber fabrics, which are usually applied unidirectionally, allow forces to be absorbed transversely to the direction of loading and thus impede the transverse strain of the loaded column. The resulting transverse pressure leads to a multi-axial stress state, which significantly increases both the load-bearing capacity and the maximum compression of the whole component.

This method combines the advantages of shotcrete and CFRP confinement. On the one hand, due to the additional concrete application, the old concrete can be repassivated. This process restores the alkaline environment, stops the carbonation of the carbonation

of the old cross-section and thus prevents further corrosion of the existing reinforcement. On the other hand, the fire protection properties can be improved in parallel with
the increase in load-bearing capacity. Another significant advantage is the corrosion
resistance of the textile reinforcement material. The concrete cover required to ensure
durability is eliminated, which both saves material and reduces the space required for the
reinforcement measure. TRC thus provides a sustainable, more environmentally friendly
and lighter option for column reinforcement.

2 Experimental Investigation

2.1 Experimental Programm

The experimental program, shown in Table 1 includes 12 steel-reinforced concrete cylindrical specimens, which are divided into two groups with different compressive strengths.
All specimen have a diameter D of 200 mm and height H of 1000 mm and have been
tested under a longitudinal force load. The main objectives of the experimental program
were (a) to analyze the possible increase in ultimate load and (b) the exploration of
the influence of the concrete compressive strength on the reinforcing effect. In order to
simulate a component in need of rehabilitation in the existing structure the core concrete
compressive strength of the first series (C20) was chosen to be low. From the literature,
it is known that the core concrete compressive strength has a great influence on the
confining pressure of the reinforcement system, which is why it was increased in the
second series (C55) up to a target compressive strength class of C55. Specimen R of each
series represent the reference unconfined specimens consisting only of the reinforced
core concrete. Specimen with the Label 2L20 are equipped with a 20 mm thick layer of
fine-grained concrete in which two layers of carbon textile are embedded.

Table 1. Experimental Program

Series	Specimen label	No. of spec	H/D	Confinement	f_{c0} [N/mm^2]	$f_{cm,m}$ [N/mm^2]
C20	R	3	5	reference	27.67	-
	2L20	3	5	2 layers, 20 mm fine-grained concrete	30.76	94.96
C55	R	3	5	reference	*	-
	2L20	3	5	2 layers, 20 mm fine-grained concrete	*	81,62

*material characteristic value to be determined

2.2 Specimen Preparation

Each test series was cast out of the same recipe designed to obtain a cylindrical compressive strength (f_{c0}) of 20 MPa and 55 MPa. The cement content was 240 kg/m^3 in
C20 and 330 kg/m^3 in C55 while the water cement ratio (w/c) was 0.75 in C20 and

0.48 in C55. The cement:sand:gravel proportions in the concrete mixtures were around 1:3.22:4.80 in C20 and 1:2.34:3.46 in C55 by weight and the maximum size of the coarse aggregate was 16 mm in both series. With the aim to determine the mechanical properties of the concrete, three cylindrical specimens measuring 150×300 mm for each series were made out of the same batch and tested on the same day as the main specimens. The compressive strength of the core f_{c0} and the fine-grained concrete fcm,m is shown in Table 1. Furthermore, the specimens were provided with a steel reinforcement content of 2.36 cm^2/m. Six rebars with a diameter of Ø12 mm as longitudinal reinforcement and every 10 cm cross-sectional reinforcement out of Ø6 mm curved rebars were used. The tensile strength of the steel rebars was 500 MPa and a young's modulus of 200 GPa. The concrete cover at all of the specimen was 15 mm.

Before any confinement works, the surface of the specimen needed to be prepared. A middle roughness of the surface of around 1 mm could be achieved by sandblasting the specimen until the aggregate with a diameter of >4 mm was visible. Furthermore 24 h before confining, the specimens were prewetted and covered with foil. The surface was wetted again and cleaned of dust 20 min before confining. To ensure uniform loading, all specimen were capped with a fine-grained concrete.

2.3   Confining Materials

The confining materials, which were used are regulated by the german general technical approval (Z-31.10-182 (2016)). It designates the use of the textile reinforcement TUDALIT-BZT1-TUDATEX, which a bidirectional warped mesh impregnated with a film-forming dispersion based on Styrene-butadiene rubber. According to the approval, only the yarns in the warp direction with the red knitting thread (Table 2) may be used for reinforcement. Important mechanical properties are summarized in Table 2.

As a mortal matrix, the fine-grained concrete TF10 CARBOrefit® is being used. This concrete has a maximum grain size of 1 mm and has been specially developed for the processing of carbon reinforcements in the hand lay-up and spray process. The concrete mixture, which is available as ready-mixed concrete, has a characteristic minimum compressive strength equal to 80 MPa. The compressive and flexural strengths was measured using three prisms for each confined specimen.

2.4   Instrumentation of the Specimens and the Experimental Setup

All of the 12 specimens were stored for more than 28 days under controlled termperature and humidity conditions (20 °C and 60% relative humidity) until the testing. The wrapping started with the application of the first layer of fine-grained of 5 mm; after, the first ply of carbon mesh was applied and slightly pressed into the mortar. The textile was then wrapped around the specimens under slight tension, and the next layer of concrete was applied in parallel (Fig. 1(a)).

These processes of wrapping and applying the fine-grained concrete were repeated until the required number of layers was achieved. As a concrete cover, a final 5 mm layer of fine-grained concrete was applied. By using additional stencils, uniform total layer thicknesses of 20 mm could be realized (Fig. 1(b)). In addition, to prevent premature debonding failure of fibers, an overlap length of 50 cm was provided in the confined

Table 2. Characteristics of the textile reinforcement TUDALIT-BZT2-V.FRAAS according to Z-31.10-182 (2016) [2]

Properties of a coated yarn	Carbon yarns in warp direction	Structure of mesh
Number of filaments per yarn	3200/3300 tex	
Fiber cross-sectional area Textile Yarn	140 mm^2/m 1.8 mm^2	
Tensile strength Mean value Characteristic value	1980 MPa 1890 MPa	
Modulus of elasticity Mean value Characteristic value	170 GPa 166 GPa	
Ultimate strain Mean value Characteristic value	1.28% 1.24%	
Coating	Styrene-butadiene rubber (SBR) - Lefasol VLT-1	

specimen. Furthermore, one layer of CFRP was applied to avoid a failure in the column head and foot (Fig. 1(c)).

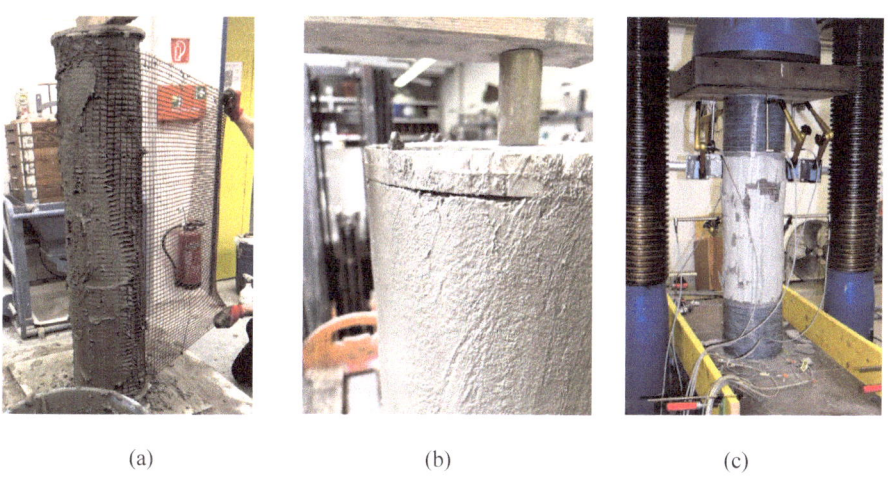

    (a)   (b)   (c)

Fig. 1. (a) wrapping process; (b) uniform total layer thicknesses; (c) test setup

All tests were performed using a servo-hydraulic compression testing machine with a maximum load carrying capacity of 6,000 MPa. The tests were done on deformation-controlled mode, primarily to allow accurate analysis of the processes of load transfer to the reinforcing layer and failure. The test speed was set at 0.01 mm/s according to empirical values. Axial and lateral displacements were measured by external linear variable differential transducers (LVDT) mounted on two opposite sides of the specimen. The test setup is shown in Fig. 1(c).

3 Experimental Results

The evaluation of the test results showed that from confining the columns with TRC an significant increase in load-bearing capacity in relation to the unconfined specimen can be attained. With identical initial conditions, specimen with a lower core concrete compressive strength achieved a higher percentage increase in load-bearing capacity. An increase by an average of 67% compared to those with a higher concrete compressive strength class with 56% could be achieved. As expected, a dependence of the core concrete compressive strength of the component to be reinforced with the effectiveness of the reinforcement becomes apparent. The test results are showed in Table 3.

By confining a column with TRC the lateral expansion can be limited by the textile absorbing the axial loads as tensile stress in hoop direction. Once activated due to volume increase during load the textile provides confining pressure, which is continuously increased with gaining axial load. If the formation of lateral expansion is limited by a higher concrete compressive strength, this results in a lower as tensile stress in textile. A lower increase in load capacity is the consequence.

Table 3. Experimental Results

Series	Label	Age	H	F_{max}	ΔF	Strength increase	COV
–	–	[d]	[cm]	[kN]	[kN]	[%]	[%]
C20	R-1	107	100.7	1125.2	1107.1	0	1,7
	R-2	107	101.0	1115.6			
	R-3	107	100.8	1080.5			
	2L20-1	106	100.3	1892.4			
	2L20-2	106	100.4	1794,9	1847.8	66,9	2.2
	2L20-3	106	100.4	1856,1			
C55	R-1	294	100,4	1845,4			
	R-2	294	100,2	1969,8	1901,1	0	2,7
	R-3	294	100,9	1888,2			
	2L20–1	295	100,6	2895,3			
	2L20–2	295	100.4	3216,4	2978,1	56,7	5,7
	2L20–3	295	100.4	2822,7			

The experimental results are shown visually in Fig. 2, while radial strain is negative and axial strain is positive. The increase in stiffness before macro cracking compared to the unreinforced specimen is noticeable. This is due to the increased cross-section and the much higher compressive strength of the fine-grained concrete.

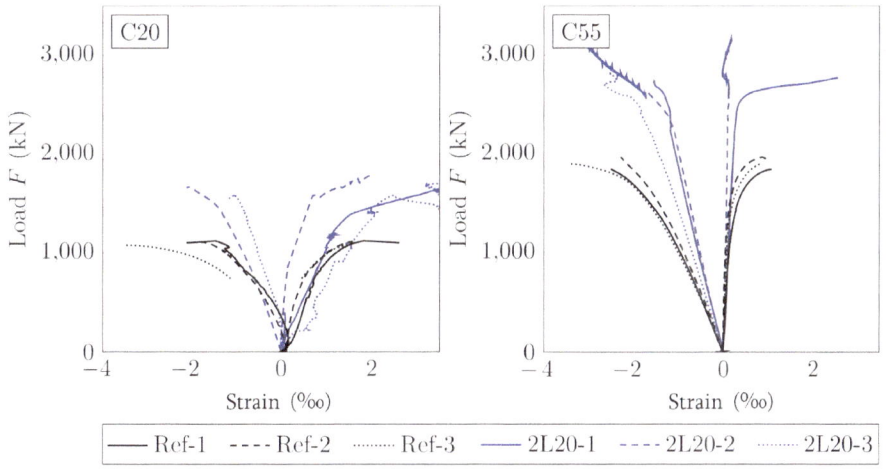

Fig. 2. Diagrams of the experimental results

4   Summary

In this paper, a part of a more extensive experimental program is showed. The retrofitting of RC columns with TRC increased the load-bearing capacity compared to the unreinforced specimens. The collected results show that with lower concrete compressive strength a higher strengthening effect can be observed. Also a typical behavior for TRC could be observed. After the axial compressive strength of the core concrete is reached, the textile can be gradually activated by reaching the tensile strength of the fine-grained concrete characterized by elongated cracks. The compressive load acting on the component can thus be converted into the textile. A renewed absorption of the load is then the consequence.

Acknowledgements. The authors would like to thank the University of Applied Sciences Leipzig (HTWK Leipzig) for the sponsorship of this research. Additionally, PAGEL Spezial-Beton GmbH & Co. KG is gratefully acknowledged for providing the fine-grained concrete mixture.

References

1. Zilch, K., Niedermeier, R., Finckh, W.: Sachstandsbericht Verstärken von Betonbauteilen mit geklebter Bewehrung. In: Deutscher Ausschuss für Stahlbeton (DAfStb), Heft 591 (2011)
2. Allgemeine bauaufsichtliche Zulassung Z-31.10-182: Verfahren zur Verstärkung von Stahlbeton mit TUDALIT (Textilbewehrter Beton). Deutsches Institut für Bautechnik (2016)

Improvement of the C-PC Properties

Evaluation of Strength and Modulus of Elasticity of Polymer-Modified Cement Concrete (PCC) Under Thermal Impact Within a Defined Service Temperature Range

Alexander Flohr[✉], Catharina Rohde, Savitha Devarajamohalla Narayana, and Andrea Osburg

F.A. Finger-Institute for Building Materials Engineering, Chair of Construction Chemistry and Polymer Materials, Bauhaus-Universität Weimar, Weimar, Germany
alexander.flohr@uni-weimar.de

Abstract. Polymer-modified cement mortars (PCM) and concretes (PCC) are mainly used in concrete repair and restoration exhibiting improved durability, suitable chemical resistance, and beneficial adhesion strength compared to unmodified cementitious materials. Due to these favorable properties, the material is increasingly implemented in construction. Commonly, the modifiers applied to cementitious binders consist of thermoplastic polymers, which feature a change in the deformation behavior under the influence of different temperatures. Despite the distinct temperature-dependent properties of the polymers, the load-dependent deformation behavior of PCM and PCC was barely investigated within a service temperature range. To make statements about the effect of polymers on the load bearing and elastic deformation behavior of PCM and PCC, the engineering properties of the material have to be experimentally assessed under thermal conditioning. Accordingly, the compressive and flexural strength as well as dynamic and static modulus of elasticity of seven different PCM mixtures were characterized while the specimens were exposed to service temperatures of −20 °C, 20 °C, and 60 °C. After the specimens were thermally conditioned in a climate chamber, the samples were transferred to the equally conditioned test machine and tested in the proposed temperature scope. The experimental results reveal influential changes in all tested mechanical attributes for the modified system within the applied service temperature range compared to an unmodified reference. This knowledge is essential to further investigate the temperature impact on the material and develop appropriate prediction models for the application of polymer-modified cementitious materials in construction and the integration in design guidelines.

Keywords: Polymer modified mortar · Service temperature · Thermal impact · Mechanical properties · Modulus of elasticity

L. Czarnecki et al. (Eds.): ICPIC 2023, 61, pp. 513–521, 2025.
https://doi.org/10.1007/978-3-031-72955-3_52

1 Introduction

Firstly, patented in the 1920s [1], the development of and research on polymer-modified cementitious materials (PCC) increasingly intensified in the 1960s and 1970s [2]. Since then, the compound material was mainly used in repair and restoration but became eventually more important also as construction material due to specific mechanical properties compared with ordinary cement mortar and concrete. Alongside with Portland cement and aggregates, the modified mortar or concrete incorporates usually thermoplastic polymers. These polymers have the underlying characteristic to be sensitive to temperature. While higher temperatures can cause a decrease of strength and deformation resistance, lower temperatures can foster a brittle state of the polymer. Whereas the difference in temperature causing to change the material from a brittle to plastic state often lies within a small temperature range of approximately 40 °C to 70 °C. Due to the necessity of application of polymer-modified mortar and concrete for refurbishment matters and the wish to be able to estimate the behavior of the material within the supposed service temperature range, multiple studies were conducted exposing the material to high-temperature conditions above 200 °C [3–8]. Some studies included elevated temperatures but do not address typical mechanical behavior such as flexural and compressive strength or the determination of the modulus of elasticity [9–13]. And a manageable number of studies were concerned with polymer-modified cementitious material exposed to cold climates [14]. Hence, the typical mechanical engineering properties of polymer-modified mortar and concrete under the thermal impact within a defined service temperature range between −20 °C to 60 °C need to be experimentally evaluated to further investigate the thermally implied impact and to develop fitting prediction models.

2 Experimental Program

To obtain adequate results, the experimental program followed consecutive steps, which were performed under alternating thermal conditions as it can be seen in Fig. 1. At a temperature of 20 °C, the material was prepared and mixed to adequate mortar specimens, fresh mortar properties such as consistency, air void content, and fresh bulk density were determined as well as the mortar specimens were stored according to the corresponding standards. While the specimens were conditioned and then hardened mortar properties were obtained at temperatures of −20 °C, 20 °C, and 60 °C.

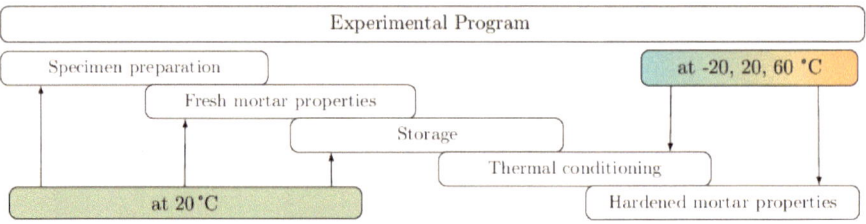

Fig. 1. Flow diagram of the experimental program

2.1 Materials

Cement. All experiments were conducted using a Portland cement (CEM I 42.5 R). The product exhibited a density of 3.11 g/cm^3, an average particle size of 14.29 μm, and a specific surface (Blaine) of 3580 cm^2/g. The chemical composition can be seen in Table 1.

Table 1. Chemical components of the used Portland cement in weight percentage

Chemical component	CaO	SiO_2	Al_2O_3	Fe_2O_3	SO_3	MgO	K_2O	Na_2O	LOI
Percentage [%]	64.1	19.5	5.0	2.9	3.2	1.5	1.0	0.2	2.2

Polymers. To modify the mortar, three different thermoplastic polymer dispersions were used: styrene-acrylate copolymer (SA), ethylene-vinyl acetate copolymer (EVA) and styrene butadiene rubber copolymer (SBR). All three polymer dispersions contained water as liquid phase and are commercially available as well as specifically suitable for the application in mortars. The material characteristics of the polymer dispersions are summarized in Table 2.

Table 2. Material properties of the polymer dispersions

Polymer	SA	EVA	SBR
Main constituents	styrene, acrylic acid ester	ethylene, vinyl acetate	styrene, butadiene
Solid particle content [%]	50.6	53.7	50.8
Density [g/cm^3]	1.03 ± 0.02	1.07	1.02
Particle size range [μm]	0.04 – 2.11	0.52 – 7.08	0.08 – 0.21
Mean particle size [μm]	0.15	1.34	0.13
pH value at 20 °C [−]	8.18	3.24	8.04
Dynamic viscosity [mPas]	50 – 200 (25 °C)	400 (23 °C)	30 – 150 (23 °C)
MFT [°C]	33	0	12
T_g [°C]	22	−6	16
T_m [°C]	380	330	130

2.2 Specimen Preparation

According to DIN EN 196-1, a sand-to-cement ratio (s/c) of 3.0 was predefined so that 450 g cement and 1350 g sand were applied to each mixture. With respect to a w/c-ratio of 0.40 and the water given due to the liquid phase of the polymer dispersions, the mix design was determined as it can be seen in Table 3. The mixing process in compliance with DIN EN 196-1 was performed utilizing a mixer with automated mix program preset. The designation of the samples is based on the following principle. REF is the abbreviation of the reference sample whilst PCM is the one for polymer-modified mortars. The following numbers 1 to 3 mark the particular modifying polymer and the last numbers the polymer content in percent based on cement weight (p/c).

Table 3. Mix design for the mortars

Sample ID	w/c	p/c	s/c	Water [g]	Polymer [g]	Cement [g]	Sand [g]
PCM_Ref	0.4	0.00	3.0	180.0	0.0	450	1350
PCM_SA05	0.4	0.05	3.0	158.0	44.5	450	1350
PCM_SA15	0.4	0.15	3.0	114.1	133.5	450	1350
PCM_EVA05	0.4	0.05	3.0	160.6	41.9	450	1350
PCM_EVA15	0.4	0.15	3.0	121.8	125.7	450	1350
PCM_SBR05	0.4	0.05	3.0	158.2	44.3	450	1350
PCM_SBR15	0.4	0.15	3.0	114.6	132.9	450	1350

2.3 Specimen Conditioning

The specimens prepared for the testing of flexural and compressive strength as well as stabilized secant modulus (Young's modulus) and dynamic modulus of elasticity were stored according to the national annex of DIN EN 12390-2. After preparation, the molded mortar specimens were protected from loss of moisture and stored for 24 h at a temperature of 20 °C. Then, the specimens were demolded and cured for six days in water at a temperature of 20 °C. Following the seventh day of preparation, the mortar specimens were stored in a draught-free environment at 20 °C and 65% relative humidity. At the day of testing, the specimens were subjected to a temperature regime of −20 °C, 20 °C, or 60 °C. Prior to the analyses at −20 °C or 60 °C, the specimens were thermally conditioned utilizing a climate chamber. To guarantee a uniform temperature distribution throughout the complete specimens, the samples were cooled or heated for 2 h until the core of the sample reached the target temperature of −20 °C or 60 °C, respectively.

2.4 Test Procedures

Fresh Mortar Properties. The fresh mortar properties can be appropriately represented by determining the consistency using the slump test according to the procedure described in DIN EN 1015-3, the air void content by means of the pressure method according to DIN EN 1015-7 and the fresh bulk density identified due to volume and weight measurements according to DIN EN 1015-6.

Hardened Mortar Properties. The hardened mortar properties of the material used within this study are represented at first by the determination of bulk density according to

DIN EN 1015-10, true density, and total porosity. The total porosity has been computed by applying a ratio of dry bulk density $\rho_{s,dry}$ and true density ρ of specimens aged 2, 7, and 28 days. Subsequently, the total porosity P was calculated to:

$$P = 1 - \frac{\rho_{s,dry}}{\rho} \tag{1}$$

Furthermore, the flexural and compressive strength according to DIN EN 1015-11, the Young's modulus following DIN EN 12390-13 (Method B), and the dynamic modulus of elasticity in accordance with DIN EN ISO 12680-1 were tested at the age of 2, 7, and 28 days at temperatures of -20 °C, 20 °C and 60 °C. Therefore, prismatic specimens were produced according to DIN EN 1015-2 with an edge length d and b of 40 mm and a length L of 160 mm. The dynamic modulus of elasticity was determined using the impulse excitation technique.

3 Results

3.1 Fresh Mortar Properties

Table 4 shows the results of the fresh mortar tests. The influence of the polymers on the fresh mortar properties is significant and existing research results could be confirmed. Generally, a liquefaction of the mortar by the polymer modification was observed. Furthermore, the mortars modified with styrene butadiene rubber copolymer (SBR) showed an increased air void content, what is also often described in relating publications especially using polymers without defoamer.

Table 4. Fresh mortar properties

Sample ID	Slump [mm]	Consistency	Air void content [%]	Fresh bulk density [g/cm^3]
PCM_Ref	102	stiff	4,5	2,30
PCM_SA05	122	stiff	4,3	2,27
PCM_SA15	199	plastic	2,8	2,24
PCM_EVA05	108	stiff	4,9	2,27
PCM_EVA15	143	plastic	5,4	2,20
PCM_SBR05	146	plastic	7,2	2,20
PCM_SBR15	217	soft	10,0	2,09

3.2 Hardened Mortar Properties

Figures 2, 3, 4, and 5 show the results of the hardened mortar tests, which are the first small steps to get an impression of the temperature-dependent behavior of the mortars.

The experimental results reveal influential changes in all tested mechanical attributes for the PCMs compared to an unmodified reference, but there is no general correlation. It was surprising that at −20 °C the flexural strength of all samples except PCM_SBR05 decreases with increasing hydration time. It is likely that the ice formation process is responsible for this effect because, especially at early ages, there is still a lot of water in the samples that can freeze. The behavior of the specimens tested at 20 °C and 60 °C is basically as expected. There is an increase in flexural strength with increasing sample age. The polymer modification leads to a slight increase in flexural strength, which is more pronounced for the high polymer contents. The SBR-modified samples behave a little bit different due to the rather high air void content. A decrease in strength with increasing temperature was observed for all mortars tested, where, contrary to expectation, the temperature dependence of the PCMs was not greater than that of the reference.

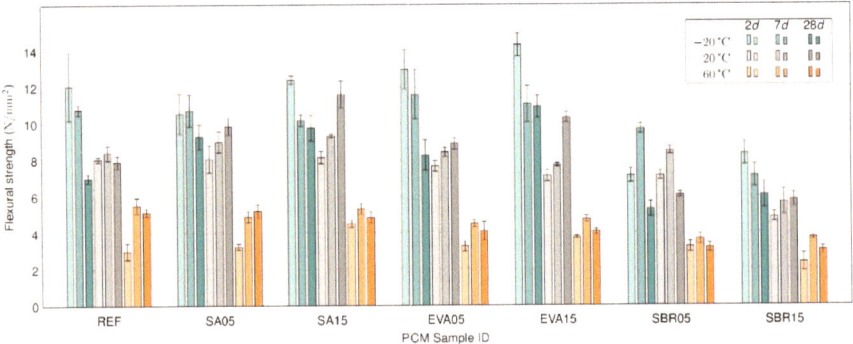

Fig. 2. Flexural strength of the mortars at 2, 7 and 28 days and −20, 20 and 60 °C

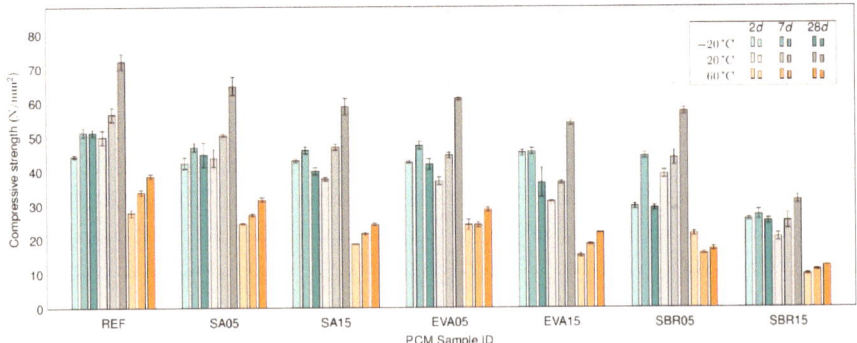

Fig. 3. Compressive strength of the mortars at 2, 7 and 28 days and −20, 20 and 60 °C

Under compressive stress, the mortars behave largely as expected. The polymer modifications cause a strength decrease of the PCMs compared to the reference. As the hydration time progresses, the strengths and deformation resistance of all mortars increase, with this behavior being most pronounced at 20 °C. Furthermore, it can be

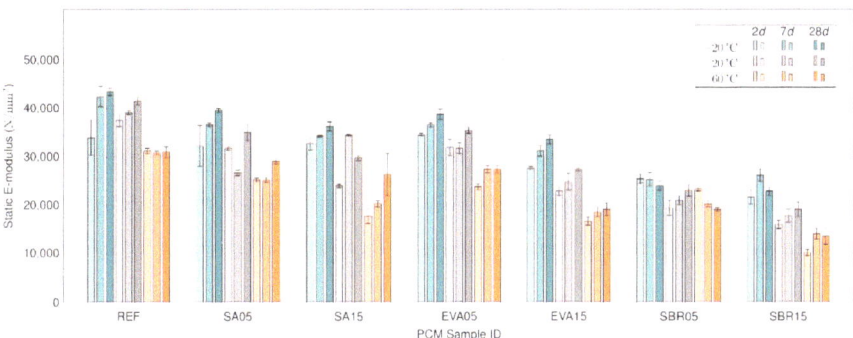

Fig. 4. Young's modulus of the mortars at 2, 7 and 28 days and −20, 20 and 60 °C

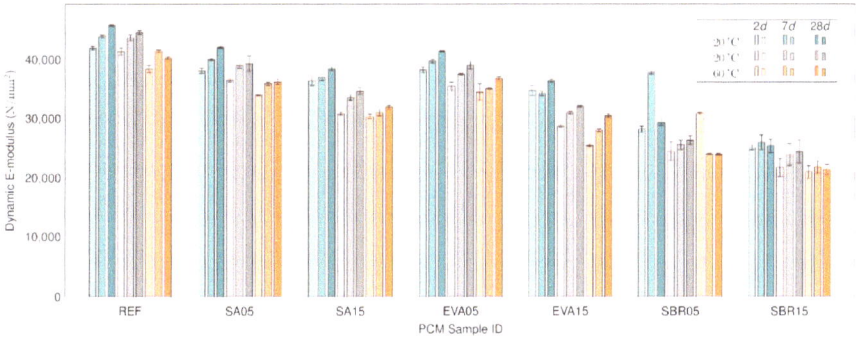

Fig. 5. Dynamic modulus of elasticity of the mortars at 2, 7 and 28 days and −20, 20 and 60 °C

observed that both compressive strength and elastic modulus are lowest at 60 °C, although the differences between −20 °C and 20 °C are hardly significant and should be considered individually for each formulation. In conclusion, it can be stated that the thermoplastic polymers do not necessarily lead to a greater temperature dependence of the hardened mortar properties. This knowledge is essential to further investigate the temperature impact on the material and develop appropriate prediction models.

4 Conclusions

The application of different service temperatures leads to significant differences in the mechanical parameters of the mortars investigated, so that temperature must be considered as a significant factor in modelling of the material behavior. The results of the presented investigations are the input parameters beside others which are to be bundled in a semi-analytical multiscale model based on the methods of continuum micromechanics. By means of a bottom-up approach, homogenized properties at the macroscale are determined using the specific microstructural behavior. The microstructure changing as a result of the temperature influence can thus be directly correlated with the macroscopic material behavior. After extending an existing multiscale model [15] by considering principles of thermo-poro-elasticity, the temperature dependence of the mechanical properties of polymermodified mortars and concretes will be predictable.

Acknowledgments. This research is supported by the German Research Foundation (DFG) via research grant for the project "Experimental investigations and microstructure-based modeling of the elastic and visco-elastic behavior of PCC depending on temperature", which is gratefully acknowledged.

References

1. Lefebure, V.: Improvements in or relating to concrete, cements, plasters and the like. Br. Pat. **217**(279), 5 (1924)
2. Ohama, Y.: Handbook of polymermodified concrete and mortars: properties and process technology. William Andrews, New Jersey (1995)
3. Won, J.-P., Choi, S.-W., Park, C.-G., Jang, C.-I.: High strength polymermodified repair cementitious composite for fire protection. Polym. Polym. Compos. **15**(5), 379–388 (2007)
4. Muhammad, B., Ismail, M., Yussuf, A., Muhammad, A.: Elastomeric influence of natural rubber latex on cement mortar at high temperatures using thermal degradation analysis. Constr. Build. Mater. **25**(5), 2223–2227 (2011)
5. Kim, H., Noguchi, T.: Burn-up characteristics of polymermodified cement mortar used for building repair. In: Proceedings of the Korea Institute of Building Construction, vol. 12, pp. 295–298 (2012)
6. Muthadhi, A., Kothandaraman, S.: Experimental investigations on polymer-modified concrete subjected to elevated temperatures. Mater. Struct. **47**(6), 977–986 (2014)
7. Song, H., Shin, H.: High temperature properties of cement mortar using EVA, EVCL redispersible polymer powder and fly ash. J. Korean Recycled Constr. Resour. Inst. **6**(4), 365–372 (2018)
8. Kim, H.-J., Park, J.-Y., Suh, H.-W., Cho, B.-Y., Park, W.-J., Bae, S.-C.: Mechanical degradation and thermal decomposition of ethylene-vinyl acetate (EVA) polymer-modified cement mortar (PCM) exposed to high-temperature. Sustainability **11**(2), 500 (2019)
9. Gad, E., El-Sukkary, M., Sayed, W., Abo-El-Enein, S.: Physicochemical characteristics of acrylicacid polymer-impregnated cement pastes. J. Chem. Technol. Biotechnol. Int. Res. Process. Environ. Clean Technol. **62**(3), 310–316 (1995)
10. Al-Gahtani, A., Rasheeduzzafar, Al-Mussallam, A.: Performance of repair materials exposed to fluctuation of temperature. J. Mater. Civ. Eng. **7**(1), 9–18 (1995)
11. Rashid, K., Ueda, T., Zhang, D., Miyaguchi, K., Nakai, H.: Experimental and analytical investigations on the behavior of interface between concrete and polymer cement mortar under hygrothermal conditions. Constr. Build. Mater. **94**, 414–425 (2015)
12. Rashid, K., Wang, Y., Ueda, T.: Influence of continuous and cyclic temperature durations on the performance of polymer cement mortar and its composite with concrete. Compos. Struct. **215**, 214–225 (2019)
13. Rashid, K., Zhang, D., Ueda, T., Jin, W.: Investigation on concrete-PCM interface under elevated temperature: At material level and member level. Constr. Build. Mater. **125**, 465–478 (2016)
14. Mirza, J., Mirza, M., Lapointe, R.: Laboratory and field performance of polymer-modified cementbased repair mortars in cold climates. Constr. Build. Mater. **16**(6), 365–374 (2002)

15. Göbel, L., Bos, C., Schwaiger, R., Flohr, A., Osburg, A.: Micromechanics-based investigation of the elastic properties of polymer-modified cementitious materials using nanoindentation and semi-analytical modeling. Cement Concr. Compos. **88**, 100–114 (2018)

Basic Study on Ultra Rapid Hardening Alkali Activated Material Using Sodium Orthosilicate

Hiroshi Yamada[1]([✉]), Mikio Wakasugi[2], Toshiyuki Kanda[2], Tomonori Seki[3], and Kazuo Ichimiya[4]

[1] Osaka Sangyo University, 3-1-1, Nakakakiuchi, Daitoushi, Japan
h-yamada@ce.osaka-sandai.ac.jp
[2] Chemical Construction, 5-5, Uozakihamamachi, Higashinadaku, Japan
[3] Sumitomo Osaka Cement, 1357-1, Ojinboucho, Funabashi, Japan
[4] National Institute of Technology, Oita College, 1666, Maki, Oita-Shi, Japan

Abstract. For the purpose of developing a repair material that contributes to decarbonization, a basic study is conducted on an alkali activated material that does not require an alkaline solution. Thin study is characterized by the use of sodium orthosilicate as an alkaline source in order to achieve powder premixing, be ultra rapid hardening. Specifically, a basic study on ultra rapid hardening property and an improvement of the length change performance of the material using only ground granulated blast furnace slag were carried out. As a result, it was clarified that sufficient compressive strength at 3 h can be achieved by increasing the alkaline usage rate, and that the length change performance can be improved by using shrinkage reducing agents and expansive agents.

Keywords: Sodium orthosilicate · Alkali active material · Ground granulated blast furnace slag · Ultra rapid hardening · Compressive strength · Length change

1 Introduction

Reduction of greenhouse gas emissions is required in order to solve global climate change problems. Alkali-activated materials (AAM) are attracting attention as construction materials that contribute to the reduction of CO_2 emissions. The AAM studied in Japan so far are mainly classified as geopolymers[1]. The geopolymer is formulated to have a low calcium content, and the general material composition is a mixture of water glass and an alkaline aqueous solution (alkaline solution), and fly ash and ground granulated blast furnace slag (GGBS) as active fillers [1]. There are cases where only GGBS is used as the active filler [2], and cases where other industrial by-products are used [3]. On the other hand, there are concerns that domestic social infrastructure, which was intensively developed during the high economic growth period, will deteriorate rapidly in the future, and there is a need to accelerate measures to deal with the deterioration. Many of the repair materials used for aging countermeasures are premixed cementitious materials, and various types are available depending on the application, from general-purpose types to ultra-rapid hardening types. However, almost no decarbonized repair

© The Author(s) 2025
L. Czarnecki et al. (Eds.): ICPIC 2023, 61, pp. 522–530, 2025.
https://doi.org/10.1007/978-3-031-72955-3_53

materials are on the market. The development of repair applications for AAM is considered to be an effective means of decarbonization, and some practical studies have begun [4]. However, these methods require the preparation of alkaline solutions on-site, which poses a problem of construction safety. In this study, we developed a premix type AAM that can be used on site by simply adding water, and developed an ultra-rapid hardening type that exhibits practical strength after 3 h.

2 Sodium Orthosilicate

To achieve premixing, powdered sodium orthosilicate (Na_4SiO_4) shown in Fig. 1 was selected as the alkaline source for AAM. Sodium orthosilicate is a substance conventionally used for degreasing and cleaning of metals, and it is considered that sodium metasilicate and NaOH are produced by hydrolysis as shown in the following formula. In other words, by using sodium orthosilicate as an alkaline source, it is possible to approach the state of using water glass and NaOH during mixing, it is the same handling as conventional cementitious repair materials.

$$Na_4SiO_4 + H_2O \rightarrow Na_2SiO_3 + 2NaOH \qquad (1)$$

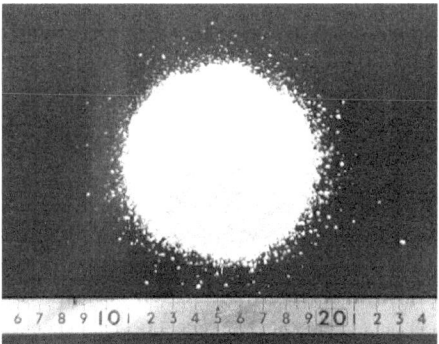

Fig. 1. Sodium orthosilicate

3 Experiment

3.1 Outline of Experiment

The initial strength development of geopolymers is affected by the alkaline concentration in the solution and the proportion of GGBS used as the active filler [1]. Therefore, as Test Series 1, the effects of the amount of sodium orthosilicate used in the solution and the amount of GGBS in the active filler on the 3-h compressive strength of AAM were investigated. Next, it has been reported that the shrinkage property of AAM using GGBS is greater than that of ordinary Portland cement [2]. Therefore, as Test Series 2, the effects of expansive agents and shrinkage reducing agents were investigated for compressive strength and length change characteristics of ultra-rapid hardening AAMs, with the aim of improving the length-change performance.

3.2   Materials, Mix Proportion and Mixing

In Test Series 1, tap water was used as the mixing water. Sodium orthosilicate, symbol AL, its density 2.39 g/cm^3 (Fig. 1) was used as the alkaline source. Fly ash equivalent to JIS Class II, symbol FA, its density 2.36 g/cm^3 and GGBS equivalent to Blaine 4000 m^2/g without gypsum, symbol GGBS, its density 2.91 g/cm^3 were used as active fillers. Dry sand for premix, symbol S, its density 2.64 g/cm^3 was used as fine aggregate. In Test Series 2, in addition to the above materials excluding FA, powdered shrinkage reducing agent, symbol SR, its density 1.20 g/cm^3 and quicklime as expansive agent, symbol EX, its density 3.34 g/cm^3, were used.

Table 1 shows the mix proportion of Test Series 1. The volume of AL in the solution and the volume of GGBS in the active filler were taken as parameters with a liquid equivalent of 401 L/m^3 as the basic condition. Since AL is water-soluble, the amount. Was set to liquid equivalent. In addition, the fine aggregate active filler ratio S/(FA + GGBS) was set to a constant volume ratio based on the mass ratio when the GGBS replacement rate was 100%. Table 2 shows the composition of Series 2. Under the basic conditions of 401 L/m^3 liquid equivalent, 17.3 vol.% alkali content, and only GGBS as the active filler, the usage rates of shrinkage reducing agent SR and expansive additive EX were used as parameters. EX was replaced as part of the active filler volume, and SR was replaced as part of the liquid equivalent. Fine aggregate by active filler ratio S/(GGBS + EX) was set to a same volume ratio.

Table 1. Mix proportion in Test Series 1

No	AL/(W + AL) (vol.%)	GGBS/(FA + GGBS) (vol.%)	Unit quantity (kg/m^3)				
			W	AL	FA	GGBS	S
1	6.49	100	375	62	0	829	829
2	9.73	100	362	93	0	829	829
3	13.0	100	349	124	0	829	829
4	17.3	100	332	166	0	829	829
5	22.7	100	310	218	0	829	829
6	17.3	66.7	332	166	224	553	829
7	17.3	33.3	332	166	448	276	829
8	17.3	0	332	166	672	0	829

※W + AL: 401 L/m^3, S/(FA + BFS): Constant volume rate based on the mass rate of the mixture with a BFS replacement rate of 100 vol.%

Using a Hobart-type mortar mixer in a room at 20 °C, the pre-mixed powder material and fine aggregate were added to the water weighed in a mixing bowl and kneaded for 2 min. Immediately after mixing, a fresh test and specimen preparation were performed. The preparation of the specimen conformed to JSCE-F 506 "Method of making cylindrical specimen of mortar or cement paste for compressive strength".

Table 2. Mix proportion in Test Series 2

No	SR usage rate (vol.%)	EX usage rate (vol.%)	Unit quantity (kg/m^3)					
			W	AL	SR	GGBS	EX	S
4	0	0	332	166	0	829	0	829
9	5	0	312	166	24	829	0	829
10	10	0	292	166	48	829	0	829
11	0	5	332	166	0	788	48	829
12	0	10	332	166	0	746	95	829

※W + AL + SR: 401 L/m^3, AL/(W + AL + SR): 17.3 vol.%, S/(BFS + EX): Constant volume ratio based on formulation No. 4

3.3   Test Method

In the property test of fresh mortar, the temperature and flow value (JIS R 5201) immediately after mixing were measured. Compressive strength test complied with JSCE-G 505 "Test method for compressive strength of mortar and cement paste using cylindrical specimen". In Test Series 1, it was tested only at an age of 3 h. In addition, demolding was performed about 2 min before the test. In Series 2 it was tested at 3 h, 1, 7, 28, 56 and 91 days. The curing conditions were sealed in a room of 20 °C. Series 2 involved a length change test. Using a formwork of φ5 × 10 cm, a simple embedded strain gauge (measurement length 60 mm) was installed in the formwork so that the center of the cylindrical specimen and the center of the measurement part coincided. The kneaded sample was cast and demolded after one day. After that, the specimen was placed in air at 20 °C and 60% RH., the length change rate was measured at 0, 1, 4, 7, 14, 28, 56 and 91 days. In order to understand the relationship between the length change rate and the mass loss rate, a specimen for mass measurement was also prepared. Mass measurements were performed on the same day as length change rate measurements. The mass loss rate was obtained by dividing the mass loss due to drying of the specimen by the mass of the specimen at the start of drying.

4   Results and Discussion

4.1   Test Series 1

Table 3 shows the fresh test results. First, focusing on the effect of the AL concentration (No. 1 to 5), the higher the AL concentration, the higher the temperature immediately after mixing and the Flow value tended to decrease. Next, focusing on the effect of the BF replacement rate (No. 4, No. 6 to 8), the higher the replacement rate, the smaller the flow value, and the temperature tended to remain almost unchanged. The flow values of No. 7 and No. 8 were both 300 or more, but No. 8 clearly showed higher fluidity.

Figure 2 shows the effect of AL concentration in solution on compressive strength at 3 h. Mix proportion No. 2 to 5 were demoldable, and the higher the AL content, the higher the compressive strength. Compressive strength increased remarkably from 13.0

to 17.3 vol% of AL concentration, and moderately increased from 17.3 to 22.7 vol%. In addition, in the case of the composition with a low AL concentration (No. 1), demolding was not possible at 3 h, but demolding was possible on the following day. Figure 3 shows the effect of GGBS replacement rate on compressive strength at 3 h. Compressive strength increased with increasing GGBS replacement rate. In addition, no hardening was observed even after the next day in the mix proportion No. 8 that used only FA as the active filler. The GGBS replacement ratio was 33.3% or more in the mix proportion that could measure the compressive strength at 3 h.

Table 3. Fresh test result in Series 1

No	AL Concentration (vol.%)	GGBS replacement rate vol.%)	Temperature (°C)	Flow value
1	6.49	100	27.3	over 300
2	9.73	100	31.6	298
3	13.0	100	35.3	253
4	17.3	100	40.0	199
5	22.7	100	41.7	133
6	17.3	66.7	39.6	282
7	17.3	33.3	39.6	over 300
8	17.3	0	39.8	over 300

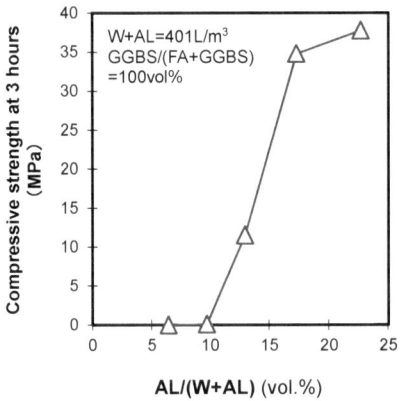

Fig. 2. Effect of AL concentration on compressive strength at 3 h

4.2 Test Series 2

Table 4 shows the test results of fresh mortar. First, the mixing temperature was almost constant regardless of the SR usage rate and EX usage rate. As for the flow value, the

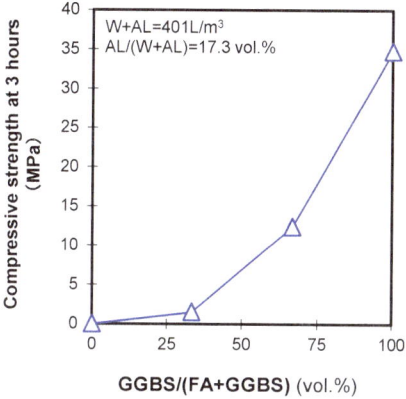

Fig. 3. Effect of GGBS replacement rate on compressive strength at 3 h

higher the SR usage rate and the EX usage rate, the smaller the flow value. In particular, when EX was used, the flow value tended to be remarkably small.

Figure 4 shows the relationship between compressive strength and age. The left figure summarizes the effect of SR usage rate, and the right figure summarizes the effect of EX usage rate. First, focusing on the left figure, the compressive strength at 3 h was almost the same for all mix proportions, and the compressive strength tended to increase with the age. However, the higher the SR content, the lower the long-term compressive strength.

Next, focusing on the right figure, the compressive strength of the mixture with an EX content of 5 vol.% was about 60% of that of the mixture with an EX content of 0 vol.% at 3 h. Although the compressive strength tended to increase with age, the rate of increase in strength was small compared to an EX content of 0 vol.%. The compressive strength of the mixture with an EX content of 10 vol.% increased until the age of 1 day, but decreased after that. After 7 days, swelling was clearly observed in the appearance.

Table 4. Fresh test result in Series 2

No	SR usage rate (vol.%)	EX usage rate (vol.%)	Temperature (°C)	Flow value
4	0	0	40.0	199
9	5	0	39.4	178
10	10	0	39.6	159
11	0	5	39.7	104
12	0	10	40.0	95

Figure 5 shows the relationship between drying time and length change. First, focusing on the left figure summarizing the effect of the SR usage rate, the length change

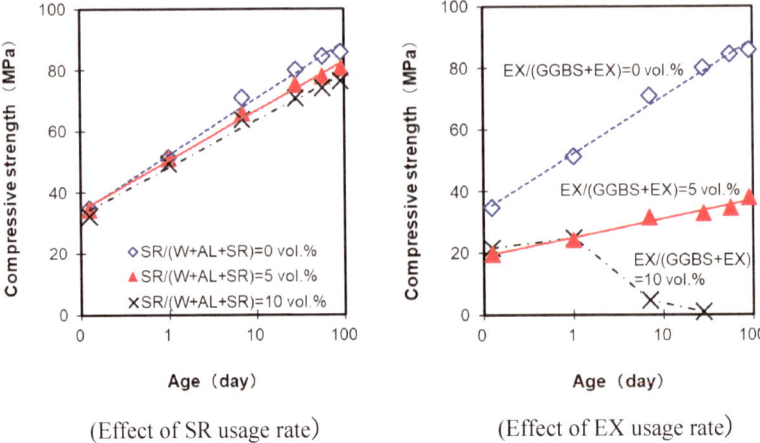

Fig. 4. Relationship between age and compressive strength

became negative from the initial stage of drying, indicating a tendency to shrink. In addition, the shrinkage rate tended to increase with the passage of drying time, regardless of the composition. Furthermore, the higher the SR usage rate, the lower the shrinkage rate, and the shrinkage rate of the mix proportion with SR usage rate of 10 vol.% decreased by about 2000×10^{-6} after drying time 91 days compared to the mix proportion without SR. Next, focusing on the right figure summarizing the effect of EX usage rate, it was confirmed that the length change became a positive value at an EX usage rate of 10 vol.%, indicating expansion. It should be noted that the expansion rate reached about 9000×10^{-6} at the drying time of 4 days, and after that it became impossible to measure. The mix proportion with 5 vol.% of EX showed a negative length change from the initial stage of drying, just like the without EX. However, that with 5 vol.% of EX showed smaller shrinkage of about 700×10^{-6}, which was almost the same after 14 days.

Figure 6 shows the relationship between and drying time and moisture loss. Focusing on the left figure, the moisture loss of the order of several percent was confirmed in hardened cement [5], but the moisture loss in this study was much smaller than that.

No clear trend due to blending was confirmed. Moreover, no long-term water loss was confirmed, and in some cases, moisture was taken in from the outside of the hardened body, resulting in a negative moisture loss. Next, focusing on the right figure, the higher the EX usage rate, the greater the negative moisture loss rate with the passage of drying time, and the tendency to take in moisture from the outside became noticeable.

From the above, it is considered that the length change of ultra-rapid hardening AAM using sodium orthosilicate may be caused by shrinkage due to polycondensation reaction as experienced in a geopolymer. In addition, it is considered that in mix proportion with EX, the continuous absorption of moisture from the outside reacts with EX and causes expansion. The shrinkage suppression mechanism and the mechanism of the curing reaction itself due to the use of SR have not been elucidated in this study, so they will be studied in the future.

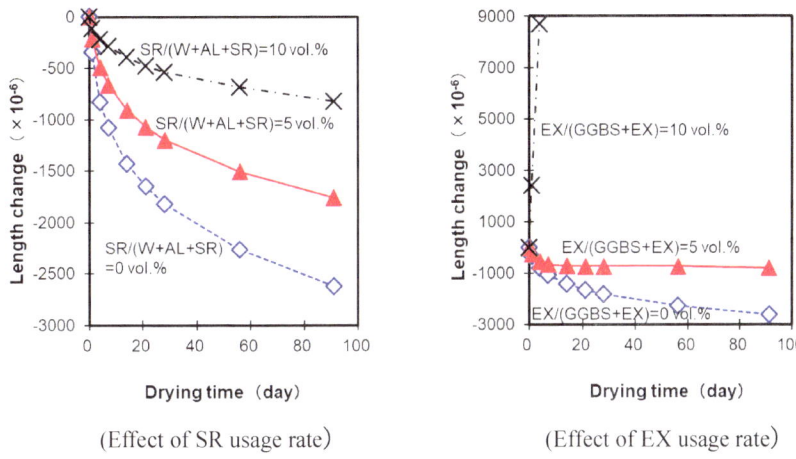

Fig. 5. Relationship between drying time and length change

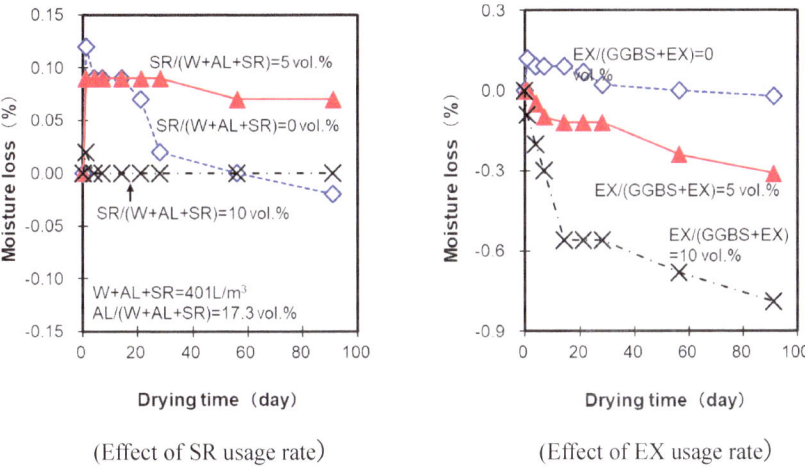

Fig. 6. Relationship between drying time and moisture loss

5   Conclusions

The results of basic studies on alkali-activated materials using sodium orthosilicate are as follows. Compressive strength at 3 h tended to increase with increasing alkali usage rate and ground granulated blast furnace slag replacement rate. The effect of powder shrinkage reducing agent on 3-h compressive strength is small. Compressive strength at 3 h tended to decrease with increasing use of expansive additive. In addition, when the expansive additive content was less than 5 vol.%, the strength increased with age. At 10 vol.% expansive additive, the compressive strength decreased after 1 day, and finally could not be measured. The use of shrinkage reducing agents and expansive additives can improve the length change performance. The mass of the hardened alkali-activated material using sodium orthosilicate did not change or tended to increase slightly even after curing in an environment of 20 °C and 60% RH.

References

1. Japan Concrete Institute: Research Committee Report on Application of Geopolymer Technology to Construction Field (2017)
2. Ichinomiya, K., Kai, M.: Fundamental physical properties of AAM mortar produced from ground granulated blast furnace slag and alkaline solution. In: Abstracts of 2022 JSCE Western Branch Conference, pp.683–684 (2022)
3. Kondo, R., Lee, J., Ikeda, O.: Study on various performances and ability to fix harmful elements of hardened geopolymer using municipal waste incineration ash. Annu. Proc. Concr. Eng. **42**(1), 1816–1821 (2020)
4. Uehara, M., Sato, T., Kosaka, Y., Yamaguchi, M.: Preparation and properties of room-temperature-curing potassium (K)-type geopolymer plastering mortar. **42**(1), 1780–1785 (2020)
5. Yamada, H., Seki, T., Yamamoto, M.: Effect of fine aggregate grain size on shrinkage characteristics of mortar. Materials **71**(7), 617–622 (2022)

Optimization of Eco-Sustainable, Form-Stable Phase Change Material to Be Incorporated in Aerial-Lime-Based Mortars

Paulina Guzmán García Lascurain[1], Alessandro Amendola[1], Mariaenrica Frigione[2(✉)], Antonella Sarcinella[2], Lucia Toniolo[1], and Sara Goidanich[1]

[1] Department of Chemistry, Materials and Chemical Engineering "Giulio Natta", Politecnico di Milano, Milan, Italy
[2] Innovation Engineering Department, University of Salento, Lecce, Italy
mariaenrica.frigione@unisalento.it

Abstract. The building sector faces a challenge to find innovative and sustainable ways to increment the energy-efficiency of buildings and reduce their environmental impact. Recently, the incorporation of phase change material (PCM), based on a polymeric active phase (PEG-1000) in waste stone aggregates, has proven to be a promising option to be used for building restoration. Mortars that include PCM aggregates demonstrated to have favorable thermal properties, that would lead to a reduction of energy requirement for heating/cooling needs. However, the inclusion of aggregates impregnated by PEG causes a reduction in the mechanical properties of the mortars possibly due to (i) a lack of compatibility between aggregate and binder, or (ii) a problem with the confinement of the PEG, causing its dispersion in the mortar. Therefore, the aim of this study was to investigate the causes associated to the reduction of the mechanical properties and propose a method to prevent it. Preliminary results showed that, given its high water solubility, the PEG 1000 included in the stone aggregates tends to be *washed away* when these aggregates are incorporated in the mortar mixture. This hypothesis was confirmed by FTIR spectroscopy. Therefore, an additional confinement method using a layer to coat the stone aggregates impregnated by PEG 1000 was proposed in this study. Different materials were tested as coating layer: powder calcium hydroxide, milk of lime (suspension of $Ca(OH)_2$ in water), pozzolana, and cocciopesto. Carbonated mortar samples using the proposed coated aggregates were, then, analyzed using FTIR to evaluate the efficiency of this encapsulation methodology. Preliminary results suggested a relevant improvement in terms of PEG confinement.

Keywords: Phase change materials (PCMs) · Lime-based mortars · Thermal energy storage (TES) · Circular economy

1 Introduction

The environmental impact of human activities is a concerning issue that has been tackled in recent years by many disciplines and industrial sectors. Particularly for the construction industry, the energetic consumption of a building during its use is considerably

© The Author(s) 2025
L. Czarnecki et al. (Eds.): ICPIC 2023, 61, pp. 531–538, 2025.
https://doi.org/10.1007/978-3-031-72955-3_54

higher than the energy required for its construction [1, 2]. This large consumption is mainly due to cooling and heating systems [3]. Thus, is fundamental to mitigate this issue by finding solutions to improve the energy efficiency of buildings. On the other hand, the renovation and restoration of historical buildings to prolong their usable life has also a positive impact on the environment. It has been estimated that the renovation of a building could save 30–60% of energy with respect to constructing a brand-new one [4]. The renovation of historical buildings also permits to implement innovative solutions to improve the building's energy consumption, for example Phase Change Materials (PCM) for passive thermal regulation.

The implementation of PCM in the mortar mixture has been proposed in the last decade [3, 5] as a suitable solution to obtain better thermal stability in the buildings. These materials are characterized by a large latent melting heat; thus, they are able to absorb/release thermal energy trough changing its physical phase [6]. PCM can act, in fact, as passive systems reducing the total energy demand required for the maintenance of comfort conditions inside a building. Even if their implementation in lime-based mortars is recent [7–9], the results show a promising improvement of the thermal properties of the PCM-based mortars. A sustainable solution was employed for the incorporation of the polymeric phase of the PCM, i.e. PEG 1000 (Poly(ethylene glycol)), in the mortars. Form-Stable method was employed using waste pieces of Lecce Stone (LS) as the support matrix for the form stable PCM (FS-PCM) [7, 10–12]. The inclusion of these aggregates, on the other hand, causes a reduction in the mechanical properties of the mortars, forcing to reformulate the mix-design in order to compensate for the loss in mechanical properties [11].

It was hypothesized that the reduction in mechanical properties may be due to (i) a lack of compatibility between aggregate and binder, or (ii) a problem with the confinement of the PEG, causing its dispersion in the mortar. Starting from these considerations, the aim of the present work was to investigate the causes associated to the reductions in mechanical properties of the PCM-based mortars and propose some new methods to prevent it.

2 Materials and Methods

2.1 Modification of the Form-Stable Phase Change Material Aggregates

The Form-Stable Phase Change Material (FS-PCM), were obtained according to the process described by Frigione et al. [7]. These composite materials were produced using Lecce stone (LS) as the matrix for supporting the PCM. The LS is composed mainly by $CaCO_3$ and was obtained from the waste of the stone cutting process from a quarry located near Lecce (Italy). The active phase for the PCM was Poly(ethylene glycol) (trade name PEG 1000) supplied by Sigma—Aldrich (Germany). This is a non-toxic, low environmental impact, low-flammable, and relatively cheap thermoplastic polymer, with a melting temperature in the range 37°–40 °C. The composite made with LS and the PEG 1000 was the FS-PCM used as aggregate in the mortar mixture. In the present work, these aggregates were modified in four different ways, indicated as: size reduction, polishing, washing, and coating.

The size reduction was performed by mechanically compressing the FS-PCM in a mortar without grinding them. For the polishing method, sandpaper of 600 to 2500 grade was used to partially remove the outermost layer of the PEG polymer composing the PCM. Water washing with deionized water was performed in two different configurations: static and manually stirred. Finally, the application of a coating was chosen in order to obtain an external layer on the composite PCM aggregates. FS-PCM was placed in a heating plate under constant stirring (500–700 rpm) at 34°C (i.e. at a temperature within the range of phase change of the PEG 1000). The coating material was added in an excess of approximately 20% with respect to the weight of the aggregates and mixed for 20–30 min. The coating materials were: powder calcium hydroxide (CH), milk of lime (suspension of $Ca(OH)_2$ in water) (ML), pozzolana (P), and cocciopesto (C). The latter being a fine powder made of clay materials mixed with lime and sand. The calcium hydroxide was provided by Fassa Bortolo Srl (Italy), the milk of lime was produced using the same calcium hydroxide. The pozzolana and cocciopesto were supplied by CTS company (Italy).

2.2 Composition of Mortar Formulations

With the aim of understanding the behavior of the coated aggregates when they are included in a mortar, a set of mortars were produced and subsequently analyzed trough FTIR. All mortars, possessing the same formulation, were based on aerial lime, provided by Fassa Bortolo Srl, and carbonatic sand of different granulometries, provided by the same company. To these mortars the modified aggregates were added. The composition of the mortars are reported in Table 1. The amount of water added to the mix corresponded to 36% by weight of total dry components (i.e. sum of all the weights added in the proportions expressed in Table 1).

Table 1. Proportions of dry components of the mix-design expressed in percentage by weight.

Binder: Hydrated lime $(Ca(OH)_2)$	FS-PCM or modified aggregates (CH, ML, P, C)	Sand (0.1 – 0.6mm)	Sand (0.6 – 1.4mm)	Total of dry components
30	40	10	20	**100**

2.3 Material Characterization

In order to characterize the surface properties of the aggregates before modification, optical microscopy observations were carried out using a Leica DM6 microscope at 5X and 10X magnifications, coupled with a Flexcam C1 camera.

Fourier-Transform Infrared Spectroscopy (FTIR) measurements were made to determine and quantify the presence of PEG 1000 in the unmodified aggregates. The same technique was also used to analyze the presence of PEG 1000 in the mortar mixture and evaluate the efficacy of the confinement method. All measurements were made using a Macro Attenuated Total Reflectance FTIR (Macro-ATR-FTIR) Thermo Nicolet iS20 spectrometer, equipped with a DTGS detector and a Smart itX diamond accessory for ATR, in the spectral range 4000 - 400 cm^{-1}, collecting 64 scans for each measurement with a 4 cm^{-1} spectral resolution (diameter of the window 2 mm).

3   Results and Discussion

The unmodified FS-PCM was characterized by a *waxy* surface clearly visible under the microscope (Fig. 1a), which indicates that PEG polymer was able to fill the porosity present in the Lecce Stone aggregates but it is also present on their surface. This observation confirmed that the reduction in mechanical properties of the PCM-based mortars can be due to a lack of compatibility between the binder and the polymer present on the aggregate, thus surface modifications might help to avoid this issue. Moreover, the average size of the aggregates is 2mm [7]. It is known that to have adequate mechanical properties, the granulometry curve of the aggregates must be compact and even [13]: as the FS-PCM aggregates are quite large, a reduction of their size could be a possible means to improve the mechanical characteristics of mortars, without modifying the PCM content.

The chemical nature of the composite aggregates was characterized by FTIR: the characteristic signals of both $CaCO_3$ stone and PEG 1000 are visible in the spectra shown in Fig. 1b. As some of the characteristic peaks of both PEG 1000 and $CaCO_3$ superpose, especially in the region of $1500-600$ cm^{-1}, the identification of the presence of PEG polymer in the composite PCM was obtained through the observation of the peak at 2885 cm^{-1}.

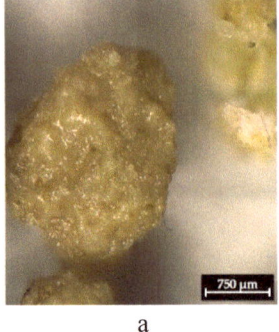
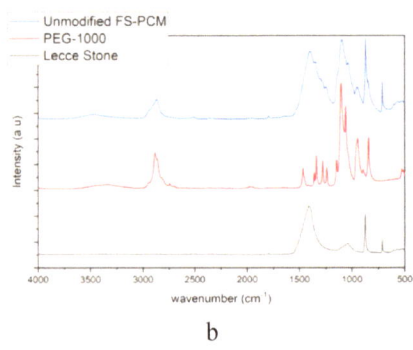

a   b

Fig. 1. (a) Optical microscopy image of a single unmodified FS-PCM aggregate and (b) FTIR spectra of the FS-PCM and of its components.

The other three modification methods implemented, i.e. polishing, water washing, and coating. Aimed at increasing the compatibility between the PCM aggregates and the mortars, by eliminating the superficial layer of PEG 1000 or by covering it.

Both size reduction and polishing modifications gave unsatisfactory results due to the reduced mechanical resistance of the impregnated Lecce Stone: when the PCM composite aggregates were either pressed or polished, they immediately shattered. The water washing method revealed that, due to the high solubility of PEG 1000 in water (value reported in product datasheet: 1500g in 30mL of H$_2$O), after 30 to 45 min of washing in static conditions, no PEG polymer remained in the PCM aggregates, confirmed by FTIR measurements (Fig. 2). The washing process carried out in stirring conditions was able to accelerate the *washing out* of PEG 1000, and almost no FTIR signal of PEG 1000 was found after a 5 min process. These results led to hypothesize that, when the FS-PCM aggregates are included in a mortar, part of PEG 1000 can be dispersed in it, possibly worsening its mechanical properties.

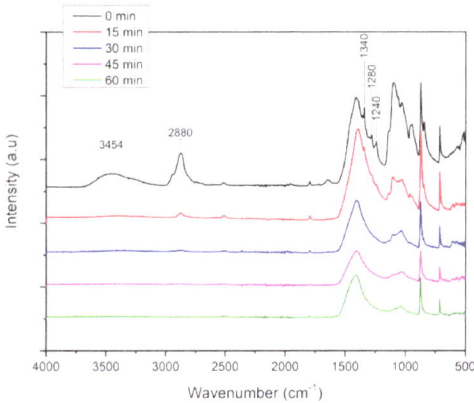

Fig. 2. FTIR spectra of the washing process under static conditions of the FS-PCM aggregates.

The coating methodology involved the application of an external layer of different materials (i.e. powder calcium hydroxide (CH), milk of lime (suspension of Ca(OH)$_2$ in water) (ML), pozzolana (P), and cocciopesto (C)) on the PCM-based aggregates, with the aim of improving the compatibility between the aggregate and the binder. In the case of the coating materials in powder form (i.e. CH, P, and C), the adhesion between the powder and the aggregate was achieved through the heating of the aggregates at a mild temperature (34 °C) in order to improve the adherence of the coating powder. The ML coating was applied at room temperature: once the aggregates were completely covered, they were left to dry for a few hours. The resulting materials are shown in Fig. 3. It is possible to observe that the color of the original FS-PCM (Fig. 3a) was changed to white by the application of both CH and LM coatings (Figs. 3b and 3c, respectively); on the other hand, the pozzolana and cocciopesto coatings modified the color into an *earth* or *brick* tone (Figs. 3d and 3e, respectively). This is an important aspect as color compatibility is an important parameter to take into account in restoration applications.

In order to evaluate the efficacy of the coating method with the different materials on the confinement of the PEG 1000 inside the stone aggregates, both FS-PCM unmodified and the modified aggregates were incorporated in a mortar mixture. The FTIR-ATR spectra obtained on the surface and on a cut section of each mortar are shown in Fig. 4. It is important to highlight that the powder analyzed was obtained by scratching the binder and carefully avoiding the aggregates. The peak corresponding to the C-H stretching of the PEG 1000 chain (2885cm^{-1}) was employed to indicate the presence of this

Fig. 3. Form stable phase change material aggregates (a) coated with (b) calcium hydroxide powder, (c) milk of lime, (d) pozzolana, and (e) cocciopesto.

polymer in the mortars, as it does not interfere with any other signal of the $CaCO_3$. From the intensity of this peak, it was concluded that, in the case of mortars containing the unmodified aggregates, PEG 1000 tends to migrate from the cross section to the surface of the mortar. On the other hand, the mortars containing the coated aggregates always exhibited a reduced intensity of the peak relative to PEG 1000, thus partly preventing the *washing-out* of this polymer. In particular, both calcium hydroxide and milk of lime coatings are able to effectively prevent the *washing-out* of PEG 1000 and, thus, its migration. Moreover, the aggregates coated with pozzolana and cocciopesto were able to partially confine PEG 1000, as its signal can be still observed in the FTIR spectra.

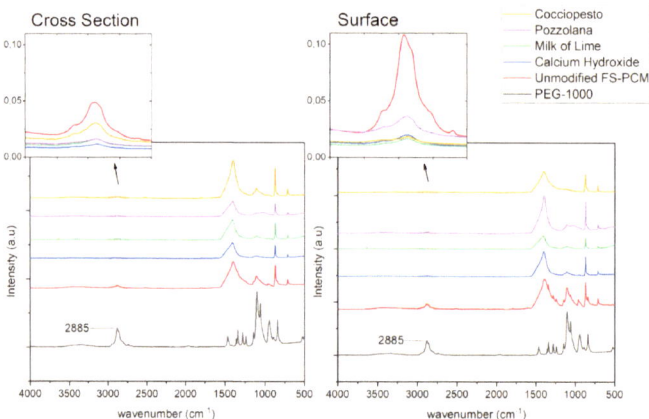

Fig. 4. FTIR-ATR measurements performed on a cross-section or surface of the mortars containing the unmodified FS-PCM or the four coated aggregates. In black, the reference spectrum of PEG 1000 is reported.

4 Conclusions

The presented study demonstrated that the impregnation method used to produce the PEG 1000-based FS-PCM was able to completely fill the porosity characteristic of the Lecce Stone inert support. On the other hand, PEG 1000 was found also on the surface of the PCM aggregates: this occurrence is probably responsible for the reduction of the adhesion between these aggregates and the binder composing the mortar. FTIR analysis revealed that the PEG 1000 composing the PCM is likely to migrate from the bulk to the surface of the mortar due to its high solubility in water.

In order to limit this phenomenon, thus improve the mechanical properties of the mortars containing a FS-PCM, different modifications were carried out on the PEG-based FS-PCM. Among others, the coating method proved to be an efficient and simple solution to confine the PEG 1000 inside the FS-PCM aggregates, avoiding its leaching. This method, in fact, was able to prevent the migration of the PEG 1000 present in the PCM to the surface of the mortar. This solution could also positively affect the thermal effectiveness of the PCM-based mortars, as more active material remains within the aggregates.

Further research is underway to assess the mechanical and thermal properties of mortars including modified FS-PCM aggregates, in order to confirm that this method is effective in limiting the reduction in mechanical properties of PCM-based mortars.

Acknowledgements. The present study was performed as part of the PhD program financed by a PhD Fellowship DOT1316197 (IMTR) CUP D45F21003710001 (Green) financed by Italian PON "Ricerca e Innovazione" 2014–2020 funding program. The research was also partly funded by PON Ricerca e Innovazione 2014–2020 Risorse React Eu - DM 1062/2021 Azione IV.4, "Dottorati e contratti di ricerca su tematiche dell'innovazione" and Azione IV.6, "Contratti di ricerca su tematiche green". The authors are grateful to Dr. Samuele Beraldo (Fassa Bortolo Srl) for providing calcium hydroxide used in this work.

References

1. Santamouris, M., Vasilakopoulou, K.: Present and future energy consumption of buildings: challenges and opportunities towards decarbonisation. E-Prime - Advances in Electrical Engineering, Electronics and Energy. **1**, 100002 (2021). https://doi.org/10.1016/J.PRIME.2021.100002
2. Ortiz, O., Castells, F., Sonnemann, G.: Sustainability in the construction industry: a review of recent developments based on LCA. Constr. Build. Mater. **23**, 28–39 (2009). https://doi.org/10.1016/J.CONBUILDMAT.2007.11.012
3. da Cunha, S.R.L., de Aguiar, J.L.B.: Phase change materials and energy efficiency of buildings: A review of knowledge. J Energy Storage. **27**, 101083 (2020). https://doi.org/10.1016/J.EST.2019.101083
4. Economidou, M., Atanasiu, B., Despret, C., Maio, J., Nolte, I., Rapf, O., Laustsen, J., Ruyssevelt, P., Staniaszek, D., Strong, D.: Europe's buildings under the microscope. A country-by-country review of the energy performance of buildings (2011)
5. Liu, L., Hammami, N., Trovalet, L., Bigot, D., Habas, J.P., Malet-Damour, B.: Description of phase change materials (PCMs) used in buildings under various climates: a review. J Energy Storage. **56**, 105760 (2022). https://doi.org/10.1016/J.EST.2022.105760
6. Akeiber, H., Nejat, P., Majid, M.Z.A., Wahid, M.A., Jomehzadeh, F., Zeynali Famileh, I., Calautit, J.K., Hughes, B.R., Zaki, S.A.: A review on phase change material (PCM) for sustainable passive cooling in building envelopes. Renew. Sustain. Energy Rev. **60**, 1470–1497 (2016). https://doi.org/10.1016/J.RSER.2016.03.036
7. Frigione, M., Lettieri, M., Sarcinella, A., Barroso de Aguiar, J.: Sustainable polymer-based phase change materials for energy efficiency in buildings and their application in aerial

lime mortars. Constr. Build. Mater. **231**, 117149 (2020). https://doi.org/10.1016/J.CONBUI
LDMAT.2019.117149

8. Cunha, S., Aguiar, J.B., Ferreira, V.M., Tadeu, A.: Influence of adding encapsulated phase
   change materials in aerial lime based mortars. Adv Mat Res. **687**, 255–261 (2013). https://
   doi.org/10.4028/WWW.SCIENTIFIC.NET/AMR.687.255

9. Ventolà, L., Vendrell, M., Giraldez, P.: Newly-designed traditional lime mortar with a phase
   change material as an additive. Constr. Build. Mater. **47**, 1210–1216 (2013). https://doi.org/
   10.1016/J.CONBUILDMAT.2013.05.111

10. Frigione, M., Lettieri, M., Sarcinella, A.: Phase change materials for energy efficiency in
    buildings and their use in mortars. Materials. **12**, 1260 (2019)

11. Frigione, M., Lettieri, M., Sarcinella, A., de Aguiar, J.L.B.: Applications of sustainable
    polymer-based phase change materials in mortars composed by different binders. Mater.
    12(21), 3502 (2019). https://doi.org/10.3390/MA12213502

12. Sarcinella, A., De Aguiar, J.L.B., Lettieri, M., Cunha, S., Frigione, M.: Thermal performance
    of mortars based on different binders and containing a novel sustainable phase change material
    (PCM). Materials **13**, 2055 (2020). https://doi.org/10.3390/MA13092055

13. Papayianni, I., Hughes, J.: Testing properties governing the durability of lime-based repair
    mortars. RILEM Tech. Lett. **3**, 135–139 (2018). https://doi.org/10.21809/RILEMTECH
    LETT.2018.81

Testing the Performance of Vinyl Ester and Polyester Polymer Concrete Following Exposure to Artificially Induced Climatic Environments

Dhruv Parbhoo Gulabbhai$^{(\boxtimes)}$, Deon Kruger, and Mayur Kishor Hira

Faculty of Engineering and the Built Environment, University of Johannesburg, Johannesburg, South Africa
dgulabbhai@gmail.com

Abstract. The exposure of polymer concrete to artificially designed environmental conditions of high-and low-temperatures, and moisture levels allowed for the assessment of strength performance and aesthetic value. Strength performance indicated the maximum capability of the product to carry a load successfully, whereas the aesthetics assessed the appearance of the product, that can be measured using spectrophotometry. In this study, materials such as water and Portland cement typically used to form traditional concrete were replaced by two polymer resins namely - vinyl ester and polyester, thus making it polymer concrete. As such, compressive strengths of cube samples were tested, the change in cubes' masses was measured using a balance - prior and post exposure to the artificially induced environments, the colour change tests (spectroscopy analysis) were performed using the spectrophotometer tests. Compressive strengths of over 75MPa were achieved, thereby justifying promising concrete strengths. Mass losses recorded were almost negligible, thereby showing toughness to conditions presented in the artificially induced environments. Minor colour changes were noticed- thereby showing a good resistance to harsh weather conditions on the surface properties. Therefore, the assessed products displayed desired characteristics for strength performance and aesthetic value, subsequently, creating a product that promotes sustainability.

Keywords: Polymer concrete · Vinyl ester resin · Polyester resin · Compressive Strength · Aesthetics

1 Introduction

Polymer concrete can be defined as the composite material formulated from the use of polymeric resins to fully replace the cement binders in conventional cement concrete [1]. The normalised use of cement in conventional concrete has thus developed the need to investigate more sustainable methods of producing concrete without compromising the durability and strength properties of the composite material, when exposed to harsh climatic environments.

© The Author(s) 2025
L. Czarnecki et al. (Eds.): ICPIC 2023, 61, pp. 539–546, 2025.
https://doi.org/10.1007/978-3-031-72955-3_55

The use of polymeric resins such as vinyl ester and polyester to replace water and conventional Portland cement in concrete, coupled with the use of recycled materials as aggregates such as polypropylene plastic and rubber crumbs, can be a sustainable way to producing polymer concrete. It can be touted to be a "greener" product by reducing the overall carbon dioxide emissions [2], achieving higher strength properties, increasing hardening rates, and upholding the aesthetic value of the composite materials.

Like conventional concrete, the impact that various environmental conditions have on polymer concrete is important to consider as these conditions will govern the sustainability, durability, strength performance and suitability of these composite materials.

Thus, to investigate the sustainability of the strength performance and aesthetic value of polymer concrete materials when exposed to artificially induced climatic environments, six different polymer concrete mix designs were formulated. These mix designs incorporated the use of vinyl ester and polyester polymer resins, which were mixed with three different aggregate systems, namely: recycled polypropylene plastic, recycled rubber crumbs from old tyres and a conventional crusher sand/crusher stone mix.

Polyester and vinyl ester resins are unsaturated polymer resins which are essential thermosetting matrix components for composite materials. Polyester and vinyl ester resins are available at a lower cost when compared to epoxy resins, however their matrix systems are used for improved strength properties and chemical resistance [3, 4], thus contributing to their economic sustainability. In this investigation, these six polymer concrete materials were exposed to artificially induced climatic environments.

2   Experimental Design

For this test, a total of one hundred and twenty cube specimens were produced. Twenty specimens were made for each mix design. From the twenty, ten underwent exposure of harsh climates for testing, whereas the other ten were used as control specimens.

2.1   Mix Designs

The two different polymer resins were incorporated with three different sets of aggregate systems. For the mixes, 1.2% of the supplier-provided catalyst was mixed with the vinyl ester resin for the three mix designs utilising vinyl ester resin, while 2% of the supplier-provided catalyst was mixed with the polyester resin for the remaining three mix designs utilising polyester resin. Table 1 details the six mix designs used for the tests.

2.2   Casting of Cube Specimens

Casting of cube specimens was done with guidelines from the South African Bureau of Standards (SANS 5861, SANS 5863 and SANS 50196) [5–7] and the American Society for Testing and Materials (ASTM C109) [8], whereby 50mm x 50mm x 50mm cube specimens were casted. The cube moulds were filled with the various mix designs and compacted. After 24 h, it was demoulded and stored at room temperature.

Table 1. Mix design compositions and volume ratios.

Mix Design Composition	Abbreviation	Volume Ratio	Ratio Explanation
Vinyl ester resin with conventional aggregates and recycled rubber crumbs	RV	1: 4.2: 1.4: 1.4	Vinyl ester: Sand: Stone: Rubber
Vinyl ester resin with conventional aggregates and recycled polypropylene plastic	PV	1: 5: 1.7: 1.7	Vinyl ester: Sand: Stone: Plastic
Vinyl ester resin with conventional aggregates	CV	1: 5: 2.5	Vinyl ester: Sand: Stone
Polyester resin with conventional aggregates and recycled rubber	RP	1: 4.2: 1.4: 1.4	Polyester: Sand: Stone: Rubber
Polyester resin with conventional aggregates and recycled polypropylene plastic	PP	1: 5: 1.7: 1.7	Polyester: Sand: Stone: Plastic
Polyester resin with conventional aggregates	CP	1: 5: 2.5	Polyester: Sand: Stone

2.3 Exposure of Cube Specimens to Artificially Induced Climatic Environments.

Sixty cube specimens (experimental group) were exposed to artificially induced climatic environments, namely: high temperatures, negative temperatures, and high moisture levels. High temperatures up to 70 °C, negative temperatures up to -13 °C and 100% moisture levels were used as parameters to expose the specimens to the induced climatic environments.

Exposure to high temperatures and high moisture levels. Sixty cube specimens were exposed to high temperatures of up to 70 °C at two-hour intervals. Thereafter, the specimens were submerged into water for a two-hour period, to replicate a harsh environment of rapid heating followed by high moisture. This cycle was replicated twenty-five times.

Exposure to low temperatures. The same sixty cube specimens were exposed to negative temperatures reaching as low as -13 °C in eight-hour intervals. After the eight-hour period, the specimens were left at room temperature for four hours to replicate a freeze-thaw action sequence. This cyclic exposure was replicated twenty-five times.

2.4 Compressive Strength Testing

Upon exposing the sixty cube specimens (experimental group) to the artificially induced climatic environments, the compressive strength of the experimental and control group was tested and recorded to ascertain the effect that the harsh conditions have on degree of degradation for strength performance. The testing of the compressive strengths was done with guidelines outlined by the South African Bureau of Standards (SANS 5863) [6].

2.5 Mass Loss

After exposing the sixty cube specimens (experimental group) to the artificially induced climatic environments, the mass of the specimens in both the experimental and control group were recorded to determine the effect that the harsh conditions have on durability and weathering resistance. Mass loss is an undesired property that compromises durability.

2.6 Aesthetic Value

The assessment of aesthetics was determined through the analysis and measurement of colour change by using spectrophotometry tests. These spectrophotometry tests were performed on all one hundred and twenty cube specimens and were done in line with the ASTM D2244 – 21 [9]. The way the different mix design constituents affect colour change was of significance to its visual aesthetic value.

3 Results and Discussion

The results of findings of the compressive strength test, mass loss test and aesthetics test are detailed and discussed in this section.

3.1 Compressive Strength

Compressive strength testing was conducted on all one hundred and twenty cube specimens, sixty of which were exposed to the artificially induced climatic environments and sixty of which acted as the control group. This was done to ascertain the effect that these harsh climatic environments have on the strength performance of the polymer concrete mix designs. The compressive strength performance is illustrated in Fig. 1.

Figure 1 shows the compressive strengths achieved for the various mix designs. A relationship can be observed between the aggregate systems utilised in each mix design, whereby the use of conventional aggregates exhibited the highest overall compressive strength, followed using recycled plastic and recycled rubber aggregates.

High compressive strengths of about 89 MPa were recorded by the polyester resin mix with conventional aggregates (CP) and the vinyl ester resin mix with conventional aggregates (CV). These compressive strengths are like that of M80 concrete that can be used for the construction of dam spillways, high-rise buildings, and bridges [10, 11]. Further, the use of polypropylene plastic for compressive strength tests results showed promising results. Vinyl ester resin with polypropylene plastic mix (PV) and polyester with polypropylene plastic mix (PP), achieved compressive strengths of just under 30 MPa. This is slightly short of achieving a M30 concrete, which is necessary for footings, slabs, patios, and strainer posts [10, 11, 13]. The mix designs utilising the recycled rubber crumbs achieved compressive strengths ranging from 7 MPa to 9 MPa, which is like the compressive strength of M7.5 and M10 concrete, which can be utilised as plain cement concrete (PCC) for the purposes of levelling course or bedding [10, 11].

In the case of recycled plastic materials incorporated into the aggregate system and like that of conventional concrete incorporated with plastic materials, a reduction of

Compressive Strength of the Mix Designs

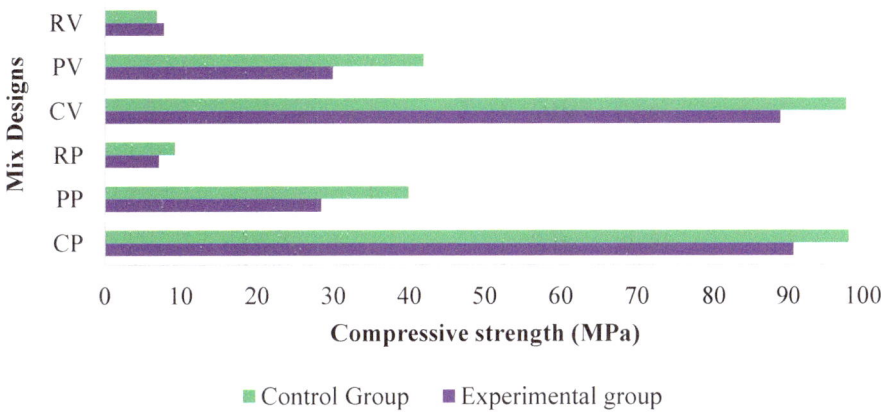

■ Control Group   ■ Experimental group

Fig. 1. Compressive strength results of the various polymer concrete mix designs before and after exposure to the various artificially induced climatic environments.

compressive strength was observed with an increased content of plastic aggregate. This is attributed to low strength properties of the plastic aggregates and poor bonding within the mix design composition [12, 13]. In the case of recycled rubber crumbs incorporated into the aggregate system and like that of conventional concrete incorporated with rubber crumbs, the lower compressive strengths observed is attributed to the low elastic modulus of rubber when compared to the conventional aggregates and poor adhesion within the mix design composition [14, 15].

3.2   Mass Loss

Mass Losses for the Mix Designs

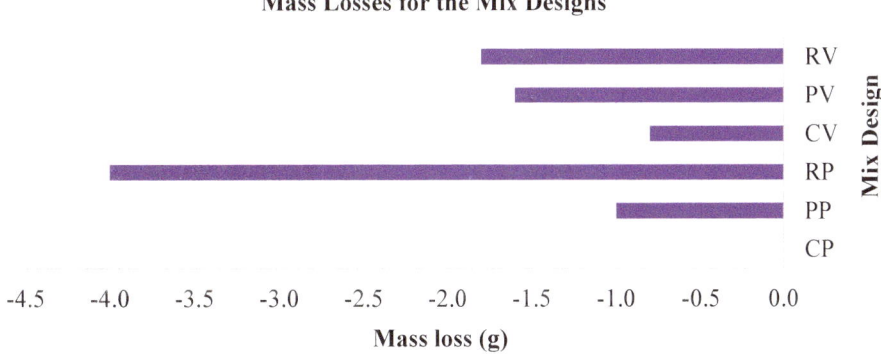

Fig. 2. Mass loss of the various polymer concrete mix designs after exposure to artificially induced climatic environments.

Figure 2 shows that the mix design incorporating the polyester resin, coupled with the conventional aggregate system (CP) displayed the best durability as it showed no changes to its average mass. Vinyl ester with normal aggregate (CV), follows as the next ideal mix design due to a small loss in mass of 0.8g. Recycled polypropylene plastic within the aggregate system also showed promising results, indicating a good level of binding strength. Rubber crumbs in the aggregate system, particularly with the use of the polyester resin (RP), indicated the highest level of overall degradation and mass change, making it an inadequate choice.

The compressive strength results and mass losses observed are largely attributed to the content of recycled materials within the mix design. The increased content of recycled materials resulted in lower strength and durability which is attributed due to poor bonding and thus contributing to poor abrasion resistance and durability thereof. However, the incorporation of recycled plastic and rubber within the mix designs at varying contents are said to improve its overall impact resistance and energy absorption [12, 14, 15].

3.3   Colour Change

The vinyl ester with conventional aggregates (CV) mix design was the next best result, that recorded a change in colour, ΔE* of 1.2. Therefore, the use of vinyl ester with plastic is recommended for esthetics of sight and environmental sustainability. The mix designs that contained polyester with either polypropylene plastic or normal aggregates yielded promising results, since the change in colour, ΔE* was less than 2. However, the use of recycled rubber was shown to yield higher changes of colour, thus making it a less visually aesthetic product, despite it being a more environmentally sustainable product. Figure 3 illustrates the degree of colour change observed.

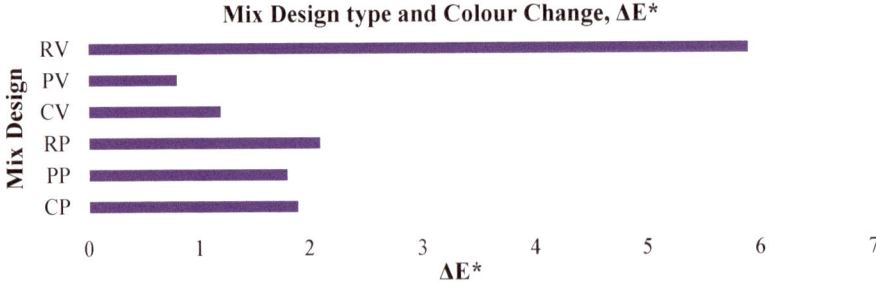

Fig. 3. Spectrophotometer test results

4   Conclusion

In this study, the performance of the polymer resins of vinyl ester and polyester used with conventional aggregates, rubber crumbs and plastic were assessed for strength, mass loss and aesthetics, upon exposure of test samples to the artificially induced climatic conditions of high temperature, low temperature, and high moisture levels. For compressive strength performance and resistance to mass loss, polyester resin with a conventional aggregate system (CP) yielded the best results. This shows that polyester resin is promising, though its compatibility with recycled aggregate substitutes needs to be better incorporated, due to it having poor adhesive properties and a low elastic modulus. The aesthetic test revealed optimistic findings for both polymers in terms of colour change resistance. Vinyl ester with polypropylene plastic (PV) revealed the best aesthetic value, since the colour change was the least. A resistance to colour change indicates that it could withstand the harsh climate. Therefore, the ideal mix design from this study is polyester resin with a conventional aggregate system (CP). That is due to it having the best compressive strength, the least mass loss, and acceptable aesthetic value properties. It is, however, recommended that future research in this study be continued for the selected polyester resins with varying content of recycled materials, and/ or different recycled materials to replace conventional aggregates when utilised in polymer concrete, to achieve similar or better performance.

References

1. Ohama, Y.: Developments in the Formulation and Reinforcement of Concrete, 1st edn. Woodhead Publishing, Cambridge (2008)
2. Kiruthika, C., Prabha, S.L., Neelamegam, M.: Different aspects of polyester polymer concrete for sustainable construction. Mater. Today Proc. **43**, 1622–1625 (2021)
3. Hodgkin, J.: Encyclopedia of Materials: Science and Technology—Thermosets: Epoxies and Polyesters. Pergamon, Oxford (2001)
4. Kandelbauer, A., Tondi, G., Zaske, O., Goodman, S.: Unsaturated Polyesters and Vinyl Esters (2014)
5. Standard, South African Bureau of Standards: South African National Standards 5861–3. Concrete tests—Part 3: Making and curing of test specimens (2006)
6. Standard, South African Bureau of Standards: South African National Standards 5863. Concrete tests—Compressive strength of hardened concrete (2006)
7. Standard, South African Bureau of Standards: South African National Standards 50196–3. Methods of testing cement—Part 3: Determination of setting times and soundness (2006)
8. Standard, American Society for Testing and Materials: C109/C109M. Compressive Strength of hydraulic Cement Mortars (2021)
9. Standard, American Society for Testing and Materials: D2244 – 21. Standard Practice for Calculation of Color Tolerances and Color Differences from Instrumentally Measured Color Coordinates (2022)
10. Civil Lead: Grade of Concrete and Their Uses—Concrete Mix Ratio. https://www.civillead. com/concrete-mix-ratio-for-various-grade-of-concrete/. Last accessed 01 Apr 2023
11. Different Grades of Concrete, Their Strength and Selection for Construction—The Constructor. https://theconstructor.org/concrete/grades-concrete-strength-selection/20570/. Last accessed 01 Apr 2023
12. Siddique, R., Khatib, J., Kaur, I.: Use of recycled plastic in concrete: A review. Waste Manage. **28**, 1835–1852 (2008)
13. Abu-Saleem, M., Zhuge, Y.Z., Hassanli, R., Ellis, M., Rahman, M., Levett, P.: Evaluation of concrete performance with different types of recycled plastic waste for kerb application. Constr. Build. Mater. **293**, 123477 (2021)
14. Karunarathna, S., Linforth, S., Kashani, A., Liu, X., & Ngo, T.: Effect of recycled rubber aggregate size on fracture and other mechanical properties of structural concrete. J. Cleaner Prod. **314**, 128230 (2021)

15. Abdelmonem, A., El-Feky, M.S., Nasr, A.R., Kohail, M.: Performance of high strength concrete containing recycled rubber. Constr. Build. Mater. **227**, 116660 (2019)

Improvement of the Thermal Shock Resistance of Epoxy Resin Coating Using Sedimented Recycled Fine Aggregate

Kamil Krzywiński[✉] and Łukasz Sadowski

Faculty of Civil Engineering, Wroclaw University of Science and Technology, Wrocław, Poland
kamil.krzywinski@pwr.edu.pl

Abstract. The floors made of epoxy resin coating that are exposed to the thermal loads are usually not durable enough to be used in industrial facilities. To enhance thermal properties of the coating recycled fine aggregate was used to reduce thermal expansions in the interphase zone in between the coating and substrate. Three different types of coating were analyzed: pure epoxy, specially homogenous, and functionally graded material. The top floor was loaded with temperature to obtain heat flux and strain results. Finite element method was used to simulate the heat transfer and heat load. Simulations show that sedimentation of the aggregate reduces heat flow to the substrate during loading. It means that properly designed aggregate can improve durability of the industrial floor which can be more resistance to thermal gradient.

Keywords: Polymer · Floor Coating · Thermal Shock

1 Introduction

The industrial floors are often exposed to the forklifts loads. The damage of the floor is usually caused by mechanical and dynamic loads [1]. However, thermal load can also damage floor structure [2]. All kind of floors can be damaged by temperature gradient [3], but the polymer coatings are one of the most sensitive [4]. The thermal durability of polymer coatings depends on thermal expansion and conductivity. Therefore, in this work authors present numerical simulation of heat flow in floor structure made of epoxy resin coating and concrete substrate. The simulation helps to analyze impact of the temperature growth on the interphase zone where the crack can be occurred during thermal loading [5]. The thermal load is simulated as a temperature growth in the place of spinning wheel of the forklift [6]. To change thermal expansion properties of the coating the epoxy resin is modified with recycled fine aggregate sourced from old concrete [7]. Various size of the filler is used to obtain different structure of the polymer composite: specially homogenous, and functionally graded material. The numerical simulations are carried out in Abaqus software. The results show that the aggregate reduces thermal expansion in the interphase zone in between the coating and substrate.

L. Czarnecki et al. (Eds.): ICPIC 2023, 61, pp. 547–554, 2025.
https://doi.org/10.1007/978-3-031-72955-3_56

2   Materials and Methods

2.1   Materials

The composite structure is designed with a 3 mm coating layer and a 97 mm concrete substrate, 100 mm in total. Three coating types were selected for the simulation. First coating type is pure epoxy resin (ER) Sto IHS BV made of component (A) bisphenol and component (B) hardener, where A:B is 100:33. Second coating type contains IHS BV epoxy resin and recycled fine aggregate which does not sediment in liquid epoxy resin matrix. The hardened composite creates specially homogenous (SH) coating. Third and last coating type is made of recycled fine aggregate which sediments in epoxy resin matrix during curing time. The sedimentation of the aggregate depends on curing temperature, curing time, viscosity and density of the epoxy resin, A:B ratio, density, roughness and diameter of the aggregate. In this particular case authors change aggrege size to create different type of the coatings. Sedimentation of the aggregate in third type of the coating creates pure layer of the epoxy resin in upper part of the coating. The structure of the middle coating contains light and small particles (graded material – 1). The lower part of the coating is made of the aggregate with the greatest size (graded material – 2). Above described layers create functionally graded material (FGM) coating that is presented in Fig. 1c.

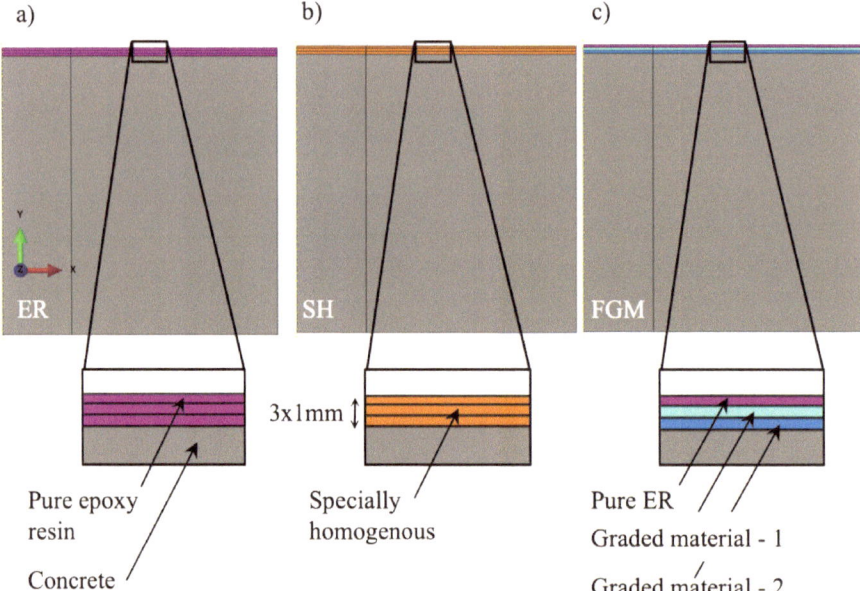

Fig. 1. The materials used in FEM simulations: a) pure epoxy resin (ER) coating; b) specially homogenous (SH) coating; c) functionally graded material (FGM) coating.

The substrate is designed of concrete C25/30. The properties of materials are summarized in Table 1. Material properties.. In simulation temperature depended properties are used. Because of lack of space in Table 1. Material properties. Are presented only properties in room temperature 20°C. In FEM model all properties are in range between 20 and 120°C.

Table 1. Material properties.

Property	Concrete	ER	SH	FGM	
				Graded mat. 1	Graded mat. 2
Density [kg/m^3]	2100	1080	1437	1488	1947
Thermal conductivity [W/(mK)]	1.333	0.188	0.467	0.585	0.704
Thermal expansion [10^{-6} 1/K]	11.00	26.49	21.07	20.29	13.32
Heat capacity [J/(kgk)]	900	1270	1141	1122	956
Young's modulus [MPa]	31000	1500	2000	2100	2500
Poisson's ratio [-]	0.20	0.32	0.27	0.28	0.22

2.2 FEM Simulation

The FEM simulation is carried out in Abaqus 2017. The composite size is 100x100 mm. The coating is divided into 3 equal horizontal stripes with 1 mm thickness. The displacement in y and x direction is fixed at the bottom and right edge, respectively (Fig. 2a). The initial temperature 20°C is applied to the whole model. The temperature load is applied to the 25 mm long top-left edge. The radiation of the top-right edge is designed with emissivity of 0.7.

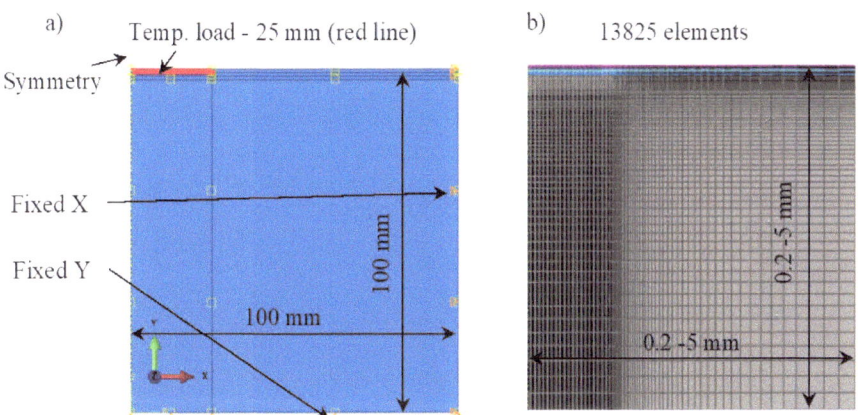

Fig. 2. The FEM model: a) highlighted load and boundary conditions; b) mesh size.

To obtain heat flux and displacement results two simulations are necessary to be created. The first simulation is heat transfer which gives heat flux results. Second model is created in static mode with thermal load imported from the results of the first model. All models are created as axisymmetric. As a result the model can be presented as a cylinder using symmetry axis on the vertical right edge.

2.3 Temperature Load

The electrical heater SKU: HA502AC (2000W) is used to obtain thermal load amplitude. The heater is installed 10 cm above the floor surface with heating temperature $T = 200°C$, and fan speed 2/5. The temperature data is collected with temperature recorder TERMIO 31 ($\pm 0,6$ °C) and 30 cm STK probe (with range up to 800°C). The end of the probe is installed in the center of heating area 3 mm above the floor surface. The collected data is presented in Fig. 3.

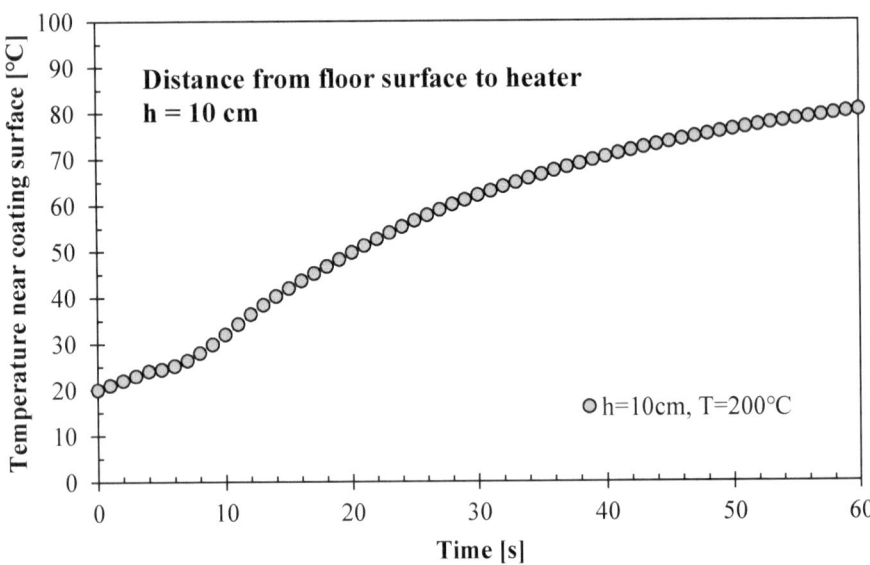

Fig. 3. The temperature load over time applied to the top surface of the composite.

3 Results and Discussion

First simulation contains heat flux for whole composite. The results are presented in Fig. 4. It can be seen that the ER coating with low thermal conductivity is slowing down heat flux. As a result, the temperature of the concrete does not increase a lot. It means that high temperature difference occurs in this model. Great temperature difference can affect thermal expansion and generates stress. Nevertheless, the other models present dipper heat propagation, except for the right side of the model where that the surface lose heat by radiation.

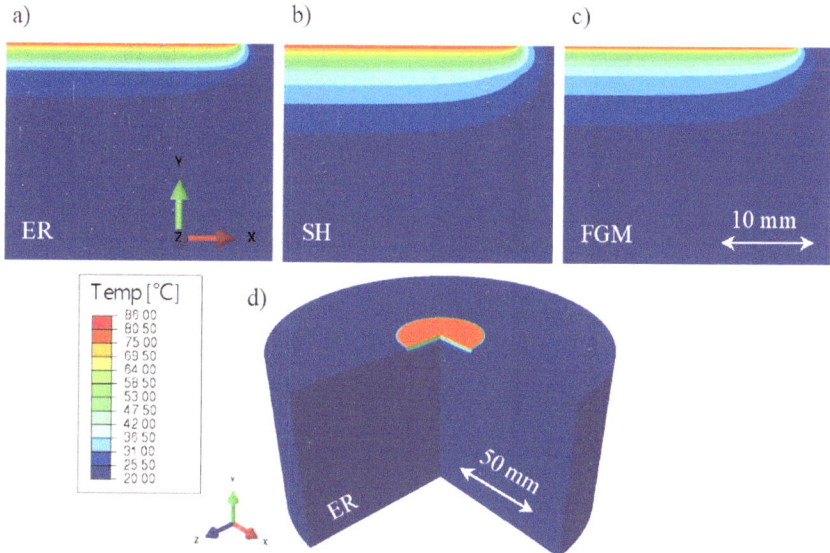

Fig. 4. The results of heat flux after 60 s of heating: a) pure epoxy resin (ER) coating; b) specially homogenous (SH) coating; c) functionally graded material (FGM) coating; d) 3D view of composite with ER coating.

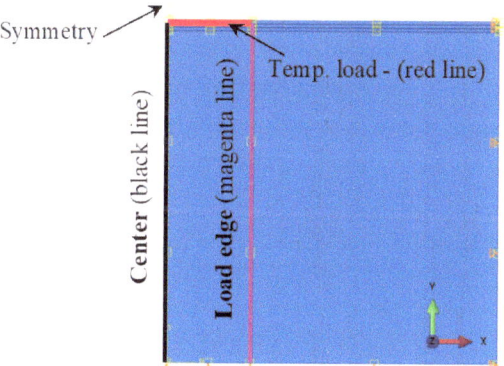

Fig. 5. The description of paths used during results generation – path no. 1 - center (black line); path no. 2 - load edge (magenta).

From Fig. 4 it can be seen that there are two areas where the results differ. Close to the symmetry and to the edge of load. In these areas two paths (Fig. 5) are created: center path (black line), and load edge path (magenta line).

Figure 6 presents nodes temperature over the height of the model. The radiation influence on heat loss at the load edge is highly visible. The temperature at the load edge is lower than in the center up to 30°C, depending on the coating. The grates temperature difference between top coating surface and concrete boundary has composite with ER coating (around 50°C). The coating with SH seems to have the lowest temperature difference between those surfaces. However, the important areas are between 3rd layer and concrete where thermal expansion of different material can generate the stress. For this particular case FGM has the lowest thermal gradient.

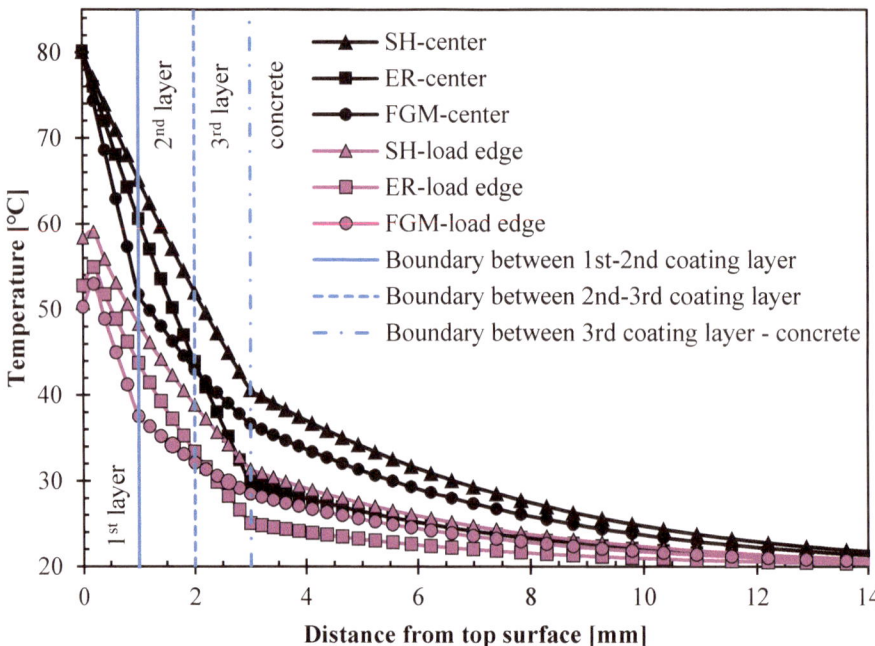

Fig. 6. The heat flux results for two paths (center and load edge) and different coatings (ER, SH, and FGM). The vertical blue lines represent boundaries between the layers.

Figure 7 shows magnitude displacement of all models. Great thermal expansion of epoxy resin cab be observed in Fig. 7a. Two other models have similar total displacement. It is because these two coatings were designed to contain the same volume value of aggregate. By having the same amount of epoxy resin matrix the behave of these two coatings should be similar in some cases.

To understand how the displacement affects models it is important to calculate the average strain between the layers $\Delta\varepsilon_i$ using Eq. 1. Where ε_{avg} is average strain in upper (j) and lower (k) layer.

$$\Delta\varepsilon_i = \varepsilon_{avg,j} - \varepsilon_{avg,k} \tag{1}$$

Figure 8 presents the strain difference between the layers in the load edge area where strain value is high in x direction. Each great value in the chart means that there is a high chance of stress generation. However, the difference between coating layers is not the most important. Because those layers have similar properties. Moreover, the temperature of the top layer can exceed glass transition temperature. It can make the material softer and more vulnerable. The most important in the case of durability should be analyzed

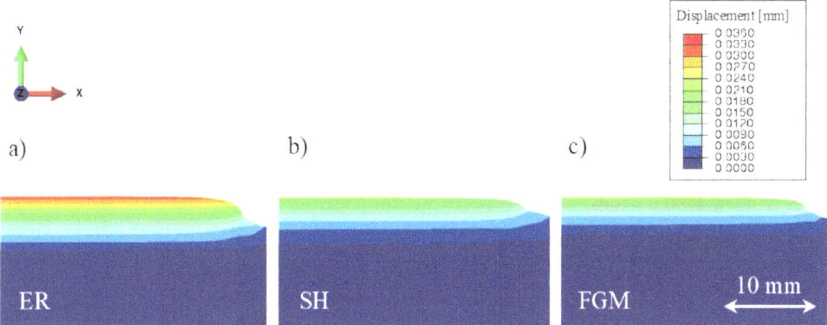

Fig. 7. The magnitude displacement of: a) pure epoxy resin (ER) coating; b) specially homogenous (SH) coating; c) functionally graded material (FGM) coating.

the strain difference between 3rd layer and concrete substrate. It is visible that FGM coating obtains the lowest value of $\Delta\varepsilon_i$.

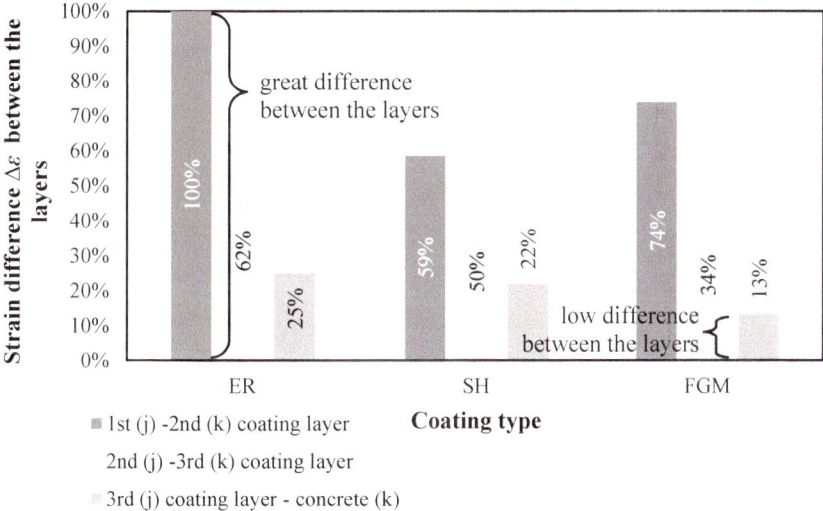

Fig. 8. The strain difference $\Delta\varepsilon$ between the layers along load edge path.

4 Conclusions

This work is focused on improvement of the thermal shock resistance of epoxy resin coating using sedimented recycled fine aggregate. Based on the performed analysis and simulations it can be stated that modification of the epoxy resin coating with recycled fine aggregate can improve the thermal shock by reducing thermal expansion in near-substrate area. However, to obtain such an improvement it is important to properly designed size of the aggregate to obtain pure epoxy resin layer at the top of the coating. This pure layer slows down heat flux into the composite and aggregates at the bottom reduce the thermal expansion. Nonetheless, FGM has almost the same abrasion resistance comparing to ER. Pure epoxy resin occurs in the top layer in both types of coating. The only composite that can enhance its abrasion resistance is SH because of filler particles that are allocated in near-surface area. Each coating type has different assets which should be taken into account during design process.

Acknowledgements. The authors received funding from the project supported by the National Centre of Science, Poland [Grant No. 2019/35/O/ST8/01546].

References

1. Chowaniec, A.: Using waste limestone powder as filler in a toxic epoxy resin coating and its influence on adhesive properties. Chem. Eng. Trans. **88**, 1063–1068 (2021). https://doi.org/10.3303/CET2188177
2. Sadowski, Ł., Hoła, J., Żak, A., Chowaniec, A.: Microstructural and mechanical assessment of the causes of failure of floors made of polyurethane-cement composites. Compos. Struct. **238** (2020). https://doi.org/10.1016/j.compstruct.2020.112002
3. Wang, Q., Chen, H., Wang, Y., Sun, J.: Thermal shock effect on the glass thermal stress response and crack propagation. Proc. Eng. (2013)
4. Carbas, R.J.C., Marques, E.A.S., Da Silva, L.F.M., Lopes, A.M.: Effect of cure temperature on the glass transition temperature and mechanical properties of epoxy adhesives. J. Adhes. (2014)
5. Krzywiński, K., Sadowski, Ł., Fedoruk, K., Sieradzki, A.: Fundamental understanding of the thermal properties of amorphous epoxy resin coatings filled with quartz-feldspar powder sourced from quarry waste. Appl. Surf. Sci. **614** (2023). https://doi.org/10.1016/j.apsusc.2022.156133
6. Sadowski, Ł, Czarnecki, S., Krzywiński, K., Moj, M., Chowaniec, A., Żak, A.: Dry spinning wear of cementitious materials: a novel testing method and mechanism. Measurement **196**, 111216 (2022). https://doi.org/10.1016/j.measurement.2022.111216
7. Krzywiński, K., Sadowski, Ł, Stefaniuk, D., Obrosov, A., Weiß, S.: Engineering and manufacturing technology of green epoxy resin coatings modified with recycled fine aggregates. Int. J. Precis. Eng. Manuf.-Green Technol. (2021). https://doi.org/10.1007/s40684-021-00377-w

Development of Mechano-Responsive CNT-Epoxy Concrete

Hye-Kyoung Jeon[1]([⊠]), Yu-Jin Jung[1], and Sung-Hwan Jang[1,2]

[1] Department of Smart City Engineering, Hanyang University ERICA, Ansan, South Korea
sj2527@hanyang.ac.kr
[2] Department of Civil and Environmental Engineering, Hanyang University ERICA, Ansan, South Korea

Abstract. Cracking in engineering concrete structures poses a significant problem as it not only accelerates the rate of deterioration but also diminishes the structural strength. The current repair materials being used are capable of short-term repairs only and may result in destroying the concrete structure due to additional damage. In this study, Carbon nanotube (CNT)-Epoxy concrete is proposed as a sensing repair material to address this problem. The authors assess the electrical properties of samples based on the CNT content and aggregates and report on the optimal mixing ratio of CNT-Epoxy concrete. The electro-mechanical characterization of CNT-Epoxy concrete was evaluated through a series of experiments such as static and cyclic loadings. CNT-Epoxy concrete exhibits accurate sensing response under compression and maintains a consistent cyclic response. The findings from the study demonstrate that CNT-Epoxy concrete is an effective material for repairing concrete cracks, offering favorable compressive strength, and showing reliable sensing ability.

Keywords: Mechano-responsive · CNT-Epoxy concrete · Repair material

1 Introduction

Engineering concrete structures such as bridges, tunnels, and dams are frequently exposed to service load, which leads to cracks initiated of the structure's surface [1]. These cracks reduce local stiffness and create material discontinuities, which eventually destroy the concrete structure as damage accumulates [2, 3]. With an increasing focus on the maintenance of concrete structures, it has become crucial to address repair work promptly to prevent futher deterioration. Currently, repair work is being performed on cracks or damaged areas in various concrete structures [4–6]. In recent years, polymer concrete has been increasingly recognized as a prominent material for repair work in various applications [7]. Polymer concrete has been widely used as a repair material because of its rapid curing property, excellent bond to cement concrete and reinforcing bars, and impressive strength and durability [8–10]. In contrast, polymer concrete can only be repaired, and research on repair materials with sensing capabilities is limited.

© The Author(s) 2025
L. Czarnecki et al. (Eds.): ICPIC 2023, 61, pp. 555–561, 2025.
https://doi.org/10.1007/978-3-031-72955-3_57

CNT is one of the additions that can be used to create multifunctional materials [11]. The structural morphology of CNT can be used as a composite mixed with different materials to impart electric-mechanical properties. The self-sensing capability of CNT nanocomposites is regarded as a key requirement for future structural health monitoring technologies and applications [12, 13].

In this study, we intend to fabricate CNT based epoxy concrete that can effectively repair cracks and enable load sensing in concrete structures. We calculated the electrical conductivity for CNT content to select the appropriate CNT content for CNT-Epoxy concrete with and without aggregate. We confirmed the mechano-responsive of CNT-Epoxy concrete when a static load was applied. We tested the cyclic load using this result and obtained a significant conclusion regarding the mechano-responsive performance.

2 Experimental

2.1 Materials

MWCNT was purchased from Nanolab, Inc. (MA, USA) and featured a diameter of 15nm, a length of 5-20μm, and a purity higher than 85wt%. Epoxy (DH-150) was provided by Daehwa Precision (Gyeonggi-do, South Korea) and was used as the base polymer matrix. Aggregate was obtained from Joomoonjin Silica sand Co., LTD (Gangwon-do, South Korea). Acetone was used as a dispersant and was provided by Samchun chemicals Co., LTD (Gyeonggi-do, South Korea) with 99.7% purity. All the materials were used as received.

2.2 Fabrication of CNT-Epoxy Concrete

Figure 1 showed the fabrication process of CNT-Epoxy concrete. A 50 ml of acetone was used as a dispersant to reduce the viscosity of the epoxy and to improve the dispersion performance of CNT [14]. The Q700CA ultrasonicator of Qsonica LLC. (CT, USA) and a TR50M three-roll mill of Trilos (CA, USA) were used to disperse the CNT in the epoxy concrete. A mixed sample of CNT, pure acetone, and epoxy resin was dispersed in an ultrasonicator using a 90% amplitude pulse model for 40 min. Then, the sample was placed on a hot plate to allow for the evaporation of acetone. Subsequently, epoxy hardener and aggregates were mixed with the sample using a three-roll mill. Note that 80wt% of aggregates were selected based on the optimal aggregate content reported by Jung et al. [8].

2.3 Characterization

The electrical resistance of CNT-Epoxy concrete was measured using a 2450 Sourcemeter of Kethley (OH, USA) with the two-probe measurement method. Electrical conductivity was calculated to select the most suitable amount of CNT to impart the desired function to CNT-Epoxy concrete. The electrical conductivity, σ, was calculated using the Eq. (1):

$$\sigma = L/AR \tag{1}$$

1) Mixing
(Acetone + Epoxy resin +
CNTs)

2) Dispersing with
ultra sonicator

3) Evaporating acetone
with hot plate

4) Mixing
(Sample + Epoxy
hardener + aggregates)

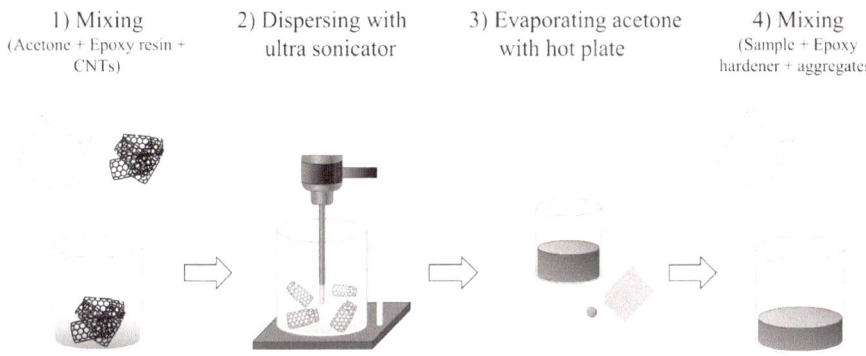

Fig. 1. Fabrication process of CNT-Epoxy concrete

where σ (S/m) was the electrical conductivity of the sample, A(m^2) was the cross-sectional area of the specimen, L(m) was the sample thickness, and R(Ω) was the evaluated electrical resistance.

The 8861 universal testing machine of Instron (MA, USA) was used for the static and cyclic load tests to evaluate the electro-mechanical properties of CNT-Epoxy concrete. Static loading was performed under displacement control of 1.3 mm/min. Cyclic loading was performed under lading frequency 0.5Hz. During the static and cyclic loading, the change in resistance of the specimens was recorded using a 2700 multimeter of Kethley (OH, USA) with a data acquisition (DAQ) system (Fig. 2).

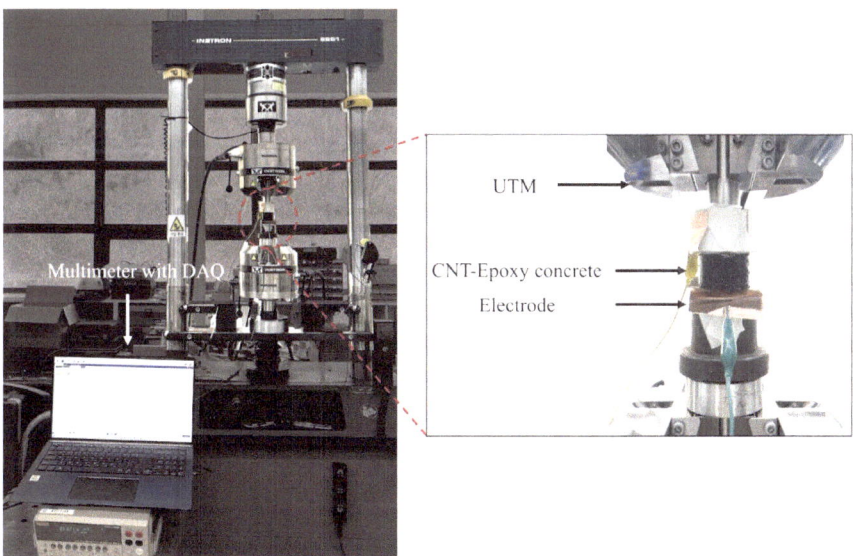

Fig. 2. Electro-mechanical test set up

3　Result and Discussion

Figure 3 showed the electrical conductivity of CNT-Epoxy concrete. The electrical conductivity of CNT-Epoxy concrete increases as increasing CNT content. This is because the electrical conductivity of CNT-Epoxy concrete was determined by the conductive path of dispersed CNTs [15]. Both CNT-Epoxy concrete with and without aggregates exhibited variations in electrical conductivity but demonstrated a similar increasing trend. CNT-Epoxy concrete with aggregate was also selected for its conductive and economic feasibility. At 0.5 wt% of CNT, the electrical conductivity started to converge, and the electrical conductivity was stable. Therefore, CNT content 0.5 wt% was selected for reasons of economic efficiency and stability.

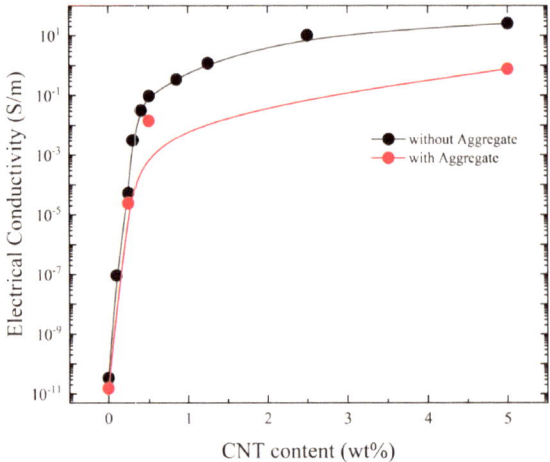

Fig. 3. Electrical conductivity of CNT-Epoxy concrete as a function of CNT content

The mechanical properties and electrical changes of CNT-Epoxy concrete under the compression are shown in Fig. 4. The compressive strength-strain behaviors of CNT-Epoxy concrete are plotted in Fig. 4(a). Compared with CNT-Epoxy concrete and South Korea repair material compressive strength standards, CNT-Epoxy concrete has a much higher compressive strength. Figure 4(b) showed the relative resistance of CNT-Epoxy concrete under the compressive loading. The sensitivity of the S was determined by

$$S = (\Delta R/R_0)/\Delta P \tag{2}$$

where ΔR was the change in the resistance, R_0 was the initial resistance when no compressive strength is applied, and ΔP was the applied compressive strength. The S was defined as the slope of the curve and is divided into two ranges based on compressive strength. A sharp decrease in resistance was observed in the range of 0–5 MPa, and then a stable decrease in resistance was observed in the range of 5–10 MPa. The compressive sensitivities of samples were calculated as 11.83 and 0.73, respectively. This is attributed to the structural variation of CNT-Epoxy concrete during the compression process, leading to different rates of change in the contact area between CNT in CNT-Epoxy concrete.

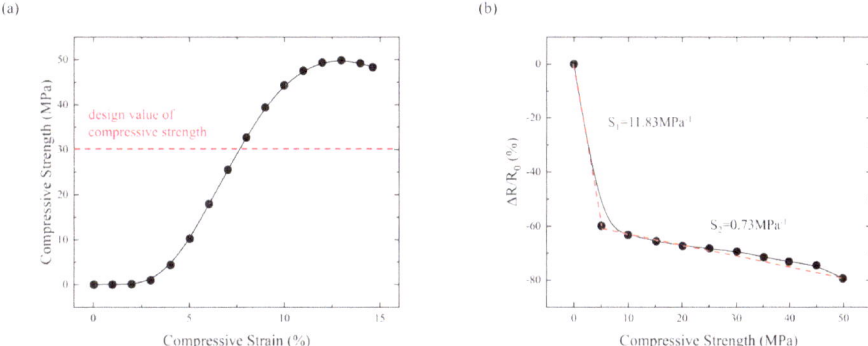

Fig. 4. CNT-Epoxy concrete in compression test (a) Mechanical properties (b) Electro-mechanical properties

Figure 5 showed the electro-mechanical properties of CNT-Epoxy concrete under the cyclic loading conditions, and CNT-Epoxy concrete under stepwise cyclic loading levels was explored as well. Figure 5(a) showed the stepwise resistance response of CNT-Epoxy concrete at intervals of 5MPa. The maximum relative resistance of samples decreased with each compressive strength level and the resistance recovered to its initial value after external compressive strength release. Similar to the resistance response shown in Fig. 4(b), the minimum resistance response for each strength exhibited decreasing trend with stepwise cyclic loading. Figure 5(b) showed the cyclic loading resistance response of CNT-Epoxy concrete. The results indicated that CNT-Epoxy concrete achieved good resilience and reproducibility for every cyclic loading level.

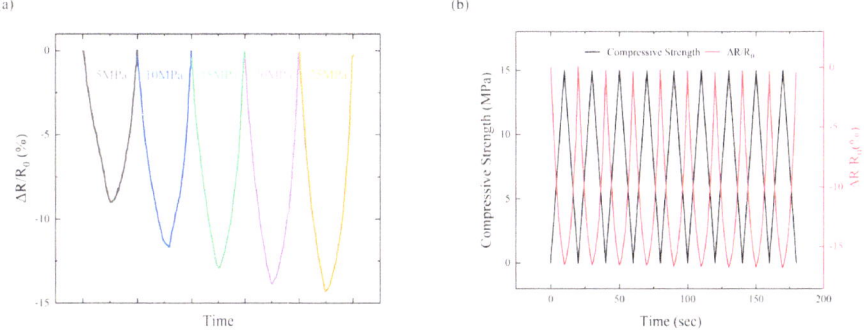

Fig. 5. Electro-mechanical properties of CNT-Epoxy concrete under cyclic loading (a) Stepwise cyclic loading response of CNT-Epoxy concrete up to 25MPa (b) Cyclic response of CNT-Epoxy concrete under 15MPa

4 Conclusions

In summary, we developed mechano-responsive CNT-Epoxy concrete by mixing CNT into epoxy concrete. CNT-Epoxy concrete showed a clear decrease in electrical resistance as increasing compressive strength and accurate and stable compressive sensing performance. Under cyclic loading conditions, the minimum resistance response of CNT-Epoxy concrete varies with the change in stepwise cyclic levels. CNT-Epoxy concrete shows a constant resistance response repeatability under repeated load and secures stability and responsivity. Therefore, CNT-Epoxy concrete exhibits the potential to be used as a multifunctional construction material for repair.

Acknowledgements. This work was supported by the Technology Innovation Program (20014127, Development of a smart monitoring system integrating 3D printed battery-free antenna sensor technology with AI optimization) funded by the Ministry of Trade, Industry & Energy (MOTIE, Korea) and the National Research Foundation of Korea (NRF) grant funded by the Korea government (MSIT) (No. 2020R1C1C1005273).

References

1. Aboudi, J.: Stiffness reduction of cracked solids. Eng Fatigue Mech **26**(5), 637–650 (1987)
2. Feist, C., Hofstetter, G.: An embedded strong discontinuity model for cracking of plain concrete. Comput. Methods Appl. Mech. Eng. **195**(52), 7115–7138 (2006)
3. Dhital, D., Lee, J.R.: A fully non-contact ultrasonic propagation imaging system for closed surface crack evaluation. Exp. Mech. **52**, 1111–1122 (2012)
4. Lukovic, M., Ye, G., Van Breugel, K.: Reliable concrete repair: A critical review. In 14th International Conference Structural Faults and Repair (2012)
5. Contamine, R., Larbi, A.S., Hamelin, P.: Identifying the contributing mechanisms of textile reinforced concrete (TRC) in the case of shear repairing damaged and reinforced concrete beams. Eng. Struct. **46**, 447–458 (2013)
6. Baharuddin, N.K., Nazri, F.M., Bakar, B.H.A., Beddu, S., Tayeh, B.A.: Potential use of ultra high-performance fibre-reinforced concrete as a repair material for fire-damaged concrete in terms of bond strength. Int. J. Integrat. Eng. **12**(9), 87–95 (2020)
7. Bedi, R., Chandra, R., Singh, S.P.: Mechanical properties of polymer concrete. J. Compos. **2013**, 1–12 (2023)
8. Jung, K.C., Roh, I.T., Chang, S.H.: Evaluation of mechanical properties of polymer concretes for the rapid repair of runways. Compos. B Eng. **58**, 352–360 (2014)
9. Czarnecki, L.: Polymer-concrete composites for the repair of concrete structures. MATEC Web Confer. **199**, 01006 (2018)
10. Hashemi, M.J., Jamshidi, M., Aghdam, J.H.: Investigating fracture mechanics and flexural properties of unsaturated polyester polymer concrete (UP-PC). Constr. Build. Mater. **163**, 767–775 (2018)
11. Camacho, M.D.C., Galao, O., Baeza, F.J., Zornoza, E., Garcés, P.: Mechanical properties and durability of CNT cement composites. Materials **7**(3), 1640–1651 (2014)
12. Jang, S.H., Kawashima, S., Yin, H.: Influence of carbon nanotube clustering on mechanical and electrical properties of cement pastes. Materials **9**(4), 220 (2016)
13. Dong, W., Li, W., Shen, L., Sun, Z., Sheng, D.: Piezoresistivity of smart carbon nanotubes (CNTs) reinforced cementitious composite under integrated cyclic compression and impact. Compos. Struct. **241**, 112106 (2020)

14. Ma, P.C., et al.: Enhanced electrical conductivity of nanocomposites containing hybrid fillers of carbon nanotubes and carbon black. ACS Appl. Mater. Interfaces. **1**(5), 1090–1096 (2009)
15. Jang, S.H., Li, L.Y.: Self-sensing carbon nanotube composites exposed to glass transition temperature. Materials **13**(2), 259 (2020)

Special Properties of Concrete

Sustainable Development Approach for 3D Concrete Printing

Maria Kaszyńska and Szymon Skibicki[(✉)]

Faculty of Civil and Environmental Engineering, West Pomeranian University of Technology in Szczecin, Szczecin, Poland
szymon.skibicki@zut.edu.pl

Abstract. In the current era, sustainability has gained significant importance within the field of civil engineering. The promising technology of 3D printing for cementitious materials addresses the mentioned challenges. This study provides a briefly overview of the sustainable approach to 3D printed concrete, covering both technological and material aspects. The paper presents a thorough analysis of the essential properties of 3D printed concrete from a sustainable perspective. Specifically, the composition of binders and aggregates is examined in relation to sustainable development. In case of technological aspects various research studies have demonstrated that the mentioned aspects of 3D concrete printing have the potential to achieve a minimum reduction of 50% in CO_2 emissions. Furthermore, modifying materials can help protect natural resources from depletion, and the use of alternative binders can further reduce CO_2 emissions. The findings presented in this work pave the way for new directions in future research endeavors.

Keywords: 3D printing · 3D concrete printing · additive manufacturing · buildability · green strength · recycled aggregate · LC3 · CSA · calcined clay · sustainable development

1 Introduction

Tha amount of publication upon 3D concrete printing increase very fast over last years. The research teams focuses on analysis the early age properties of 3D printable mix, which are crucial for fulfill the requirements of printing process [1–6]. In other side the research team try to improve hardened and durability properties of 3D printed mixes [7–11]. Nevertheless mentioned works evaluate the properties of 3D printed structures.. It should be noted that initially, 3D printing relied on high-performance concretes as the primary material choice [12, 13], characterized by a significant proportion of cement that detrimentally affects the environment [14, 15]. Nevertheless the increased presence of binders in the printing mixture has a favorable impact on its rheological properties, which is a crucial parameter in the process [1]. The production processes involved in the cement industry result in substantial amounts of CO_2 being emitted, making it a major contributor to anthropogenic CO_2 emissions. This connection arises from the unique nature of the production processes, which generate large volumes of CO_2 [16].

© The Author(s) 2025
L. Czarnecki et al. (Eds.): ICPIC 2023, 61, pp. 565–576, 2025.
https://doi.org/10.1007/978-3-031-72955-3_58

In 2019, in cement and concrete industry it was approximated that about 10% of global CO_2 emissions associated with energy use, transportation, production, and demolition activities were attributable to factors such as fuel combustion, power consumption, and carbonate decomposition [17, 18]. According to IPCC report [19], cement industry itself plays a vital role as a major contributor to anthropogenic carbon dioxide (CO_2) emissions, accounting for approximately 7% of global emissions.

The mentioned environmental problem do not limit to cement industry, the additional factor is related to the aggregate. The 3D printed concrete usually need to used only fine aggregate due to pumping process. The mentioned type of aggregate is also on the brink of depletion. It should be noted that aggregate is the most widely used material in concrete in terms of volume [20]. The excavation of this type of material has led to irreversible environmental changes [21].

Nowadays, sustainability factors are becoming increasingly important for the entire civil engineering sector. This notion compels the industry and research community to offer sustainable solutions for 3D printing materials. Existing research has, in many instances, tackled these aforementioned issues [22–26]. However, there is still a need to draw conclusions and explore new research directions. This paper provides a brief summary of the sustainable approach for 3D printed concrete, encompassing both technology and material aspects. The presented work sets a new course for future research. The paper focuses exclusively on the extrusion-type printing process of cementitious materials.

2 Crucial 3D Printed Materials and Technological Properties

Finding a suitable mix and machines that meet the requirements of 3D concrete printing is one of the most challenging problems in this technology. Many research teams consider various aspects related to these problems [27]. These aspects can be related to technology issues, such as the pumping system and robot controls, or to materials issues associated with the appropriate mix design. It should be noted that both technological and materials aspects are closely interconnected. A mix that is suitable for printing with one type of pump may be impossible to print with another type of pump [28]. The example of successfully printing process of cementitious mix is presented on Fig. 1.

The parameter that describes the transportation of fresh concrete from the pump to the extrusion nozzle is defined as **pumpability/deliverability** [12]. One of the main issues concerning these properties is the possible segregation within the hose. This can occur due to an inappropriate mixture composition (e.g., very high viscosity) or insufficient mixing before the printing process begins. Another problem highlighted by research teams is the potential segregation of concrete particles within the hose [9]. Additionally, the friction between the mixture and the hose wall can increase viscosity, making pumping impossible [29]. To ensure adequate pumpability, many research teams use mixtures with a high cement content, which reduces the friction between the mixture and the hose walls. Several research teams have attempted to determine the pumpability parameter for their specific equipment.

Extrudability and printability are parameters that determine the ability to extrude a mixture through the printing nozzle while maintaining the desired shape (without significant deformations) and an acceptable level of path discontinuity [9]. The extruded mixture must not exhibit structural damage, should not be segregated, and its consistency should remain consistent during construction. In addition mentioned parameters can be defined as the ability to arrange the mixture-nozzle system to print a path of desired quality with appropriate rheological properties [9]. For example to measure the quality of extrudability, the shape retention factor (SRF) was introduced by Panda et al. [30]. It is an indicator that assesses the disparity between the printed path and the printing nozzle. The SRF is calculated by dividing the width of the printed layer by the width of the nozzle used. When the SRF value increases, it indicates unsatisfactory shape retention of the mixture. A value of one for SRF signifies that the printed mixture maintains the shape corresponding to the size of the nozzle.

In addition to mentioned parameters, some research teams [27] defined **print quality** as well. This property evaluate the defects of printed layers, which can also influence the mechanical and durability properties.

The duration of proper workability, which allows for printing, is known as **open time** [12]. It refers to the period during which a concrete mixture can be pumped, meets quality requirements, and is capable of achieving adequate adhesion to the surface it is being placed on. Open time is particularly crucial for geopolymer or other alkali-activated materials, as their open time is typically very limited.

For mainly a materials point of view, it is important to design a mixture that has the ability to carry loads and maintain the same shape under its own weight and the weight of subsequent layers without excessive deformation. These requirements allow for the printing of high and complex structures. To evaluate these crucial properties, many testing methods have been proposed, starting with basic rheological parameters such as slump flow, yield stress, or structurization rate [1]. These parameters can be evaluated using traditional rheological methods. A better way to evaluate the "green" (non-hardened) mechanical properties of the mix is uniaxial unconfined compression testing (UUCT) or similar tests. There are two ways to perform the test: (1) at constant intervals [2, 15], and (2) at a constant displacement rate [4, 31]. This type of testing was initially adopted by Perrot et al. [2], and later modified by Wolfs et al. [4], who used a triaxial compression test. Casagrande et al. [5] also performed various assumptions by examining different test setups, including casting procedures, different displacement rates, etc. However, the UUCT test does not accurately evaluate the ability to carry loads [15]. **Buildability** or **shape retention** during printing is one of the best ways to examine the mentioned aspects. This research procedure involves direct printing tests, usually until collapse, to evaluate the real load-carrying capacity of the printed mixes. Different research teams print different structures for the purpose of this test and analyze various aspects such as changing the height of layers [32], complete collapse [15], and the type of collapse [33, 34]. It should be noted that these assumptions can be influenced by stability aspects [33]. There are also other indirect techniques to evaluate the green properties of the mixture, including penetration tests [35] or ram extrusion tests [36]. However, due to the limitations of this study, the topic will not be further expanded.

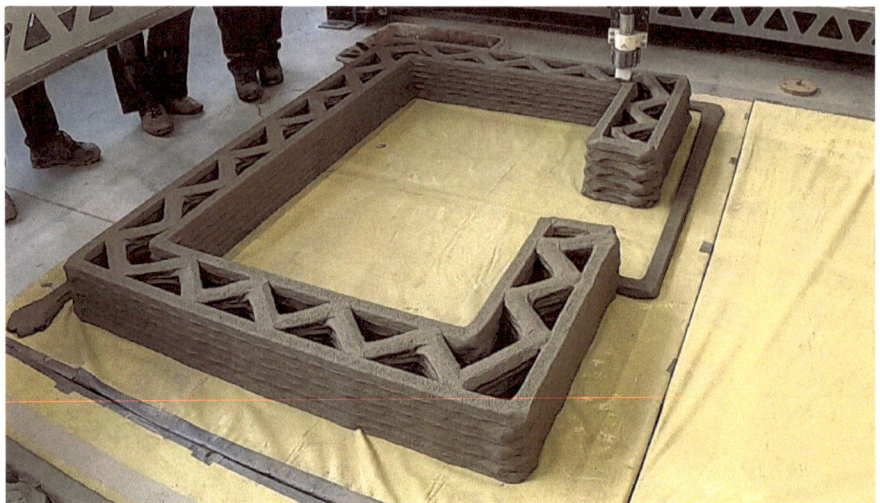

Fig. 1. The wall during printing process.

Table 1. Chosen environmental aspects related to 3D concrete printing.

Sustainable aspect	Potential solution	Chosen reference
Aggregate composition	Copper slag	[37]
	Recycled sand	[6, 38]
	Recycled glass aggregate	[39]
	Artificial / plastic aggregate	[40]
	Tyre waste / recycled rubber aggregate	[41]
Binder composition	Alternative binders or SCM	[42]
	Chemical admixtures	[43]
	Inert microfillers	[31, 44]
	Alternative binder composition	[45–48]
Technology aspects	Speed up the construction process	[49–51]
	Reducing the labor cost	
	Reducing the amount of material used	
	Reducing the amount of construction waste	

3   Sustainable Approach for 3D Printed Mixes

The sustainable aspects mentioned in point 1 can be categorized into three main groups: aggregate composition, binder composition and technological aspects. Table 1 presents these aspects along with potential solutions related to 3D printing technology. These groups play a crucial role in the sustainable development of 3D printing technology.

In the case of binder composition, a high amount of binder is typically used for 3D printed concrete [52], but this aspect needs to be addressed and changed. As for aggregate composition, it should be noted that the use of coarse aggregate is limited in 3D concrete printing. This limitation necessitates the use of high-quality quartz aggregate, which leads to the depletion of limited natural resources. Technological aspects should provide additional advantages for 3D printing technology, thereby improving the overall construction process. In the following sections of the study, each of these groups will be reviewed in detail.

4   Aggregate Composition

The potential solution for aggregate composition could involve the use of alternative and recycled aggregates. Recycled sand shows promise for 3D printed concrete, as it often exhibits higher water absorption, which can enhance the load-bearing capacity of the printed structure [38]. However, research indicates that the origin of the aggregate plays a crucial role in the final performance of the mixture. For instance, recycled aggregate obtained from crushing and screening 100% waste concrete [38] has a positive impact on the green strength of concrete. On the other hand, aggregates derived from post-abrasive wear of steel and engineering plastics (spent garnet) have a negative influence on green properties [6]. A comparison of these effects is presented in Fig. 2.

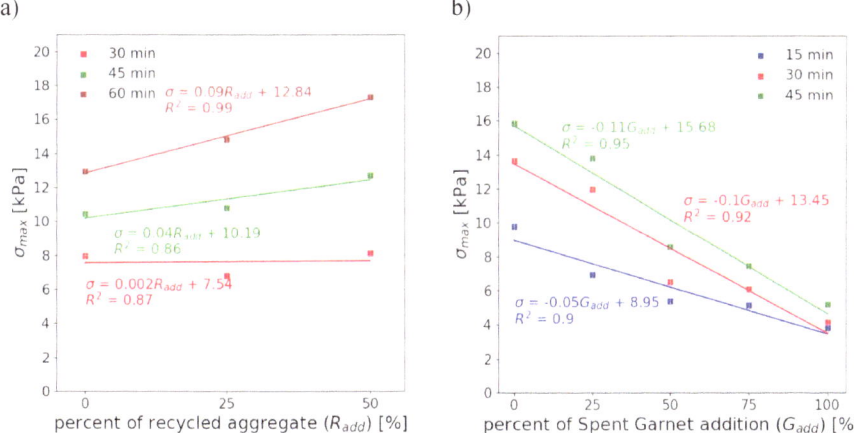

Fig. 2. The comparison of the influence of recycled aggregate on the green compressive strength depends on its source. The two sources considered are: a) recycled aggregate obtained by crushing and screening 100% of waste concrete (based on [38]); b) recycled aggregate obtained as post-abrasive wear of steel and engineering plastics (spent garnet) (reproduced from [6])

The artificial / plastic recycled aggregate (PET) was successfully used in 3D printing technology by Skibicki et al. [40]. The research proves that this type of material does not influence the rheological properties of the mixture, and the modified reference mix with the mentioned aggregate meets all the printing requirements. However, PET granules hinder the hardened properties. Additionally, exposure to freeze-thaw cycles and high temperatures significantly reduces the material's properties, up to 82.4% and 68.8%, respectively. The research concludes that adding up to 10% by volume results in a reasonable reduction in durability properties.

It should be noted that recycled glass aggregates [39], tyre waste, and recycled rubber aggregate [41, 53] do not significantly influence the rheological properties. However, data on the utilization of these materials in 3D printing of cementitious materials is limited. Evaluation of the hardened and durability properties is needed.

Promising results for buildability evaluation were obtained by adding 10–15% of copper slag [37]. This procedure led to an increase in buildability, although a higher amount of slag significantly deteriorates the print quality.

5   Binder Composition

The potential solution for binder composition in 3D printed mixes has been evaluated in many research papers [42, 54]. For this brief study, specific assumptions were chosen for review. Changing the binder can be favorable for the environment. Cement production results in the production of 0.91 kg of CO_2 for every 1 kg of cement [55], but the use of supplementary cementitious materials helps limit CO_2 production. Silica fume and fly ash have a minimal environmental footprint [22] due to their waste nature. However, the availability of silica fume and especially fly ash is becoming limited according to EU policy. In relation to 3D printing technology, silica fume improves buildability [42], although there is a lack of research on high amounts of silica fume addition. Fly ash, on the other hand, reduces buildability but can act as a retarder, especially when used with alternative binders such as Calcium Sulfoaluminate Cement (CSA) or Calcined Clay Cement (LC3) [56]. Both materials have positive effects on long-term mechanical properties.

Another assumption to reduce CO_2 emissions is the use of alternative binders such as Magnesium Cement (MG), Calcium Sulfoaluminate Cement (CSA), and Calcined Clay Cement (LC3), which have approximately 45% [57], 35% [45] and 73% [55] lower CO2 production than ordinary cement, respectively. The MG [58] does not contribute the properties requires for 3D printing, but existing literature data only analyze pastes, and further research is needed. The CSA and LC3 cement [36, 47, 48] greatly improve green strength properties but significantly reduce the open time due to their fast hardening process. Both materials require efficient retardation to control the hardening process.

A sustainable and promising idea for 3D printing materials is the use of inert micro-fillers as partial replacements for aggregate [31, 44]. In cited research, successful replacement of up to 30% by volume was achieved. Figure 3 present the influence of limestone powder (LP) to binder (B) ratio on the green strength of 3D printed mixes. The presented comparison is based on findings from the West Pomeranian University of Technology [15, 31]. Five mixes with different amounts of binder were compared (in each case, the binder consists of 70% cement, 20% fly ash, and 10% silica fume). The results clearly indicate that the green strength increases with an increase in the amount of limestone powder in the mix. This phenomenon allows for a possible reduction in the binder content within the mix. Additionally, it should be noted that the effectiveness of limestone powder, in most cases, increases with the binder amount. Nevertheless, the presented research shows that it is possible to reduce the amount of binder and achieve the same green strength. In summary, this approach helps maintain similar green strength and rheological properties required for printing, even for mixes with reduced binder. However, it may slightly reduce the hardened properties [44]. Additionally, due to the higher amount of binder, some cement remains unhydrated, and partial replacement by inert microfillers does not significantly impact the hardened mechanical properties [31].

Binder composition holds great promise in preserving the natural environment. However, due to the limitations of this study, only the main findings have been highlighted.

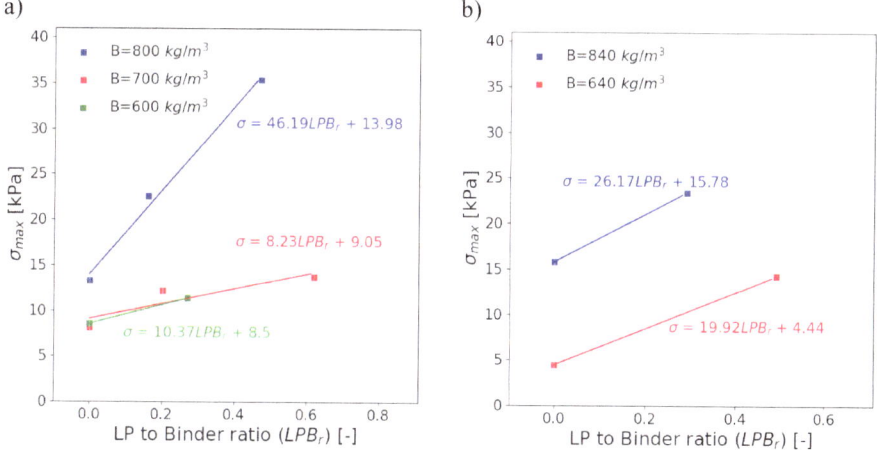

Fig. 3. The comparison of the influence of limestone powder (LP) to binder (B) ratio on the green strength of 3D printed mixes: a) the mixes with w/b = 0.3 (based on [31]), the results for the 600 kg/m^3 binder mix are limited to only two LP to binder ratios. This limitation is due to the impossibility of testing the mix with the highest limestone content; b) the mixes with w/b = 0.28 (based on [15])

6 Technological Aspects

Different varieties of 3D printers are currently employed, such as Cartesian robots, robotic manipulators, and Delta robots [51]. Numerous items have already been successfully manufactured utilizing this technology, serving both as demonstrations and for practical applications [51].

Taking into account that the expenses related to formwork can make up 35–54% of the overall costs involved in constructing a concrete structure [59] the utilization of additive manufacturing yields significant benefits. The greatest reduction in construction costs and completion time occurs with elements of complex geometry (e.g., curved walls, arches, etc.). In such situations, traditional methods often lead to significant cost and time increases, whereas in 3D printing technology, the shape of the printed structure does not have a major impact on the overall cost of the project. This innovative approach not only enables the creation of structures without the need for formwork, but it also leads to a reduction in total production time, costs, and labor. Additionally, this technology enhances the safety of construction site workers, minimizes waste generation, and utilizes raw materials with a low level of embodied energy [59].

Some studies have analyzed the mentioned problems using a quantitative approach [49, 50, 60]. Weng et al. [50] conducted a comparison between the construction processes of precast concrete bathrooms using 3D concrete printing technology and traditional methods. The results indicate that certain costs were significantly reduced with 3D printing technology (e.g., a 47% reduction in material costs), while electricity costs increased by over 25 times. However, when considering the overall cost, the comparison demonstrates that 3D concrete printing technology leads to a construction cost reduction of over 25%. Furthermore, the analysis of CO_2 emissions reveals that this technology achieves a reduction of over 85%. Similar results were obtained by Mohammad et al. [49], providing evidence that 3D concrete printing can reduce CO_2 emissions by over 50% when compared to four different construction scenarios involving concrete walls, two using traditional methods and two utilizing 3D printing technology.

7 Conclusions

Based on the own research and experiences with 3D printing, and the international research achievements presented in this brief study, the following conclusion remarks can be made:

- Replacing natural aggregates with recycled materials can have a positive effect on the properties of 3D printed concrete. However, the impact of this phenomenon depends on the specific materials used. Plastic aggregates do not significantly affect the green strength and rheological properties, but adding more than 10% of these materials can reduce the hardened and durability properties. The source of recycled sand also plays a crucial role in determining the final effect, but there is limited data on this topic, necessitating further research.
- Substituting the binder is one of the most effective ways to reduce negative environmental impacts. Significant reductions in CO_2 emissions can be achieved by using waste materials like silica fume or fly ash. However, fully replacing the binder with alternative binders such as CSA, MG, or LC3 is even more environmentally favorable. Limited data is available on the use of MG in 3D printing technology, and further research is needed, even in relation to mortar usage. CSA and LC3 have shown improvements in early age properties, but careful control of the hardening process is required due to their short open time.
- Existing research has pointed out that 3D concrete printing technology could be favorable for the environment. Some studies have shown that if this technology is fully adopted by the industry, it could reduce CO_2 emissions by a minimum of 50%.

To sum up, there are promising approaches for sustainable 3D printing materials. However, most of these solutions require further research to fully evaluate all the necessary properties. It should be noted that, particularly for alternative binders, the commercial usage of these materials is still uncertain due to their short open time. Nonetheless, the rapid development of 3D printing research holds the potential for new solutions that meet both printing and sustainability requirements.

References

1. Roussel, N.: Rheological requirements for printable concretes. Cem. Concr. Res. **112**, 76–85 (2018)
2. Perrot, A., Rangeard, D., Pierre, A.: Structural built-up of cement-based materials used for 3D-printing extrusion techniques. Mater. Struct. **49**(4), 1213–1220 (2016)
3. Wolfs, R.J.M., Bos, F.P., Salet, T.A.M.: Early age mechanical behaviour of 3D printed concrete: numerical modelling and experimental testing. Cem. Concr. Res. **106**, 103–116 (2018)
4. Wolfs, R.J.M., Bos, F.P., Salet, T.A.M.: Triaxial compression testing on early age concrete for numerical analysis of 3D concrete printing. Cement Concr. Compos. **104**, 103344 (2019)
5. Casagrande, L., Esposito, L., Menna, C., Asprone, D., Auricchio, F.: Effect of testing procedures on buildability properties of 3D-printable concrete. Constr. Build. Mater. **245**, 118286 (2020)
6. Skibicki, S., Jakubowska, P., Kaszyńska, M., Sibera, D., Cendrowski, K., Hoffmann, M.: Early-age mechanical properties of 3D-printed mortar with spent garnet. Materials (Basel) **15**(1), 100 (2021)
7. Nodehi, M., Aguayo, F., Nodehi, S.E., Gholampour, A., Ozbakkaloglu, T., Gencel, O.: Durability properties of 3D printed concrete (3DPC). Autom. Constr. **142**, 104479 (2022)
8. Wu, P., Wang, J., Wang, X.: A critical review of the use of 3-D printing in the construction industry. Autom. Constr. **68**, 21–31 (2016)
9. Buswell, R.A.: Leal de Silva WR, Jones SZ, Dirrenberger J:3D printing using concrete extrusion: a roadmap for research. Cem. Concr. Res. **112**, 37–49 (2018)
10. Skibicki, S., Techman, M., Federowicz, K., Olczyk, N., Hoffmann, M.: Experimental study of hardened young's modulus for 3D printed mortar. Materials (Basel) **14**(24), 76433 (2021)
11. Hoffmann, M., Żarkiewicz, K., Zieliński, A., Skibicki, S., Marchewka, Ł.: Foundation piles—a new feature for concrete 3D printers. Materials **14**(10), 2545 (2021)
12. Le, T.T., Austin, S.A., Lim, S., Buswell, R.A., Gibb, A.G.F., Thorpe, T.: Mix design and fresh properties for high-performance. Mater. Struct. **45**, 1221–1232 (2012)
13. Le, T.T., Austin, S.A., Lim, S., Buswell, R.A., Law, R., Gibb, A.G.F., et al.: Hardened properties of high-performance printing concrete. Cem. Concr. Res. **42**(3), 558–566 (2012)
14. Kaszynska, M., Skibicki, S.: Influence of Eco-Friendly Mineral Additives on Early Age Compressive Strength and Temperature Development of High-Performance Concrete. IOP Conf. Ser.: Earth Environ. Sci. **95**, 42060 (2017)
15. Kaszyńska, M., Skibicki, S., Hoffmann, M.: 3D concrete printing for sustainable construction. Energies **13**(23), 6351 (2020)
16. Deja, J., Uliasz-Bochenczyk, A., Mokrzycki, E.: CO2 emissions from Polish cement industry. Int. J. Greenhouse Gas Control **4**(4), 583–588 (2010)
17. Cao, Z., Masanet, E., Tiwari, A., Akolawala, S.: Decarbonizing concrete: deep decarbonization pathways for the cement and concrete cycle in the United States, India, and China (2021)
18. Belaïd, F.: How does concrete and cement industry transformation contribute to mitigating climate change challenges? Resour. Conserv. Recycl. Adv. **15**(1), 200084 (2022)
19. Metz, B., Davidson, O., Coninck, Hd., Loos, M., Meyer, L.: IPCC special report on carbon dioxide capture and storage. Cambridge University Press, New York, NY (United States) (2005)

20. Ersan, Y.C., Gulcimen, S., Imis, T.N., Saygin, O., Uzal, N.: Life cycle assessment of lightweight concrete containing recycled plastics and fly ash. Eur. J. Environ. Civil Eng., 1–14 (2020)
21. Safiuddin, M., Alengaram, U.J., Rahman, M.M., Salam, M.A., Jumaat, M.Z.: Use of recycled concrete aggregate in concrete: a review. J. Civ. Eng. Manag. **19**(6), 796–810 (2013)
22. Bhattacherjee, S., Basavaraj, A.S., Rahul, A.V., Santhanam, M., Gettu, R., Panda, B., et al.: Sustainable materials for 3D concrete printing. Cement Concr. Compos. **122**(8), 104156 (2021)
23. Markin, V., Krause, M., Otto, J., Schröfl, C.: Mechtcherine V:3D-printing with foam concrete: From material design and testing to application and sustainability. J. Building Eng. **43**, 102870 (2021)
24. Teixeira, J., Schaefer, C.O., Rangel, B., Alves, J.L., Maia, L., Nunes, S., et al.: Development of 3D printing sustainable mortars based on a bibliometric analysis. Proc. Inst. Mech. Eng. Part L J. Mater. Des. Appl. **235**(6), 1419–1429 (2021)
25. Khan, S.A., Koç, M., Al-Ghamdi, S.G.: Sustainability assessment, potentials and challenges of 3D printed concrete structures: a systematic review for built environmental applications. J. Clean. Prod. **303**(2), 127027 (2021)
26. de Brito, J., Kurda, R.: The past and future of sustainable concrete: a critical review and new strategies on cement-based materials. J. Clean. Prod. **281**(8), 123558 (2021)
27. Kazemian, A., Yuan, X., Cochran, E., Khoshnevis, B.: Cementitious materials for construction-scale 3D printing: laboratory testing of fresh printing mixture. Constr. Build. Mater. **145**, 639–647 (2017)
28. Skibicki, S., Kaszynska, M., Federowicz, K., Techman, M., Zielinski, A., Olczyk, N., et al.: Druk 3D kompozytów betonowych metodą przyrostową - doświadczenia zespołu szczecińskiego. Inżynieria i Budownictwo **77**(7), 328–333 (2021)
29. Wangler, T., Roussel, N., Bos, F.P., Salet, T.A.M., Flatt, R.J.: Digital concrete: a review. Cem. Concr. Res. **123**, 105780 (2019)
30. Panda, B., Sonat, C., Yang, E.-H., Tan, M.J., Unluer, C.: Use of magnesium-silicate-hydrate (M-S-H) cement mixes in 3D printing applications. Cement Concr. Compos. **117**(3), 103901 (2021)
31. Skibicki, S., Kaszyńska, M., Wahib, N., Techman, M., Federowicz, K., Zieliński, A., et al.: Properties of Composite Modified with Limestone Powder for 3D Concrete Printing. In: Bos, F.P., Lucas, S.S., Wolfs, R.J.M., Salet, T.A.M. (eds.) Second RILEM International Conference on Concrete and Digital Fabrication, pp. 125–134. Springer International Publishing, Cham (2020)
32. Chougan, M., Ghaffar, S.H., Sikora, P., Chung, S.-Y., Rucinska, T., Stephan, D., et al.: Investigation of additive incorporation on rheological, microstructural and mechanical properties of 3D printable alkali-activated materials. Mater. Des. **202**, 109574 (2021)
33. Suiker, A.S.J., Wolfs, R.J.M., Lucas, S.M., Salet, T.A.M.: Elastic buckling and plastic collapse during 3D concrete printing. Cem. Concr. Res. **135**(3), 106016 (2020)
34. Wolfs, R.J.M., Suiker, A.S.J.: Structural failure during extrusion-based 3D printing processes. Int. J. Adv. Manuf. Technol. **104**(1–4), 565–584 (2019)
35. Pott, U., Stephan, D.: Penetration test as a fast method to determine yield stress and structural build-up for 3D printing of cementitious materials. Cement Concr. Compos. **121**, 104066 (2021)
36. Chen, Y., Li, Z., Figueiredo, C., Çopuroğlu, O., Veer, F., Schlangen, E.: Limestone and calcined clay-based sustainable cementitious materials for 3D concrete printing: a fundamental study of extrudability and early-age strength development. Appl. Sci. **9**(9), 1809 (2019)
37. Ma, G., Sun, J., Wang, L., Aslani, F., Liu, M.: Electromagnetic and microwave absorbing properties of cementitious composite for 3D printing containing waste copper solids. Cement Concr. Compos. **94**(5), 215–225 (2018)

38. Ding, T., Xiao, J., Qin, F., Duan, Z.: Mechanical behavior of 3D printed mortar with recycled sand at early ages. Constr. Build. Mater. **248**(2005), 118654 (2020)
39. Ting, G.H.A., Tay, Y.W.D., Tan, M.J.: Experimental measurement on the effects of recycled glass cullets as aggregates for construction 3D printing. J. Clean. Prod. **300**(5), 126919 (2021)
40. Skibicki, S., Pułtorak, M., Kaszyńska, M., Hoffmann, M., Ekiert, E., Sibera, D.: The effect of using recycled PET aggregates on mechanical and durability properties of 3D printed mortar. Constr. Build. Mater. **335**(2), 127443 (2022)
41. Liu, J., Setunge, S.: Tran P:3D concrete printing with cement-coated recycled crumb rubber: compressive and microstructural properties. Constr. Build. Mater. **347**(2), 128507 (2022)
42. Lu, B., Weng, Y., Li, M., Qian, Y., Leong, K.F., Tan, M.J., et al.: A systematical review of 3D printable cementitious materials. Constr. Build. Mater. **207**, 477–490 (2019)
43. Sikora, P., Chougan, M., Cuevas, K., Liebscher, M., Mechtcherine, V., Ghaffar, S.H., et al.: The effects of nano- and micro-sized additives on 3D printable cementitious and alkali-activated composites: a review. Appl. Nanosci. **6**(112), 100135 (2021)
44. Skibicki, S., Kaszyńska, M., Techman, M.: Maturity testing of 3D printing concrete with inert microfiller. MATEC Web Conf. **219**(1), 3008 (2018)
45. Hanein, T., Galvez-Martos, J.-L., Bannerman, M.N.: Carbon footprint of calcium sulfoaluminate clinker production. J. Clean. Prod. **172**(4), 2278–2287 (2018)
46. Mohan, M.K., Rahul, A.V., van Dam, B., Zeidan, T., de Schutter, G., van Tittelboom, K.: Performance criteria, environmental impact and cost assessment for 3D printable concrete mixtures. Resour. Conserv. Recycl. **181**, 106255 (2022)
47. Mohan, M.K., Rahul, A.V., Tao, Y., de Schutter, G., van Tittelboom, K.: Hydration re-initiation of borated CSA systems with a two-stage mixing process: an application in extrusion-based concrete 3D printing. Cem. Concr. Res. **159**, 106870 (2022)
48. Tambara Júnior, L.U.D., de Matos, P.R., Lima, G.S., Silvestro, L., Rocha, J.C., Campos, C.E.M., et al.: Effect of the nanosilica source on the rheology and early-age hydration of calcium sulfoaluminate cement pastes. Constr. Build. Mater. **327**(12), 126942 (2022)
49. Mohammad, M., Masad, E.: Al-Ghamdi SG:3D concrete printing sustainability: a comparative life cycle assessment of four construction method scenarios. Buildings **10**(12), 245 (2020)
50. Weng, Y., Li, M., Ruan, S., Wong, T.N., Tan, M.J., Ow Yeong, K.L., et al.: Comparative economic, environmental and productivity assessment of a concrete bathroom unit fabricated through 3D printing and a precast approach. J. Clean. Prod. **261**, 121245 (2020)
51. Hoffmann, M., Skibicki, S., Pankratow, P., Zieliński, A., Pajor, M., Techman, M.: Automation in the construction of a 3D-printed concrete wall with the use of a lintel gripper. Materials **13**(8), 1800 (2020)
52. Pacheco, J., Santos, K., Sikora, P., Skibicki, S., Techman, M., Federowicz, K., et al.: Recycled Aggregates and 3D printing technology: production requirements, printability and way forward (Recycl3D project report D.1.1.) (1.0). Zenodo (2023)
53. Aslani, F., Dale, R., Hamidi, F., Valizadeh, A.: Mechanical and shrinkage performance of 3D-printed rubberised engineered cementitious composites. Constr. Build. Mater. **339**(1), 127665 (2022)
54. Panda, B., Paul, S.C., Mohamed, N.A.N., Tay, Y.W.D., Tan, M.J.: Measurement of tensile bond strength of 3D printed geopolymer mortar. Measurement **113**, 108–116 (2018)
55. Gettu, R., Pillai, R.G., Santhanam, M., Basavaraj, A.S., Rathnarajan, S.: Dhanya BS:Sustainability-based decision support framework for choosing concrete mixture proportions. Mater. Struct. **51**(6), 69 (2018)
56. Xu, Z., Zhang, D., Li, H., Sun, X., Zhao, K., Wang, Y.: Effect of FA and GGBFS on compressive strength, rheology, and printing properties of cement-based 3D printing material. Constr. Build. Mater. **339**(3), 127685 (2022)

57. Ruan, S., Unluer, C.: Comparative life cycle assessment of reactive MgO and Portland cement production. J. Clean. Prod. **137**(5), 258–273 (2016)
58. Douba, A., Badjatya, P., Kawashima, S.: Enhancing carbonation and strength of MgO cement through 3D printing. Constr. Build. Mater. **328**(1), 126867 (2022)
59. Paul, S.C., van Zijl, G.P.A.G., Tan, M.J., Gibson, I., Campbell, R.I.: A review of 3D concrete printing systems and materials properties: current status and future research prospects. Rapid Prototyping J. **23**(4), 0 (2018)
60. Yang, H., Chung, J.K.H., Chen, Y., Li, Y.: The cost calculation method of construction 3D printing aligned with internet of things. J. Wireless Com. Network **2018**(1), 65 (2018)

Evolution of Early-Age Mechanical and Failure Behavior of 3D Printed Polymer Concrete

Mohammad Amin D. Najvani, Daniel Heras Murcia, and Mahmoud Reda Taha$^{(\boxtimes)}$

Department of Civil, Construction & Environmental Engineering, University of New Mexico, Albuquerque, NM 87131, USA
mrtaha@unm.edu

Abstract. The increasing interest in 3d printing of concrete for infrastructure applications necessitates having a design for this process. Previous research has mostly focused on 3D printable cement-based concrete mixes, with less attention given to 3D printed polymer concrete (PC). PC is a concrete type that uses polymer instead of cement as a binder. It offers improved compressive and tensile strengths, crack resistance and bond strengths, and superior durability than traditional Portland cement concrete, making it an excellent material for 3D printing. This study aims to understand the evolution of the early-age mechanical properties of fresh polymer concrete and its potential failure during printing. Unconfined uniaxial compression and direct shear tests were performed on fresh polymer concrete for the first 110 min after mixing to determine the evolution of mechanical and failure characteristics with time. Such characteristics include compressive strength, modulus of elasticity, cohesive strength, and friction angle. A time-dependent early-age Mohr-Coulomb failure envelope is established to describe the mechanical and failure behavior of 3D printed polymer concrete.

Keywords: Polymer Concrete · Green Strength · 3D Printing · Rheology

1 Introduction

3D printing is a computer-controlled method of sequentially layering materials to create three-dimensional shapes. This process helps fabricate complex geometries and produce prototypes. The construction industry's gradual shift towards automation has prompted growing research to explore the potential of this technology. 3D concrete printing, in particular, enables architects and builders to produce intricate and asymmetric patterns while minimizing human error leading ways to innovative designs [1, 2].

Concrete made from polymers as a binder, called Polymer Concrete (PC), has been used in various field applications since the 1950s. Such applications include precast architectural façades, underground utilities, wastewater pipes and tanks, manholes, machine foundations, bridge deck overlays, and closures [3]. PC exhibits superior properties such as high tensile and bond strength and outstanding durability [4, 5]. However, its use is often restricted due to its comparatively higher cost [6, 7]. The enhanced characteristics of PC to being impermeable concrete with very high durability [7, 8].

© The Author(s) 2025
L. Czarnecki et al. (Eds.): ICPIC 2023, 61, pp. 577–584, 2025.
https://doi.org/10.1007/978-3-031-72955-3_59

The unique features of PC and its numerous industrial applications make it a desirable material for 3D printing [9, 10]. However, it is crucial to understand PC's early age rheological, mechanical, and failure characteristics to ensure successful printing and product quality and prevent printing failures. This study examines the time-dependent evolution of fresh polymer concrete's mechanical and failure characteristics, referred to as "green strength" properties, such as modulus of elasticity, Poisson's ratio, and failure criteria. One crucial property of 3D printing is material stability during the printing process [11–13]. Green strength refers to the ability of fresh concrete to retain its shape and support its weight immediately after compaction or mixing, which results from inter-particle cohesion and friction. The Mohr-Coulomb model represents the green material's failure, a simple and widely used approach to describe the material's strength and deformation properties under different loading conditions [14–18]. The use of Mohr-Coulomb failure criteria can be supported by the similarly of fresh PC with geomaterials. We aim to characterize the 3D printing process for PC by investigating how its mechanical and failure characteristics change over time. To achieve this, we conducted uniaxial unconfined compression and direct shear tests to measure the mechanical properties of fresh PC. Testing was performed over 110 min to realize the time-dependent evolution of the properties. These observations shall help model and predict the behavior of the 3D printed PC.

2   Materials and Methods

2.1   Materials

PC was produced using Polyester resin cured at room temperature using a Methylene ethyl ketone peroxide hardener. The hardener content was adjusted to achieve a 2-h setting time. High-quality silica sand with a nominal maximum size of 2.36 mm was used as the main aggregate. In addition, a filler combining silica fume, type-F fly ash and fumed silica was used to control material thixotropy. The PC mix design is presented in Table 1. The PC was cured in dry conditions and ambient temperature of 22 °C for 7 days of age and showed a mean compressive strength of 59.8 MPa and a mean modulus of elasticity of 7.0 GPa.

Table 1. PC mix design of 1 cubic meter and 7-day mechanical properties

Material	Weight (kg/m^3)
Polymer resin	390
Hardeners	3.0
Aggregate	1537
Fillers (Silica Fume, fumed silica, and fly ash)	162
7-day compressive strength	59.8 MPa
7-day modulus of elasticity	7.0 GPa

2.2   Test Methods

Strength assessment of green PC was executed at different time intervals, including t = 0,30,60,90, and 110 min, with t = 0 being the earliest point after compaction, demolding, placing the sample in the testing arrangement, and initiating the examination, as suggested by previous researchers [15]. The process mentioned above takes around 10 to 15 min. The time span from 0 to 110 min was chosen to match the polymer concrete's setting time and the typical length of the printing process.

For the uniaxial unconfined compression test, as per ASTM D2166-13 [19], a cylinder made of PC material is subjected to axial loading until it fails under compression. The maximum load is divided by the specimen's cross-sectional area to determine the unconfined compressive strength. The cylindrical specimens have a height of 101.6 mm and a diameter of 50.8 mm. Before placing the PC mix in the cylindrical molds, a thin Teflon sheet was laid on both sides, and a lubricant oil spray was used, ensuring the sample's easy removal from the molds. The PC mix was added in three layers, each compacted for 10s on a 30 Hz vibration table to obtain a uniform sample. When the sample was carefully removed from the molds, the Teflon sheet was removed as well. A universal testing machine with a 25.0 kN load capacity and a 1.0 N resolution was used for testing. A steel head with a similar diameter was used to transfer the load to the specimen, while a hardened 3D-printed PC plate was placed beneath the sample to mimic the printing conditions of the layer's bottom in contact with the hardened PC (Fig. 1.a). To replicate the loading rate during printing and avoid the impacts of thixotropic build-up [15], a displacement-controlled test was conducted at a rate of 30 mm/min. At each age, five fresh PC samples were tested. To record the PC specimens' vertical and lateral deformation during loading, a non-contact digital image correlation measurement system [20] was used.

The direct shear test, described in ASTM D3080-98 [21], is a method used to investigate the relationship between shear strength and normal stress. The test involves placing a circular disc of material into a shear box mold composed of two steel plates with a 63 mm diameter opening and a specimen height of 36 mm. The plates can slide laterally with respect to each other and are held together with two locking screws to prevent specimen extrusion. Prior to testing, the molds are lightly greased with a lubricant oil spray to minimize side friction. PC was placed in the mold in three layers and compacted for 10 s on a 30 Hz vibration table to ensure homogeneity. The specimen is then tested using the direct shear device equipped with a 450N load cell and data logger (Fig. 1.b). The bottom plate is moved laterally at a constant rate of 10 mm/min while normal vertical stress, σ_n, is applied using steel weights. Three normal loads of 0N, 10N, and 20N are applied, corresponding to normal loads of 1.1N, 10.9N, and 20.7N, respectively, considering the self weight of PC, on the shear plane located in the middle of the sample. A thin layer of grease is applied to their surface to reduce friction between the plates. Stresses are computed with an updated cross-section due to the considerable deformations during testing. Five specimens are tested at each ages underthree normal load combinations.

The two main parameters that determine the failure of a printed element are strength and stiffness growth over time which are produced by the thixotropic build-up of PC [22–24]. The Mohr-Coulomb failure criterion was used to describe the shear behavior

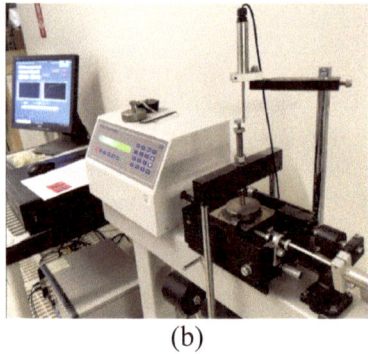

(a)   (b)

Fig. 1. (a) Test set up for the uniaxial unconfined compression test of fresh PC (b) Direct shear test set up for testing fresh PC.

of green printed PC and can be expressed by Eq. (1):

$$\tau_y(t) = C(t) + \sigma_n tan\phi(t) \tag{1}$$

where $C(t)$ is the time-dependent cohesion strength of green PC, σ_n is the normal stress, and $\phi(t)$ is the time-dependent friction angle of the fresh PC layer placed against another fresh PC layer. Finally, $\tau_y(t)$ is the time-dependent shear strength between two green PC layers.

To understand the evolution of mechanical and fracture characteristics of PC with time, time-temperature measurements of the polymer were made and analyzed using the analysis of temperature derivative (ADT) method [25]. In this method, the polymer resin and hardener were mixed with the same ratio used to produce PC and stirred for one minute in a glass cup. Two thermocouples were prepared with twisted junctions to minimize thermal inertia and directly submerged into the polymer mixture. The thermocouples were positioned near the center, and data were collected every minute.

3 Results and Discussions

Under compressive stresses and from 0 to 30 min, the green PC exhibited a mean compressive strength of 2.1 kPa, gradually increasing to 274 kPa at 110 min. The growth of compressive strength of the fresh polymer concrete up to 110 min is shown in a linear fit in Fig. 2(a). The elastic modulus of green PC starts at 8.1 kPa at t = 0 min and shows linear growth up to 11.6 kPa at t = 30 min, as shown in Fig. 2(b). Subsequently, the elastic modulus increased rapidly and reached 4354.9 kPa at t = 110 min. The polymerization reaction of the polymer in the PC seems to start at 30 min (Fig. 4) and might be responsible for the initiation of an observable elastic modulus and compressive strength after 30 min of placing.

In shear testing, there was a noticeable linear growth in strength and stiffness during the first 110 min of PC age. The Mohr-Coulomb failure parameters for PC were determined based on direct shear tests, as shown in Fig. 3, Fig. 3 (a) shows that PC cohesion increases slowly and linearly during the first 60 min. A significant change in the rate of

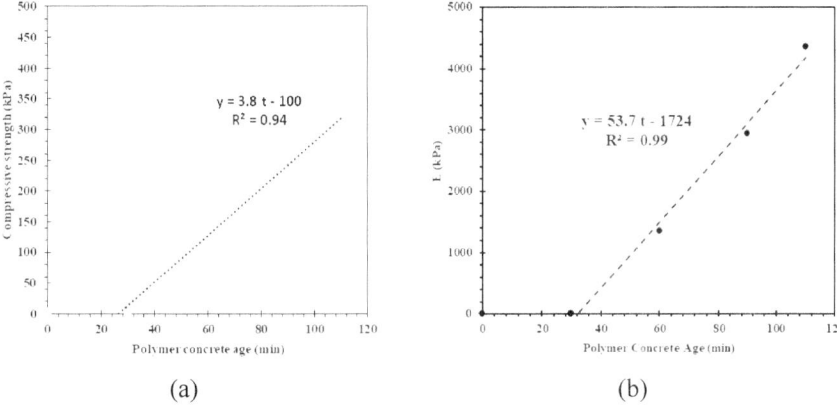

Fig. 2. a) Time-dependent compressive strength of fresh PC up to 110 min; (b) Time-dependent elastic modulus of fresh PC up to 110 min.

cohesion increase was observed after 60 min of the PC age. This might be attributed to polymer gelation (gel transition) at this time. A piecewisefunction (Fig. 3.a) is used to describe the evolution of PC cohesion with time. On the other hand, Fig. 3 (b) shows that the angle of friction of PC seems to be constant at 57 degrees.

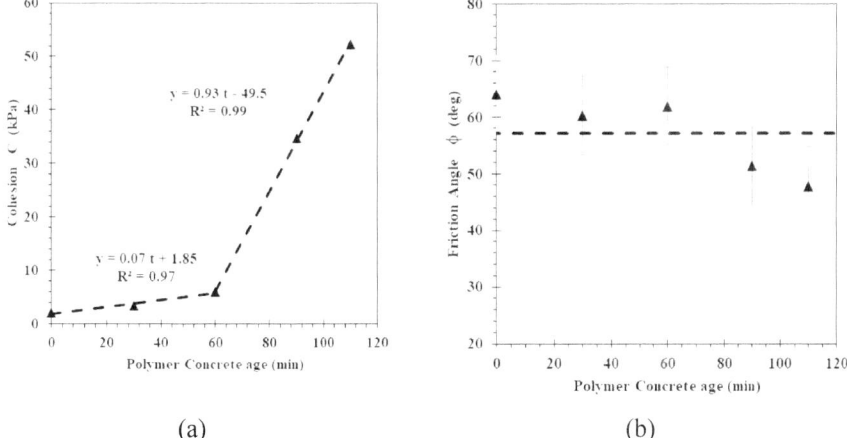

Fig. 3. (a) PC cohesion evolution with time (b) PC internal friction angle of PC of 57° seems constant over time.

The polymer temperature measurements are presented in Fig. 4. Temperature time derivative was calculated using Origin base functions and curves smoothed using the Savitzky-Golay method. It is apparent that polymer gelation, identified by the peak of the first derivative of the temperature-time curve, takes place at 60 min. This can explain the significant change in the PC cohesion growth rate after 60 min of mixing. This is critical information for the 3D printing process.

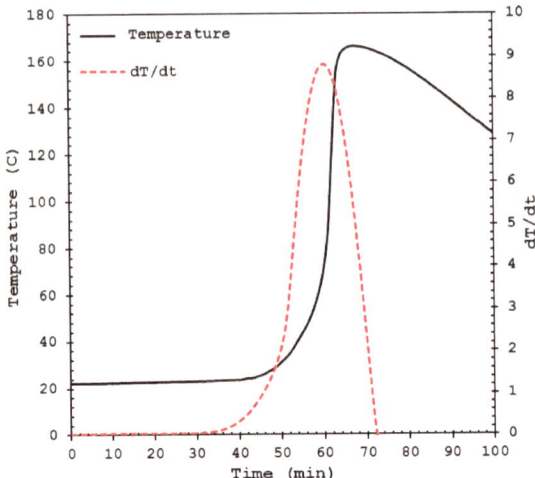

Fig. 4. Thermal analysis of the polymer showing polymer gelling to take place close to 60 min.

The above investigation showed the time evolution of the mechanical properties of 3D printed PC, including the compressive strength and the axial stiffness. The tests also showed the change in the failure criteria of 3D printed PC, including the cohesion and friction angle with time. Further research is underway to use this data for developing a computational model to simulate the 3D printing process. Such modeling is important to ensure the stability of the 3D printed PC structure.

4 Conclusions

The work presented herein demonstrates the extraction of green strength and failure criteria of fresh PC used for 3D printing. The extracted parameters can be used to develop a computational model that simulates the 3D printing process of PC. The compression strength, the elastic modulus, and the shear strength of fresh PC mix were measured at 0, 30, 60, 90, and 110 min of age. The failure behavior can be described by a time-dependent Mohr-Coulomb failure criterion and linear stress-strain behavior up to failure. The green strength parameters and failure criteria of PC, evolves over time. PC's mean green compressive strength increased from 2.1 kPa at printing time to 274 kPa at 110 min. The mean green elastic modulus increased from 8.1 kPa at time of printing to 4355 kPa at 110 min. PC failure criteria parameters are evidently time-dependent. The PC starts to develop an appreciable strength and stiffness after 30 min of placing due to the start of the polymerization reaction. Furthermore, cohesion showed to have a piecewise function relationship with time with a significant change of slope at 60 min. Temperature analysis of the polymer showed this change coincides with the polymer gelation where the solid polymer formation starts to take place. On the other hand, PC internal friction angle was found to have a constant value of 57 degrees.

References

1. Casagrande, L., Esposito, L., Menna, C., Asprone, D., Auricchio, F.: Effect of testing procedures on buildability properties of 3D-printable concrete. Constr. Build. Mater. **245**, 118286 (2020). https://doi.org/10.1016/j.conbuildmat.2020.118286
2. Lim, S., Buswell, R.A., Le, T.T., Austin, S.A., Gibb, A.G.F., Thorpe, T.: Developments in construction-scale additive manufacturing processes. Autom. Constr. **21**, 262–268 (2012). https://doi.org/10.1016/j.autcon.2011.06.010
3. Keith, C., et al. Assessment of moisture-tolerant coatings for decreasing open top construction time (2017)
4. Gorninski, J.P., Dal Molin, D.C., Kazmierczak, C.S.: Study of the modulus of elasticity of polymer concrete compounds and comparative assessment of polymer concrete and portland cement concrete. Cem. Concr. Res. **34**, 2091–2095 (2004). https://doi.org/10.1016/j.cemconres.2004.03.012
5. Abdel-Fattah, H., El-Hawary, M.M.: Flexural behavior of polymer concrete. Constr. Build. Mater. **13**, 253–262 (1999). https://doi.org/10.1016/S0950-0618(99)00030-6
6. 548.6R-19: Polymer Concrete: Guidelines for Structural Applications, n.d.:5
7. Handbook_of_Polymer-Modified_Concrete_and_Mortars_1221893.pdf, n.d.
8. Taha, M.M.R.: Nano-modified polymer concrete: a new material for smart and resilient structures. In: Taha, M.M.R. (ed.) International Congress on Polymers in Concrete (ICPIC 2018), pp. 61–73. Springer International Publishing, Cham (2018). https://doi.org/10.1007/978-3-319-78175-4_6
9. Taha, M.M.R. (ed.): International Congress on Polymers in Concrete (ICPIC 2018). Springer International Publishing, Cham (2018) https://doi.org/10.1007/978-3-319-78175-4
10. Douba, A., Emiroglu, M., Kandil, U.F., Reda Taha, M.M.: Very ductile polymer concrete using carbon nanotubes. Constr. Build. Mater. **196**, 468–477 (2019). https://doi.org/10.1016/j.conbuildmat.2018.11.021
11. Lloret Kristensen, E., Gramazio, F., Kohler, M., Langenberg, S.: Complex Concrete Constructions – Merging Existing Casting Techniques with Digital Fabrication, pp. 613–622. Singapore (2013). https://doi.org/10.52842/conf.caadria.2013.613
12. Voigt, T., Mbele, J.-J., Wang, K., Shah, S.P.: Using fly ash, clay, and fibers for simultaneous improvement of concrete green strength and consolidatability for slip-form pavement. J. Mater. Civ. Eng. **22**, 196–206 (2010). https://doi.org/10.1061/(ASCE)0899-1561(2010)22:2(196)
13. Heras Murcia, D., Genedy, M., Reda Taha, M.M.: Examining the significance of infill printing pattern on the anisotropy of 3D printed concrete. Constr. Build. Mater. **262**, 120559 (2020). https://doi.org/10.1016/j.conbuildmat.2020.120559
14. Wolfs, R.J.M., Bos, F.P., Salet, T.A.M.: Triaxial compression testing on early age concrete for numerical analysis of 3D concrete printing. Cem. Concr. Compos. **104**, 103344 (2019). https://doi.org/10.1016/j.cemconcomp.2019.103344
15. Wolfs, R.J.M., Bos, F.P., Salet, T.A.M.: Early age mechanical behaviour of 3D printed concrete: Numerical modelling and experimental testing. Cem. Concr. Res. **106**, 103–116 (2018). https://doi.org/10.1016/j.cemconres.2018.02.001
16. Hüsken, G., Brouwers, H.J.H.: On the early-age behavior of zero-slump concrete. Cem. Concr. Res. **42**, 501–510 (2012). https://doi.org/10.1016/j.cemconres.2011.11.007
17. Alexandridis, A., Gardner, N.J.: Mechanical behaviour of fresh concrete. Cem. Concr. Res. **11**, 323–339 (1981). https://doi.org/10.1016/0008-8846(81)90105-8
18. Investigation into Yield Behavior of Fresh Cement Paste: Model and Experiment. ACI Mater J. **107** (2010). https://doi.org/10.14359/51663459
19. ASTM D 2166 – 00 _RDIXNJY_.pdf, n.d.

20. Padilla, A., Najvani, M.A., Knight, E., Rougier, E., Stormont, J., Taha, M.M.R.: Correlating damage and cracking with air (gas) permeability in concrete using the Brazilian tension test. Constr. Build. Mater. **348**, 128616 (2022). https://doi.org/10.1016/j.conbuildmat.2022.128616
21. D18 Committee: Test Method for Direct Shear Test of Soils Under Consolidated Drained Conditions. ASTM International; n.d. https://doi.org/10.1520/D3080_D3080M-11
22. Murcia, D.H., Abdellatef, M., Genedy, M., Taha, M.M.: Rheological characterization of three-dimensional-printed polymer concrete. ACI Mater J. **118** (2021)
23. Roussel, N.: A thixotropy model for fresh fluid concretes: Theory, validation and applications. Cem. Concr. Res. **36**, 1797–1806 (2006). https://doi.org/10.1016/j.cemconres.2006.05.025
24. Roussel, N., Ovarlez, G., Garrault, S., Brumaud, C.: The origins of thixotropy of fresh cement pastes. Cem. Concr. Res. **42**, 148–157 (2012). https://doi.org/10.1016/j.cemconres.2011.09.004
25. Smoleń, J., Olesik, P., Gradoń, P., Chudy, M., Mendala, B., Kozioł, M.: The use of the ATD technique to measure the gelation time of epoxy resins. Materials **14**, 6022 (2021). https://doi.org/10.3390/ma14206022

Physical and Mechanical Characterization of Cement Boards with Incorporation of Free Phase Change Materials

Sandra Cunha[✉], Ingried Aguiar, and José Aguiar

Centre for Territory, Environment and Construction (CTAC), University of Minho, Campus de Azurém, Guimarães, Portugal
sandracunha@civil.uminho.pt

Abstract. Currently it is necessary to give a new direction to the construction sector, making it essential to change the way that buildings are constructed and rehabilitated, with the aim of obtaining a construction with greater sustainability value. To minimize energy consumption, it is important to take advantage of renewable energy sources like solar power. Phase change materials (PCM) can help reduce building energy consumption due to their energy storage capacity. PCM has been studied in different solutions for walls, ceilings, and floors, essentially using encapsulation techniques. The use of PCM in building materials has social, environmental, and economic benefits, including increased thermal comfort, decreased energy consumption from non-renewable sources, and reduced air conditioning needs and costs. The development of cement boards incorporating PCM brings a new option for the thermal improvement of buildings, which can be used in new buildings and rehabilitation operations. The main objective of this study was the development of cement boards with the incorporation of pure and free PCM, through direct incorporation, consisting of a simple, low-cost, and very promising incorporation technique. The results showed that the technique was easy to use in manufacturing cement boards for interior coatings of building walls and ceilings. It was also possible to observe a decrease in the water/cement ratio with the incorporation of PCM and a consequent decrease in the porosity, which resulted in a slight reduction in its mechanical strengths, without ever compromising the necessary performance for its function.

Keywords: Phase change materials · Boards · Physical and mechanical performance

1 Introduction

Construction is responsible for 40% of energy consumption, carbon dioxide emissions, and natural raw material consumption [1, 2]. Efforts must be made to change the paradigm of construction, which is dependent on-air conditioning systems and building materials with high embodied energy, to a more holistic and sustainable approach [1].

Europe is currently experiencing significant energy supply challenges, which have been exacerbated by recent international conflicts in natural gas and oil producing and

© The Author(s) 2025
L. Czarnecki et al. (Eds.): ICPIC 2023, 61, pp. 585–593, 2025.
https://doi.org/10.1007/978-3-031-72955-3_60

exporting countries. As a result, there is an increasing need for European Union countries to explore options for achieving greater energy independence. One potential solution is to invest in renewable energy sources, which could help to reduce reliance on traditional energy sources. Additionally, the incorporation of functional construction materials, such as those with thermal storage capacity, has the potential to enhance energy efficiency and comfort in both residential and commercial buildings.

The use of phase change materials (PCMs) has been employed as a measure to regulate energy consumption in buildings. PCMs have been studied for building applications due to their thermal storage capacity based on solar energy, with many of the studied applications focused on building interiors [2–6]. Regarding PCM incorporation into building materials, the encapsulation technique is the most widely used and the most expensive [7, 8]. In this study, the direct incorporation technique of PCM into mortars was used, which is a low-cost technique [3, 9, 10].

The main objective of this study was to develop and characterize interior panels with direct incorporation of PCM. The composition of the panels was based on the composition of the mortars previously developed. Four distinct compositions were developed with different PCM contents (0%, 5%, 10%, and 20%). The behavior of the panels was evaluated in the hardened state, based on open pore volume, real and apparent density, total porosity, water absorption by immersion, water absorption by capillarity, and flexural strength.

2 Experimental Program

2.1 Materials

The cement used was CEM II/B-L 32.5 N with a density of 3030 kg/m³. The fly ash used was produced in a Portuguese coal-fired power plant and had a density of 2420 kg/m³. The aggregate used was composed of two different types of sand (A and B). Sand A has a minimum dimension of 0.125 mm and a maximum dimension of 0.5 mm, with an average particle size of 439.9 μm and a density of 2600 kg/m³. Sand B is composed of particles with a minimum dimension of 0.125 mm and a maximum dimension of 8 mm, with an average particle size of 762 μm and a density of 2569 kg/m³. The PCM used is non-encapsulated, of organic nature, composed of a paraffin with a transition temperature between 20 °C and 23 °C, enthalpy of 200 kJ/kg, solid state density of 760 kg/m³ and liquid state density of 700 kg/m³.

2.2 Test Methods

Regarding the characterization of the boards with incorporation of PCM, they were evaluated considering their physical and mechanical behavior. Considering the absence of specific harmonized standards for this type of material, it was decided to adapt the standards for natural stone boards, NP EN 1469 [11]. The physical behavior was based on real and apparent density, total and open porosity, and water absorption by capillarity and immersion. The mechanical behavior was based on flexural strength. The determination of real and apparent density, and total porosity was performed according to the specification NP EN 1936 [12]. The determination of real density and total porosity was carried out according to the Le Chatelier method described in the specification NP EN 1936 [12]. Water absorption by immersion was performed according to the specification NP EN 13755 [13], and water absorption by capillarity was determined according to NP EN 1925 [14].Finally, the flexural behavior of the boards with the incorporation of PCM was determined based on the specification NP EN 12372 [15] with load control and a speed of 50 N/s.

2.3   Boards Prototype

The determination of the geometry and dimension of the prototype of the PCM-enhanced boards was based on market research, which revealed a wide availability of natural stone boards and composites with different dimensions suitable for various project needs. Therefore, it was decided to follow the geometrical indications provided in specification NP EN 12372 [15], considering a geometry that allows for easy handling and industrial production, with a prismatic geometry of dimensions 100x100x20 mm^3 selected.

2.4   Formulations

The development of these formulations was based on previous works carried out by the authors regarding the incorporation of phase change materials (PCMs) into mortars for interior coatings [3, 9, 10]. Four different compositions of cement mortars were developed by incorporating different levels of unencapsulated PCMs (Table 1). A reference mortar was produced without the addition of PCM (C0PCM), and three mortars with varying incorporation levels of 5% PCM (C5PCM), 10% PCM (C10PCM), and 20% PCM (C20PCM) relative to the mass of the aggregate. The binder used comprised 500 kg/m^3, consisting of 40% cement and 60% fly ash, while the aggregate consisted of a mixture of 50% Sand A and 50% Sand B. The aggregates were used in dry form.

Table 1. Mortars composition (kg/m^3).

Composition	Cement	Fly ash	Sand A	Sand B	PCM	Water
C0PCM	200	300	684.9	684.9	0	280
C5PCM	200	300	583.6	583.6	58.4	275
C10PCM	200	300	519.1	519.1	103.8	260
C20PCM	200	300	423.7	423.7	169.5	240

3   Test Results and Discussion

3.1   Open Pores Volume

The results presented in Fig. 1 indicate that the presence of a higher content of phase change material (PCM) in the composite board leads to an increase in the open pore volume. Comparison with reference boards, which did not incorporate PCM, revealed that the incorporation of 5% PCM resulted in a more than 30% increase in the open pores volume. This behavior can be attributed to the geometry of the PCM composite board prototype, which has a larger exposed surface area, facilitating water evaporation during the curing process. Moreover, PCM exhibits a higher tendency to migrate towards the surface, which also contributes to an increase in the open pore volume. In this case, PCM can move from its initial location, but the quantity of PCM involved in this effect is not significant. This phenomenon is known as the "board effect".

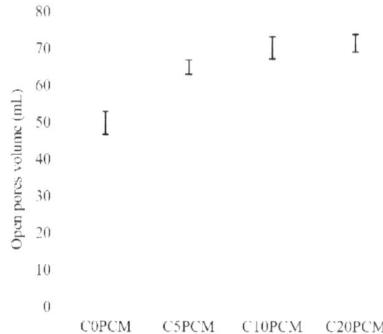

Fig. 1. Open pores volume of the developed mortars.

3.2   Real and Apparent Density

The results of Fig. 2 indicates that the incorporation of higher PCM content into the mortars led to a decrease in both apparent and real density. Specifically, a decrease of over 11% in apparent density was observed with the incorporation of 5% PCM, compared to the reference boards. In contrast, the decrease observed in real density was less significant, around 5%.

The behavior of apparent density is strongly influenced by the volume of open pores, as higher values of open pores volume (Fig. 1) lead to lower values of apparent density. In contrast, the behavior of real density can be explained by the lower density of PCM compared to the sand used. Additionally, this behavior could also be observed due to the decrease in the water/binder ratio in the mortars that was demonstrated in previous works [3, 9, 10].

3.3   Total Porosity

The Fig. 3 depicts the total porosity of the PCM composite boards, and it is evident that the total porosity increases with the incorporation of higher PCM contents. Comparing the board without PCM incorporation (0% PCM) to the board with 5% PCM incorporation, a significant increase in total porosity of 48% was observed. Furthermore, the increase in total porosity was 65% and 76% for an incorporation of 10% and 20% of PCM, respectively. The observed increase in total porosity can be attributed to the higher ease

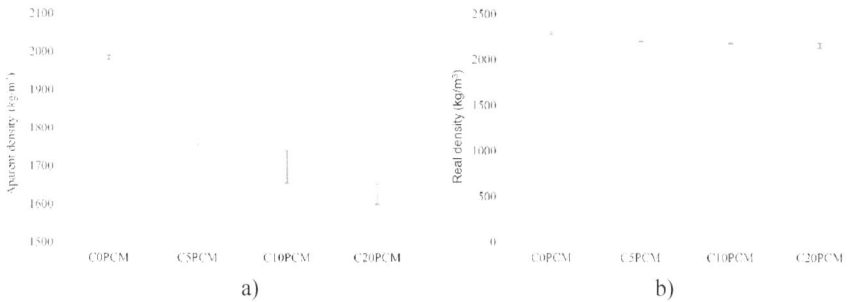

Fig. 2. Boards density: (a) Apparent density; (b) Real density.

of water evaporation, resulting from the higher exposed surface area of the boards, which is associated with the "board effect" described earlier. This effect mainly affects the macroporosity of the boards.

Fig. 3. Total porosity of the developed mortars.

3.4   Water Absorption by Immersion

The presented results in Fig. 4 show the water absorption by immersion of the developed boards. It was observed that the water absorption tends to increase with the presence of higher PCM contents. An increase of approximately 3%, 7%, and 21% in water absorption by immersion was observed for boards with 5%, 10%, and 20% PCM incorporation, respectively. This behavior could also be explained by the increase in the liquid/binder ratio in the mortars that was demonstrated in previous works [3, 9, 10]. Also, due to the increase in open pore volume (Fig. 1) and total porosity (Fig. 3) in the boards with PCM incorporation. This increase in pore volume and total porosity is associated with the presence of a higher macroporosity in the boards.

Fig. 4. Water absorption by immersion of the developed mortars.

3.5 Water Absorption by Capillarity

The capillary absorption coefficient of the boards provides information about their micro-porosity. Figure 5a shows the capillary absorption coefficient of the developed boards. The boards with 5% and 10% PCM incorporation had a similar capillary water absorption coefficient. However, for boards with 20% PCM incorporation, there was a decrease of about 25% in the water absorption coefficient by capillarity. Figure 5b shows the amount of water absorbed by capillarity. The incorporation of higher PCM content led to lower water absorption. This behavior can be explained by the more compact microstructure of the mortars with PCM incorporation, which leads to a decrease in the microporosity of the mortars [9] and consequently of the boards.

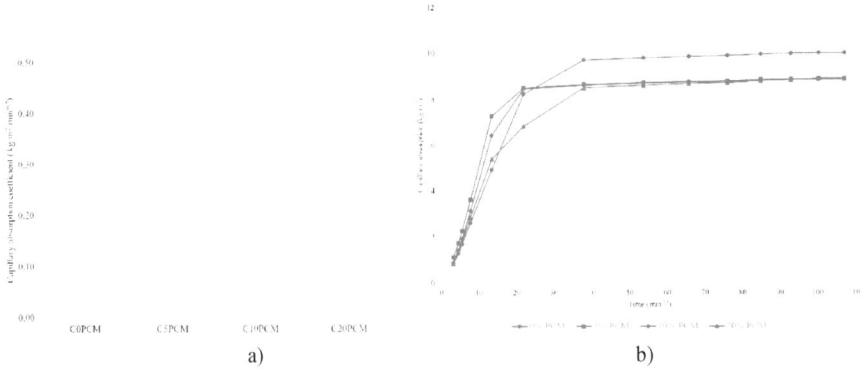

a)   b)

Fig. 5. Water absorption by capillarity: (a) Capillary absorption coefficient; (b) Water absorption by capillary for each PCM composite board.

3.6 Flexural Strength

Based on Fig. 6, a decrease in the flexural strength of the boards was observed with the incorporation of PCM. The inclusion of 5% of non-encapsulated PCM resulted in a decrease in flexural strength of over 15%. This reduction in strength is attributed to the increase in liquid content in the mortars [3, 9, 10] and higher macroporosity, which is caused by the greater ease of water evaporation due to the "board effect" explained in this study.

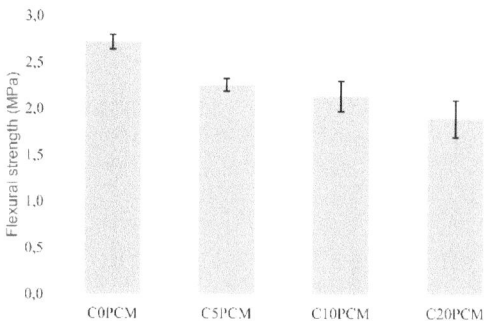

Fig. 6. Flexural strength of the developed mortars.

4 Conclusions

This study demonstrates that it is possible to use mortars with the incorporation of non-encapsulated PCM to produce boards. However, the incorporation of non-encapsulated PCM in mortars and boards alters their physical and mechanical properties. Regarding the characterization of the boards, the following aspects were observed:

- The volume of open pores, total porosity, and water absorption by immersion decrease with the incorporation of a higher content of PCM. This behavior can be justified by the geometry of the board, since the exposed surface during the curing process is larger than the surface exposed in mortar specimens, facilitating the evaporation of chemically uncombined water, a phenomenon known as the "board effect".
- The apparent and real density decreased with the incorporation of PCM, which can be attributed to the higher macroporosity of the boards caused by the presence of PCM and the "board effect".
- The capillary water absorption decreased with the incorporation of PCM due to the existence of a more compact microstructure of PCM mortars with respect to smaller pores.
- The flexural strength decreased with the incorporation of non-encapsulated PCM, which is associated with an increase in the liquid/binder ratio in mortars and the higher macroporosity caused by the "board effect" described in this study.

In summary, the behavior of boards with PCM incorporation is largely influenced by the properties of the mortars that originate them. However, in terms of mechanical behavior, the performance of the boards was not affected by their geometry, which proves that the selected dimensions constitute a possible solution for the prefabrication of construction materials obtained based on functional mortars.

Acknowledgements. This work was supported by FCT/MCTES through national funds (PID-DAC) under the R&D Unit Centre for Territory, Environment and Construction (CTAC) under reference UIDB/04047/2020.

References

1. Matos, A., Delgado, J., Guimarães, A.: Linking energy poverty with thermal building regulations and energy efficiency policies in Portugal. Energies **15**(1), e329 (2022)
2. Cunha, S., Aguiar, J., Tadeu, A.: Thermal performance and cost analysis of mortars made with PCM and different binders. Constr. Build. Mater. **122**, 637–648 (2016)
3. Cunha, S., Leite, P., Aguiar, J.: Characterization of innovative mortars with direct incorporation of phase change materials. J. Energy Storage **30**, e101439 (2020)
4. Rebelo, F., Figueiredo, A., Vicente, R., Ferreira, V.: Study of a thermally enhanced mortar incorporating phase change materials for overheating reduction in buildings. J. Energy Storage **46**, e 103876 (2022)
5. Salgueiro, T., Samagaio, A., Gonçalves, M., Figueiredo, A., Labrincha, J., Silva, L.: Incorporation of phase change materials in an expanded clay containing mortar for indoor thermal regulation of buildings. J. Energy Storage **36**, e102385 (2021)
6. Cabeza, L., Navarro. R., Pisello, A., Oliveri, L., Bartolomé, C., Sánchez, J., Álvarez., S., Tenorio, J.: Behaviour of a concrete wall containing micro-encapsulated PCM after a decade of its construction. Solar Energy **200**, 108–113 (2020)
7. Bahrar, M., Djamai, Z., Mankibi, M., Larbi, A., Salvia, M.: Numerical and experimental study on the use of microencapsulated phase change materials (PCMs) in textile reinforced concrete panels for energy storage. Sustain. Cities Soc. **41**, 455–468 (2018)
8. Saxena, R., Rakshit, D., Kaushik, S.: Phase change material (PCM) incorporated bricks for energy conservation in composite climate: a sustainable building solution. Sol. Energy **183**, 276–284 (2019)
9. Cunha, S., Silva, M., Aguiar, J.: Behavior of cementitious mortars with direct incorporation of non-encapsulated phase change material after severe temperature exposure. Constr. Build. Mater **230**, e 117011 (2020)
10. Cunha, S., Lima, M., Aguiar, J.: Influence of adding phase change materials on the physical and mechanical properties of cement mortars. Constr. Build. Mater. **127**, 1–10 (2016)
11. Portuguese Quality Institute (IPQ). NP EN 1469. Natural stone products. Slabs for classing. Requirements (2015)
12. Portuguese Quality Institute (IPQ). NP EN 1936. Natural stone test methods. Determination of real density and apparent density, and of total and open porosity (2008)
13. Portuguese Quality Institute (IPQ). NP EN 13755. Natural stone test methods. Determination of water absorption at atmospheric pressure (2008)
14. Portuguese Quality Institute (IPQ). NP EN 1925. Natural stone test methods. Determination of water absorption coefficient by capillarity (2000)
15. Portuguese Quality Institution (IPQ). NP EN 12372. Natural stone test methods. Determination of flexural strength under concentrated load (2008)

Novel Sustainable Polymer-Based Phase Change Materials (PCMs) for Mortars Based on Different Binders for the Energy Efficiency of Buildings Located in Different Climatic Regions

Antonella Sarcinella[1]($\boxtimes$), José Luís Barroso de Aguiar[2], Sandra Cunha[2], and Mariaenrica Frigione[1]

[1] Department of Innovation Engineering, University of Salento, Lecce, Italy
antonella.sarcinella@unisalento.it
[2] Department of Civil Engineering, University of Minho, Guimarães, Portugal

Abstract. A possible solution to reduce the consumption of fossil fuel and energy demand to power heating and cooling devices is represented by Phase Change Materials (PCMs). They can absorb, store and release energy according to their physical state that changes with the environmental temperature. In this work, novel eco-sustainable PCMs were developed through the form-stable method. Through this process, it was possible to create composite PCMs consisting of a natural inert matrix (i.e., a very porous stone obtained from processing waste) and an eco-friendly polymer, i.e., Poly-Ethylene Glycol (PEG). The composite PCMs were used to replace aggregates in mortars based on different binders (i.e., hydraulic lime, and cement). A complete characterization was performed on the new PCMs assessing their thermal stability and thermal efficiency. The study of the properties of the PCM-based mortars, in their fresh and hardened states, allowed to identify those with suitable mechanical properties. These latter were, then, subjected to a further investigation to assess their thermal behavior in response to different climatic loads. Encouraging results were achieved that allowed to establish the effectiveness of the novel PCMs in thermo-regulating an indoor environment.

Keywords: Phase Change Material (PCM) · Poly-Ethylene Glycol (PEG) · Mortars · Circular Economy (CE) · Energy efficiency

1 Introduction

To handle high energy demand, fossil fuel depletion and climate change, it is imperative to invest in the development of renewable energy systems as an alternative to conventional energy systems. Currently, energy storage systems are becoming increasingly well-known because their ability to improve the energy efficiency and energy savings was verified [1].

© The Author(s) 2025
L. Czarnecki et al. (Eds.): ICPIC 2023, 61, pp. 594–602, 2025.
https://doi.org/10.1007/978-3-031-72955-3_61

In this area, Phase Change Materials (PCMs) play a key role as they can store excess energy by making it available when needed. Their integration into a building, through a suitable construction material (such as mortars) is one of the possible ways to improve the building energy efficiency while ensuring a good thermal comfort for the inhabitants [2]. Furthermore, a mortar occupies a large surface area in a building, and this allows a great storage capability [3].

Starting from these premises, two sustainable PCMs were developed with the aim to integrate them into plaster mortars. Low-cost, non-flammable, non-toxic polymers, i.e., Poly-Ethylene Glycol 1000 (PEG1000) and Poly-Ethylene Glycol 800 (PEG800) were selected as active PCM phases to be included in a porous matrix. The latter, a stone obtained in the form of small pieces from processing waste, was sieved to gain granules of an appropriate grain size. These granules were, then, vacuum-impregnated with the each PEG in liquid state, following the form-stable method [4]. Two different composite PCMs (i.e., LS/PEG800 and LS/PEG1000) were obtained, that were used as aggregates in mortars based on different binders (hydraulic lime and cement). The physical properties (i.e., workability and mechanical properties) of the developed mortars were evaluated. As expected, when one of the two form-stable PCM was included into a mortar, its mechanical performance decreased if compared to the same mortar without any PCM. Nevertheless, by adjusting the quantities of the mortar components, adequate mechanical characteristics were achieved. Further analyses were carried out on the optimized mortars to evaluate their thermal performance, in order to assess whether the presence of a PEG-based PCM can assure a proper thermal indoor comfort, saving energy and costs.

2 Materials and Methods

2.1 Materials

Two polymers with a different molecular weight were selected to produce the composite PCMs: Poly-Ethylene Glycol 1000 (PEG 1000), possessing a greater molecular weight and provided by Sigma-Aldrich Company (Germany) and Poly-Ethylene Glycol 800 (PEG 800) with a lower molecular weight and supplied by Wuhan Fortuna Chemical Co. (China). The different molecular weights of PEGs correspond to a different range of phase change temperatures: these two PEGs (i.e., PEG 1000 and PEG 800) were, in fact, selected on the basis of their phase change temperatures which should ensure the manufacture of two PCMs suitable for different climates. In addition, a third PCM was produced by combining the previous PCMs, to obtain a 50% mix, in order to extend the temperature range in which the melting/solidification process takes place. The thermal properties of the two single PCMs were measured through Differential Scanning Calorimetry (DSC), and the relative thermograms are shown in Fig. 1.

PEG 800 exhibits two endothermic peaks during the melting process: one pointed at around 18 °C and a smaller one at 25 °C and, at the same time, two exothermic peaks during the solidification process: one centered at around 13 °C and the other at 9 °C. The melting/solidification enthalpy is approximately of 150 J/g. The PCM based on this polymer was, then, selected for cold climates (i.e., continental temperatures). On the other hand, PEG 1000 shows an endothermic peak nearly at 43 °C and an exothermic peak close to 23 °C, with a latent heat in both heating and cooling stages of about 129 J/g.

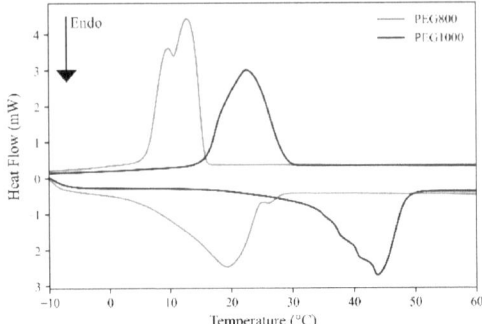

Fig. 1. Thermograms recorded through DSC on pure PEG 800 and pure PEG 1000.

The PCM manufactured with PEG 1000 was considered suitable for warm climates (i.e., Mediterranean/Tropical temperatures). The mix of the two PCMs, i.e. that based on PEG 800 and that produced starting from PEG 800, was expected to enlarge the range of phase change temperatures in order to be suitable for a wide variety of climate zones.

Lecce Stone (LS) was employed as matrix to contain the polymeric phase of the PCMs, i.e., each PEG, since it has a highly porous structure [5]. LS is a limestone very common in southern Italy and in fact it was supplied at km-0 from a local quarry. The processing waste of this stone (in the form of small flakes) were used with the aim of recycling this material in a circular economy perspective. The pieces of stone were, then, sieved to obtain appropriate size granules, i.e. between 1.6 and 2.0 mm.

Different binders were used to manufacture the mortar compositions, i.e., aerial lime, hydraulic lime, gypsum, and cement, as reported in [6–8]. However, the focus was pointed out on hydraulic lime and cement-based PCM containing mortars since they exhibited the greatest mechanical properties. A natural hydraulic-lime (HL), with a density of 2700 kg/m^3, was supplied by Cimpor (Lisbon, Portugal); a cement (C, CEM I 42.5 R), with a density of 3030 kg/m^3, was purchased by SECIL (Lisbon, Portugal). A polyacrylate (MasterGlenium SKY 627), supplied by BASF, was used as superplasticizer (SP) to keep low the water content with the aim of improving the mechanical properties of the mortars.

2.2 Preparation of the Form-Stable PCMs and Their Inclusion in Mortars

Two composite PCMs were produced starting by the selected PEGs (i.e., PEG 800 and PEG 1000) using LS as support matrix through the so-called *form-stable* method, by vacuum impregnation of the stone, as described in [5, 9]. In Fig. 2a, a PEG-based PCM is shown. The final PCMs obtained, indicated as LS/PEG800 and LS/PEG1000, were used to replace inert fine aggregates (usually sand) in the mortar formulations.

The composite PCMs were, then, thermally characterized to confirm their capacity to serve as thermal energy storage/release materials in the form of aggregates. The thermal tests were also employed to measure their melting/solidification temperature ranges and the relative latent heats. The results of DSC test performed on the LS/PEG800 and LS/PEG1000 PCMs are shown in Fig. 2b. The melting peak temperature for LS/PEG800

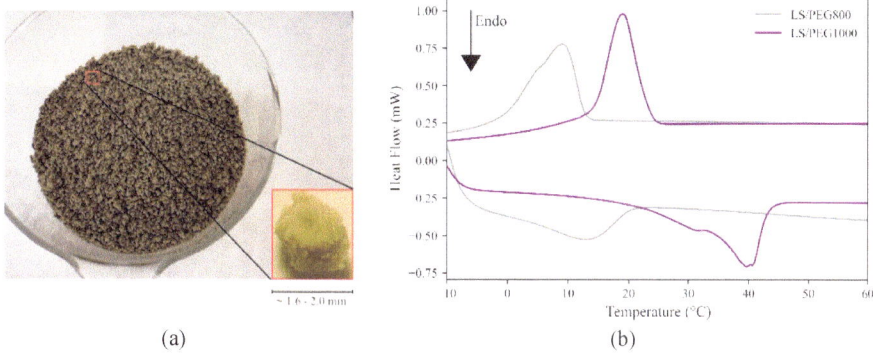

(a)   (b)

Fig. 2. (a) Composite PCMs used as aggregates for mortars. (b) Thermograms recorded through DSC on the PCMs LS/PEG800 and LS/PEG1000.

is about 13 °C and the crystallization peak temperature about 9 °C; for the LS/PEG1000 composite, the melting temperature peak is about 40 °C and the crystallization one about 19 °C. For both PCMs, the enthalpy measured in both heating and cooling stages is about 29 J/g. The results of these tests confirmed that both PEG-based PCMs possess an adequate thermal behavior. They were, then, incorporated into the mortars based on both binders. Reference mortar formulations were also manufactured using LS particles as aggregate. The compositions of the different mix designs are summarized in Table 1.

As it can be seen from Table 1, an increased binder content (from 800 kg/m^3 to 1000 kg/m^3) and a higher amount of the SP (from 15 kg/m^3 to 20 kg/m^3) were employed in the study in order to improve the mechanical properties of the final mortars. The water at saturation point was used to completely impregnate the pure LS granules, i.e., to prevent them to absorb the water required to the mortars.

2.3 Methods

According to the European standard EN 1015-3, the workability of all the developed mortars reported in Table 1 was evaluated to determine their flowability in their fresh state, using the flow-table test. According to European standard EN 1015-11, the mechanical characteristics of the developed mortars were studied (in both flexural and compressive mode) employing a Lloyd dynamometer equipment. The mechanical tests were performed at a speed of 6 μm/s in flexural configuration and at a speed of 12 μm/s in compressive one, respectively. The results obtained on three specimens (at least) for each mortar were, finally, averaged.

3 Results and Discussion

The results of the tests performed on all the produced mortars (i.e., workability and mechanical characterizations) are presented in Table 2. In the same Table, the classification of each mortar, according to the standard NP EN 998-1, is also reported.

Table 1. Mortar formulations (indicated as kg/m^3)

| Sample | Binder Content | Aggregates | | | SP | Water Saturation | Water | Water/Binder |
		LS	LS/PEG800	LS/PEG1000				
HL$_{800}$_LS	800	1092	0	0	15	275	320	0.40
HL$_{800}$_LS/PEG1000	800	1729	0	398	15	0	375	0.47
HL$_{1000}$_LS	1000	682	0	0	20	171	380	0.38
HL$_{1000}$_LS/PEG800	1000	1082	249	0	20	0	320	0.32
HL$_{1000}$_LS/PEG800_LS/PEG1000	1000	1082	124	124	20	0	320	0.32
C$_{800}$_LS	800	1070	0	0	15	269	296	0.37
C$_{800}$_LS/PEG1000	800	1347	0	310	15	0	360	0.45
C$_{1000}$_LS	1000	772	0	0	20	194	390	0.39
C$_{1000}$_LS/PEG800	1000	1307	301	0	20	0	300	0.30
C$_{1000}$_LS/PEG800_LS/PEG1000	1000	1307	150	150	20	0	300	0.30

Table 2. Physical properties of the mortar formulations in fresh state (to determine the workability) and hardened state (to assess the mechanical properties).

Sample	Workability (mm)	Flexural Strength (MPa)	Compressive Strength (MPa)	Mortar Classification
HL$_{800}$_LS	165 ± 2.0	2.8 ± 0.5	17.0 ± 0.2	CSIV
HL$_{800}$_LS/PEG1000	175 ± 2.0	0.4 ± 0.1	1.5 ± 0.1	CSI – CSII
HL$_{1000}$_LS	175 ± 1.0	5.2 ± 1.5	11.7 ± 0.5	CSIV
HL$_{1000}$_LS/PEG800	170 ± 3.0	2.3 ± 0.4	3.5 ± 0.2	CSII – CSIII
HL$_{1000}$_LS/PEG800_LS/PEG1000	163 ± 2.0	2.1 ± 0.5	3.6 ± 0.5	CSII – CSIII
C$_{800}$_LS	160 ± 1.0	9.2 ± 0.9	26.3 ± 0.4	CSIV
C$_{800}$_LS/PEG1000	178 ± 3.0	1.9 ± 0.3	3.4 ± 0.8	CSII
C$_{1000}$_LS	180 ± 0.5	11.8 ± 1.1	65.6 ± 6.1	CSIV
C$_{1000}$_LS/PEG800	170 ± 1.0	2.1 ± 0.1	3.9 ± 1.2	CSII – CSIII
C$_{1000}$_LS/PEG800_LS/PEG1000	170 ± 4.0	2.0 ± 0.2	4.4 ± 0.7	CSII – CSIII

According to the results shown in Table 2, all the mortars presented excellent workability values, i.e. falling within the established range (160-180mm). Analyzing the results of the mechanical tests, as expected, the presence of a PCM caused reductions in the mechanical properties of mortars [10], irrespective to the type of PCM included in that mortar. However, it was possible to achieve suitable values of compressive strength when the both the binder (from 800 kg/m^3 to 1000 kg/m^3) and the superplasticizer (from 15 kg/m^3 to 20 kg/m^3) contents were increased. The developed mortars were, in fact, classified as CSII, CSIII, and CSIV, according to the EN 998-1 standard. These values resulted appropriate for the intended applications, i.e., the development of plaster mortars. The mortars were also subjected to DSC analysis to assess the presence of any PCM and its thermal characteristics. All the mortars containing a PEG-based PCM were found to be suitable from a thermal point of view for the intended climatic zone, in terms of melting/crystallization range of temperatures, as reported in [11]. The investigation moved, then, to the analysis of the thermal behavior of the mortars when subjected to a controlled temperature program. An experimental setup able to study the thermal behavior of the mortars applied on the wall of a small box in the four seasons of the year was realized, as illustrated in Fig. 3. The simulation of the four seasons took place in a climatic chamber, selecting two climatic conditions: those characteristics of Mediterranean regions and those typical of Continental area. Each mortar composition, reported in Table 1, was applied (thickness of 1cm) on the inner walls of a scaled test cell (20x20x20cm^3) composed by an insulating material (i.e., expanded polystyrene, XPS). A thermocouple (Type K) was placed in the center of the cell to measure the internal temperature during the experiment.

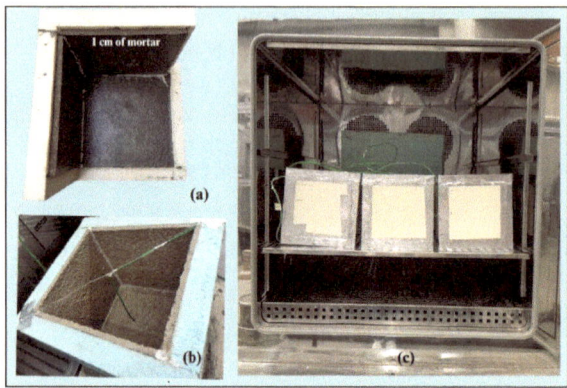

Fig. 3. Experimental setup employed to evaluate thermal performances: (a) section of a small-scale box with 1 cm-thick of mortar; (b) thermocouple centered inside the small-scale box; (c) climatic chamber where the small-scale boxes were placed to conduct the test.

The results obtained in the experiments performed on some of the PCM-based mortars, i.e., those containing PEG 1000, have been reported in [12]: they confirmed that the presence of the LS/PEG1000 composite PCM is able to positively influence the thermal behavior of the mortars. It was observed, in fact, that the maximum/minimum peaks were reduced/increased (especially in warmer seasons), with a consequent decrease in the energy needed for the indoor cooling/heating. The same tests have been recently carried out on mortars containing LS/PEG800 or LS/PEG800_LS/PEG1000 PCMS: the relative results has been just published in [13].

4   Conclusions

When a PCM is selected, it is extremely important to know its characteristic phase change temperatures (i.e. melting/crystallization) and latent heat capacity. In this work, two polymers with different molecular weights were selected to produce two PCMs: PEG1000 and PEG800. These PEGs were chosen for their sustainability characteristics and for their melting and crystallization range if temperatures considered appropriate for the intended applications, i.e. to produce indoor plaster mortars suitable for climatic conditions typical of Mediterranean regions and Continental areas. The polymers were included in a porous inert matrix (i.e., LS) through a vacuum impregnation process, obtaining as a result two composite PCMs, i.e. LS/PEG800 and LS/PEG1000. The PCMs were, then, used as aggregates in different mortars, i.e. based on hydraulic lime or cement. The addition of the PEG-based PCMs in mortars affected their physical properties in both fresh and solid states. However, with a suitable selection of the mortar composition, it was possible to achieve an appropriate workability and adequate mechanical properties. On the mortars that achieved the best mechanical performance the thermal performance was evaluated in order to assess their attitude to reduce the need for cooling and heating, from a building energy efficiency perspective.

Acknowledgements. This research was funded by PON Ricerca e Innovazione 2014–2020 Risorse React Eu - DM 1062/2021 Azione IV.4, "Dottorati e contratti di ricerca su tematiche dell'innovazione" and Azione IV.6, "Contratti di ricerca su tematiche green".

References

1. Xu, J., Zhang, X., Zou, L.: A review: progress and perspectives of research on the functionalities of phase change materials. J. Energy Storage **54**, 105341 (2022). https://doi.org/10.1016/j.est.2022.105341

2. Aridi, R., Yehya, A.: Review on the Sustainability of Phase-Change Materials Used in Buildings. Energy Convers. Manag.: X. **15**, 100237 (2022). https://doi.org/10.1016/j.ecmx.2022.100237

3. Frigione, M., Lettieri, M., Sarcinella, A.: Phase change materials for energy efficiency in buildings and their use in mortars. Materials **12**, 1260 (2019). https://doi.org/10.3390/ma12081260

4. Chinnasamy, V., Heo, J., Jung, S., Lee, H., Cho, H.: Shape stabilized phase change materials based on different support structures for thermal energy storage applications–a review. Energy **262**, 125463 (2023). https://doi.org/10.1016/j.energy.2022.125463

5. Frigione, M., Lettieri, M., Sarcinella, A., Barroso de Aguiar, J.: Sustainable polymer-based phase change materials for energy efficiency in buildings and their application in aerial lime mortars. Constr. Build. Mater. **231**, 117149 (2020)https://doi.org/10.1016/j.conbuildmat.2019.117149

6. Frigione, M., Lettieri, M., Sarcinella, A., Barroso de Aguiar, J.L.: Applications of sustainable polymer-based phase change materials in mortars composed by different binders. Materials **12**, 3502 (2019). https://doi.org/10.3390/ma12213502

7. Sarcinella, A., de Aguiar, J.L.B., Frigione, M.: Physical properties of eco-sustainable form-stable phase change materials included in mortars suitable for buildings located in different continental regions. Materials **15**, 2497 (2022). https://doi.org/10.3390/ma15072497

8. Frigione, M., Sarcinella, A., Barroso de Aguiar, J.L.: Development and performance of eco-sustainable form-stable Phase Change Materials (PCMs) for mortars to be applied in buildings located in different climatic areas. Coatings **13**, 258 (2023). https://doi.org/10.3390/coatings13020258

9. Sarcinella, A., Aguiar, J'.L.B. de, Frigione, M.: Physical properties of an eco-sustainable, form-stable phase change material included in aerial-lime-based mortar intended for different climates. Materials **15**, 1192, (2022). https://doi.org/10.3390/ma15031192

10. Li, M., Shi, J.: Review on micropore grade inorganic porous medium based form stable composite phase change materials: preparation, performance improvement and effects on the properties of cement mortar. Constr. Build. Mater. **194**, 287–310 (2019). https://doi.org/10.1016/j.conbuildmat.2018.10.222

11. Sarcinella, A., de Aguiar, J.L.B., Frigione, M.: Use of sustainable Phase Change Material (PCM) in mortars for building energy efficiency. J. Phys.: Conf. Ser. **2385**, 012009, (2022). https://doi.org/10.1088/1742-6596/2385/1/012009

12. Sarcinella, A., De Aguiar, J.L.B., Lettieri, M., Cunha, S., Frigione, M.: Thermal performance of mortars based on different binders and containing a novel sustainable Phase Change Material (PCM). Materials **13**, 2055 (2020). https://doi.org/10.3390/ma13092055

13. Sarcinella, A., Barroso de Aguiar, J.L., Jesus, C., Frigione, M.: Thermal properties of PEG-based form-stable Phase Change Materials (PCMs) incorporated in mortars for energy efficiency of buildings. J. Energy Storage **67**, 107545 (2023). https://doi.org/10.1016/j.est.2023.107545

The Effects of Modified Andreassen Particle-Packing Model on Polymer Modified Self-Leveling Heavy-Weight Mortar

A. Orhun Erüz[1,1(✉)], M. Hulusi Özkul[2], Özlem Akalın[3], and Muhammed Maraşlı[4]

[1] Building Materials and Construction Chemistry, TU Berlin, Berlin, Germany
orhun.eruz@hotmail.com
[2] Department of Civil Engineering, Beykent University, Istanbul, Turkey
[3] Research and Development Center, PLUSTECHNO, Istanbul, Turkey
[4] Research and Development Center, Fibrobeton, Istanbul, Turkey

Abstract. Heavy-weight concretes are known for their high unit weight inherited from the aggregates and developed mainly for radiation shielding. Therefore, minimal porosity besides the high unit weight is a desired property. Numerous particle-packing theories were put forward to decrease the porosity by an ideal reference curve; modified Andreassen model based on the size distribution of ingredients to adjust the fineness. This research investigates the effect of the mentioned method on heavy-weight mortars. Cement, micro and nano silica combinations, and their polymer additive mixtures were used as binders in the specimens, along with barite and finely ground magnetite aggregates. In this work, the aggregate size limit selected 3 mm and the fineness factor of q was chosen as 0.22 and 0.25, depending on the mixture. To achieve a self-levelling consistency, the w/c ratio was kept constant at 0.40, and a superplasticizer was added to maintain the workability. Consequently, the specimens were examined for unit weight, compressive strength, and capillary water absorption. The collected results were analyzed, and the difference between groups was compared according to their composition.

Keywords: Heavy-weight Mortar · Polymer-modified Heavy-weight Mortar · Particle-Packing Theory

1 Introduction

Concrete is the most produced material due to its ability to be tailored to meet specific needs. Heavy-weight concretes have a unit weight above 2600 kg/m3 in the dry state achieved by using heavy-weight aggregates. Such heavy particles can be from natural or artificial sources. Examples of natural aggregates are barite, hematite and magnetite [1]. Heavy-weight concretes were initially introduced as a safer measurement for constructions that tended to slide, such as retaining walls. However, with nuclear technology's emergence, heavy-weight concrete production shifted to radiation shielding [2].

Particle packing models aim to fit the highest possible number of solid particles within a unit volume. The increase of the solid ratio leads to a lower ratio of the water and

L. Czarnecki et al. (Eds.): ICPIC 2023, 61, pp. 603–610, 2025.
https://doi.org/10.1007/978-3-031-72955-3_62

trapped air. Thus, if desired, the approach might lead to reduced water-to-solid and water-to-cement ratios. This impacts the various properties, including porosity, volumetric stability, and low segregation. In the early 20th century, researchers Fuller and Thomson came up with the idea of using an optimization curve for particle packing. Their curve is essentially an equation that factors in the sizes of particles used and can be tuned based on fineness with an adjustable distribution exponent (q).

In the 1980s Japan, a durable and underwater castable concrete, Self-Leveling Concrete (SLC), was developed [3]. Even though the SLC has similar proportioning to conventional concrete, one of its unique properties is self-levelling under its own weight without external force and segregation. However, the coarse aggregate content of SLC is limited compared to traditional concrete. Furthermore, powders not only consist of cement but also mineral admixtures or filler materials, which are used to increase the segregation resistance of concrete. To reduce the water content and increase the segregation resistance, implementing superplasticizers is crucial. The ability of superplasticizers to adsorb on cement grains and repel the solid particles further apart by steric hindrance provides a higher flowability in the fresh state.

Polymer Modified Concrete (PMC) contains polymer along with cementitious particles. Depending on the polymer type, the increase in strength can continue for several years due to the continuation of curing. The PMC is preferable for its high flexural and compressive strength and high durability properties. Even though polymer is added to the cement mixture in slight amounts due to partially economic reasons, the properties are enhanced to the desired level [4]. Furthermore, the polymer in a modified cement system might lead to a decrease in permeability due to the inter-bounding of polymer particles.

The aim of this study is to evaluate the influence of various factors, including heavy-weight aggregates, micro silica, nano silica, polymer admixture and particle size distribution (PSD) of these materials in self-levelling heavy-weight mortar with unit weight over 2600kg/m^3 and flowability diameter of 175 mm. As one of the main uses of this type of mortar is shielding against radiation, minimal porosity and homogeneity are essential aims. To realize these with the segregating nature of fine ground heavy-weight aggregates, modified Andreassen model is used to modify the PSD of ingredients for homogeneity, whereas self-levelling concrete technology is used to increase the segregation resistance of the mortar. The tests are carried out in both fresh and hardened states according to the standards to obtain results.

1.1 Modified Andreassen Model

The Modified Andreassen model is an optimization design that succeeds the Fuller and Andreassen models [5]. The research on the properties of concrete by determining the role of the size distribution of the aggregates is also known as particle packing models (PPM). The pursuit for optimal packing started in 1907 with Thompson and Fuller [6]. Later, the first PPM based on the void ratio of binary mixes was developed by Furnas in 1929. The initial models lacked a calculation procedure that required the determination of the theoretical packing density of a given combination of particles by the PSD of each class. Moreover, the models were insufficient to include some factors such as shape factor, interparticle forces, water demand and strength. Due to the unincorporated factors, the empirical model does not deliver the most optimized proportioning alone and requires data based on trial and error.

The particle density can be defined as the solid volume in the total volume of the mixture. Calculating by porosity is also possible in this situation in Eq. (1) or by the void ratio in Eq. (2):

$$\pi = 1 - \Phi \tag{1}$$

$$\Phi = 1 - \varepsilon \tag{2}$$

where π represents porosity, Φ packing density and ε void ratio.

The modified Andreassen model is designed to bring limitations in both the lower size and distribution factor of q. The parameter q varies between 0.21 and 0.36 for optimal packing. As Eq. (3) suggests, any change in the q affects the fineness and, consequently, the flowability and workability properties of the mixture. According to Emma software guidebook [8], a q value lower than 0.25 and 0.28 are beneficial for the flowability of mortar and concrete, respectively.

$$P(d) = \frac{d^q - d_{min}^q}{d_{max}^q - d_{min}^q} \tag{3}$$

In Eq. (3), d represents particle diameter considered in mm, P(d) size cumulative distribution function of considered d diameter, d_{max} coarsest particle diameter in the mixture in mm, d_{min} finest particle in the mixture in mm and the q as a parameter.

2 Experimental Studies

2.1 Materials

Heavy-weight aggregates of barite and magnetite which were selected smaller than 3 mm, fine ground CEM II 42,5R ordinary Portland cement (OPC), micro and nano silicas, water, and HR50 superplasticizer and a polymer commercially known as Bettolatex, which is a vinyl-acrylic copolymer dispersion-based polymer [7] admixtures were used in the research. In order to achieve SLC-type mortar, a distribution exponent of q in the modified Andreassen between 0.21 and 0.25 is suitable [8, 9]. Micro and nano silicas were added to mixtures to reduce the porosity and increase the solidity, as all the solid particles were considered in the PSD. The superplasticizer to obtain an SLC type is undeniable and was implemented to reduce water demand, enhance flowability and increase the mortar's strength and unit weight. Also, polymer admixture was used in the specimens to minimize water absorption of the mortar.

2.2 Mixtures of Mortars

3 groups consisting of 6 subgroups were prepared for the study. The materials in Table 1 were used to reach close proximity to optimum curves, as shown in Fig. 1 and 2. These groups were named according to their group and subgroup. The subgroups PS and PP

Table 1. Materials, PSDs, and unit weights

Class ID	Size Class (micron)	Density, ρ (g/cm^3)
Magnetite	1–600	4.9
Barite	1–3000 (8 different size groups)	4.2
Micro silica	1–30	1.3
Nano silica	0,02–0,24	2.21
Cement	1–40	3.14

were the reference specimens which contained only superplasticizer but no pozzolan admixture, with PP having polymer admixture. This pattern continues in subgroups MS and MP, which stand for micro silica and subgroups NS and NP, which were micro-silica and nano-silica combinations. All mixtures have 0.05 superplasticizer to cement ratio. Figure 1 is the optimal curve of P and M groups; meanwhile, Fig. 2 is for the N group due to the involvement of nanoparticles (Table 2).

Table 2. Volumetric usage percentage of materials in each group

Material	PS	PP	MS	MP	NS	NP
Cement	17,99	17,88	18,20	18,08	17,95	17,84
Magnetite	7,33	7,29	7,42	7,37	5,23	5,20
Barite	49,51	49,22	48,24	47,92	50,62	50,32
Water	22,45	21,48	22,26	21,20	20,42	19,50
Micro silica	-	-	1,18	1,17	1,16	1,15
Nano silica	-	-	-	-	1,97	1,96
Superplasticizer	2,72	2,71	2,70	2,68	2,66	2,65
Polymer	-	1,42	-	1,58	-	1,39

3   Tests and Results

3.1   Unit Weight Test in Fresh and Hardened States

The unit weight was calculated by division of the weight of the mortar in a premeasured container in fresh state. In hardened state, it is calculated by division of the weight of samples which were 28 days cured and oven dried for 40 h at 70 °C by their measured volume. The test results are illustrated in Fig. 3.

3.2   Slump Flow Tests

The slump flow test is more appropriate for the SCC rather than the slump test due to the effectiveness of superplasticizers to cause much higher flowability. The flow tests were

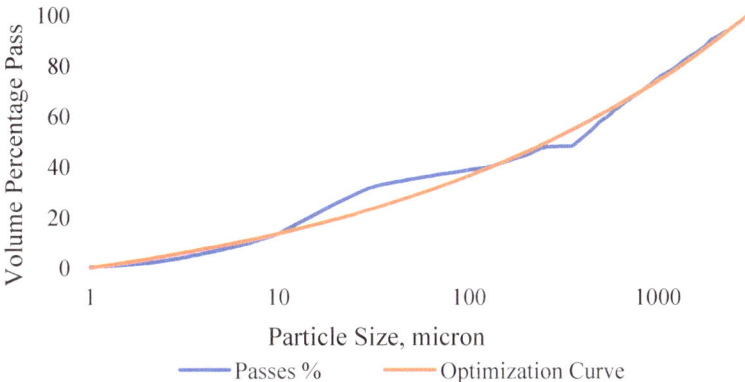

Fig. 1. Optimization Curve for PS, PP, MS, and MP q = 0.22 with 1 micron minimum and 3000 microns maximum

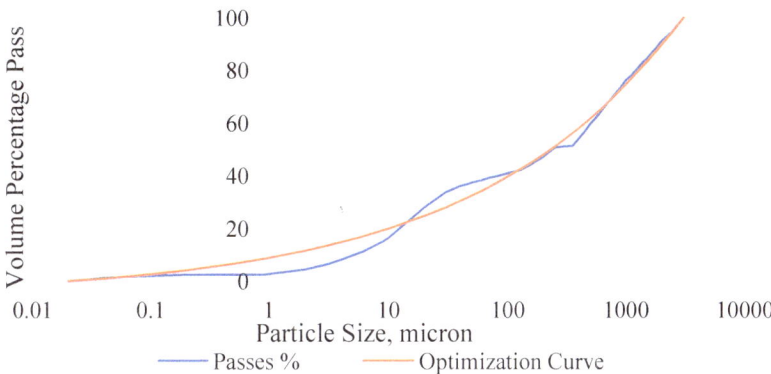

Fig. 2. Optimization curve for NS and NP q = 0.25 with 0,01 micron minimum and 3000 microns maximum

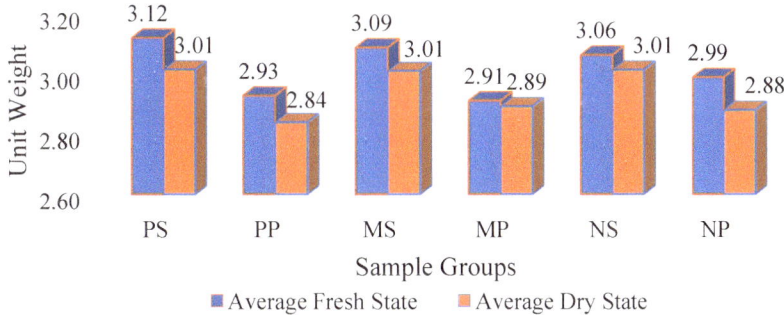

Fig. 3. Unit Weights of Mortars in Fresh and Hardened States

conducted according to the guidance of the TS EN 1170–1[10]. The results are shown in Table 3.

Table 3. Flow values of fresh mortar of each groups

Sample Group	Average (mm)	Standard Deviation
PS	192.5	4.7
PP	185	2.4
MS	195	1.2
MP	172.5	4.7
NS	190	2.4
NP	172.5	2.4

3.3 Compressive Strength Tests

4x4x4 cm3 volume of the samples were subjected to compressive strength examination. Mean values of the test results were calculated and demonstrated in Table 4.

Table 4. Compressive strength values of hardened mortar of each groups

Sample Group	Average Compressive Strength (MPa)	Standard Deviation
PS	56.3	1.1
PP	43.6	0.6
MS	65.7	1.4
MP	50.2	0.9
NS	60.3	5.5
NP	50.5	0.4

3.4 Test for Water Absorption by Capillary

The oven dried and cooled samples were coated on the sides with paraffin to avoid any other water absorption excluding the bottom surface. The samples were placed in the water and the weight measured at predetermined times and absorption was calculated according to ASTM C1585 [11]. The results are shown in the Fig. 4.

4 Conclusions

The aim of this study is to evaluate the influence of various factors in self-levelling heavy-weight mortar.

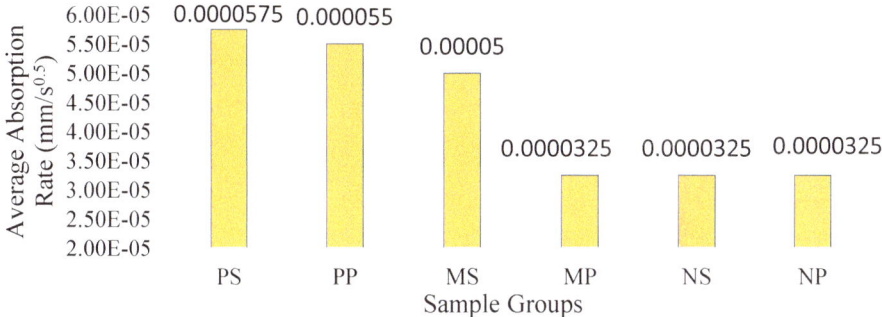

Fig. 4. Water Absorption Rates in mm/sn$^{0.5}$

By standards, the classification of heavy weight is any mortar heavier than 2600 kg/m3 in oven dried state. Figure 3 demonstrates that the entirety of the sample groups was above this limit. Among the sample groups, polymer modified groups were shown lower unit weight. This could be linked to the volume that was filled with lighter polymer particles and probable air entraining properties of the polymer admixture.

The flowability of the mixtures showed that two thirds of the samples surpassed the target of 175 mm and only MP and NP groups were below. As a general trend among the polymer modified groups to have lower flowability, it can be deducted that vinyl acrylic copolymer inhibits the property.

Compressive strength values were the highest at MS and NS groups with 65.7 and 60.3 MPa, respectively, which utilize micro silica and a combination of micro silica and nano silica mixtures likely due to pozzolanic reactions and filler effect of the materials. However, polymer admixture commonly reduces the lower compressive strengths compared to ones without polymer admixture. Another study shows results where the acrylic dispersion polymer admixture increases the compressive strength [12]. Nonetheless, these results might stem from the difference of type between polymer admixtures.

Figure 4 shows that the water absorption rates decreased for both polymer and mineral additions. In the case of micro silica and nano silica addition, the absorptions decreased significantly with the latter being more effective. The fineness or structures of pozzolans might have the influence behind the superior impermeability of the mixtures. As it is intended, the polymer admixture diminished the absorption rates in PP and MP groups, whereas did not affect NP group any further. Provided that the polymeric bound network effectively fills capillary voids, the uniformity in nano silica incorporated groups indicates that the capillary voids were sealed off by reaction outputs already and did not require any further polymer supplementation. In general, polymer subgroups mostly lowered the water absorption, except for N group which shows no significant difference.

It is planned in future to carry out research on self-compacting heavy-weight mortar with further mechanical tests and various radiation permeability tests.

References

1. ASTM C 637, "Standard specification for aggregates for radiation-shielding concrete," ASTM Int., vol. 4, no. 100, pp. 1–4 (2015). https://doi.org/10.1520/C0637-20.2
2. Esen, Y., Yilmazer, B.: Investigation of some physical and mechanical properties of concrete produced with barite aggregate (2010)
3. Okamura, H., Ouchi, M.: Self-compacting concrete, J. Adv. Concr. Technol. **1**(1), April 2003," pp. 5–15 (2003)
4. Martínez-Barrera, G., Vigueras-Santiago, E., Gencel, O., Hagg Lobland, H.E.: Polymer concretes: A description and methods for modification and improvement. J. Mater. Educ. **33**(1–2), 37–52 (2011)
5. Andersen, P., Johansen, V.: Particle packing and concrete properties. Mater. Sci. Concr. II Skaln. J Mindess S (Edited),The Am. Ceram. Soc. Inc.,Westerville, Ohio, pp. 111–147, 1991
6. Fuller, W.B., Thompson, S.E.: THE LAWS OF PROPORTIONING CONCRETE (1907)
7. Betton Yapı Kimyasalları, "Bettolatex." https://www.betton.com.tr/tr/bettolatex (accessed Jun. 14, 2023)
8. "User Guide – EMMA," Elkem Materials,, [Online]. Available: www.elkem.com/en/Silicon-materials/Support/Software-EMMA/ Software EMMA
9. Snehal, K., Das, B.B.: Application of andreassen and modified andreassen model on cementitious mixture design: A review. Recent Developments in Sustainable Infrastructure Select Proceedings of ICRDSI **2019**, 729–750 (2019)
10. TS EN 1170 - 1, "Precast concrete products-test method for glass-fibre reinforced cement-part 1: measuring the consistency of the matrix 'sulump test' method," Turkish Standards Institution, Ankara (1999)
11. ASTM C1585–20, Standard test method for measurement of rate of absorption of water by hydraulic cement concretes. ASTM Int., **41**(147), 1–6 (2013). https://doi.org/10.1520/C1585-20.2
12. Zalegowski, K., Piotrowski, T., Garbacz, A.: Influence of polymer modification on the microstructure of shielding concrete. Materials **13**(3), 498. MDPI AG. (2020). Retrieved from https://doi.org/10.3390/ma13030498

Neutron Radiation Shielding of C-PC and Geopolymers

Tomasz Piotrowski[✉], Piotr Prochoń, and Magdalena Wojtkowska

Faculty of Civil Engineering, Warsaw University of Technology, Warsaw, Poland
tomasz.piotrowski@pw.edu.pl

Abstract. The aim of the paper is to evaluate neutron shielding efficiency of ordinary concrete, heavy-weight concrete and geopolymer concrete modified with epoxy additive. Evaluation was based on neutron shielding efficiency calculations. Since the commonly used fast neutron effective removal cross-section calculation does not take into account thermal neutron absorption reactions, a method adopted from the fast neutron effective removal cross-section calculation based on the macroscopic cross-section calculation for compositions is therefore applied. The results confirmed that an efficient neutron shield requires a balanced mixture of light and heavy nuclei and polymer modification is a proper way for increasing neutron shielding efficiency.

Keywords: Shielding · Neutron Radiation · Concrete · Epoxy · Geopolymers

1 Introduction

Concrete is one of the most common material used for radiation shielding purposes. It involves using the material's density and atomic composition to protect both against gamma and neutrons and reduce their harmful effects on individuals and equipment in the vicinity of a radiation source. While the protection against the high energy electromagnetic gamma rays is rather simple, the shielding against neutrons is more complex. For gamma rays density is the main factor influencing the effectiveness of the shielding. For neutron shielding, density is of marginal importance due to the different nature of the neutrons. When neutrons are emitted from a radiation source, as they have no charge, they can travel through until they interact with the nuclei of a target atom. The absorption of fast, high energy neutrons, is practically impossible, so to stop a fast neutron, first its energy has to be decreased to the level of a thermal neutron.

Following the laws of mechanics if a neutron of a limited mass hits a large nucleus, it will lose a small part of its energy. But if it collides with a nuclei whose mass is close to the mass of a nucleus, the energy loss will be large. That is why the best moderators are light elements in concrete like hydrogen which is present in the chemically bounded water [1] and oxygen that is from aggregate and cement used to make concrete. The other elements contributed by aggregate have relatively smaller effectiveness but due to big total mass can create an important share in total neutron attenuation efficiency. The second step of neutron shielding is to absorb thermal neutrons. It is independent from

© The Author(s) 2025
L. Czarnecki et al. (Eds.): ICPIC 2023, 61, pp. 611–620, 2025.
https://doi.org/10.1007/978-3-031-72955-3_63

the atomic number of the target nuclei or any other simple relation. The best absorbers are neither the light elements (moderators) nor the heavy atoms which are efficient in gamma radiation shielding. The best neutron absorbers in concrete are chloride (Cl) and iron (Fe). Unfortunately, chloride as a component of reinforced concrete is not recommended as it can induce the corrosion of the reinforcing steel. Iron is not ideal either, as thermal neutron absorption in iron results in the emission of high-energy secondary gamma radiation and can cause significant activation of the concrete. For this reason, it is beneficial to use other elements such as gadolinium (Gd), cadmium (Cd), boron (B) or cobalt (Co).

Recently a number of researches have been performed for geopolymers [2]. Geopolymers are a type of inorganic polymer material that can be synthesized from various industrial waste materials such as fly ash, slag, and clay. They are formed by the reaction of these waste materials with alkaline activators such as sodium silicate and sodium hydroxide. Geopolymers have a unique chemical structure that gives them properties such as high strength, low porosity, and fire resistance, making them a promising alternative to traditional cement and concrete. The chemical structure of geopolymers consists of a three-dimensional network of tetrahedral and octahedral units, which are linked together by covalent bonds. This structure is similar to that of natural minerals such as zeolites and feldspars. The use of industrial waste materials to produce geopolymers not only provides a sustainable solution for waste management but also reduces the environmental impact of traditional cement and concrete production. Geopolymers have a wide range of applications, including as a binder in construction materials, such as concrete, mortar, and tiles, as well as in the production of refractory materials, composites, and coatings. They also have potential for use in waste immobilization, catalysis, and energy storage.

The aim of the paper is to evaluate the neutron shielding efficiency of concretes and geopolymers modified by epoxy resin. It has been assumed that polymer is added in 5, 10, 15 and 20% in relation to cement or geopolymer precursors mass [3]. Such content of epoxy can create a separate phase in composite so they can be named C-PC (Cement-Polymer Composite) and G-PC (Geopolymer-Polymer Composite). The evaluation is based on neutron shielding efficiency calculation of fast neutron effective removal cross-section and additionally on the macroscopic cross-section calculation for compositions - Evaluation was based on neutron shielding efficiency calculations – a method adopted from the fast neutron effective removal cross-section calculation allowing for evaluation of fast neutron attenuation and thermal neutron absorption reactions separately.

2 Materials and Methods

The calculation has been performed on ordinary concrete with dolomite aggregate (density 2520 kg/m^3), heavy-weight concrete with magnetite aggregate (density 4090 kg/m^3) and geopolymer concrete with silica based aggregate (density 2000 kg/m^3). Cement for concrete was CEM III/B 42,5 N-LH/SR and its oxide composition was obtained from XRF analysis. Dolomite aggregate was 87% pure dolomite with 10% of calcite and 3% of iron sulfide. Magnetite aggregate contained more that 90% of Fe$_3$O$_4$ with minor content of Silica, Calcium and Aluminum oxides. Geopolymer precursors were fly ash, zeolite, metakaolin and calcium hydroxide. The secondary components were limestone powder, cenospheres, perlite powder. The silica based aggregate and silica glass with NaOH solution in water has been used in geopolymers as well.

The neutron shielding efficiency of compounds can be compared based on an equivalent absorption cross-section called a *fast neutron effective removal cross-section*, Σ_R [4, 5]. It is a linear attenuation coefficient given in cm^{-1} and is defined as a probability that a fast energy neutron undergoes on the first collision, which removes it from the group of penetrating, uncollided neutrons. The concept of this phenomena is based on the presence of hydrogen as it is the main moderator that dominates the attenuation of neutrons. Calculation of *fast neutron effective removal cross-sections* is by analogy to the calculation of mass attenuation coefficients of gamma-ray according to the equation:

$$\Sigma_R = \sum_i W_i \cdot (\Sigma_R/\rho)_i \tag{1}$$

where: W_i – partial density of ith constituent, $(\Sigma_R/\rho)_i$ – fast neutron mass removal coefficient of ith constituent

The *fast neutron mass removal cross-section* of constituents is related to the microscopic nuclear properties and varies smoothly with the atomic weight. The value can be calculated using the empirical equations or measured. For most elements and some compounds, experimental and theoretical values of the fast neutron mass removal cross-sections have been published [6–8]. In order to estimate the neutron shielding efficiency of concrete in a more detailed way and do not describe not fast neutron attenuation only, a method based on macroscopic cross-sections for a different interaction can be used [9]. In this method, instead of fast neutron attenuation cross-sections, it uses a database of neutron scattering lengths and cross-sections that includes the thermal neutron microscopic cross-section as well [10]. Thus the *macroscopic neutron scattering cross-section* Σ_s, or the *macroscopic thermal neutron absorption cross-section* Σ_a, and finally their sum named the *total macroscopic neutron cross-section* Σ_T, can be calculated using equation:

$$\Sigma_j = \sum_i W_i \cdot \left(\Sigma_j/\rho\right)_i \tag{2}$$

where: W_i – partial density of ith constituent, $\left(\Sigma_j/\rho\right)_i$ – neutron mass attenuation coefficient of ith constituent for a specific interaction (j).

The key issue for shielding efficiency evaluation is atomic composition determination. It is presented in elemental weight fractions and partial densities. For concretes it was assumed that only 20% of the water is chemically bound water by cement hydration. Later weight fractions for specific elements of the concrete and the partial densities from oxide composition of the constituents were calculated. In case of geopolymers the calculations were simpler as all constituents with full mass have been taken into account. Weight fractions of concretes and geopolymer has been modified by addition in 5, 10, 15 and 20% of by epoxy resin composed of Carbon (76%), Oxygen (17%) and Hydrogen (7%) (Tables 1 and 2).

Dominating elements in concrete are mainly the income form aggregates that contribution in total unit mass is more than 80%. For ordinary concrete (OC) they were calcite

Table 1. Basic weight fraction of concretes

Element	H	C	O	Na	Mg	Al	Si	P	S	Cl	K	Ca	Ti	Fe
Ordinary concrete	0,0033	0,1031	0,4958	0,0000	0,0989	0,0075	0,0199	0,0000	0,0107	0,0002	0,0006	0,2421	0,0005	0,0174
Magnetite concrete	0,0020	0,0000	0,2803	0,0018	0,0028	0,0070	0,0259	0,0045	0,0013	0,0001	0,0017	0,0474	0,0003	0,6248

Table 2. Basic weight fraction of geopolymer

Element	H	C	O	Na	Mg	Al	Si	P	S	Cl	K	Ca	Ti	Fe	Cr	Mn
geopolymer mortar	0,0262	0,0037	0,5749	0,0440	0,0047	0,0493	0,2366	0,0005	0,0000	0,0000	0,0086	0,0363	0,0023	0,0126	0,0001	0,0002

(CaCO₃) and Mg/Ca oxides from dolomite aggregate and for magnetite concrete (MC) it was Fe oxide. In case of geopolymer (GP) due to presence of sand Silica and oxide content is the highest. It can be also found that hydrogen content in geopolymer mortar is f one order of magnitude bigger than in cement based concretes. These remarks from atomic composition can be directly transferred to analysis of contribution in neutron shielding efficiency parameters (Figs. 1, 2, and 3). The highest contribution to fast neutron mass removal cross-section of all composite's results from aggregates (Fig. 1). It can be also observed much bigger contribution of hydrogen in geopolymer (GP) in comparison to both concretes (OC and MC). This effect is even more visible on macroscopic neutron scattering cross-section values (Fig. 2). It proves that the hydrogen in concrete is the most important component regarding fast neutron shielding efficiency. The next in importance are oxide (O) due to big contribution in total mass and iron (Fe) that is present in magnetite concrete. Iron is also the key factor contributing thermal neutron absorption cross-section (Fig. 3). It is the only peak for magnetite concrete (MC) and for ordinary one (OC) although it is less than 2% in total mass its contribution is visible. Finally, contribution to total macroscopic neutron cross-section (Fig. 4) is similar to macroscopic neutron scattering cross-section (Fig. 2) as the cross-section values for absorption are of two magnitudes less than for scattering and they do not change the total value significantly. The last remark from the contribution analysis is that an addition of epoxy to every composite slightly increases hydrogen and carbon content and thus their importance in the shielding efficiency parameters.

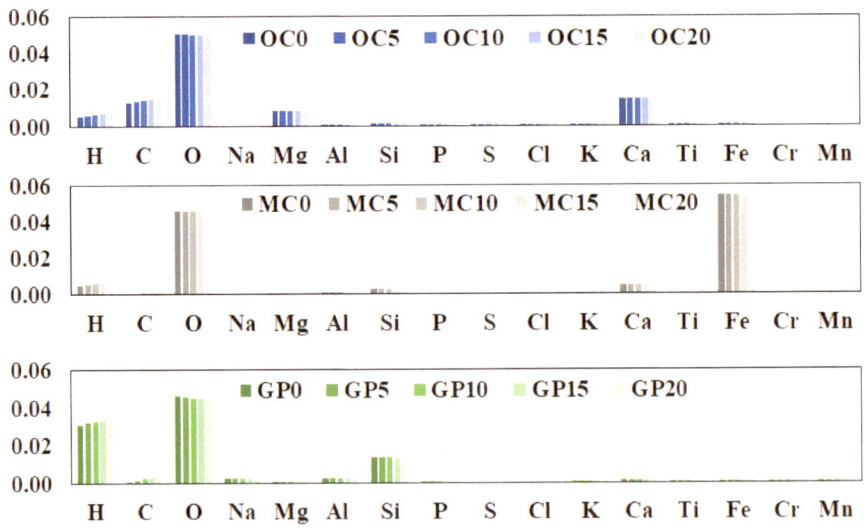

Fig. 1. Contribution to fast neutron mass removal cross-section

The comparison of radiation shielding parameters absolute values shows that ordinary concrete (OC) is the worst from every point of view (Fig. 5). Geopolymer mortar (GP) is the best when fast neutron scattering and total macroscopic neutron cross section

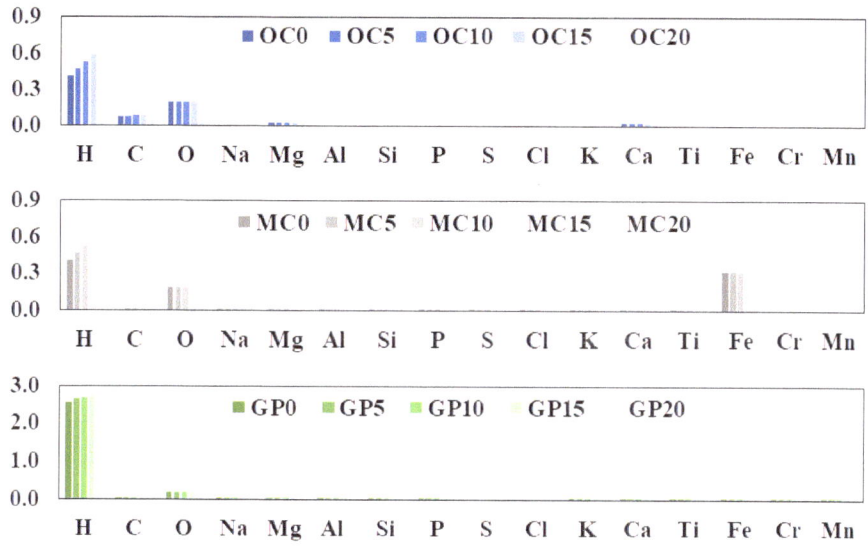

Fig. 2. Contribution to macroscopic neutron scattering cross-section

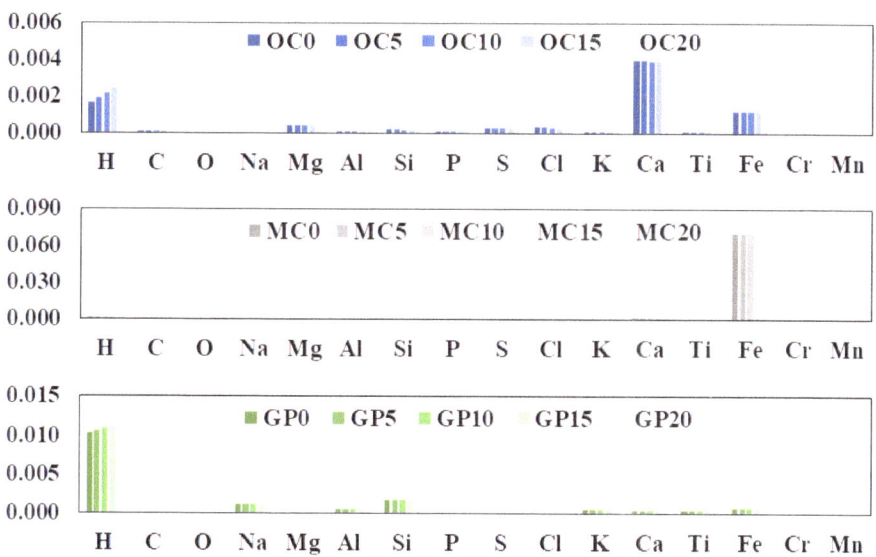

Fig. 3. Contribution to macroscopic thermal neutron absorption cross-section

is compared. Magnetite concrete (MC) has the highest values of thermal neutron absorption cross-section and surprisingly fast neutron mass removal cross-section as well. It is probably due to the fact that in fast neutron mass removal cross-section calculation issues other than simple fast neutron scattering also have some influence. They simply decide about fast neutron mass removal coefficient value of each atomic constituent.

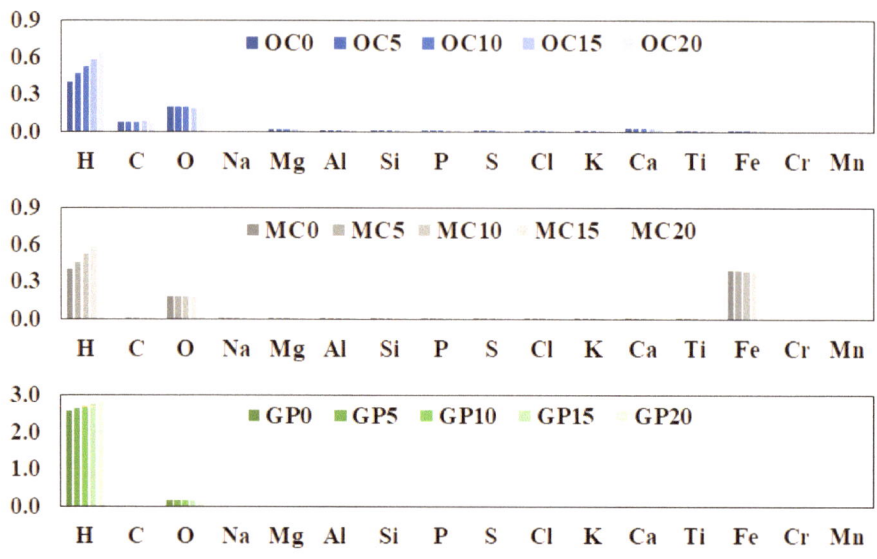

Fig. 4. Contribution to total macroscopic neutron cross-section

The positive influence of epoxy addition on every parameter is clearly visible by positive value of slope coefficients in trend line equations. It doesn't differ much between composites but changes regarding the type of parameter. The smallest influence it has on thermal neutron absorption (Fig. 5c) and the highest for fast neutron scattering (Fig. 5b) and total neutron macroscopic neutron cross-section (Fig. 5d). Moderate effect is noted in case of fast neutron mass removal cross-section (Fig. 5a).

3 Conclusions

The following conclusions from the research described above can be drawn:

- Fast neutron mass removal cross-section computational method is a proper one for general neutron shielding efficiency evaluation. If more detailed analysis distinguishing fast neutron scattering and thermal neutron absorption is needed, the method based on macroscopic neutron cross-section is recommended
- Neutron shielding efficiency is depending strongly on atomic composition and the biggest influence is due to aggregate type used
- Hydrogen content is the main factor influencing neutrons shielding efficiency especially fast neutron attenuation described by fast neutron mass removal cross-section and macroscopic neutron scattering cross-section
- Magnetite concrete is an effective material in radiation shielding due to its density that is crucial for gamma shielding and iron influencing neutron attenuation and absorption
- Geopolymers seems to be promising material for neutron radiation shielding purposed especially due to high hydrogen content

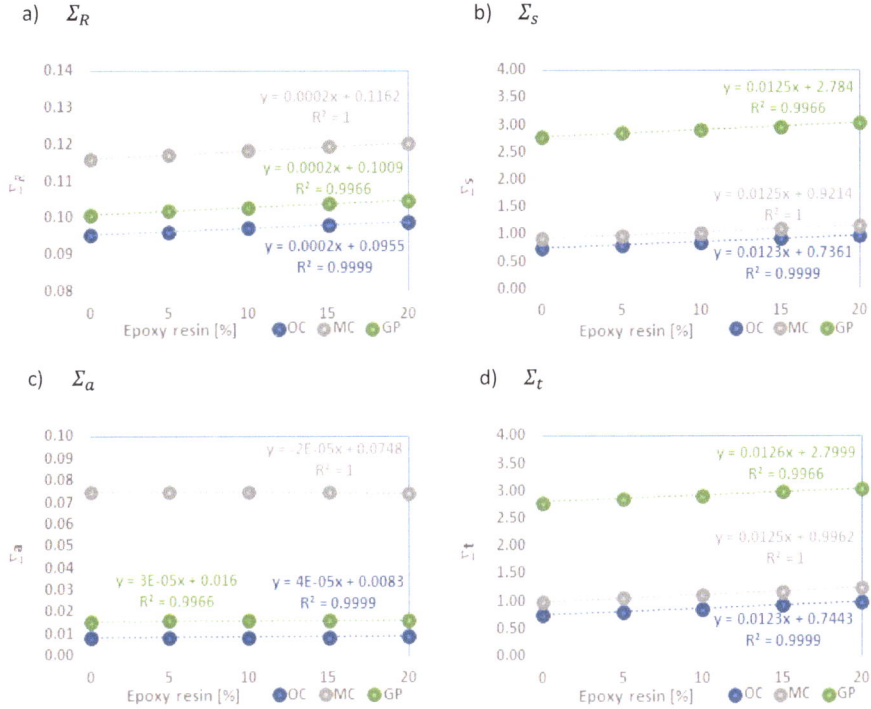

Fig. 5. Neutron shielding efficiency parameters

- Modification by polymers e.g. epoxy resin increases shielding efficiency due to increase of hydrogen input into atomic composition.

Acknowledgements. This work was supported by NCBR , Project Lider No LIDER/49/0187/L-11/19/NCBR/2020 "Sustainable materials – alkali-activated mortars - for specific shielding performance in civil engineering".

References

1. Prochoń, P., et al.: The effect of cement and aggregate type and w/c ratio on the bound water content and neutron shielding efficiency of concretes. Constr. Build. Mater. **264**, 120210 (2020). https://doi.org/10.1016/j.conbuildmat.2020.120210
2. Prochoń, P., et al.: Influence of activators on mechanical properties of modified fly ash based geopolymer mortars. Materials **13**(5), 1033 (2020). https://doi.org/10.3390/ma13051033
3. Zalegowski, K., et al.: Influence of polymer modification on the microstructure of shielding concrete. Materials **13**(3), 498 (2020). https://doi.org/10.3390/ma13030498
4. El-Khayatt, A.M.: Calculation of fast neutron removal cross-sections for some compounds and materials. Ann. Nucl. Energy **37**(2), 218–222 (2010). https://doi.org/10.1016/j.anucene.2009.10.022

5. El-Khayatt, A.M., El-Sayed Abdo, A.: MERCSF-N: a program for the calculation of fast neutron removal cross sections in composite shields, , Ann. Nucl. Energy **36**(6), 832–836 (2009). https://doi.org/10.1016/j.anucene.2009.01.013
6. Kaplan, M.F.: Concrete Radiation Shielding. John Wiley & Sons Inc, New York (1989)
7. Profio, A.E.: Radiation Shielding and Dosimetry. John Wiley & Sons Inc, New York (1979)
8. Chilton, A.B., et al.: Principles of Radiation Shielding. Prentice-Hall, Englewood Cliffs, New Jersey (1984)
9. Piotrowski, T.: Neutron shielding evaluation of concretes and mortars: a review. Constr. Build. Mater. **277**, 122238 (2021). https://doi.org/10.1016/j.conbuildmat.2020.122238
10. Sears, V.F.: Special feature section of neutron scattering lengths and cross sections of the elements and their isotopes. Neutron News **3**(3), 29–37 (1992)

Correction to: Towards the Use of Waste Limestone Powder as a Filler for Epoxy Coatings in Floors: Research on Mechanical Properties

Agnieszka Chowaniec-Michalak, Sławomir Czarnecki, and Łukasz Sadowski

Correction to:
Chapter 36 in: L. Czarnecki et al. (Eds.): *Concrete-Polymer Composites in Circular Economy*, **Springer Proceedings in Materials,**
https://doi.org/10.1007/978-3-031-72955-3_36

The original version of the book was inadvertently published with incorrect funding information in chapter 36. The book and the chapter have been updated with these changes.

The updated version of this chapter can be found at
https://doi.org/10.1007/978-3-031-72955-3_36

Author Index